Electromagnetic Shielding

WILEY SERIES IN MICROWAVE AND OPTICAL ENGINEERING

KAI CHANG, Editor
Texas A&M University

A complete list of the titles in this series appears at the end of this volume.

Electromagnetic Shielding

SALVATORE CELOZZI
RODOLFO ARANEO
GIAMPIERO LOVAT

Electrical Engineering Department "La Sapienza" University
Rome, Italy

IEEE Press

A JOHN WILEY & SONS, INC., PUBLICATION

Published by John Wiley & Sons, Inc., Hoboken, New Jersey

Published simultaneously in Canada

Library of Congress Cataloging-in-Publication Data

Celozzi, Salvatore.
Electromagnetic shielding / Salvatore Celozzi, Rodolfo Araneo, Giampiero Lovat.
p. cm.
Includes bibliographical references and index.
ISBN 978-0-470-05536-6
1. Shielding (Electricity) 2. Magnetic shielding. I. Araneo, Rodolfo.
II. Lovat, Giampiero. III. Title.
TK454.4.M33C45 2008
621.3–dc22

2008006387

Printed in the United States of America
10 9 8 7 6 5 4 3 2 1

Contents

Preface

This book might have been titled *Introduction to Electromagnetic Shielding*, since every chapter, that is to say, every section, could well be the subject of a book. Thus the goal of this book is to provide a first roadmap toward a full understanding of the phenomena at the core of the complex and fascinating world of electromagnetic shielding.

The book is organized in twelve chapters and three appendixes. A word of explanation about this choice is due. Consistent with the title of the Series to which this book belongs, electrostatic and magnetic shielding are relegated to Appendixes A and B, respectively. While these discussions could possibly have merited being presented in the first chapters, our final decision about their collocation in the appendixes was also influenced by the fact that they could be sufficiently contracted and are familiar to most readers approaching a book on electromagnetic shielding. The third appendix covers standards and measurement procedures. Its location at the end of the book is due to the rapid obsolescence of any material covering a subject of this type. Standards proliferation is a real problem for almost any engineer, and the field of electromagnetic shielding does not represent an exception to this modern-day "disease."

The chapters are organized as follows: First some introductory remarks are presented on the electromagnetics of shielding (Chapter 1), followed by a description of the arsenal of conventional and less-conventional materials (Chapter 2). A brief review of the figures of merit suitable for a quantitative and comparative analysis of shielding performance is offered in Chapter 3, and this chapter forms the initial point of an ongoing (and possibly endless) discussion on crucial issues at the root of a number of further considerations. The core of the analysis methods available for electromagnetic shielding starts at Chapter 4, where the subject of a stratified medium illuminated by a plane wave and the analogy between governing equations and transmission-line equations are covered at some length. Chapter 5 deals with numerical methods suitable for the analysis of actual shielding problems. Its content follows from both the number of numerical techniques available to solve this class of configurations and the need of presenting, for each of them, pros and cons and examples. Chapter 6 is entirely devoted to apertures and to their effects on shielding performance, since apertures are generally considered to be the most important coupling path between the shielded and source regions. The book continues with a thorough analysis of enclosures, since the shielded volume is often a closed region. The special case of cables is considered in Chapter 8. Cables are the subject of excellent textbooks. Although, at least in principle, a possible choice was to refer to the existing literature without any attempt of inclusion, their omission would have been a serious deficiency. So our compromise was to present the very basics of shielded cables.

Conceptually cables can be viewed as systems to be shielded as well as components of a shielding configuration, which are the subject of Chapter 9. These details are much more important than might appear at a first glance and are often decisive in the achieved level of performance. The reader should always refer to manufacturers' specifications, bearing in mind that shielding components are often not directly comparable, because different test fixtures can yield different measured data.

The last three chapters cover some distinctive issues in frequency-selective shielding, shielding design procedures, and uncommon ways of shielding. The interested readers will find there several starting points for fields that are still subject to much exploration.

Before acknowledgments, apologies: we are perfectly aware that not all the contributors to this research and technical field have been mentioned. Space limitations have further imposed some omissions reflecting our different personal views. The authors will be grateful to anybody who will bring to their attention any relevant omission. We are ready to include also in a future edition references worthy of being cited.

The authors are indebted to all those who nurtured their education, to ancestors who instilled the value of education, parents and relatives, to fathers of today's philosophical and scientific culture, to pioneers in electromagnetics, to their very special educators and colleagues. In particular, they wish to acknowledge the patience and the competence of Dr. Paolo Burghignoli. The first author is particularly grateful to the late Dr. Motohisa Kanda, whose friendship and encouragement in focusing on the specific topic of electromagnetic shielding ultimately led to the writing of this book.

Final thanks go to Dr. Chang, Editor of this Series, for having encouraged us to contribute to this prestigious Series and to the editorial staff at Wiley for their definitive merit in the improvement of the quality of the original manuscript.

Salvatore Celozzi
Rodolfo Araneo
Giampiero Lovat
Rome, Italy
July 2007

CHAPTER ONE

Electromagnetics behind Shielding

Shielding an electromagnetic field is a complex and sometimes formidable task. The reasons are many, since the effectiveness of any strategy or technique aimed at the reduction of the electromagnetic field levels in a prescribed region depends largely upon the source(s) characteristics, the shield topology, and materials. Moreover, as it often happens, when common terms are adopted in a technical context, different definitions exist. In electromagnetics *shielding effectiveness* (SE) is a concise parameter generally applied to quantify shielding performance. However, a variety of standards are adopted for the measurement or the assessment of the performance of a given shielding structure. Unfortunately, they all call for very specific conditions in the measurement setup. The results therefore are often useless if the source or system configurations differ even slightly. Last among the difficulties that arise in the solution of actual shielding problems are the difficulties inherent in both the solution of the boundary value problem and the description of the electromagnetic problem in mathematical form.

1.1 DEFINITIONS

To establish a common ground, we will begin with some useful definitions. An electromagnetic shield can be defined as [1]:

> [A] housing, screen, or other object, usually conducting, that substantially reduces the effect of electric or magnetic fields on one side thereof, upon devices or circuits on the other side.

Electromagnetic Shielding by Salvatore Celozzi, Rodolfo Araneo and Giampiero Lovat

This definition is restrictive because it implicitly assumes the presence of a "victim." The definition is also based on a misconception that the source and observation points are in opposite positions with respect to the shield, and it includes the word "substantially" whose meaning is obscure and introduces an unacceptable level of arbitrariness.

Another definition [2] of electromagnetic shielding that is even more restrictive is:

> [A] means of preventing two circuits from electromagnetic coupling by placing at least one of the circuits in a grounded enclosure of magnetic conductive material.

The most appropriate definition entails a broad view of the phenomenon:

> [A]ny means used for the reduction of the electromagnetic field in a prescribed region.

Notice that no reference to shape, material, and grounding of the shield is necessary to define its purpose.

In general, electromagnetic shielding represents a way toward the improvement of the electromagnetic compatibility (EMC, defined as the capability of electronic equipment or systems to be operated in the intended electromagnetic environment at design levels of efficiency) performance of single devices, apparatus, or systems. Biological systems are included, for which it is correct to talk about health rather than EMC. Electromagnetic shielding also is used to prevent sensitive information from being intercepted, that is, to guarantee communication security.

Electromagnetic shielding is not the only remedy for such purposes. Some sort of electromagnetic shielding is almost always used in apparatus systems to reduce their electromagnetic emissions and to increase their electromagnetic immunity against external fields. In cases where the two methodologies for reducing the source levels of electromagnetic emission or strengthening the victim immunity are not available or are not sufficient to ensure the correct operation of devices or systems, a reduction of the coupling between the source and the victim (present or only potentially present) is often the preferred choice.

The immunity of the victims is generally obtained by means of filters that are analogous to electromagnetic shielding with respect to conducted emissions and immunity. The main advantage of filters is that they are "local" devices. Thus, where the number of sensitive components to be protected is limited, the cost of filtering may be much lower than that of shielding. The main disadvantage of using a filter is that it is able to arrest only interferences whose characteristics (e.g., level or mode of transmission) are different from that of the device, so the correct operation in the presence of some types of interference is not guaranteed. Another serious disadvantage of the filter is its inadequacy or its low efficiency for the prevention of data detection.

In general, the design of a filter is much simpler than that of a shield. The filter's designer has only to consider the waveform of the interference (in terms of voltage or current) and the values of the input and output impedance [3], whereas the shield's designer must include a large amount of input information and constraints, as it will be discussed in the next chapters.

Any shielding analysis begins by an accurate examination of the shield topology [4–6]. Although the identification of the coupling paths between the main space regions is often trivial, sometimes it deserves more care, especially in complex configurations. A shield's complexity is associated with its shape, apertures, the components identified as the most susceptible, the source characteristics, and so forth. Decomposition of its configuration into several subsystems (each simpler than the original and interacting with the others in a definite way) is always a useful approach to identify critical problems and find ways to fix and improve the overall performance. This approach is based on the assumption that each subsystem can be analyzed and its behavior characterized independently of the others components/subsystems. For instance, in the frequency domain and for a linear subsystem, for each coupling path and for each susceptible element, it is possible to investigate the transfer function $T(\omega)$ relating the external source input $S(\omega)$ and the victim output $V(\omega)$ characteristics as $V(\omega) = U(\omega) + T(\omega)S(\omega)$, where $U(\omega)$ represents the subsystem output in the absence of external-source excitation. In the presence of multilevel barriers, the transfer function $T(\omega)$ may ensue from the product of the transfer functions associated with each barrier level.

The foregoing approach can be generalized for a better understanding of the shielding problem in complex configurations. However, it is often sufficient for attention to be given to the most critical subsystems and components, on one hand, and the most important coupling paths, on the other hand, for the fixing of major shielding problems and the improvement of performance [7]. The general approach is obviously opportune in a design context. The complete analysis of the relations between shielding and grounding is left to the specific literature.

1.2 NOTATION, SYMBOLOGY, AND ACRONYMS

The abbreviations and symbols used throughout the book are briefly summarized here in order to make clear the standard we have chosen to adopt. Of course, we will warn the reader anytime an exception occurs.

Scalar quantities are shown in italic type (e.g., V and t), while vectors are shown in boldface (e.g., $\mathbf{e}$ and $\mathbf{H}$); dyadics are shown in boldface with an underbar (e.g., $\underline{\boldsymbol{\varepsilon}}$ and $\underline{\mathbf{G}}$). A physical quantity that depends on time and space variables is indicated with a lowercase letter (e.g., $\mathbf{e}(\mathbf{r}, t)$ for the electric field). The Fourier transform with respect to the time variable is indicated with the corresponding uppercase letter (e.g., $\mathbf{E}(\mathbf{r}, \omega)$), while the Fourier transform with respect to the spatial variables is indicated by a tilde (e.g., $\tilde{\mathbf{e}}(\mathbf{k}, t)$); when the Fourier transform with respect to both time and spatial variables is considered, the two symbologies are combined (e.g., $\tilde{\mathbf{E}}(\mathbf{k}, \omega)$).

The sets of spatial variables in rectangular, cylindrical, and spherical coordinates are denoted by (x, y, z), (ρ, ϕ, z), and (r, ϕ, θ), respectively. The boldface latin letter $\mathbf{u}$ is used to indicate a unit vector and a subscript is used to indicate its direction: for instance, $(\mathbf{u}_x, \mathbf{u}_y, \mathbf{u}_z)$, $(\mathbf{u}_\rho, \mathbf{u}_\phi, \mathbf{u}_z)$, and $(\mathbf{u}_r, \mathbf{u}_\phi, \mathbf{u}_\theta)$ denote the unit vectors in the rectangular, cylindrical, and spherical coordinate system, respectively.

We will use the "del" operator notation ∇ with the suitable product type to indicate gradient $(\nabla[\cdot])$, curl $(\nabla \times [\cdot])$, and divergence operators $(\nabla \cdot [\cdot])$; the Laplacian operator is indicated as $\nabla^2[\cdot]$. The imaginary unit is denoted with $j = \sqrt{-1}$ and the asterisk $*$ as a superscript of a complex quantity denotes its complex conjugate. The real and imaginary parts of a complex quantity are indicated by $\mathrm{Re}[\cdot]$ and $\mathrm{Im}[\cdot]$, respectively, while the principal argument is indicated by the function $\mathrm{Arg}[\cdot]$. The base-10 logarithm and the natural logarithm are indicated by means of the $\log(\cdot)$ and $\ln(\cdot)$ functions, respectively.

Finally, throughout the book, the international system of units SI is adopted, electromagnetic is abbreviated as EM, and shielding effectiveness as SE.

1.3 BASIC ELECTROMAGNETICS

1.3.1 Macroscopic Electromagnetism and Maxwell's Equations

A complete description of the macroscopic electromagnetism is provided by Maxwell's equations whose validity is taken as a postulate. Maxwell's equations can be used either in a differential (local) form or in an integral (global) form, and there has been a long debate over which is the best representation (e.g., David Hilbert preferred the integral form but Arnold Sommerfeld found more suitable the differential form, from which the special relativity follows more naturally [8]). When stationary media are considered, the main difference between the two representations consists in how they account for discontinuities of materials and/or sources. Basically, if one adopts the differential form, some boundary conditions at surface discontinuities must be postulated; on the other hand, if the integral forms are chosen, one must postulate their validity across such discontinuities [9,10].

Maxwell's equations can be expressed in scalar, vector, or tensor form, and different vector fields can be considered as fundamental. A full description of all these details can be found in [8]. In this book we assume the following differential form of the Maxwell equations:

$$\begin{aligned}
\nabla \times \mathbf{e}(\mathbf{r}, t) &= -\frac{\partial}{\partial t}\mathbf{b}(\mathbf{r}, t),\\
\nabla \times \mathbf{h}(\mathbf{r}, t) &= \mathbf{j}(\mathbf{r}, t) + \frac{\partial}{\partial t}\mathbf{d}(\mathbf{r}, t),\\
\nabla \cdot \mathbf{d}(\mathbf{r}, t) &= \rho_{\mathrm{e}}(\mathbf{r}, t),\\
\nabla \cdot \mathbf{b}(\mathbf{r}, t) &= 0.
\end{aligned} \tag{1.1}$$

From these equations the continuity equation can be derived as

$$\nabla \cdot \mathbf{j}(\mathbf{r}, t) = -\frac{\partial}{\partial t}\rho_{\mathrm{e}}(\mathbf{r}, t). \tag{1.2}$$

In this framework the EM field—described by vectors **e** (electric field, unit of measure V/m), **h** (magnetic field, unit of measure A/m), **d** (electric displacement, unit of measure C/m^2), and **b** (magnetic induction, unit of measure Wb/m^2 or T)—arises from sources **j** (electric current density, unit of measure A/m^2) and ρ_e (electric charge density, unit of measure C/m^3). Further, except for static fields, if a time can be found before which all the fields and sources are identically zero, the divergence equations in (1.1) are a consequence of the curl equations [8], so under this assumption the curl equations can be taken as independent.

It can be useful to make the Maxwell equations symmetric by introducing fictitious magnetic current and charge densities **m** and ρ_m (units of measure V/m^2 and Wb/m^3, respectively) which satisfy a continuity equation similar to (1.2) so that (1.1) can be rewritten as

$$\begin{aligned}
\nabla \times \mathbf{e}(\mathbf{r},t) &= -\frac{\partial}{\partial t}\mathbf{b}(\mathbf{r},t) - \mathbf{m}(\mathbf{r},t),\\
\nabla \times \mathbf{h}(\mathbf{r},t) &= \mathbf{j}(\mathbf{r},t) + \frac{\partial}{\partial t}\mathbf{d}(\mathbf{r},t),\\
\nabla \cdot \mathbf{d}(\mathbf{r},t) &= \rho_e(\mathbf{r},t),\\
\nabla \cdot \mathbf{b}(\mathbf{r},t) &= \rho_m(\mathbf{r},t).
\end{aligned} \tag{1.3}$$

As it will be shown later, the equivalence principle indeed requires the introduction of such fictitious quantities.

It is also useful to identify in Maxwell's equations some "impressed" source terms, which are independent of the unknown fields and are instead due to other external sources (magnetic sources can be only of this type). Such "impressed" sources are considered as known terms in Maxwell's differential equations and indicated by the subscript "i." In this connections, (1.3) can be expressed as

$$\begin{aligned}
\nabla \times \mathbf{e}(\mathbf{r},t) &= -\frac{\partial}{\partial t}\mathbf{b}(\mathbf{r},t) - \mathbf{m}_i(\mathbf{r},t),\\
\nabla \times \mathbf{h}(\mathbf{r},t) &= \mathbf{j}(\mathbf{r},t) + \frac{\partial}{\partial t}\mathbf{d}(\mathbf{r},t) + \mathbf{j}_i(\mathbf{r},t),\\
\nabla \cdot \mathbf{d}(\mathbf{r},t) &= \rho_e(\mathbf{r},t) + \rho_{ei}(\mathbf{r},t),\\
\nabla \cdot \mathbf{b}(\mathbf{r},t) &= \rho_{mi}(\mathbf{r},t).
\end{aligned} \tag{1.4}$$

The impressed-source concept is well known in circuit theory. For example, independent voltage sources are voltage excitations that are independent of possible loads.

Although both the sources and the fields cannot have true spatial discontinuities, from a modeling point of view, it is useful to consider additionally sources in one or two dimensions. In this connection, surface- and line-source densities can be introduced in terms of the Dirac delta distribution δ, as (singular) idealizations of actual continuous volume densities [8,11].

Finally, in the frequency domain, Maxwell's curl equations are expressed as

$$\begin{aligned} \nabla \times \mathbf{E}(\mathbf{r},\omega) &= -j\omega \mathbf{B}(\mathbf{r},\omega) - \mathbf{M}_{\mathrm{i}}(\mathbf{r},\omega), \\ \nabla \times \mathbf{H}(\mathbf{r},\omega) &= \mathbf{J}(\mathbf{r},\omega) + j\omega \mathbf{D}(\mathbf{r},\omega) + \mathbf{J}_{\mathrm{i}}(\mathbf{r},\omega), \end{aligned} \tag{1.5}$$

where the uppercase quantities indicate either the Fourier transform or the phasors associated with the corresponding time-domain fields.

1.3.2 Constitutive Relations

By direct inspection of Maxwell's curl equations in (1.1), it is immediately clear that they represent 6 scalar equations with 15 unknown quantities. With fewer equations than unknowns no unique solution can be identified (the problem is said to be indefinite). The additional equations required to make the problem definite are those that describe the relations among the field quantities **e**, **h**, **d**, **b**, and **j**, enforced by the medium filling the region where the EM phenomena occur. Such relations are called *constitutive relations*, and they depend on the properties of the medium supporting the EM field.

In nonmoving media, with the exclusion of bianisotropic materials, the **d** field depends only on the **e** field, **b** depends only on **h**, and **j** depends only on **e**. These dependences are expressed as constitutive relations, with the **e** and **h** fields regarded as causes and the **d**, **b**, and **j** fields as effects.

If a linear combination of causes (with given coefficients) produces a linear combination of effects (with the same coefficients), the medium is said to be *linear* (otherwise nonlinear). In general, the constitutive relations are described by a set of constitutive parameters and a set of constitutive operators that relate the above-mentioned fields inside a region of space. The constitutive parameters can be constants of proportionality between the fields (the medium is thus said *isotropic*), or they can be components in a tensor relationship (the medium is said anisotropic). If the constitutive parameters are constant within a certain region of space, the medium is said *homogeneous* in that region (otherwise, the medium is inhomogeneous). If the constitutive parameters are constant with time, the medium is said *stationary* (otherwise, the medium is nonstationary).

If the constitutive operators are expressed in terms of time integrals, the medium is said to be *temporally dispersive.* If these operators involve space integrals, the medium is said *spatially dispersive.* Finally, we note what is usually neglected, that the constitutive parameters may depend on other nonelectromagnetic properties of the material and external conditions (temperature, pressure, etc.).

The simplest medium is vacuum. In vacuum the following constitutive relations hold:

$$\begin{aligned} \mathbf{d}(\mathbf{r},t) &= \varepsilon_0 \mathbf{e}(\mathbf{r},t), \\ \mathbf{b}(\mathbf{r},t) &= \mu_0 \mathbf{h}(\mathbf{r},t), \\ \mathbf{j}(\mathbf{r},t) &= \mathbf{0}. \end{aligned} \tag{1.6}$$

The quantities $\mu_0 = 4\pi \cdot 10^{-7}$ H/m and $\varepsilon_0 = 8.854 \cdot 10^{-12}$ F/m are the free-space magnetic permeability and dielectric permittivity, respectively. Their values are related to the speed of light in free space c through $c = 1/\sqrt{\mu_0 \varepsilon_0}$ (in deriving the value of ε_0, the value of c is assumed to be $c = 2.998 \cdot 10^8$ m/s).

For a linear, homogeneous, isotropic, and nondispersive material the constitutive relations can be expressed as

$$\begin{aligned} \mathbf{d}(\mathbf{r},t) &= \varepsilon \mathbf{e}(\mathbf{r},t), \\ \mathbf{b}(\mathbf{r},t) &= \mu \mathbf{h}(\mathbf{r},t), \\ \mathbf{j}(\mathbf{r},t) &= \sigma \mathbf{e}(\mathbf{r},t), \end{aligned} \tag{1.7}$$

where μ and ε are the magnetic permeability and dielectric permittivity of the medium, respectively. These quantities can be related to the corresponding free-space quantities through the dimensionless relative permeability μ_r and relative permittivity ε_r, such that $\mu = \mu_r \mu_0$ and $\varepsilon = \varepsilon_r \varepsilon_0$. The dimensionless quantities $\chi_m = \mu_r - 1$ and $\chi_e = \varepsilon_r - 1$ (known as magnetic and electric susceptibilities, respectively) are also used. The third equation of (1.7) expresses Ohm's law, and σ is the conductivity of the medium (unit of measure S/m). If the medium is inhomogeneous, μ, ε, or σ are quantities that depend on the vector position $\mathbf{r}$. If the medium is anisotropic (but still linear and nondispersive) the constitutive relations can be written as

$$\begin{aligned} \mathbf{d}(\mathbf{r},t) &= \underline{\boldsymbol{\varepsilon}} \cdot \mathbf{e}(\mathbf{r},t), \\ \mathbf{b}(\mathbf{r},t) &= \underline{\boldsymbol{\mu}} \cdot \mathbf{h}(\mathbf{r},t), \\ \mathbf{j}(\mathbf{r},t) &= \underline{\boldsymbol{\sigma}} \cdot \mathbf{e}(\mathbf{r},t), \end{aligned} \tag{1.8}$$

where $\underline{\boldsymbol{\varepsilon}}$, $\underline{\boldsymbol{\mu}}$, and $\underline{\boldsymbol{\sigma}}$ are called the permittivity tensor, the permeability tensor, and the conductivity tensor, respectively (they are space-dependent quantities for inhomogeneous media).

For linear, inhomogeneous, anisotropic, stationary, and temporally dispersive materials, the constitutive relation between $\mathbf{d}$ and $\mathbf{e}$ is expressed by a convolution integral as

$$\mathbf{d}(\mathbf{r},t) = \int_{-\infty}^{t} \underline{\boldsymbol{\varepsilon}}(\mathbf{r},t-t') \cdot \mathbf{e}(\mathbf{r},t')\mathrm{d}t'. \tag{1.9}$$

The constitutive relations for other field quantities have similar expressions. Causality is implied by the upper limit t in the integrals (this means that the effect cannot depend on future values of the cause). If the medium is nonstationary, $\underline{\boldsymbol{\varepsilon}}(\mathbf{r},t,t')$ has to be used instead of $\underline{\boldsymbol{\varepsilon}}(\mathbf{r},t-t')$. The important concept expressed by (1.9) is that the behavior of $\mathbf{d}$ at the time t depends not only on the value of $\mathbf{e}$ at the same time t but also on its values at all past times, thus giving rise to a time-lag between cause and effect. In the frequency domain the constitutive relation (1.9) is expressed as

$$\mathbf{D}(\mathbf{r}, \omega) = \underline{\boldsymbol{\varepsilon}}(\mathbf{r}, \omega) \cdot \mathbf{E}(\mathbf{r}, \omega), \tag{1.10}$$

where, with a little abuse of notation, $\underline{\boldsymbol{\varepsilon}}(\mathbf{r}, \omega)$ indicates the Fourier transform of the corresponding quantity in the time domain. The important point to note here is that, in the frequency domain, temporal dispersion is associated with complex values of the constitutive parameters; causality establishes a relationship between their real and imaginary parts (known as the Kramers–Kronig relation) [8] for which neither part can be constant with frequency. In addition it can be shown that the nonzero imaginary part of the constitutive parameters is related to dissipation of EM energy in the form of heat.

Finally, if the medium is also spatially dispersive (and nonstationary), the constitutive relation takes the form

$$\mathbf{d}(\mathbf{r}, t) = \iiint_V \left[\int_{-\infty}^{t} \underline{\boldsymbol{\varepsilon}}(\mathbf{r}, \mathbf{r}', t, t') \cdot \mathbf{e}(\mathbf{r}, \mathbf{r}', t') \mathrm{d}t' \right] \mathrm{d}V', \tag{1.11}$$

where V indicates the whole three-dimensional space; as before, similar expressions hold for the constitutive relations of other field quantities as well. The integral over the volume V in (1.11) expresses the physical phenomenon for which the effect at the point $\mathbf{r}$ depends on the value of the cause in all the neighboring points $\mathbf{r}'$. An important point is that if the medium is spatially dispersive but homogeneous, the constitutive relations involve a convolution integral in the space domain. Therefore the constitutive relations in a linear, homogeneous, and stationary medium for the Fourier transforms of the fields with respect to both time and space can be written as

$$\tilde{\mathbf{D}}(k, \omega) = \tilde{\underline{\boldsymbol{\varepsilon}}}(\mathbf{k}, \omega) \cdot \tilde{\mathbf{E}}(\mathbf{k}, \omega). \tag{1.12}$$

Very often, in the frequency domain, the contributions in Maxwell's equations (1.5) from the conductivity current and the electric displacement are combined in a unique term by introducing an equivalent complex permittivity. For simplicity, we consider isotropic materials for which complex permittivity is a scalar quantity, defined as $\varepsilon_C = \varepsilon - j\sigma/\omega$. Thus we can rewrite (1.5) in a dual form as

$$\begin{aligned} \nabla \times \mathbf{E}(\mathbf{r}, \omega) &= -j\omega\mu(\mathbf{r}, \omega)\mathbf{H}(\mathbf{r}, \omega) - \mathbf{M}_i(\mathbf{r}, \omega), \\ \nabla \times \mathbf{H}(\mathbf{r}, \omega) &= j\omega\varepsilon_C(\mathbf{r}, \omega)\mathbf{E}(\mathbf{r}, \omega) + \mathbf{J}_i(\mathbf{r}, \omega). \end{aligned} \tag{1.13}$$

Finally, it is important to note that for the study of electromagnetism in matter, the EM field can be represented by four vectors other than **e**, **h**, **d**, and **b** (provided that the new vectors are a linear mapping of these vectors). In particular, the common alternative is to use vectors **e**, **b**, **p**, and **m** (not to be confused with the magnetic current density), where the new vectors **p** and **m** are called polarization and magnetization vectors, respectively, and Maxwell's equations are consequently written as

$$\begin{aligned}
&\nabla \times \mathbf{e}(\mathbf{r},t) = -\frac{\partial}{\partial t}\mathbf{b}(\mathbf{r},t),\\
&\nabla \times \left[\frac{\mathbf{b}(\mathbf{r},t)}{\mu_0} - \mathbf{m}(\mathbf{r},t)\right] = \mathbf{j}(\mathbf{r},t) + \frac{\partial}{\partial t}[\varepsilon_0\mathbf{e}(\mathbf{r},t) + \mathbf{p}(\mathbf{r},t)],\\
&\nabla \cdot [\varepsilon_0\mathbf{e}(\mathbf{r},t) + \mathbf{p}(\mathbf{r},t)] = \rho_e(\mathbf{r},t),\\
&\nabla \cdot \mathbf{b}(\mathbf{r},t) = 0.
\end{aligned} \tag{1.14}$$

From (1.1) and (1.14), it follows that

$$\begin{aligned}
\mathbf{p}(\mathbf{r},t) &= \mathbf{d}(\mathbf{r},t) - \varepsilon_0\mathbf{e}(\mathbf{r},t),\\
\mathbf{m}(\mathbf{r},t) &= \frac{\mathbf{b}(\mathbf{r},t)}{\mu_0} - \mathbf{h}(\mathbf{r},t);
\end{aligned} \tag{1.15}$$

or, in the frequency domain,

$$\begin{aligned}
\mathbf{P}(\mathbf{r},\omega) &= \mathbf{D}(\mathbf{r},\omega) - \varepsilon_0\mathbf{E}(\mathbf{r},\omega),\\
\mathbf{M}(\mathbf{r},\omega) &= \frac{\mathbf{B}(\mathbf{r},\omega)}{\mu_0} - \mathbf{H}(\mathbf{r},\omega).
\end{aligned} \tag{1.16}$$

Next we introduce the equivalent polarization current density $\mathbf{j}_P = \partial\mathbf{p}/\partial t$, the equivalent magnetization current density $\mathbf{j}_M = \nabla \times \mathbf{m}$, and the equivalent polarization charge density $\rho_P = -\nabla \cdot \mathbf{p}$ so that the Maxwell equations take the form

$$\begin{aligned}
&\nabla \times \mathbf{e}(\mathbf{r},t) = -\frac{\partial}{\partial t}\mathbf{b}(\mathbf{r},t),\\
&\nabla \times \frac{\mathbf{b}(\mathbf{r},t)}{\mu_0} = \mathbf{j}(\mathbf{r},t) + \mathbf{j}_P(\mathbf{r},t) + \mathbf{j}_M(\mathbf{r},t) + \varepsilon_0\frac{\partial\mathbf{e}(\mathbf{r},t)}{\partial t},\\
&\varepsilon_0\nabla \cdot \mathbf{e}(\mathbf{r},t) = \rho_e(\mathbf{r},t) + \rho_P(\mathbf{r},t),\\
&\nabla \cdot \mathbf{b}(\mathbf{r},t) = 0.
\end{aligned} \tag{1.17}$$

1.3.3 Discontinuities and Singularities

As was mentioned in the previous section, in the absence of discontinuities, Maxwell's equations in differential form are valid everywhere in space; nevertheless, for modeling purposes, discontinuities of material parameters or singular sources are often considered. In such cases other field relationships must be postulated (alternatively, they can be derived from Maxwell's equations in the integral form if such integral forms are postulated to be valid also across the discontinuities).

Let us consider the presence of either (singular) electric and magnetic source densities (electric $\mathbf{j}_S$ and ρ_{eS} and magnetic $\mathbf{m}_S$ and ρ_{mS}) distributed over a surface S, which separates two regions (region 1 and region 2, respectively), or discontinuities in the material parameters across the surface S; the EM field in each region is indicated by the subscript 1 or 2. Let $\mathbf{u}_n$ be the unit vector normal to the surface S directed from region 2 to region 1. In such conditions the following jump conditions hold:

$$\begin{aligned}\mathbf{u}_n \times (\mathbf{h}_1 - \mathbf{h}_2) &= \mathbf{j}_S, \\ \mathbf{u}_n \times (\mathbf{e}_1 - \mathbf{e}_2) &= -\mathbf{m}_S, \\ \mathbf{u}_n \cdot (\mathbf{d}_1 - \mathbf{d}_2) &= \rho_{eS}, \\ \mathbf{u}_n \cdot (\mathbf{b}_1 - \mathbf{b}_2) &= \rho_{mS},\end{aligned} \tag{1.18}$$

and

$$\begin{aligned}\mathbf{u}_n \cdot (\mathbf{j}_1 - \mathbf{j}_2) &= -\nabla_S \cdot \mathbf{j}_S - \frac{\partial \rho_{eS}}{\partial t}, \\ \mathbf{u}_n \cdot (\mathbf{m}_1 - \mathbf{m}_2) &= -\nabla_S \cdot \mathbf{m}_S - \frac{\partial \rho_{mS}}{\partial t},\end{aligned} \tag{1.19}$$

where $\nabla_S[\cdot] = \nabla[\cdot] - \mathbf{u}_n \partial[\cdot]/\partial n$. It is clear that when $\mathbf{j}_S$ and $\mathbf{m}_S$ are zero, the tangential components of both electric and magnetic fields are continuous across the surface S. In particular, if discontinuities in the material parameters are present, the electric surface current density $\mathbf{j}_S$ may be different from zero at the boundary of a perfect electric conductor (PEC, within which $\mathbf{e}_2 = \mathbf{0}$), and the magnetic surface current density $\mathbf{m}_S$ may be different from zero at the boundary of a perfect magnetic conductor (PMC, within which $\mathbf{h}_2 = \mathbf{0}$). Then the jump conditions at the interface between the conventional medium and the PEC are written as

$$\begin{aligned}\mathbf{u}_n \times \mathbf{h} &= \mathbf{j}_S, \\ \mathbf{u}_n \times \mathbf{e} &= \mathbf{0}, \\ \mathbf{u}_n \cdot \mathbf{d} &= \rho_{eS}, \\ \mathbf{u}_n \cdot \mathbf{b} &= 0, \\ \mathbf{u}_n \cdot \mathbf{j} &= -\nabla_S \cdot \mathbf{j}_S - \frac{\partial \rho_{eS}}{\partial t}, \\ \mathbf{u}_n \cdot \mathbf{m} &= 0.\end{aligned} \tag{1.20}$$

Likewise, at the interface between a conventional medium and a PMC, the results are

$$\begin{aligned}\mathbf{u}_n \times \mathbf{h} &= \mathbf{0}, \\ \mathbf{u}_n \times \mathbf{e} &= -\mathbf{m}_S, \\ \mathbf{u}_n \cdot \mathbf{d} &= 0, \\ \mathbf{u}_n \cdot \mathbf{b} &= \rho_{mS}, \\ \mathbf{u}_n \cdot \mathbf{j} &= 0, \\ \mathbf{u}_n \cdot \mathbf{m} &= -\nabla_S \cdot \mathbf{m}_S - \frac{\partial \rho_{mS}}{\partial t}.\end{aligned} \tag{1.21}$$

In these jump conditions the $\mathbf{u}_n$ unit vector points outside the conductors.

Finally, some other singular behaviors of fields and currents worth mentioning occur in correspondence to the edge of a dielectric or conducting wedge and to the tip of a dielectric or conducting cone. The solution of the EM problem in such cases is not unique, unless the singular behavior is specified. The order of singularity can be determined by requiring that the energy stored in the proximity of the edge or of the tip remains finite. Further details can be found in [11] and [12].

1.3.4 Initial and Boundary Conditions

As was noted earlier Maxwell's equations together with the constitutive relations represent a set of partial differential equations. However, it is well known that in order to obtain a solution for this set of equations, both initial and boundary conditions must be specified. The initial conditions are represented by the constraints that the EM field must satisfy at a given time, while boundary conditions are, in general, constraints that the EM field must satisfy over certain surfaces of the three-dimensional space, usually surfaces that separate regions of space filled with different materials. In these cases the boundary conditions coincide with the jump conditions illustrated in the previous section. Other important examples of boundary conditions that can easily be formulated in the frequency domain are the *impedance boundary condition* and *radiation condition at infinity.* The impedance boundary condition relates the component $\mathbf{E}_t$ of the electric field tangential to a surface S with the magnetic field as

$$\mathbf{E}_t = Z_S(\mathbf{u}_n \times \mathbf{H}), \tag{1.22}$$

where Z_S (surface impedance) is a complex scalar quantity. The radiation condition at infinity (also known as the Sommerfeld radiation condition or the Silver–Müller radiation condition) postulates that in free space, in the absence of sources at infinity, there results

$$\lim_{r \to +\infty} r \left[\mathbf{E} - \sqrt{\frac{\mu_0}{\varepsilon_0}} (\mathbf{H} \times \mathbf{u}_r) \right] = \mathbf{0}. \tag{1.23}$$

1.3.5 Poynting's Theorem and Energy Considerations

For simplicity, in what follows we will refer to time-harmonic fields and sources, and we will use the phasor notation in the frequency domain. It is understood that this is an idealization, since true monochromatic fields cannot exist. However, the simplicity of the formalism and the fact that a monochromatic wave is an elemental component of the complete frequency-domain spectrum of an arbitrary time-varying field make the assumption of monochromatic fields an invaluable tool for the investigation of the EM-field theory. Nevertheless, great care must be given to the use of such an assumption because it can lead to nonphysical consequences:

a classical example consists in determining the energy stored in a lossless cavity. An infinite value is actually obtained, since the cavity stores energy starting from a remote instant $t = -\infty$. The problem can be overcome by considering time-averaged quantities, but some other problems can arise when the filling material is dispersive.

A fundamental consequence of Maxwell's equations is the Poynting theorem by which an energetic interpretation is made of some field quantities. In particular, it can be shown that given a region V bounded by a surface S, from Maxwell's equations the following identity holds:

$$\oint\!\!\!\oint_S \mathbf{u}_n \cdot \mathbf{\Pi} \mathrm{d}S + \iiint_V p_\mathrm{d} \mathrm{d}V + \iiint_V (p_H + p_E) \mathrm{d}V = \iiint_V (p_\mathrm{i} + p_\mathrm{mi}) \mathrm{d}V. \tag{1.24}$$

Equation (1.24) expresses the Poynting theorem. The real part of the right-hand side of (1.24) (where $p_\mathrm{i} = -\mathbf{J}_\mathrm{i}^* \cdot \mathbf{E}/2$ and $p_\mathrm{mi} = -\mathbf{M}_\mathrm{i} \cdot \mathbf{H}^*/2$) represents the time-averaged power furnished by the impressed sources to the EM field, and the left-hand side of (1.24) represents the destination of such a power. The Poynting vector $\mathbf{\Pi}$ is defined as

$$\mathbf{\Pi} = \frac{1}{2} \mathbf{E} \times \mathbf{H}^*. \tag{1.25}$$

The real part of its flux across the surface S (first addend in equation (1.24)) represents the time-averaged power radiated through the surface S. The second addend in (1.24) (where $p_\mathrm{d} = \mathbf{J}^* \cdot \mathbf{E}/2$) represents the time-averaged dissipated Joule power. The terms $p_H = j\omega \mathbf{H}^* \cdot \mathbf{B}/2$ and $p_E = -j\omega \mathbf{E} \cdot \mathbf{D}^*/2$ have a clear physical meaning only for nondispersive media. In particular, for simple isotropic materials (with complex constitutive parameters μ and ε_C), the Poynting theorem can also be expressed as

$$\begin{aligned} \nabla \cdot \mathrm{Re}\{\mathbf{\Pi}\} - \omega \mathrm{Im}\{\varepsilon_\mathrm{C}\} \frac{|E|^2}{2} - \omega \mathrm{Im}\{\mu\} \frac{|H|^2}{2} &= -\mathrm{Re}\left\{ \frac{\mathbf{J}_\mathrm{i}^* \cdot \mathbf{E}}{2} + \frac{\mathbf{M}_\mathrm{i} \cdot \mathbf{H}^*}{2} \right\}, \\ \nabla \cdot \mathrm{Im}\{\mathbf{\Pi}\} + 2\omega (\mathrm{Re}\{\mu\} \frac{|H|^2}{4} - \mathrm{Re}\{\varepsilon_\mathrm{C}\} \frac{|E|^2}{4}) &= -\mathrm{Im}\left\{ \frac{\mathbf{J}_\mathrm{i}^* \cdot \mathbf{E}}{2} + \frac{\mathbf{M}_\mathrm{i} \cdot \mathbf{H}^*}{2} \right\}. \end{aligned} \tag{1.26}$$

In the first equation of (1.26) the terms involving the imaginary parts of μ and ε_C correspond to time-averaged power densities dissipated through a conduction current or for different mechanisms (magnetic and dielectric hysteresys); moreover for media that cannot transfer energy (mechanical or chemical) into the field (i.e., *passive media*) such imaginary parts must be nonpositive. In general, we will refer to *lossless* isotropic media as those materials having the imaginary parts of μ and ε_C identically zero. It can be shown that lossless anisotropic media are characterized by complex tensor permeability and permittivity which are both Hermitian.

For nondispersive media the term into the brackets in the second equation of (1.26) represents the difference between the time-averaged magnetic and electric energy densities. The right-hand side is called *reactive power density.* As the second equation of (1.26) shows, such reactive power density (divided by 2ω) represents a sort of energy exchange between the external and the internal region.

1.3.6 Fundamental Theorems

Three fundamental theorems with applications to EM theory are briefly recalled in this section. They are the *uniqueness*, *reciprocity*, and *equivalence theorems.*

Uniqueness Theorem As in any other problem of mathematical physics, the uniqueness property is a fundamental condition for a problem to be well-posed. First of all, a uniqueness theorem establishes the mandatory information that one needs to obtain the solution of the problem. Second, it is of critical importance to know that the solution that one can obtain through different techniques is also unique. Third, the uniqueness theorem is a fundamental tool for the development of other important theorems, such as the equivalence theorem and the reciprocity theorem. In the rest of the chapter we will refer to frequency-domain problems.

The uniqueness theorem can be formulated as follows: There exists a unique EM field that satisfies Maxwell's equations and constitutive relations in a lossy region provided that the tangential component of **E** over the boundary, or the tangential component of **H** over the boundary, or the former over part of the boundary and the latter over the remaining part of the boundary, are specified. In the case of regions of infinite extent, the boundary conditions for the tangential components of the field are replaced by the radiation condition at infinity.

It is important to know that the proof of the theorem is strictly valid only for lossy media (and, in turn, this restriction is a consequence of the ideal time-harmonic assumption). However, the lossless case can be obtained in the limit of vanishing losses [13].

Reciprocity Theorem Another important theorem of electromagnetism is the reciprocity theorem, which follows directly from Maxwell's equations. In fact, given a set of sources $\{\mathbf{J}_{i1}, \mathbf{M}_{i1}\}$ that produce the fields $\{\mathbf{E}_1, \mathbf{H}_1, \mathbf{D}_1, \mathbf{B}_1\}$ and a second set of sources $\{\mathbf{J}_{i2}, \mathbf{M}_{i2}\}$ that produce the fields $\{\mathbf{E}_2, \mathbf{H}_2, \mathbf{D}_2, \mathbf{B}_2\}$, from (1.6) the following identity can be obtained

$$\begin{aligned}\nabla \cdot (\mathbf{E}_1 \times \mathbf{H}_2 - \mathbf{E}_2 \times \mathbf{H}_1) &= j\omega(\mathbf{H}_1 \cdot \mathbf{B}_2 - \mathbf{H}_2 \cdot \mathbf{B}_1 - \mathbf{E}_1 \cdot \mathbf{D}_2 + \mathbf{E}_2 \cdot \mathbf{D}_1) \\ &\quad + (\mathbf{H}_1 \cdot \mathbf{M}_{i2} - \mathbf{H}_2 \cdot \mathbf{M}_{i1} - \mathbf{E}_1 \cdot \mathbf{J}_{i2} + \mathbf{E}_2 \cdot \mathbf{J}_{i1}).\end{aligned} \tag{1.27}$$

The media for which the first term in the right-hand side of (1.27) is zero are called *reciprocal*. It can be shown that this is the case for isotropic media and also for anisotropic media provided that both the tensor permittivity and permeability are symmetric; examples of nonreciprocal media are lossless gyrotropic materials (for which the tensor constitutive parameters are Hermitian but not symmetric). Therefore, from (1.27), for reciprocal media there results

$$\nabla \cdot (\mathbf{E}_1 \times \mathbf{H}_2 - \mathbf{E}_2 \times \mathbf{H}_1) = (\mathbf{H}_1 \cdot \mathbf{M}_{i2} - \mathbf{H}_2 \cdot \mathbf{M}_{i1} - \mathbf{E}_1 \cdot \mathbf{J}_{i2} + \mathbf{E}_2 \cdot \mathbf{J}_{i1}). \quad (1.28)$$

By integrating (1.28) over a finite volume V bounded by a closed surface S, we obtain the *Lorentz reciprocity theorem*, that is,

$$\oint\!\!\oint_S (\mathbf{E}_1 \times \mathbf{H}_2 - \mathbf{E}_2 \times \mathbf{H}_1) \cdot \mathbf{u}_n dS = \iiint_V (\mathbf{H}_1 \cdot \mathbf{M}_{i2} - \mathbf{H}_2 \cdot \mathbf{M}_{i1} - \mathbf{E}_1 \cdot \mathbf{J}_{i2} + \mathbf{E}_2 \cdot \mathbf{J}_{i1}) dV. \quad (1.29)$$

A system for which the integral at the left-hand side of (1.29) vanishes is said to be *reciprocal*. It can be shown that this is the case if the region V is source free or if an impedance boundary condition holds over the surface S. In such cases, from (1.29), the *reaction theorem* can be obtained, which is expressed by

$$\iiint_V (\mathbf{H}_1 \cdot \mathbf{M}_{i2} - \mathbf{E}_1 \cdot \mathbf{J}_{i2}) dV = \iiint_V (\mathbf{H}_2 \cdot \mathbf{M}_{i1} - \mathbf{E}_2 \cdot \mathbf{J}_{i1}) dV. \quad (1.30)$$

These results can be extended to infinite regions if the impedance boundary condition is replaced by the radiation condition at infinity.

The usefulness of the reciprocity theorem can be understood by considering the EM problem of an elemental electric dipole $\mathbf{J}_{i1} = \mathbf{u}_1 \delta(\mathbf{r} - \mathbf{r}_1)$ placed in free space and a second elemental electric dipole $\mathbf{J}_{i2} = \mathbf{u}_2 \delta(\mathbf{r} - \mathbf{r}_2)$ placed inside a metallic enclosure having an aperture in one of its walls ($\mathbf{u}_1$ and $\mathbf{u}_2$ are the unit vectors along two arbitrary directions). According to the discussion above, the system is reciprocal, and from (1.30) we obtain

$$\mathbf{u}_2 \cdot \mathbf{E}_1(\mathbf{r}_2) = \mathbf{u}_1 \cdot \mathbf{E}_2(\mathbf{r}_1). \quad (1.31)$$

Equation (1.31) expresses the fact that the component along $\mathbf{u}_2$ of the electric field radiated by the dipole (placed in free space) at the point $\mathbf{r}_2$ inside the cavity is equal to the component along $\mathbf{u}_1$ of the electric field radiated by the dipole (placed inside the enclosure) at the point $\mathbf{r}_1$ in free space.

The reciprocity theorem can also be used to show that the EM field produced by an electric surface current density distributed over a PEC surface is identically zero (and, dually, that produced by a magnetic surface current density distributed over a PMC surface). Other applications of the reciprocity theorem regard the mode excitation in waveguides and cavities and receiving and transmitting properties of antennas [13].

Equivalence Principle The equivalence principle is a consequence of Maxwell's equations and the uniqueness theorem. Basically, it allows the original EM problem to be replaced with an equivalent problem whose solution coincides with that of the original problem in a finite region of space. To be effective, the equivalent problem should be easier to solve than the original one.

The first form of the equivalence principle (also known as the *Love equivalence principle*) establishes that the EM field $\{\mathbf{E}, \mathbf{H}\}$ outside a region V bounded by a surface S enclosing the sources $\{\mathbf{J}_\mathrm{i}, \mathbf{M}_\mathrm{i}\}$ is equal to that produced by the equivalent sources $\{\mathbf{J}_S, \mathbf{M}_S\}$ distributed over the surface S and given by

$$\begin{aligned} \mathbf{J}_S &= \mathbf{u}_n \times \mathbf{H}_S, \\ \mathbf{M}_S &= -\mathbf{u}_n \times \mathbf{E}_S, \end{aligned} \tag{1.32}$$

where $\{\mathbf{E}_S, \mathbf{H}_S\}$ is the EM field $\{\mathbf{E}, \mathbf{H}\}$ in correspondence of the surface S and $\mathbf{u}_n$ is the unit vector normal to S pointing outside the region V. It can be shown that the field produced by the equivalent sources inside V is identically zero. The equivalent sources are considered as known terms in the formulation of the problem. However, they depend on the field $\{\mathbf{E}, \mathbf{H}\}$, which is unknown. In practice, there are many problems for which approximate expressions can be found for the equivalent currents, and in any case they are extensively used to formulate exact integral equations for the considered problem.

A second form of the equivalence principle is known as the *Schelkunoff equivalence principle*. It is based on the fact that, according to the Love equivalence principle, the field produced by the equivalent sources inside V is identically zero. It differs from the Love equivalence principle since the medium filling the region V is replaced with a PEC (the boundary conditions on S are not changed). This way, the equivalent magnetic current $\mathbf{M}_S$ is the only radiating source. Dually, the region V could be replaced with a PMC; in this case, $\mathbf{J}_S$ would be the only radiating source. However, it must be pointed out that the two situations considered by Love and Schelkunoff are different: in the latter, the presence of a PEC (or PMC) body must be explicitly taken into account.

1.3.7 Wave Equations, Helmholtz Equations, Electromagnetic Potentials, and Green's Functions

In linear, homogeneous, and isotropic media, we can take the curl of Maxwell's equations in (1.13) and obtain the electric field and the magnetic field *wave equations* as

$$\begin{aligned} \nabla \times \nabla \times \mathbf{E}(\mathbf{r}) - k^2\mathbf{E}(\mathbf{r}) &= -j\omega\mu\mathbf{J}_\mathrm{i}(\mathbf{r}) - \nabla \times \mathbf{M}_\mathrm{i}(\mathbf{r}), \\ \nabla \times \nabla \times \mathbf{H}(\mathbf{r}) - k^2\mathbf{H}(\mathbf{r}) &= -j\omega\varepsilon_\mathrm{C}\mathbf{M}_\mathrm{i}(\mathbf{r}) + \nabla \times \mathbf{J}_\mathrm{i}(\mathbf{r}), \end{aligned} \tag{1.33}$$

where $k^2 = \omega^2\mu\varepsilon_\mathrm{C}$ is called *medium wavenumber* and the dependence on frequency of the fields is assumed and suppressed. From the vector identity

$\nabla \times \nabla \times [\cdot] = \nabla\nabla \cdot [\cdot] - \nabla^2[\cdot]$ applied to (1.33), the Maxwell divergence equations, and the equation of continuity, the *vector Helmholtz equations* for the electric and magnetic fields can be derived as

$$\begin{aligned} \nabla^2\mathbf{E}(\mathbf{r}) + k^2\mathbf{E}(\mathbf{r}) &= j\omega\mu\mathbf{J}_\mathrm{i}(\mathbf{r}) - \frac{\nabla\nabla \cdot \mathbf{J}_\mathrm{i}(\mathbf{r})}{j\omega\varepsilon_\mathrm{C}} + \nabla \times \mathbf{M}_\mathrm{i}(\mathbf{r}), \\ \nabla^2\mathbf{H}(\mathbf{r}) + k^2\mathbf{H}(\mathbf{r}) &= j\omega\varepsilon_\mathrm{C}\mathbf{M}_\mathrm{i}(\mathbf{r}) - \frac{\nabla\nabla \cdot \mathbf{M}_\mathrm{i}(\mathbf{r})}{j\omega\varepsilon_\mathrm{C}} - \nabla \times \mathbf{J}_\mathrm{i}(\mathbf{r}). \end{aligned} \tag{1.34}$$

Both the vector wave equations and the vector Helmholtz equations are inhomogeneous differential equations whose forcing terms can be quite complicated functions. Therefore auxiliary quantities (known as *potentials*) are usually introduced to simplify the analysis. Different choices are possible, although the most common are the magnetic and electric (vector and scalar) potentials $\{\mathbf{A}, \mathbf{F}, V, W\}$ in the Lorentz gauge, which are defined as solutions of the following equations:

$$\begin{aligned} \nabla^2\mathbf{A}(\mathbf{r}) + k^2\mathbf{A}(\mathbf{r}) &= -\mu\mathbf{J}_\mathrm{i}(\mathbf{r}), \\ \nabla^2\mathbf{F}(\mathbf{r}) + k^2\mathbf{F}(\mathbf{r}) &= -\varepsilon_\mathrm{C}\mathbf{M}_\mathrm{i}(\mathbf{r}), \\ \nabla^2 V(\mathbf{r}) + k^2 V(\mathbf{r}) &= -\frac{\rho_\mathrm{e}(\mathbf{r})}{\varepsilon_\mathrm{C}}, \\ \nabla^2 W(\mathbf{r}) + k^2 W(\mathbf{r}) &= -\frac{\rho_\mathrm{m}(\mathbf{r})}{\mu}. \end{aligned} \tag{1.35}$$

The Lorentz gauge implies that $\nabla \cdot \mathbf{A} = -j\omega\mu\varepsilon_\mathrm{C} V$ and $\nabla \cdot \mathbf{F} = -j\omega\mu\varepsilon_\mathrm{C} W$. The electric and magnetic fields are expressed in terms of the potentials as

$$\begin{aligned} \mathbf{E}(\mathbf{r}) &= -j\omega\mathbf{A}(\mathbf{r}) - \nabla V(\mathbf{r}) - \frac{1}{\varepsilon_\mathrm{C}}\nabla \times \mathbf{F}(\mathbf{r}), \\ \mathbf{H}(\mathbf{r}) &= -j\omega\mathbf{F}(\mathbf{r}) - \nabla W(\mathbf{r}) + \frac{1}{\mu}\nabla \times \mathbf{A}(\mathbf{r}). \end{aligned} \tag{1.36}$$

All the equations in (1.35) are again inhomogeneous (vector or scalar) Helmholtz equations, but now with a simple forcing term. They can be solved by means of the Green function method. Basically the scalar Helmholtz equation can be written as an operator equation of the kind

$$L[f(\mathbf{r})] = h(\mathbf{r}), \tag{1.37}$$

where $L[\cdot] = \nabla^2[\cdot] + k^2[\cdot]$, f is the unknown function, and h is the forcing term (and appropriate boundary conditions must be specified). The scalar Green function

$G(\mathbf{r}, \mathbf{r}')$ is thus defined as the solution of the equation

$$L[G(\mathbf{r}, \mathbf{r}')] = -\delta(\mathbf{r} - \mathbf{r}') \tag{1.38}$$

subjected to the same boundary conditions. This way it can be shown that the function $f(\mathbf{r})$ can be expressed through a superposition integral in terms of the Green function and the forcing term as

$$f(\mathbf{r}) = \int_V G(\mathbf{r}, \mathbf{r}')h(\mathbf{r}')\mathrm{d}V'. \tag{1.39}$$

In particular, for a scalar Helmholtz equation in free space (subjected to the radiation condition at infinity) it results

$$G(\mathbf{r}, \mathbf{r}') = \frac{e^{-jk|\mathbf{r}-\mathbf{r}'|}}{4\pi|\mathbf{r} - \mathbf{r}'|} \tag{1.40}$$

If vacuum is considered, $k = k_0 = \omega\sqrt{\mu_0\varepsilon_0}$ is the free-space wavenumber.

In free space the vector Helmholtz equations in (1.35) can be easily separated in three scalar Helmholtz equations, each characterized by the same scalar Green function (1.40). The electric and magnetic fields can be expressed by using the scalar Green function for the potentials in terms of electric and magn-etic sources:

$$\begin{aligned}
\mathbf{E}(\mathbf{r}) &= j\omega\mu \int_V \underline{\mathbf{G}}_{\mathrm{e}}(\mathbf{r}, \mathbf{r}') \cdot \mathbf{J}_{\mathrm{i}}(\mathbf{r}')\mathrm{d}V' + \int_V \underline{\mathbf{G}}_{\mathrm{m}}(\mathbf{r}, \mathbf{r}') \cdot \mathbf{M}_{\mathrm{i}}(\mathbf{r}')\mathrm{d}V', \\
\mathbf{H}(\mathbf{r}) &= \int_V \underline{\mathbf{G}}_{\mathrm{m}}(\mathbf{r}, \mathbf{r}') \cdot \mathbf{J}_{\mathrm{i}}(\mathbf{r}')\mathrm{d}V' - j\omega\varepsilon_{\mathrm{C}} \int_V \underline{\mathbf{G}}_{\mathrm{e}}(\mathbf{r}, \mathbf{r}') \cdot \mathbf{M}_{\mathrm{i}}(\mathbf{r}')\mathrm{d}V'.
\end{aligned} \tag{1.41}$$

The free-space *electric dyadic Green function* $\underline{\mathbf{G}}_{\mathrm{e}}$ is

$$\underline{\mathbf{G}}_{\mathrm{e}}(\mathbf{r}, \mathbf{r}') = \left(\underline{\mathbf{I}} + \frac{\nabla\nabla}{k^2}\right) G(\mathbf{r}, \mathbf{r}'), \tag{1.42}$$

and the free-space *magnetic dyadic Green function* $\underline{\mathbf{G}}_{\mathrm{m}}$ is

$$\underline{\mathbf{G}}_{\mathrm{m}}(\mathbf{r}, \mathbf{r}') = \nabla G(\mathbf{r}, \mathbf{r}') \times \underline{\mathbf{I}}, \tag{1.43}$$

where $\underline{\mathbf{I}}$ is the identity 3×3 tensor.

1.4 BASIC SHIELDING MECHANISMS

EM shielding may be pursued by any of the following main strategies, or by a combination of them:

- Interposition of a "barrier" between the source and the area (volume) where the EM field has to be reduced.
- Introduction of a mean capable of diverging the EM field from the area of interest.
- Introduction of an additional source whose effect is the reduction of the EM field levels in the prescribed area with respect to a situation involving the original source or source system.

The choice of strategy is made according to the characteristics of the source (electromagnetic or physical) and to the characteristics of the area to be protected. Of course, several other factors such as costs or insensitivity to source variations must be accounted for as well.

The interposition of a barrier is particularly effective in reducing the EM field levels when the shield material is highly conducting or when it is characterized by constitutive parameters such that the level of attenuation of the field propagating through the shield is high. A simple situation, which will be studied in detail in Chapter 4, may help clarify this point. A uniform plane wave propagating in a medium with permeability μ, permittivity ε, and conductivity σ, has a propagation constant expressed by $\gamma = \sqrt{j\omega\mu\,(\sigma + j\omega\varepsilon)} = \alpha + j\beta$, where α is the attenuation constant and β the phase constant. Any combination of the values μ, ε, and σ giving rise to a high value of the attenuation constant α is suitable for shielding purposes.

An EM field can be diverted by means of an alternative path, not necessarily enclosing the area to be shielded. Such a path may offer better propagation characteristics to the electric field (by means of highly conducting materials), the electric induction (by means of high permittivity materials), the magnetic induction (high permeability materials).

Generally speaking, at relatively low frequencies (e.g., below tens of MHz) the dominant coupling mechanism is related to pass-through cables and connectors. Above such an approximate threshold the propagation of EM waves through apertures and shield discontinuities becomes more and more important.

Discontinuities treatment is a major issue in shielding theory and practice. Chapters 6 and 9 provide a sound approach and practical solutions, respectively, to this very special key point around which the whole shielding problem turns.

1.5 SOURCE INSIDE OR OUTSIDE THE SHIELDING STRUCTURE AND RECIPROCITY

In general, the techniques for introducing a shield that excludes EM interference from a certain region are identical to those used for confining an EM field in the neighborhood of the source. This is an immediate consequence of the reciprocity theorem as formulated in (1.30) or (1.31). The simplest shield consists in an infinite planar screen that divides two regions of space, region 1 and region 2. When a source is placed in $\mathbf{r}_1$ (in region 1), it produces a certain field at $\mathbf{r}_2$ (in region 2). If the assumptions of the reciprocity theorem are fulfilled, it is easy to see that such a field is the same as that produced at $\mathbf{r}_1$ by the *same* source placed at $\mathbf{r}_2$. There is no difficulty in generalizing these considerations to the more involved case of a source in the presence of an enclosure. In this case region 1 is the interior of the enclosure while region 2 is the external region: thus the field radiated at $\mathbf{r}_2$ by a source placed in $\mathbf{r}_1$ is the same as that radiated at $\mathbf{r}_1$ by the same source placed at $\mathbf{r}_2$. This also means, for example, that the shielding performance of a shielding structure can be calculated (or measured) either by placing the source outside the structure and determining the field inside it or by placing the source inside the structure and determining the field outside it.

Although the previous considerations are quite simple, the basic assumptions must be clear. First of all, the reciprocity theorem (or some of its modifications that account also for nonreciprocal media [12]) must hold: this implies, for example, that the above described conclusions for linear media are not valid in the presence of nonlinear media. The two considered situations (with source inside and outside the shielding structure) must be identical. In particular, the same source must be used in both situations (i.e., same orientation, same amplitude, and same frequency). Finally, from a practical point of view, since a sensor is always used to measure the field at a point, its interactions with the rest of the system (and, in particular, with the field source) must be negligible (at least, within the accuracy limits of measurements).

REFERENCES

[1] IEEE 100. *The Authoritative Dictionary of IEEE Standards Terms*, 7th ed. New York: IEEE Press, 2000.

[2] C. Morris, ed. *Academic Press Dictionary of Science and Technology.* San Diego: Academic Press, 1992.

[3] A. I. Zverev. *Handbook of Filter Synthesis.* Hoboken, NJ: Wiley, 2005.

[4] E. F. Vance. "Shielding and grounding topology for interference control." *AFWL Interaction Note 306*, Apr. 1977.

[5] T. Karlsson. "The topological concept of a generalized shield." *AFWL Interaction Note 461*, Jan. 1988.

[6] K. S. H. Lee, ed. *EMP Interaction: Principles, Techniques, and Reference Data.* Washington, DC: Hemisphere Publishing, 1986, pp. 50–68.

[7] K. S. Kunz. "Interleaving cavity resonances for shielding enhancement in topologically definable spaces." *IEEE Trans. Electromagn. Compat.*, vol. 24, no. 1, pp. 61–64, Feb. 1982.

[8] E. J. Rothwell and M. J. Cloud. *Electromagnetics.* Boca Raton: CRC Press, 2001.

[9] S. A. Schelkunoff. "On teaching the undergraduate electromagnetic theory." *IEEE Trans. Educ.*, vol. E-15, pp. 15–25, Feb. 1972.

[10] C.-T. Tai. "On the presentation of Maxwell's theory." *Proc. IEEE*, vol. 60, no. 8, pp. 936–945, Aug. 1972.

[11] J. Van Bladel. *Singular Electromagnetic Fields and Sources.* New York: IEEE Press, 1996.

[12] R. E. Collin. *Field Theory of Guided Waves*, 2nd ed. Piscataway, NJ: IEEE Press, 1991.

[13] R. F. Harrington. *Time-Harmonic Electromagnetic Fields.* Piscataway, NJ: IEEE Press, 2001.

CHAPTER TWO

Shielding Materials

Today, because the synthesis of new materials is a very active field of research and industrial development, the arsenal of materials available for the realization of shielding structures is always increasing. This chapter provides a review of the properties of materials whose technology is mature enough that they may be considered almost on the shelf. Materials that are still ongoing development or whose present costs discourage widespread use are considered in the last section, with the caution that can be inferred when a situation is destined to change over time.

2.1 STANDARD METALLIC AND FERROMAGNETIC MATERIALS

Most shielding structures are fabricated by means of standard (i.e., nonmagnetic), conductive materials or by means of ferromagnetic materials, which are often preferred for their mechanical properties rather than their ferromagnetic behavior. Moreover it is noteworthy that in most ferromagnetic materials the magnetic permeability decreases with frequency, generally for values close to one at frequencies exceeding a few tens of kHz. Thus, in the frequency range dealt with in the following chapters (electrostatic, magnetostatic, and low-frequency shielding are discussed in Appendixes A and B) and with the purpose of shielding considerations, the main characteristic is represented by the conductivity, which may be strongly affected by the temperature and oxidation of material surfaces. A cautionary word is necessary on the fact that commercial materials are not pure and any variation in their chemical composition is able to modify their conductivity. In addition the most popular reference handbook on materials' properties [1] highlights some slight differences even for pure bulk materials, which are generally (but not always) negligible from an engineering point of view. Moreover anomalous conductive behavior can occur in the

Electromagnetic Shielding by Salvatore Celozzi, Rodolfo Araneo and Giampiero Lovat

TABLE 2.1 Electrical Conductivity of the Most Common Conductive Materials

Conductive Material	Conductivity σ [S/m]
Silver	$6.3 \cdot 10^7$
Copper	$5.9 \cdot 10^7$
Industrial copper	$5.8 \cdot 10^7$
Gold	$4.5 \cdot 10^7$
Aluminum	$3.8 \cdot 10^7$
Industrial aluminum	$3.7 \cdot 10^7$
Lead	$4.8 \cdot 10^6$
Phosphor bronze	$4 \cdot 10^6$
Aluminum nickel bronze	$2 \cdot 10^6$
Tin	$9.2 \cdot 10^6$
Brass	$1.5 \cdot 10^7$–$3 \cdot 10^7$
Steel	$5 \cdot 10^6$–10^7

frequency range from a few tens of GHz to the THz level [2–3]. Table 2.1 lists the conductivity of commonly used shielding materials at room temperature (20°C).

Ferromagnetic materials are paramagnetic materials. Below the Curie temperature ferromagnetic materials show spontaneous magnetization, and this means that the spin moments of neighboring atoms in a microscopically large region (called *domain*) result in a parallel alignment of moments. The application of an external magnetic field changes the domains, and the moments of different domains then tend to line up together. When the applied field is removed, most of the moments remain aligned, which gives rise to significant permanent magnetization. It is notable that other paramagnetic materials show antiparallel aligment of moments (antiferromagnetic materials): if the net magnetic moment is different from zero, the material is called *ferrimagnetic*.

For a review of ferromagnetism beyond the scope of this book, the reader is referred to references in [4,5]. The following discussion recalls some basic concepts about the hysteresis loop.

Hysteresis loops can take different shapes, but a few parameters allow the properties of loops to be characterized. The first type of loop encountered is the *major hysteresis loop*, which is obtained by applying to a specimen a cyclic magnetic field $\mathbf{H}$ (with amplitude H) with values large enough to saturate the material. The ensuing change of the magnetization vector $\mathbf{M}$, or the magnetic flux density $\mathbf{B} = \mu_0(\mathbf{H} + \mathbf{M})$, is recorded along the field direction (components B and M, respectively). The section of the loop from the negative to the positive saturation is called the *ascending major curve*; the other half is called the *descending major curve*. The largest achievable amplitude of magnetization (in the limit $H \to \infty$) is called *saturation magnetization*, M_S. The magnetization amplitude that remains in the specimen after a large field is applied and then reduced to zero is the *remanence*, M_r. The *coercive field* (or *coercitivity*) H_C is the magnetic-field amplitude needed to bring the magnetization from the remanence value M_r to zero; it measures the strength of the field that must be applied to a material in order to cancel out its magnetization. While saturation inductions are comparable in almost all the common materials, coercitivities span an astonishingly wide interval, varying from between 0.2 and 100 A/m for soft magnetic materials to between 200 and 2000 kA/m for hard magnetic materials [6]. The *differential susceptibility* $\chi = \partial M/\partial H$ is defined by the slope of the hysteresis curve at the considered point (i.e., for a certain value of H).

Because of hysteresis, a given point in the *H–M* plane (as shown in Figure 2.1) can be reached in an infinite number of different ways, depending on the previous field history. Two ways of particular importance are via the *return branch* and the *minor loop*. If at some point of the major loop the field is reversed, the locus of points on the *H–M* plane will enter into the hysteresis loop. Such a point is called the *turning point*, and the new curve is called the *first-order return branch*. Another reversal from this curve will originate a *second-order return branch*, and so on. When a cyclic field of variable amplitude is applied to the demagnetized specimen, a set of *minor loops* is obtained. Roughly speaking, a minor loop is formed by any pair of higher order return branches. The line obtained by starting from the demagnetized state and going directly to saturation is termed the *initial magnetization curve*. The curve is not unique because it depends on how the material was previously demagnetized. The initial magnetization curve allows us to compute the *initial susceptibility* χ_{in} as its slope at the origin and, consequently, the *initial permeability* $\mu_{in} = \mu_0(1 + \chi_{in})$. The simplest way to model a ferromagnetic material is by treating it as linear with a constant permeability equal to the initial permeability.

The wrong impression that the hysteresis loop might give is that it constitutes a unique distinctive feature of a ferromagnetic material. On the contrary, hysteresis entails a more complex structure. The hysteresis-loop representation is incomplete for anisotropic materials. This is because only the magnetization component along the field is represented as a function of the applied-field intensity and nothing is

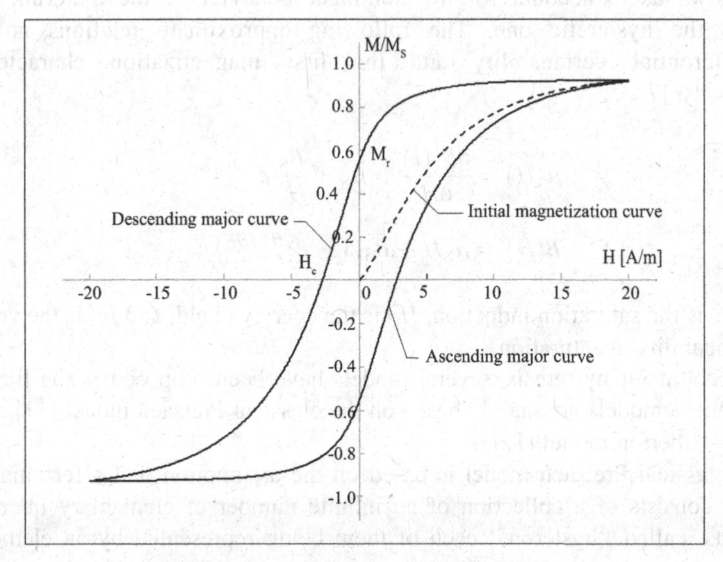

FIGURE 2.1 Example of a hysteresis loop.

said about the behavior of the magnetization components transverse to the field. Thus no anisotropy is identified. Because the magnetization vector is the result of an average process over many domains that depends on the scale of the specimen, the hysteresis loop is affected by the geometry and the dimensions of the specimen, even though the material is the same. Finally, the hysteresis shows a nonlocal memory. This is because **H** and **M** alone are not sufficient to give a complete characterization of the system, so the past time history is needed. The conclusion is that there exists nothing that can be straightforwardly called the hysteresis loop of a material [5].

Magnetic materials can be divided in soft materials (low-coercive field) and hard materials (high-coercive field), which are primarily used for permanent magnets. Soft magnetic materials may be further divided into four broad families: electrical steels, iron-nickel (FeNi) or iron-cobalt (FeCo) alloys, ferrites, and amorphous metals. In the shielding of power-frequency magnetic fields, Fe–(Ni, Co) are generally used; noticeable commercial members of the family are Mumetal, Permalloy, and Permendur.

The hysteretic behavior of ferromagnetic materials is described by separating the linear part of the $B(H)$ characteristic from the magnetization $M(H)$ as

$$B(H) = \mu_0[H + M(H)], \tag{2.1a}$$

$$\frac{\mathrm{d}B(H)}{\mathrm{d}H} = \mu_0\left[1 + \frac{\mathrm{d}M(H)}{\mathrm{d}H}\right]. \tag{2.1b}$$

The main topic when dealing with ferromagnetic materials is the correct modeling of their nonlinear hysteretic behavior.

A first way to proceed is to model only the first magnetization characteristic. This allows us to account for the nonlinear behavior of the material, but it neglects the hysteretic one. The following approximate relations hold for the differential permeability and the first magnetization characteristic, respectively [7]:

$$\mu(H) = \frac{\mathrm{d}B(H)}{\mathrm{d}H} = \mu_S + \frac{B_S}{H_C}e^{-|H|/H_C}, \tag{2.2a}$$

$$B(H) = \mu_S H + B_S(1 - e^{-|H|/H_C}), \tag{2.2b}$$

where B_S is the saturation induction, H_C is the coercive field, and μ_S is the value of the permeability at saturation.

To account for hysteresis, several models have been proposed in the literature [8–10]; these models are mainly based on the classical Preisach model [11] and on the Jiles–Atherton model [12].

The classical Preisach model is based on the assumption that a ferromagnetic material consists of a collection of an infinite number of elementary interacting fragments, called "hysteron," each of them being represented by an elementary rectangular hysteresis loop of two statistically distributed parameters (switching up h_A and switching down h_B fields, with $h_A \geq h_B$). All the hysterons are assumed to have the same saturation magnetization, M_S. If an external magnetic field is increased up to H_1, all the fragments whose up-switching fields h_A are smaller than or equal to the external field switch their magnetization up, whereas if the magnetic field is decreased to H_2 all the hysterons whose down-switching fields h_B are larger than the external field are switched down. A Preisach plane can be constructed by the switching fields (h_A, h_B), and each hysteron has its own exclusive point on the plane. The plane is divided into two parts (a part with switched-up hysterons and a part with switched-down hysterons) by a staircase line that gives the prehistory of the material. It is assumed that for each material a distribution of hysterons exists that does not change during the magnetization process. The distribution is characterized by the Preisach probability distribution function $P(h_A, h_B)$ whose values are nonnegative on the whole plane and shows even symmetry, meaning $P(h_A, h_B) = P(-h_A, -h_B)$. The knowledge of the distribution function gives the full description of the magnetization process. The total magnetization of the system with hysteresis can in fact be computed as

$$M = M_S \int_{-\infty}^{+\infty} dh_A \int_{-\infty}^{h_A} P(h_A, h_B) Q(h_A, h_B) dh_B, \tag{2.3}$$

where $Q(h_A, h_B)$ is the Preisach state function equal to $+1$ or -1, depending on the applied field. The mathematical expressions for the differential susceptibility are

$$\frac{dM}{dH} = 2 \int_{H_1}^{H} P(H, h_B) dh_B, \tag{2.4a}$$

$$\frac{dM}{dH} = 2 \int_{H}^{H_2} P(H, h_B) dh_B, \tag{2.4b}$$

where H_1 and H_2 are the largest previous minimum and the largest previous maximum, respectively. The first equation holds when the applied field increases $(dH/dt > 0)$, and the second one when it decreases $(dH/dt < 0)$. In order to give the complete description of the hysteresis process, two conditions are necessary to be fulfilled: the delation property (erasing the history of the material) and the congruency property (all the minor loops calculated between the same field limits must be congruent). It is known that the congruency property is usually not obeyed by real systems. Consequently a huge literature has tried to assess how far the real magnetization process of materials is from that predicted by the classical Preisach model and has proposed improved modifications of this classical model.

The Jiles–Atherton model is a physically based phenomenological model. The model starts from the definition of the anhysteretic magnetization M_{an} of a material, which represents a global energy minimum toward which the magnetization M strives but cannot reach because of the impedance to domain wall motion. The anhysteretic magnetization curve corresponds to the constitutive law of the material when no losses (i.e., no hysteresis) are considered, and it can be described by a modified Langevin function:

$$M_{\mathrm{an}}(H_{\mathrm{e}}) = M_{\mathrm{S}}\left[\coth\left(\frac{H_{\mathrm{e}}}{a}\right) - \frac{a}{H_{\mathrm{e}}}\right], \tag{2.5}$$

where $H_{\mathrm{e}} = H + \alpha M(H)$ is the effective magnetic field experienced by the domains, H is the external applied magnetic field, a is the domain density, and α describes the domain coupling (the latter are two parameters of the model). The magnetization process is the sum of the contributions from the irreversible magnetization M_{irr} and the reversible magnetization M_{rev}. The reversible component represents the reversible domain wall bowing and translation and the reversible rotation of the magnetic domain. The irreversible component represents the processes in the wall domains and, consequently, the energy loss in the hysteresis material. Their equations are

$$M_{\mathrm{irr}} = M_{\mathrm{an}} + k\delta\frac{\mathrm{d}M_{\mathrm{irr}}}{\mathrm{d}H_{\mathrm{e}}}, \tag{2.6a}$$

$$M_{\mathrm{rev}} = c(M_{\mathrm{an}} - M_{\mathrm{irr}}), \tag{2.6b}$$

where k is a parameter related to the hysteresis losses, c is the reversibility coefficient that belongs to the interval $[0, 1]$, and δ is $+1$ when $\mathrm{d}H/\mathrm{d}t > 0$, otherwise -1. By recalling that $M = M_{\mathrm{rev}} + M_{\mathrm{irr}}$, from (2.6), it is easy to obtain

$$\frac{\mathrm{d}M}{\mathrm{d}H} = (1-c)\frac{M_{\mathrm{an}} - M_{\mathrm{irr}}}{k\delta - \alpha(M_{\mathrm{an}} - M_{\mathrm{irr}})} + c\frac{\mathrm{d}M_{\mathrm{an}}}{\mathrm{d}H}. \tag{2.7}$$

From this expression, after differentiating (2.5) with respect to H and noting that $M_{\mathrm{irr}} = (M - cM_{\mathrm{an}})/(1-c)$, the final nonlinear differential equation can be written as

$$\frac{\mathrm{d}M}{\mathrm{d}H} = \frac{(1-c)[M_{\mathrm{an}}(H) - M(H)]}{k(1-c)\delta - \alpha[M_{\mathrm{an}}(H) - M(H)]} - \frac{cM_{\mathrm{S}}}{a}\mathrm{cosech}^2[H_{\mathrm{e}}(H)] + \frac{caM_{\mathrm{S}}}{[H_{\mathrm{e}}(H)]^2}. \tag{2.8}$$

In Table 2.2, the conductivity at room temperature (20°C) and the range of the static relative magnetic permeability of basic ferromagnetic materials are reported [4,13]. Since most of the commercially available ferromagnetic materials are patented, a number of commercial names denote similar materials with slightly different compositions and properties.

TABLE 2.2 Conductivity and Range of the Relative Magnetic Permeability of the Three Most Common Ferromagnetic Materials

Ferromagnetic Material	Conductivity [S/m]	Range of Relative Magnetic Permeability
Cobalt	$1.6 \cdot 10^7$	70–250
Nickel	$\sim 1.2 \cdot 10^7$	110–600
Iron	$\sim 1 \cdot 10^7$	150–200000

2.2 FERRIMAGNETIC MATERIALS

Ferrimagnetic materials represent a class of ceramic materials that are widely used in high-frequency applications. They are divided in garnets and ferrites, the latter being much more important in shielding applications because of their larger values of the relative magnetic permeability and of their larger losses. Ferrites have a crystal structure and are sintered by means of various metal oxides, obtaining high values of the magnetic permeability (up to 5000 in absence of saturation), of the resistivity (up to 10^{-6} Ωm at 20°C), and a relative dielectric permittivity up to 15.

Different choices for the units of the relevant parameters characterizing the ferrimagnetic behavior of these materials are possible, and care is necessary in dealing with them. In the absence of an applied static field, demagnetized ferrites are isotropic materials exhibiting a magnetic permeability that is a scalar frequency-dependent quantity. In the presence of an applied static magnetic field, ferrites are nonreciprocal materials, exhibiting an anisotropic magnetic behavior, described by the magnetic-permeability tensor $\underline{\boldsymbol{\mu}}$. In particular, if the ferrite is magnetized to saturation by a static magnetic field $\mathbf{H}_0$ (e.g., directed along the z axis of a Cartesian coordinate system so that $\mathbf{H}_0 = H_0\mathbf{u}_z$), in the presence of a small sinusoidal field $\mathbf{H}$ (such that $|\mathbf{H}| \ll H_0$), it exhibits gyromagnetic properties and the magnetic-permeability tensor can be expressed as [14,15]

$$\underline{\boldsymbol{\mu}} = \begin{bmatrix} \mu_1 & j\kappa & 0 \\ -j\kappa & \mu_1 & 0 \\ 0 & 0 & \mu_2 \end{bmatrix}, \tag{2.9}$$

where

$$\mu_1 = \mu_0(1 + \chi_\mathrm{m}), \tag{2.10a}$$

$$\chi_\mathrm{m} = \frac{\mu_0^2\gamma^2 H_0 M_0}{\mu_0^2\gamma^2 H_0^2 - \omega^2}, \tag{2.10b}$$

$$\kappa = -\frac{\omega\mu_0^2\gamma M_0}{\mu_0^2\gamma^2 H_0^2 - \omega^2}, \tag{2.10c}$$

$$\mu_2 = \mu_0. \tag{2.10d}$$

The gyromagnetic ratio γ is a quantity typical of each ferrite chemical composition, ω is the angular frequency of the applied sinusoidal field $\mathbf{H}$, and M_0 is related to H_0.

It appears that the entries of the magnetic-permeability tensor may diverge for a specific value of the angular frequency; the introduction of a damping factor α accounts for losses and is adequate for macroscopic analyses. Various models have been proposed for its inclusion and one possibility is [14]

$$\chi_{\mathrm{m}} = \frac{\mu_0 \gamma M_0 (\mu_0 \gamma H_0 + j\omega\alpha)}{(\mu_0 \gamma H_0 + j\omega\alpha)^2 - \omega^2}, \tag{2.11a}$$

$$\kappa = -\frac{\omega \mu_0 \gamma M_0}{(\mu_0 \gamma H_0 + j\omega\alpha)^2 - \omega^2}. \tag{2.11b}$$

In shielding applications, ferrites find their primary application for absorbing an incident EM field in consideration of their large values of losses: this is very useful in reducing the reflected field inside cavity-like structures once the field is penetrated through them.

2.3 FERROELECTRIC MATERIALS

Properties of ferroelectric materials have been known since the seventeenth century (in the form called *Rochelle salt*), although they were named in this way only around 1940. The industrial development and use of ferroelectric materials is even more recent and often limited to electronic devices. These materials present three main characteristics:

- Anisotropy
- Power losses
- Sensitivity to temperature variations of the main parameters

From a chemical point of view, ferroelectric materials may be ceramic, monomeric, or polymeric metallophthalocyanines. Depending on their structure, composition, and on the process used for their synthesis, the tensor $\underline{\varepsilon}_{\mathrm{r}}$ representing the relative permittivity is often diagonal, possibly exhibiting two entries with the same value (typically very large, in the range up to 3000). The frequency dependence of each entry may be accounted for by means of various models. The most adopted of them gives the electric susceptibility $\chi_{\mathrm{e}}(\omega)$, which is related to the relative permittivity by the known expression:

$$\varepsilon_{\mathrm{r}ii}(\omega) = 1 + \chi_{\mathrm{e}ii}(\omega), \qquad i = x, y, z \tag{2.12}$$

as a function of frequency and are due to Debye [16], Cole–Cole [17], Davidson–Cole [18], and Havriliak–Negami [19]. The relevant expressions for the components of the susceptibility are

$$\chi_e(\omega) \propto \frac{1}{1+(j\omega/\omega_p)} \quad \text{(Debye)}, \tag{2.13a}$$

$$\chi_e(\omega) \propto \frac{1}{1+(j\omega/\omega_p)^{1-\alpha}} \quad \text{(Cole–Cole)}, \tag{2.13b}$$

$$\chi_e(\omega) \propto \frac{1}{\left[1+(j\omega/\omega_p)\right]^{\beta}} \quad \text{(Davidson–Cole)}, \tag{2.13c}$$

$$\chi_e(\omega) \propto \frac{1}{\left[1+(j\omega/\omega_p)^{1-\alpha}\right]^{\beta}} \quad \text{(Havriliak–Negami)}, \tag{2.13d}$$

where ω_p, α, and β are characteristic parameters of each material. These parameters can be determined by starting from the chemical composition of the material but, more often, through measurements. Other adopted models can be found in [20].

Power losses are usually expressed in terms of a loss angle δ, or its tangent $\tan\delta$, or by means of its reciprocal, also called the *quality factor*, Q:

$$Q(\omega) = \frac{1}{\tan\delta(\omega)} \tag{2.14}$$

being

$$\tan\delta(\omega) = \frac{\varepsilon_r''(\omega)}{\varepsilon_r'(\omega)}, \tag{2.15}$$

where $\varepsilon_r'(\omega)$ and $\varepsilon_r''(\omega)$ represent the real and imaginary parts of each entry of the permittivity tensor, respectively. It is important recognizing that the Kramers–Kronig relationships [21,22] establish the link between the real and the imaginary parts of each entry of the permittivity tensor, that is,

$$\mathrm{Re}\left[\frac{\varepsilon(\omega)}{\varepsilon_0}\right] = 1 + \frac{1}{\pi}\mathrm{PV}\int_{-\infty}^{+\infty} \frac{\mathrm{Im}[\varepsilon(\omega')/\varepsilon_0]}{\omega'-\omega}\,d\omega', \tag{2.16a}$$

$$\mathrm{Im}\left[\frac{\varepsilon(\omega)}{\varepsilon_0}\right] = -\frac{1}{\pi}\mathrm{PV}\int_{-\infty}^{+\infty} \frac{\mathrm{Re}[\varepsilon(\omega')/\varepsilon_0]-1}{\omega'-\omega}\,d\omega', \tag{2.16b}$$

where PV stands for the principal value of the improper integral.

Another important parameter is represented by the temperature sensitivity. Temperature can considerably affect the practical use of some ferroelectric materials for some specific applications.

TABLE 2.3 Real Part of the Permittivity and Losses of Common Ferroelectric Materials

Ferroelectric Material	Range for Real Part of Relative Permittivity	Typical Loss Tangent (tan δ)
Barium titanate ($BaTiO_3$)	2500–3000	< 0.015
Barium strontium titanate($Ba_{1\text{-}x}Sr_x$)TiO_3	150–400	0.001–0.003
Ammonium dihydrogen phosphate (ADP) ($NH_4H_2PO_4$)	20–35	
Lead titanate ($PbTiO_3$)	100–200	0.045
Lead zirconate ($PbZrO_3$)	100–150	0.01–0.03
Sodium niobate ($NaNbO_3$)	200–1700	0.025–0.2
Terbium molybdate (TMO)($Tb_2(MoO_4)_3$)	100	0.0007–0.001
Triglycine sulfate (TGS)($(NH_2CH_2COOH)_3.H_2SO_4$)	200–300	0.00002–0.004
Boracite ($Mg_3B_7O_{13}Cl$)	20–50	0.0003–0.0012

In shielding problems, ferroelectric materials may be used to diverge the electric-displacement vector from an assigned region. Their use exploiting this characteristic is limited to very special applications, especially when weight considerations make metals unsuitable for shielding purposes. More often ferroeletric materials are used for the absorption of the incident EM energy, in consideration of the large values of the permittivity. In this case losses can contribute favorably to the attenuation of the EM field propagating through the material, the same as for ferrites [23].

The data on presently available ferroelectric materials are reported in Table 2.3. The range for the relative permittivity accounts for the minimum–maximum values exhibited by the various entries of the permittivity tensor.

2.4 THIN FILMS AND CONDUCTIVE COATINGS

Present technology has made available for shielding purposes materials that may be formed as thin films. The thicknesses of these materials range from about 1 μm to tens of μm and thus offer great advantages in terms of weight and often of cost, in comparison with thick barriers. Sometimes this shielding solution is chosen because it facilitates grounding or offers conductive paths to electrostatic discharges. The shielding performance of thin conductive layers is generally acceptable only at frequencies higher than tens of MHz. A thin film may take the shape of a conductive coating applied in various ways on a housing, internally or externally with respect to the expected source position. An important parameter for shielding considerations is the conductive layer's *surface resistivity* R_S, expressed in Ω and defined as

$$R_S = \frac{1}{\sigma t}, \tag{2.17}$$

where t denotes the thickness of the considered material layer. Often, in order to clearly indicate that the conductive layer has surface resistivity, technical units are adopted, and this parameter is given in terms of Ω/sq or Ω/□.

Another parameter used for thin layers is the *point-to-point resistance*, expressed in Ω, which can be easily measured also in nonflat configurations. Point-to-point resistance may be useful to place in evidence continuity problems at junctions, corners, and similar situations where the definition of surface resistivity is not applicable. However, it should be noted that measurement of the point-to-point resistance is prone to be strongly sensitive to contact pressure.

Four main technologies are currently used:

- Electroless plating
- Conductive painting
- Vacuum metallizing
- Electrolytic deposition

Electroless coatings are generally deposited on plastic housings or enclosures in the form of a double substrate. The first substrate (adherent to the structure) is often copper, and the second substrate (external) is usually nickel; its function is to prevent the oxidation of the copper layer and to provide mechanical protection. The thicknesses range between 1 and 5 μm for the copper layer and are generally less than 1 μm for the nickel one. This technology presents the main advantages of a very low weight increase and of a very uniform and continuous coating deposition. Further it does not require any particular treatment of the surface before the application. Electroless nickel coating has also been applied to wood particles in the production of wooden particleboards [24].

Conductive paints are the most diffuse coating technology for shielding applications, at least when only one layer is applied. The most common types of paint are based on silver, silver-coated copper, or a hybrid of the two types; the diluent may be a solvent or water. The thickness of the paint is easily varied and typically ranges between few μm and few tens of μm. The versatility as concerns thickness, shape, and the advantages of uniformity and low cost is counterbalanced by the need of a preliminary base treatment and by the presence of some limitations in the line-of-sight characteristics of the housing or enclosure to be painted. Some applications of this technique exist for coating [25], for instance, a concrete wall with a colloidal graphite powder, allowing for a moderate shielding performance at low cost.

Vacuum metallizing is generally applied by using aluminum or copper (in the latter case possibly coated with a chromium, nickel, or tin layer). This technology allows for the deposition of thin layers whose thickness is of the order of few μm. The main limitation in its use is represented by penetration issues in deep recesses such as those typical of ventilation ducts and apertures.

Electroplated coatings, usually in copper, are suitable for shielding applications requiring thicknesses up to 25 μm. Electroplating offers better shielding performance with respect to the other three techniques. The availability of various

TABLE 2.4 Surface Resistivity and Thickness of Coatings

Coating Technique	Typical Range for the Resistivity, ρ ($\Omega \cdot m$)	Typical Thickness Range (μm)
Electroless plating	0.01–0.03	1–100
Conductive painting	0.02–0.05	10–75
Vacuum metallizing	0.05–0.1	5–10
Electrolytic deposition	0.007–0.02	10–25

decorative finishes adds a degree of freedom that may enlarge the field of application of electrolytic deposition.

In Table 2.4 the main features of the different coating techniques are summarized.

2.5 OTHER MATERIALS SUITABLE FOR EM SHIELDING APPLICATIONS

Materials whose main function is not that of shielding EM fields have also been modified in their chemical composition or structure in order to provide them some EM performance while maintaining their original features. The most prevalent classes are as follows:

- Structural materials (concrete, plastics, etc.)
- Conductive polymers
- Conductive glasses and transparent materials
- Conductive (and ferromagnetic or ferrimagnetic) papers

2.5.1 Structural Materials

Various types of carbon or metallic (generally steel or nickel) fibers or filaments have been added to cement without varying the mechanical properties of the concrete but considerably improving the shielding performance, especially at frequencies above 1 GHz. Moreover the inclusion of filaments or fibers in various types of thermoplastic matrices has been shown to be effective against EM waves. In both cases (i.e., cement and thermoplastic matrices) the lengths of the fibers or filaments are of the order of few millimeters, and the volume fraction of material added to the basic component is typically in the range between 1% and 4% in cement pastes [26] and up to 20% in plastic pastes [27].

2.5.2 Conductive Polymers

Conducting polymers are promising materials to shield EM radiation. They can reduce or eliminate electromagnetic interference (EMI) because of their relatively large values of conductivity and dielectric permittivity and ease of control of these parameters through chemical processing [28]. Recently a high shielding efficiency of highly conducting doped polyaniline, polypyrrole, and polyacetylene in comparison with that of copper has been reported [29]. Very thin conductive polymer samples have high and weakly temperature-dependent shielding efficiencies. The easy tuning of intrinsic properties by chemical processing suggests that such polymers, especially polyaniline, are good candidates for low-frequency shielding applications. Conductivity depends on doping and on the geometry of the polymer, which can be shaped in spun fibers, especially for the polyaniline [30]. The use of this class of materials is frequent, such as in shielding gaskets, as described in Chapter 9.

2.5.3 Conductive Glasses and Transparent Materials

Different types of optically transparent sheets (i.e., transparent at frequencies having a wavelength in the range between 400 and 700 nm), obtained by using either glasses or plastics, are suitable for EM shielding. The main alternatives consist in a very thin metallic or semiconductor film over a transparent medium, or in a chemical composition capable of preserving the light transmittance and offering some level of attenuation toward the incident EM field. Metallic arrays of conductive shapes embedded in dielectric media are generally included in the classes of artificial materials (e.g., metamaterials) or composites.

Transparency to light spectrum can also be achieved by means of thin metallic films. Thin films are formed from various metals or semiconductors, such as gold [31], silver alloys [32], and indium zinc or tin or cerium oxides [33,34]. Typical thicknesses range between 10 and 100 nm, although values outside this interval are also possible. Typical values of surface resistivity range between 0.5 Ω/□ and 10 Ω/□. The light transmittance is generally higher than 70% and often close to the threshold of 90%.

2.5.4 Conductive (and Ferromagnetic or Ferrimagnetic) Papers

Conductive papers are obtained by mixing wood or synthetic pulp with metallized polyester fibers whose surfaces are coated with nickel or copper and nickel [35]. This way, by means of an amount of metal up to 15 g/m^2, acceptable performance are obtained at frequencies above 30 MHz [36]. Ferromagnetic or ferrimagnetic paper can be obtained by similar formation processes [37].

2.6 SPECIAL MATERIALS

2.6.1 Metamaterials and Chiral Materials

Artificial materials are composite structures consisting of inclusions periodically embedded in a host matrix. When the size of the inclusions and the spatial periods are small compared with the wavelength of the EM field generated by a source, such artificial materials can be homogenized; that is, they can be described as homogeneous materials with effective constitutive parameters that depend on the geometrical and physical properties of the inclusions and of the host medium, and on how the inclusions are placed in the host matrix. If the homogenized artificial materials present EM properties that conventional materials do not possess, they are also called *metamaterials* [38–40]. Other artificial materials based on periodic structures, such as electromagnetic-band-gap (EBG) structured materials and complex surfaces (e.g., high-impedance ground planes and artificial magnetic conductors), involve distances and dimensions of the order of the wavelength or more; they are strongly inhomogeneous and need to be described by the periodic-media formalism (e.g., see Chapter 10).

The most popular class of metamaterials includes structures for which the values of the (effective) permittivity and permeability are simultaneously negative, as considered in [41]. A material having this property can be used to obtain a series of surprising effects, such as backward-wave propagation in the material, a negative index of refraction, or a reversed Doppler effect. The most famous application suggests the possibility of fabricating a superlens providing spatial resolution beyond the diffraction limit [42]. Up to now there is not a common terminology used to designate such metamaterials. Some of the various terms used instead are as follows:

- Left-handed (LH) materials
- Backward-wave (BW) materials
- Negative-index (NI) or negative-refractive index (NRI) materials
- Double-negative (DNG) materials

The term "left-handed" was used in the groundbreaking paper by Veselago [41], and has been widely used since. It highlights a difference with respect to the well-known "right-hand rule" for the direction of the Poynting vector as a function of the electric and magnetic fields' directions. An objection to its use is that "LH" is also used in classifying chiral media. The term BW is not as much used because backward waves can be excited in other types of structures. The NRI term seems to be appropriate when dealing with two- or three-dimensional structures, but it is not meaningful for one-dimensional structures where propagation angles are not involved.

The terms above represent in each case a property resulting from a wave propagating within the metamaterial structure. The term DNG is instead a consequence of the properties of the (effective) constitutive parameters of the material itself (whose permittivity and permeability have both negative values). For ordinary materials these values are both positive, with the noticeable exception represented by ferrimagnetic materials (although usually negative values of the permeability occur in a very narrow band) and plasma. Moreover the reason behind the acronym DNG can be used to define double-positive (DPS) or single-negative (SNG) materials, as only-ε-negative (ENG) and only-μ-negative (MNG).

Many metamaterial structures have been designed and fabricated over the last few years, and their performances have been verified by measurements. However, this topic is still relatively new and in so rapid development that any attempt at giving limits of applicability, quantitative orders of magnitude of the characteristic parameters, and so forth, is prone to be surpassed by new discoveries, technologies, or applications.

The first DNG metamaterial structure was proposed, built, and measured in [43] for microwave applications, by suitably combining the ENG and MNG structures proposed in [33] and [34] (although it was not obvious at all that the resulting material were DNG). The ENG structure consists of a periodic arrangement of parallel metal wires with spatial periods much smaller than the operating wavelength; if the incident electric field is polarized along the wire axis, the relative effective permittivity is a purely scalar frequency-dependent quantity and takes the form

$$\varepsilon_r(\omega) = 1 - \frac{\omega_{pe}^2}{\omega^2 + j\omega\zeta_e}, \tag{2.18}$$

where ω_{pe} depends on the spatial period and on the wire radius, whereas ζ_e depends also on the wire conductivity (the relative effective permeability of the structure is $\mu_r = 1$) [44].

The MNG structure consists of a periodic arrangement of split rings (still with the dimensions and period much smaller than the operating wavelength). If the incident magnetic field is orthogonal to the rings plane, the relative effective permeability is a purely scalar frequency-dependent quantity and takes the form

$$\mu_r(\omega) = 1 - \frac{F\omega^2}{\omega^2 - \omega_{0m}^2 + j\omega\zeta_m}, \tag{2.19}$$

where F and ω_{0m} depend on the geometrical parameters of the structure, whereas ζ_m depends also on the metal losses (the relative effective permittivity of the structure is $\varepsilon_r = 1$ if the rings are in air) [45]. From equations (2.18) and (2.19) it is immediate to see that the real part of the effective permittivity is negative for $\omega^2 < \omega_{pe}^2 - \zeta_e^2$, while the real part of the effective permeability is negative inside a frequency range that in the lossless case reduces to $\omega_{0m}^2 < \omega^2 < \omega_{0m}^2/(1-F)$. Of course, for arbitrarily polarized fields the effective constitutive parameters are anisotropic (or bianisotropic), although much effort is being spent in order to obtain a fully isotropic metamaterial. It is worth mentioning that both the ENG and the MNG structures described above are known since 1950s and 1960s (e.g., [46] and [47], respectively).

So far several metamaterial applications have been developed (especially for guided-wave and antenna applications), and although EM shielding has not yet been fully explored, some properties useful for shielding purposes have already become apparent, such as its selectivity with respect to the frequency (e.g., see Chapter 12).

Chiral media are special metamaterials whose internal structure causes macroscopic effects. The most important of these effects is rotation of the polarization of the propagating field plane [48]. In the optical frequency range there are natural chiral materials that are also recognized to be optically active materials. Moreover chiral materials can be artificially obtained, for example, by randomly embedding a large number of small helices in a host medium. The constitutive relations of isotropic (or bi-isotropic) chiral materials are

$$\begin{aligned} \mathbf{D}(\mathbf{r}) &= \varepsilon \mathbf{E}(\mathbf{r}) - j\kappa\sqrt{\mu_0\varepsilon_0}\,\mathbf{H}(\mathbf{r}), \\ \mathbf{B}(\mathbf{r}) &= \mu \mathbf{H}(\mathbf{r}) + j\kappa\sqrt{\mu_0\varepsilon_0}\,\mathbf{E}(\mathbf{r}), \end{aligned} \tag{2.20}$$

where κ is an adimensional real parameter (the so-called Pasteur parameter) that accounts for the chirality. Some possible applications of chiral materials to EM shielding are demonstrated in [49] and [50]; bi-anisotropic structures (in particular, omega media) that have additional degrees of freedom are discussed in [51] for the design of antireflection coverings.

2.6.2 Composite Materials

Composite materials (especially the widely adopted fiber-reinforced composites) are generally laid up with several plies with fibers with different orientations. The aim is to provide high structural strength and mechanical behavior that is close to isotropy. For EM shielding performance, the fibers or filaments are usually made with materials exhibiting high conductivity, such as conductive carbon fibers [52,53]. The graphite structure's conductivity is achieved by lowering the electrical resistivity of highly crystalline carbon fibers through intercalation, that is, by insertion of guest atoms and molecules between single carbon layers of the graphite structure. The class of intercalates that has been proved to form the most stable compounds with graphite fibers are the alogens, bromine (Br_2), chlorine iodide (ICl), and bromine iodide (BrCl). Graphite fibers are usually embedded in isocyanate or epoxy matrices, although cyanate esters and polyimides have also been attempted. Because the orientation patterns influence the transmission and reflection properties of the composite materials, the continuous carbon fiber (CCF) composites are woven in pre-designed patterns to provide a continuously conductive network before being laminated into composites.

Casey [54] provides a detailed study of the shielding properties of both graphite-epoxy and screened boron-epoxy composite laminates. Interestingly, although anisotropic, the graphite-epoxy composite is modeled as an isotropic, homogeneous, nonmagnetic, and conductive medium (the anisotropy is assumed to have minor effects in the material characterization). The screened boron-epoxy composite is, on the other hand, modeled as an isotropic and homogeneous dielectric layer with a bonded wire-mesh screen in one surface (both bonded and unbonded wire meshes are studied in detail by Wait [55]). Casey's study shows that the two composites behave very differently as EM shielding devices: the graphite-epoxy composite proved to be a low-pass filter, whereas the screened boron-epoxy composite proved to be a high-pass filter.

The most frequently used materials for the wires are stainless steel, phosphor bronze, and copper; the wire diameters are generally in the range of 10 to 100 μm. Often the shielding properties are given in terms of the number of wires per unit length, with the number of openings per inch (OPI) being the most used measure (see also Chapter 9).

2.6.3 Nanomaterials

Nanomaterials are materials constituted by particles or fibers with dimensions typically smaller than 100 nm. Nanomaterials have shown some promise as shielding materials. In particular, carbon nanotubes (CNTs, an allotrope form of carbon, taking the form of cylindrical carbon molecules) exhibit extraordinary strength and unique electrical properties, besides being efficient heat conductors [56]. Because of the symmetry and unique electronic structure of the graphite layers, defined as graphene, the structure of a nanotube strongly determines its electrical properties. Each nanotube must be treated as a distinct molecule with a unique structure. The so-called chiral indexes n and m are used to specify the unique manner in which the single layer of graphite is rolled up in a seamless carbon nanotube. For example, if $2n + m = 3q$ (where q is an integer), the nanotube is metallic but with electric current densities higher than those of metals such as silver and copper; otherwise, the nanotube is a semiconductor. Therefore variations of nanotube diameters less than 0.1% can considerably change their electrical behavior.

Nanomaterials are being studied for possible EM applications, by way of exploiting their tunable constitutive parameters and/or selective properties. An important class of nanomaterials consists of periodic nanostructures (which may be periodic in one, two, or three dimensions). Their main application is in the design of metallo-dielectric multilayers that act as opaque screens in the radio-frequency range and as transparent screens in the optical range [57], in the development of plasmonic devices, and in the design of nanocorrugated screens that allow an extraordinary transmission of EM radiation through apertures with dimensions smaller than the operating wavelength [58]. Other nanomaterials are the so-called nano-mixtures, formed by a random arrangement of inclusions with dimensions of the order of nm embedded in a host matrix. These materials could be employed in the design of frequency-selective materials or EM absorbers with dimensions much smaller than those of conventional ones [59].

2.6.4 High-Temperature Superconductors

High-temperature superconductors (HTSCs) were discovered in 1986 by Georg Bednorz and Alex Muller while studying the conductivity of a lanthanum–barium–copper oxide ceramic, whose critical temperature (30 K) was the highest measured to date. However, this discovery started a surge of activity that unraveled superconducting behavior at temperatures as high as 135 K. The HTSC is formed as layers of copper oxide interspaced by layers containing barium and other atoms.

While the yttrium compound has a regular crystal structure, the lanthanum compound is classified as a solid solution. Very effective shielding has been observed for HTSC materials compared with the above-mentioned conventional materials, especially at very low frequencies down to DC applications [60,61]. However, some dependence of the shielding properties of these materials on the surface properties has also been observed; in particular, the use of HTSC in the form of thin films prepared by laser ablation has shown better performance compared with powdered HTSC material [62].

REFERENCES

[1] P. S. Neelakanta. *Handbook of Electromagnetic Materials: Monolithic and Composite Versions and their Applications*. Boca Raton: CRC Press, 1995.

[2] S. Lucyszyn. "Investigation of anomalous room temperatures conduction losses in normal metals at terahertz frequencies." *IEE Proc., Pt. H: Microw. Antennas Propagat.*, vol. 151, no. 4, pp. 321–329, Aug. 2004.

[3] S. Lucyszyn. "Investigation of Wang's model for room-temperature conduction losses in normal metals at Terahertz frequencies." *IEEE Trans. Microwave Theory Tech.*, vol. 53, no. 4, pp. 1398–1403, Apr. 2005.

[4] R. M. Bozorth. *Ferromagnetism*. New York: IEEE Press, 1993.

[5] G. Bertotti. *Hysteresis in Magnetism*. San Diego: Academic Press, 1998.

[6] G. E. Fish. "Soft magnetic materials." *Proc. IEEE*, vol. 78, no. 6, pp. 947–971, Jun. 1990.

[7] D. E. Merewether. "Electromagnetic pulse transmission through a thin sheet of saturable ferromagnetic material of infinite surface area." *IEEE Trans. Electromagn. Compat.*, vol. 11, no. 4, pp. 139–143, Nov. 1969.

[8] E. Della Torre. *Magnetic Hysteresis*. New York: IEEE Press, 1994.

[9] I. D. Mayergoyz. *Mathematical Models of Hysteresis*. New York: Springer-Verlag, 1991.

[10] D. C. Jiles and D. L. Atherton. "Ferromagnetic hysteresys." *IEEE Trans. Magn.*, vol. 19, no. 5, pp. 2183–2185, Sep. 1983.

[11] F. Preisach. "Über die Magnetische Nachwirkung." *Zeitschrift für Physik*, vol. 94, no. 5/6, pp. 277–302, May 1935.

[12] D. C. Jiles and D. L. Atherton. "Theory of ferromagnetic hysteresys." *J. Appl. Phys.*, vol. 55, no. 6, 2115–2120, Mar. 1984.

[13] A. Goldman. *Handbook of Modern Ferromagnetic Materials*, Boston: Kluwer, 1999.

[14] A. J. Baden Fuller. *Ferrites at Microwave Frequencies*. London: Peregrinus, 1987.

[15] D. Polder. "On the phenomenology of ferromagnetic resonance." *Phys. Rev.*, vol. 73, pp. 1120–1121, 1948.

[16] P. Debye. "Einige resultate eine kinetischen theorie der isolatoren." *Phisik. Zeits.*, vol. 13, pp. 97–100, 1912.

[17] K. S. Cole and R. H. Cole. "Dispersion and absorption in dielectrics." *J. Chem. Phys.*, vol. 9, pp. 341–351, Apr. 1941.

[18] D. W. Davidson and R. H. Cole. "Dielectric relaxation in glycole, propylene glycol, and *n*Propanol." *J. Chem. Phys.*, vol. 19, pp. 1484–1490, 1951.

[19] S. Havriliak and S. Negami. "A complex plane representation of dielectric and mechanical relaxation processes in some polymers." *Polymer*, vol. 8, pp. 161–210, 1967.

[20] H. S. Nalwa, ed. *Handbook of Low and High Dielectric Constant Materials and Their Applications*. San Diego: Academic Press, 1999.

[21] H. A. Kramers. "La diffusion de la lumière par les atomes." *Atti Congr. Int. Fis.*, vol. 2, pp. 545–557, 1927.

[22] R. de L. Kronig. "On the theory of dispersion of x-rays." *J. Opt. Soc. Am.*, vol. 12, no. 6, pp. 547–557, Jun. 1926.

[23] V. M. Petrov and V. V. Gagulin. "Microwave absorbing materials." *Inorganic Materials*, vol. 37, no. 2, pp. 93–98, Feb. 2001.

[24] C. Nagasawa, Y. Kumagai, K. Urabe, and S. Shinagawa. "Electromagnetic shielding particleboard with nickel-plated wood particles." *J. Porous Mater.*, vol. 6, no. 3, pp. 247–254, May 1999.

[25] J. Wu and D. D. L. Chung. "Improving colloidal graphite for electromagnetic interference shielding using 0.1 μm diameter carbon filaments." *Carbon*, vol. 41, no. 6, pp. 1313–1315, 2003.

[26] S. Wen and D. D. L. Chung. "Electromagnetic interference shielding reaching 70 dB in steel fiber cement." *Cem. Concr. Res.*, vol. 34, no. 2, pp. 329–332, Feb. 2004.

[27] D. D. L. Chung. "Electromagnetic interference shielding effectiveness of carbon materials." *Carbon*, vol. 39, no. 2, pp. 279–285, Feb. 2001.

[28] B. Kim, V. Koncar, E. Devaux, C. Dufour, and P. Viallier. "Electrical and morphological properties of PP and PET conductive polymer fibers." *Synth. Met.*, vol. 146, no. 2, pp. 167–174, Oct. 2004.

[29] M. S. Kim, H. K. Kim, S. W. Byun, S. H. Jeong, Y. K. Hong, J. S. Joo, K. T. Song, J. K. Kim, C. J. Lee, and J. Y. Lee. "PET fabric/polypyrrole composite with high electrical conductivity for EMI shielding." *Synth. Met.*, vol. 126, no. 2, pp. 233–239, Feb. 2002.

[30] S. Courric and V. H. Tran. "The electromagnetic properties of poly(p-phenylene-vinylene) derivatives." *Polym. Adv. Technol.*, vol. 11, no. 6, pp. 273–279, Jun. 2000.

[31] R. C. Hansen and W. T. Pawlewicz. "Effective conductivity and microwave reflectivity of thin metallic films." *IEEE Trans. Microwave Theory Tech.*, vol. 30, no. 11, pp. 2064–2066, Nov. 1982.

[32] W. M. Kim, D. Y. Ku, I.-k. Lee, Y. W. Seo, B.-k. Cheong, T. S. Lee, I.-h- Kim, and K. S. Lee. "The electromagnetic interference shielding effect of indium-zinc oxide/silver alloy multilayered thin films." *Thin Solid Films*, vol. 473, no. 2, pp. 315–320, Feb. 2005.

[33] S. W. Kim, J.-H. Lee, J. Kim, and H.-W. Lee. "Investigations of electrical properties in silica/indium tin oxide two-layer film for effective electromagnetic shielding." *J. Am. Ceram. Soc.*, vol. 87, no. 12, pp. 2213–2217, Dec. 2004.

[34] M.-M. Bagheri-Mohagheghi and M. Shokooh-Saremi. "The influence of Al doping on the electrical, optical and structural properties of SnO_2 transparent conducting films deposited by the spray pyrolysis technique." *J. Phys. D: Appl. Phys.*, vol. 37, pp. 1248–1253, 2004.

[35] S. Shinagawa, Y. Kumagai, and K. Urabe. "Conductive paper containing metallized polyester fibers for electromagnetic interference shielding." *J. Porous Mater.*, vol. 6, no. 3, pp. 185–190, May 1999.

[36] S. Shinagawa, Y. Kumagai, H. Umehara, and P. Jenvanitpanjakul. "Conductive papers for electromagnetic shielding." *Proc. Internat. Conf. Electromagn. Interfer. Compat. '99*, 6–8 Dec. 1999, pp. 372–375, 1999.

[37] J. A. Carrazana-Garcia, M. A. Lopez-Quintela, and J. R. Rey. "Ferrimagnetic paper obtained by in situ synthesis of substitued ferrites." *IEEE Trans. Magn.*, vol. 31, no. 6, pp. 3126–3130, Nov. 1995.

[38] C. Caloz and T. Itoh. *Electromagnetic Metamaterials, Transmission Line Theory and Microwave Applications*. Hoboken, NJ: Wiley-IEEE Press, 2005.

[39] G. V. Eleftheriades and K. G. Balmain, ed. *Negative-Refraction Metamaterials: Fundamental Principles and Applications*. Hoboken, NJ: Wiley-IEEE Press, 2005.

[40] R. W. Ziolkowski and N. Engheta, ed. *Electromagnetic Metamaterials: Physics and Engineering Explorations*. Hoboken, NJ: Wiley-IEEE Press, 2006.

[41] V. Veselago. "The electrodynamics of substances with simultaneously negative values of ε and μ." *Soviet Physics Uspekhi*, vol. 10, no. 4, pp. 509–514, Feb. 1968.

[42] J. B. Pendry. "Negative refraction makes a perfect lens." *Phys. Rev. Lett.*, vol. 85, no. 18, pp. 3966–3969, Oct. 2000.

[43] D. R. Smith, W. J. Padilla, D. C. Vier, S. C. Nemat-Nasser, and S. Schultz. "Composite medium with simultaneously negative permeability and permittivity." *Phys. Rev. Lett.*, vol. 84, no. 18, pp. 4184–4187, May 2000.

[44] J. B. Pendry, A. J. Holden, D. J. Robbins, and W. J. Stewart. "Low frequency plasmons in thin-wire structures." *J. Phys. Condens. Matter*, vol. 10, no. 22, pp. 4785–4809, June 1998.

[45] J. B. Pendry, A. J. Holden, D. J. Robbins, and W. J. Stewart. "Magnetism from conductors and enhanced nonlinear phenomena." *IEEE Trans. Microwave Theory Tech.*, vol. 47, no. 11, pp. 2075–1084, Nov. 1999.

[46] W. Rotman. "Plasma simulation by artificial and parallel plate media." *IRE Trans. Antennas Propagat.*, vol. AP-10, no. 1, pp. 82–95, Jan. 1962.

[47] S. A. Schelkunoff and H. T. Friis. *Antennas: Theory and Practice*. New York: Wiley, 1952.

[48] I. V. Lindell, A. H. Sihvola, S. A. Tretyakov, and A. J. Viitanen. *Electromagnetic Waves in Chiral and Bi-isotropic Media*. Norwood, MA: Artech House, 1994.

[49] D. L. Jaggard and N. Engheta. "Chiroshield: A Salisbury/Dallenbach shield alternative." *Electron. Lett.*, vol. 26, no. 17, pp. 1332–1334, Aug. 1990.

[50] A. H. Sihvola and M. E. Ermutlu. "Shielding effect of hollow chiral sphere." *IEEE Trans. Electromagn. Compat.*, vol. 39, no. 3, pp. 219–224, Aug. 1997.

[51] S. A. Tretyakov and A. A. Sochava. "Proposed composite material for non-reflecting shields and antenna radomes." *Electron. Lett.*, vol. 29, no. 12, pp. 1048–1049, June 1993.

[52] D. D. L. Chung. "Electromagnetic interference shielding effectiveness of carbon materials." *Carbon*, vol. 39, no. 2, pp. 279–285, Feb. 2001.

[53] A. Kaynak, A. Polat, and U. Yilmazer. "Some microwave and mechanical properties of carbon-fiber-polypropilene and carbon black -polypropilene composites." *Mater. Res. Bull.*, vol. 31, no. 10, pp. 1195–1206, Oct. 1996.

[54] K. F. Casey. "EMP penetration through advanced composite skin panels." *AFWL Interaction Note 315*, Dec. 1976.

[55] J. R. Wait. "Theories of scattering from wire grid and mesh structures." in *Electromagnetic Scattering*, P. L. E. Uslenghi, ed., New York: Academic, 1978, pp. 253–287.

[56] R. Saito, G. Dresselhaus, and M. S. Dresselhaus. *Physical Properties of Carbon Nanotubes*. London: Imperial College Press, 1998.

[57] M. Scalora, M. J. Bloemer, A. S. Pethel, J. P. Dowling, C. M. Bowden, and A. S. Manka. "Transparent, metallo-dielectric, one-dimensional, photonic band-gap structures." *J. Appl. Phys.*, vol. 83, no. 5, pp. 2377–2383, Mar. 1998.

[58] H. J. Lezec, A. Degiron, E. Devaux, R. A. Linke, L. Martin-Moreno, F. J. Garcia-Vidal, and T. W. Ebbesen. "Beaming light from a subwavelength aperture." *Science*, vol. 297, pp. 820–822, Aug. 1997.

[59] J. R. Liu, M. Itoh, T. Horikawa, K. Machidaa, S. Sugimoto, and T. Maeda. "Gigahertz range electromagnetic wave absorbers made of amorphous-carbon-based magnetic nanocomposites." *J. Appl. Phys.*, vol. 98, pp. 054305-1–054305-7, 2005.

[60] T. Nurgaliev. "Theoretical study of superconducting film response to AC magnetic field," *Physica C*, vol. 249, no. 1, pp. 25–32, Jul. 1995.

[61] K. A. Müller, M. Takashige, and J. G. Bednorz. "Flux trapping and superconductive glass state in La_2CuO_4-y:Ba." *Phys. Rev. Lett.*, vol. 58, no. 11, pp. 1143–1146, Mar. 1987.

[62] V. Vidyalal, K. Rajasree, and C. P. G. Vallabhan. "Measurements of electromagnetic shielding effect using HTSC materials." *Mod. Phys. Lett. B*, vol. 10, no. 7, 293–297, 1996.

CHAPTER THREE

Figures of Merit for Shielding Configurations

Before entering into the complex mathematical world that is inherent in almost any actual shielding problem, it is worth to embark on a detailed, thorough consideration about the purposes of shielding structures, and consequently about the ways to quantify how much the goal has been achieved by means of a given configuration. In other words, in this chapter priority is given to the definition of adequate figures of merit used in setting up an analysis or design problem and in comparing the performance of existing shielding structures. Unfortunately, there is not a consensus [1] on the methods used to measure the effectiveness of shielding configurations or even that of single shielding components, such as panels and films, gaskets, and shielded windows. We leave to Chapter 9 the subject of single, commonly used components perfomance. Henceforth attention is focused on the figures of merit adopted or adoptable to quantify the shielding effects of ideal situations (e.g., planar panels of infinite extent or cylindrical tubes) and of actual configurations (especially enclosures).

In a broad sense, the *shielding effectiveness* is a measure of the reduction or attenuation of the EM field at a given point in space caused by the insertion of a shield between the source and that point. Several subtleties are hidden in such a generic definition, as it will be discussed in this chapter. Starting from the generally adopted parameters, the key issues and the current trends toward the definition of more meaningful figures of merit are presented and discussed.

3.1 (LOCAL) SHIELDING EFFECTIVENESS

The shielding effectiveness (SE) of any configuration is defined as a ratio, usually expressed in decibels, between two suitable EM power, electric-field, or magnetic-field values. The three ratios—the ratio involving the power and the two ratios concerning the electric and magnetic fields—are numerically coincident only under very special circumstances. Other definitions will be presented and discussed in the next sections. The most popular choices are described in the following:

First and preferred choice [2]. The SE is defined as the ratio between the absolute value of the electric (or magnetic) field $\mathbf{E}(\mathbf{r})$ (or $\mathbf{H}(\mathbf{r})$) that is present at a given point $\mathbf{r}$ beyond the shield and the absolute value of the electric (or magnetic) field that would have been present at the same point in the absence of the shield itself. By definition, the latter is the incident field $\mathbf{E}^{\text{inc}}(\mathbf{r})$ (or $\mathbf{H}^{\text{inc}}(\mathbf{r})$). The SE is very often expressed in dB. Moreover, in order to obtain positive values in normal situations, the reciprocal of the previous definition is considered, namely

$$\text{SE}_E = 20 \log \frac{|\mathbf{E}^{\text{inc}}(\mathbf{r})|}{|\mathbf{E}(\mathbf{r})|} \tag{3.1a}$$

for the electric shielding effectiveness and

$$\text{SE}_H = 20 \log \frac{|\mathbf{H}^{\text{inc}}(\mathbf{r})|}{|\mathbf{H}(\mathbf{r})|} \tag{3.1b}$$

for the magnetic shielding effectiveness.

The third possibility involving the power before and after the shield installation reads as

$$\text{SE}_P = 10 \log \frac{|\text{Re}\{\mathbf{\Pi}^{\text{inc}}(\mathbf{r})\}|}{|\text{Re}\{\mathbf{\Pi}(\mathbf{r})\}|} \tag{3.1c}$$

where $\mathbf{\Pi}^{\text{inc}}$ and $\mathbf{\Pi}$ are the Poynting vectors of the field in the absence and in the presence of the shield, respectively. Figures 3.1 shows the configurations leading to the three evaluation of SE.

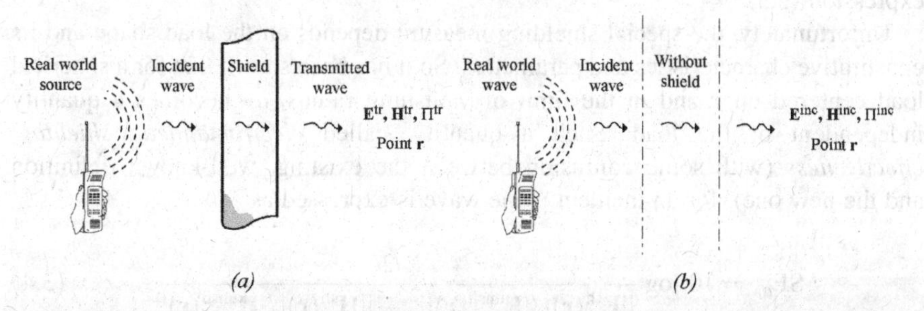

FIGURE 3.1 Configuration with (*a*) and without (*b*) the shield for the evaluation of the SE.

Second and less used choice. In this case the SE is obtained considering the ratio between the absolute values of the electric (or magnetic) field at two different points $\mathbf{r}_1$ and $\mathbf{r}_2$, placed just above and just below the screen, respectively,

$$\mathrm{SE}_E = \frac{|\mathbf{E}(\mathbf{r}_1)|}{|\mathbf{E}(\mathbf{r}_2)|} \tag{3.2a}$$

or, in terms of magnetic fields and power, respectively,

$$\mathrm{SE}_H = \frac{|\mathbf{H}(\mathbf{r}_1)|}{|\mathbf{H}(\mathbf{r}_2)|}, \tag{3.2b}$$

$$\mathrm{SE}_P = \frac{|\mathrm{Re}\{\mathbf{\Pi}(\mathbf{r}_1)\}|}{|\mathrm{Re}\{\mathbf{\Pi}(\mathbf{r}_2)\}|}. \tag{3.2c}$$

Sometimes the name *shielding factor* (or *field attenuation* [3]) instead of *shielding effectiveness* is adopted, although the term shielding factor is used to indicate expressions (3.1) as well. Usually the quantities (3.2) are expressed in dB units.

Another local figure of merit is the *special shielding measure*, a_P. This term was introduced by Klinkenbusch for the time-harmonic case [4] and defined as the ratio between the time-averaged EM power received by a load in the absence of the shield, P^{unsh}, and the EM power received by the same load when the shield is present, P^{sh}. Therefore, in dB units, at any given position $\mathbf{r}$, a_p is defined as

$$a_P = 10 \log \frac{P^{\mathrm{unsh}}(\mathbf{r})}{P^{\mathrm{sh}}(\mathbf{r})}. \tag{3.3}$$

It is evident that any standard for the measurement of the shielding performance conforms better to such a definition rather than to the classical definition. Nevertheless, in both cases there is a contrast between the finite physical dimensions of the load (i.e., the receiving antenna) and the single point of reference for a_P in the expression (3.3).

Unfortunately, the special shielding measure depends on the load shape and its constitutive characteristics and parameters. So it has been shown that for a spherical load centered on $\mathbf{r}$ and in the limit of vanishing radius, a_P becomes a quantity independent of the load. Such a quantity, called *electromagnetic shielding effectiveness* (with some confusion between the existing, well-known definition and the new one), for an incident plane wave is expressed as

$$\mathrm{SE}_{a_P} = 10 \log \frac{2}{[|\mathbf{E}^{\mathrm{sh}}(\mathbf{r})|/|\mathbf{E}^{\mathrm{unsh}}(\mathbf{r})|]^2 + [|\mathbf{H}^{\mathrm{sh}}(\mathbf{r})|/|\mathbf{H}^{\mathrm{unsh}}(\mathbf{r})|]^2}. \tag{3.4}$$

The figures of merit described above can be generalized also to *transient* incident plane waves [4] by considering the energy delivered, in a prescribed time interval, to a load located in $\mathbf{r}$ in the absence, W^{unsh}, or in the presence, W^{sh}, of a given shielding configuration. The relevant expression for the special shielding measure is

$$a_W = 10 \log \frac{W^{\mathrm{unsh}}(\mathbf{r})}{W^{\mathrm{sh}}(\mathbf{r})}. \tag{3.5}$$

Again, in the limit of vanishing load dimensions, when a transient plane wave with spectral density distribution $S(\omega)$ is considered as incident field, the following expression for the transient electromagnetic shielding effectiveness is obtained:

$$\mathrm{SE}_{a_P} = 10 \log \frac{2 \int_0^{\infty} |S(\omega)|^2 \omega \, \mathrm{d}\omega}{\int_0^{\infty} |S(\omega)|^2 \{[|\mathbf{E}^{\mathrm{sh}}(\mathbf{r})|/|\mathbf{E}^{\mathrm{unsh}}(\mathbf{r})|]^2 + \{[|\mathbf{H}^{\mathrm{sh}}(\mathbf{r})|/|\mathbf{H}^{\mathrm{unsh}}(\mathbf{r})|]^2\} \omega \, \mathrm{d}\omega}, \tag{3.6}$$

which is clearly independent of the load.

It should be noted that all the previous definitions are *local* in the sense that they provide an information only at a specific point and nothing can be said about the remaining of the shielded volume. In particular, when in the volume there are spatial resonances due to standing waves or when the field penetrating through the various paths behaves in an unpredicted or unpredictable way, the definitions above can provide useless and even misleading information. In fact the position and the spatial orientation of sensitive equipment are generally not known in advance, neither is it known what their sensitivity to EM-field levels and to the power delivered to them are. Moreover all the definitions, when applied to closed shielded volumes, require an empty enclosure to be tested, while it is recognized that the load can considerably influence the EM-field distribution and intensity. Despite such drawbacks, the great advantages of the definitions above are their simplicity and intuitiveness.

3.2 THE GLOBAL POINT OF VIEW

For practical purposes, reference will be made henceforth to closed shielded volumes, generally delimited by a shielding enclosure. Two fundamentally different philosophies are available to characterize the global performance of a given shielding structure. The first is among the recommended practices for measurement of SE, and it suggests that the average and minimum values be given on the basis of measured data at several different positions [5]. The aim is to assess the worst-case performance. But such a conservative approach can be misleading (or yield meaningless data) in the presence of resonances. The second approach is aimed at gathering a more fruitful information, preserving the synthetic form of a figure of merit. In this connection it should be noted that complex configurations, such as those dealt with in shielding problems, must be described under a number of different perspective angles, thus requiring a lot of information. The search for figures of merit capable of retaining both a synthetic form and significant information could be futile if the compromise between the two opposite requirements is not judiciously addressed. Unfortunately, any increase in the quality or in the quantity of information calls for additional work at both the design stage (in terms of computational complexity) and the experimental stage (necessary for validation or compliance purposes).

At least in principle the search for the maximum value of the field in a shielded volume is not a trivial task, and the conservative approach has relevance from an engineering point of view (opening interesting scenarios about the first philosophy). As regards the second approach, the main issues that will concern us are basically two:

- Which are the most important quantities that qualify a shielding configuration.
- How to account for the field distribution in the shielded volume.

As to the first point, classical and recent definitions are based on the assumption (both implicit and explicit) that the delivery of power or energy is the major concern in shielding problems. This assumption is adequate only in a limited set of circumstances, where there exist a number of induced effects that are related to the flux over a surface, or to a line integral, according to Maxwell's equations. In our information technology society it is more likely for undesired effects to be associated with spatial variations in these quantities rather than with the power (energy) transferred to susceptible components.

As to the second point, Maxwell's equations offer the way to obtain meaningful parameters: although some spatial derivatives of the field components are directly linked with induced, undesired effects, the characterization of the influence of the shielding enclosure on field uniformity provides useful information about the real effectiveness of the structure.

3.3 OTHER PROPOSALS OF FIGURES OF MERIT

Figures of merit corresponding to the last two comments have been introduced as *global shielding efficiency* (*GSE*) and *disuniformity reduction efficiency* (*DRE*), respectively [6,7].

Because both surface dimensions and shape as well as its orientation in space are arbitrary, the information on the reduction of the EM field at one specific position is not adequate to represent the actual behavior of an enclosure and, in particular, its performance, which is interpreted as its capability in reducing the coupling between the EM sources and the environment (closed or open) to be protected. Thus a parameter adequate for a general characterization could be represented by a suitable integral of the field quantities over the protected/shielded volume. This way the possible presence of small regions where the shielding performances are poor ("hot spots") is not revealed. However a partial recovery of this information is enabled by the second proposed parameter, *DRE*.

For the reasons above the comparison of the shielding performance of different enclosures may be facilitated by the introduction of a new parameter named *global shielding efficiency* (GSE) and defined as the integral, performed over the whole volume protected by the structure under test, of the considered field quantity:

$$\mathrm{GSE}_{\mathrm{dB}}^{E} = 20 \log \frac{\iiint_{U_{\mathrm{encl}}} |\mathbf{E}^{\mathrm{inc}}(\mathbf{r})| dV}{\iiint_{U_{\mathrm{encl}}} |\mathbf{E}(\mathbf{r})| dV} \tag{3.7a}$$

and

$$\mathrm{GSE}_{\mathrm{dB}}^{H} = 20 \log \frac{\iiint_{U_{\mathrm{encl}}} |\mathbf{H}^{\mathrm{inc}}(\mathbf{r})| dV}{\iiint_{U_{\mathrm{encl}}} |\mathbf{H}(\mathbf{r})| dV}, \tag{3.7b}$$

where U_{encl} is the internal volume of the enclosure, and $\mathbf{E}^{\mathrm{inc}}$ (or $\mathbf{H}^{\mathrm{inc}}$) and $\mathbf{E}$ (or $\mathbf{H}$) are the electric (magnetic) field in the absence and in the presence of the shield, respectively. Although the motivation of GSE is not due to the assessment of the power inside the enclosure volume, the factor 20 is retained for an easier comparison with traditional and widely used standards.

In the expressions above it has been assumed that the source of the EM field is outside the enclosure while the sensitive devices to be protected are inside. Of course, the reverse is often true, and additionally compliance with standards of many products requires the dual configuration. The reciprocity theorem (when applicable) can help in studying such a configuration, when the GSE is treated as the contribution in a point outside the enclosure due to a source placed inside at all the possible locations. Other subtleties should be fully analyzed as well, especially as concerns the practical measurements of the proposed quantities. In the following, reference is made to configurations with sources located outside the enclosure.

Several observations are noted below about some further relevant aspects concerning measurements:

- Measurement of the integral quantity (3.7a) and/or (3.7b)
- Measurement or evaluation of the field in the absence of the shielding enclosure
- Antenna configurations in the various frequency bands of interest

About the first issue, it should be clear that the output of real EM antennas is generally related to an integral of the field quantity to be measured and the antenna factor accounts for the conversion from the integral to the local quantity. However, in the case of GSE such a consideration cannot be the key for measurement; the simplest solution resides in resorting to classical expressions [8] for the approximation of an integral when the function is to be integrated at several discrete points. The regular shape of commonly used enclosures considerably simplifies this task.

The second issue concerns a well-known critical fact about shielding measurement, especially for "large" enclosures (i.e., those falling into the field of application of IEEE Std 299–1997): the fields are measured not in the absence of the enclosure, but with doors and apertures open. This implies that in the numerators of (3.7a) and (3.7b) an incorrect reference value is used because of the effects of the shielding walls.

The usual care should be given to the constancy of the power driven to the transmitting antenna in the presence and in the absence of the shield. Antennas of small dimensions are recommended because enclosures of physical dimensions smaller than those to which IEEE Std 299 is applicable are frequently encountered and the revision of the above-mentioned standard is under preparation. A small loop for the LF range and a tunable microstrip antenna may be appropriate for such a measurement. The location of antennas will depend on both the dimensions of the enclosure and the number and locations of discontinuities on the shield surfaces (apertures, seams, etc.).

Again, Maxwell's equations can mitigate the loss of information on the presence of local "hot spots" inherent in the definition of the GSE. Induced effects are due also to spatial variations of EM-field components. Thus the influence of the shielding enclosure on field uniformity can provide useful information about the real effectiveness of the structure. The proposed coefficient accounting for this aspect of the shield behavior is the *disuniformity reduction efficiency* (*DRE*):

$$\mathrm{DRE_{dB}} = 20\log\frac{\iiint_{U_{\mathrm{encl}}} f^{\mathrm{inc}}(\mathbf{r})\mathrm{d}\mathbf{V}}{\iiint_{U_{\mathrm{encl}}} f(\mathbf{r})\mathrm{d}\mathbf{V}}, \tag{3.8}$$

where

$$f^{\mathrm{inc}} = \left|\frac{\partial H_x^{\mathrm{inc}}}{\partial y}\right| + \left|\frac{\partial H_x^{\mathrm{inc}}}{\partial z}\right| + \left|\frac{\partial H_y^{\mathrm{inc}}}{\partial x}\right| + \left|\frac{\partial H_y^{\mathrm{inc}}}{\partial z}\right| + \left|\frac{\partial H_z^{\mathrm{inc}}}{\partial x}\right| + \left|\frac{\partial H_z^{\mathrm{inc}}}{\partial y}\right| \tag{3.9a}$$

and

$$f = \left|\frac{\partial H_x}{\partial y}\right| + \left|\frac{\partial H_x}{\partial z}\right| + \left|\frac{\partial H_y}{\partial x}\right| + \left|\frac{\partial H_x}{\partial z}\right| + \left|\frac{\partial H_z}{\partial x}\right| + \left|\frac{\partial H_z}{\partial y}\right|, \tag{3.9b}$$

or similar spatial derivatives of the electric field components. All the relevant spatial derivatives have been included because of the degrees of freedom existing in the placement of sensitive components and equipment inside the enclosure.

Note that only the spatial variations in Maxwell's curl equations appear in (3.9). This is because they are expected to be the causes of induced effects when a possible loading is inserted inside the enclosure. The properties of the new integral parameters are shown in Figure 3.2. Figure 3.2*a* shows the enclosure under test with a rectangular aperture on one side illuminated by a plane wave impinging perpendicularly on the aperture, with the electric field directed along the shortest side of the aperture. The results concerning the GSE are shown in Figure 3.2*b* where

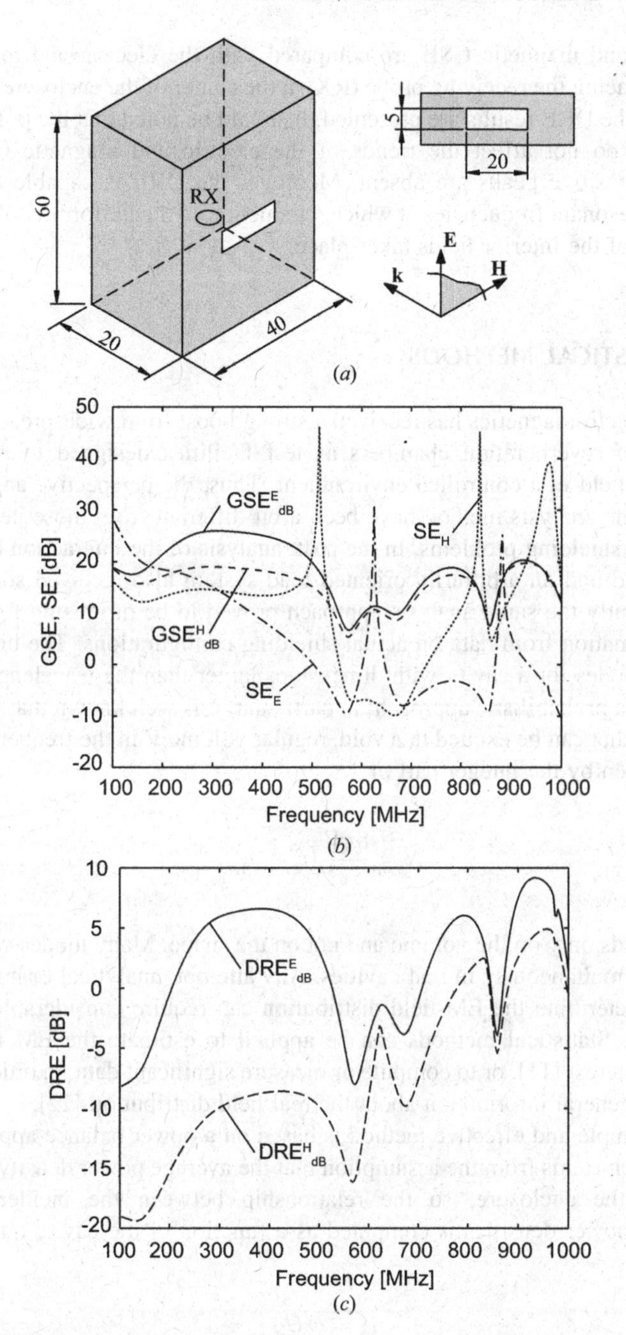

FIGURE 3.2 Geometry of the enclosure with a rectangular aperture illuminated by a plane wave (dimensions in cm) (*a*). Comparison among the electric and magnetic GSE and the classical electric and magnetic SE (*b*). Electric and magnetic DRE (*c*).

the electric and magnetic GSE are compared with the electric and magnetic SE computed placing the receiving probe (RX) at the center of the enclosure. Finally, in Figure 3.2*c* the DRE results are presented. It should be noted that the possible zeros of the field do not affect the trends of the electric and magnetic GSE where misleading positive peaks are absent. Moreover the DRE is capable to partially resolve the resonant frequencies at which a reduction of the uniformity of the spatial distribution of the interior fields takes place.

3.4 STATISTICAL METHODS

Statistical electromagnetics has received a strong boost from widespread interest in the utility of reverberation chambers as test facilities designed to simulate an "expected" field in a controlled environment. Thus, the perspective angle and the purpose of the analysis method have been a bit different from those necessary for approaching shielding problems. In the past, analysis of the interaction between an arbitrary field and an arbitrarily oriented load system also received some interest [9,10]. Recently the same analysis approach proved to be promising for gathering useful information from data on actual shielding configurations. The high spectral density of modes for a cavity with dimensions larger than the wavelength is at the root of such a probabilistic approach; in particular, it is well known that the number N of modes that can be excited in a void, regular volume V in the frequency interval $[f_1, f_2]$ is given by the integer part of

$$N = \frac{8\pi V}{3\,c^3}(f_2^3 - f_1^3), \tag{3.10}$$

which depends only on the volume and not on the shape. Many modes can in effect be excited simultaneously in real cavities. Any attempt, analytical or numerical, to accurately determine the EM-field distribution can require considerable computational effort. Statistical methods can be applied to estimate the EM field in the volume of interest [11], or to compute or measure significant data, in order to gather some more general information about the real field distribution [12].

A very simple and effective method is based on a power balance approach [13]. This approach stems from the assumption that the average power density is uniform throughout the enclosure, so the relationship between the incident and the transmitted power densities is computed as a function of the cavity quality factor Q, defined as

$$Q = \frac{\omega\,U_S}{P_d}, \tag{3.11}$$

where ω is the angular frequency of the excitation, U_S denotes the steady-state energy in the cavity, and P_d represents the dissipated power. By the main assumption of power-density uniformity, the steady-state energy U_S is proportional to the volume of the cavity V through the average energy density W, that is,

$$U_S = W\,V. \tag{3.12}$$

The dissipated power P_d is related to four terms,

$$P_d = P_{d1} + P_{d2} + P_{d3} + P_{d4}, \tag{3.13}$$

expressing the power dissipated in the walls of the enclosure, the power absorbed by the objects loading the enclosure, the power exiting from apertures and leakage paths, and that delivered to receiving antennas, respectively. It follows immediately that four partial quality factors can be introduced:

$$Q^{-1} = Q_1^{-1} + Q_2^{-1} + Q_3^{-1} + Q_4^{-1}. \tag{3.14}$$

The following approximate expressions for the four partial quality factors have been derived [13]:

$$Q_1 = \frac{3\,V}{2\mu_{rw}\,S\,\delta_w}, \tag{3.15a}$$

$$Q_2 = \frac{2\pi V}{\lambda_0\langle\sigma_o\rangle}, \tag{3.15b}$$

$$Q_3 = \frac{4\pi V}{\lambda_0\langle\sigma_a\rangle}, \tag{3.15c}$$

$$Q_3 = \frac{16\pi^2 V}{m\lambda_0^3}, \tag{3.15d}$$

where μ_{rw} is the relative magnetic permeability of the enclosure walls, S is the inner cavity surface, $\delta_w = \sqrt{2/(\omega\mu_0\mu_{rw}\sigma_w)}$ is the penetration depth of the EM field in the wall material (and σ_w is the conductivity of the enclosure walls), λ_0 is the free-space wavelength, $\langle\sigma_o\rangle$ is the sum of the absorption cross sections of the lossy objects averaged over all the angles of incidence and over all the polarizations, and $\langle\sigma_a\rangle$ is the sum of the transmission cross sections of the apertures averaged over all the angles of incidence and over all the polarizations. Finally, m is the impedance mismatch factor defined as $m = 4RR_L/|Z + Z_L|^2$, where $Z = R + jX$ and $Z_L = R_L + jX_L$ are the receiving-antenna and load impedances, respectively. By assuming that the main penetration of the EM field is through an aperture (whose total transmission cross section is σ_t), the power transmitted into the enclosure P_t can be obtained as a function of the incident power density p_{inc}. In particular,

$$P_t = \sigma_t p_{inc}. \tag{3.16}$$

The total transmission cross section σ_t is a function of frequency, polarization, angle of incidence, and shape of the aperture. Moreover, in case of multiple apertures, σ_t is assumed to be given by the sum of the individual contributions. For steady-state conditions, the power penetrating into the enclosure must be equal to the total dissipated power, and the power density in the enclosure p_{encl} is

$$p_{encl} = \frac{\sigma_t \lambda Q}{2\pi V} p_{inc}. \tag{3.17}$$

The assumption of power-density uniformity inside the enclosure allows for the SE evaluation as

$$SE = 10 \log \frac{2\pi V}{\sigma_t \lambda Q}. \tag{3.18}$$

An electrically small circular aperture has a total transmission cross section, averaged for all the angles of incidence and polarizations, given by

$$\langle \sigma_t \rangle = \frac{16}{9\pi} k^4 a^6, \tag{3.19}$$

where $k = \omega/c$ is the wavenumber and a is the aperture radius.

The methods for the determination of the statistical distribution of the EM field in regular enclosures are rather complex and beyond the purposes of this book [11,14–16]. Various approximations may be introduced that lead to much simpler formulations. Research is presently still in progress to avoid, for example, the failure of the Kolmogorov-Smirnov goodness-of-fit test occurring with most of the various cumulative distribution functions that have been adopted.

3.5 ENERGY-BASED, CONTENT-ORIENTED DEFINITION

Recently another figure of merit has been proposed for the measurement of the shielding performance of enclosures [17]. It is based on the ratio between the power absorbed by a shielded load P_A (expressed in W), and the incident power density p_{inc} (in W/m^2). This figure of merit, which has the dimensions of a surface area, has been called the *shielding aperture* (SA) of the enclosure under test, that is,

$$SA = \frac{P_A}{p_{inc}}. \tag{3.20}$$

Although the approach is not far from that proposed by Klinkenbusch, the solution to the issue concerning the *representative content* (RC) (i.e., the load to be considered) is very different; while in [4] the limit for vanishing load dimensions is considered, in [17] the adoption of a standard RC is proposed. In particular, a thin block of a special foam is considered and investigated and compared with an actual printed circuit board. The key issue appears to be the selection of RCs representative of actual configurations, and in practice, the assessment of their characteristics, in order to account for possible differences in the EM behavior of RCs used in different test facilities. A second less important issue concerns the effective measurement/evaluation of the power delivered to each RC.

3.6 PERFORMANCE OF SHIELDED CABLES

Definition (3.1) can be applied to shielded cables but with some difficulty because of the actual dimensions of common cables, which make difficult the insertion of field probes inside the shield: the wire(s) inside the shield could be considered antennas capable of detecting the EM field. Further the measurement suffers from several drawbacks: the two setups (with and without the shield) will differ in their terminal conditions because of the influence of the shield on the line characteristics, namely the propagation constant and the characteristic impedance. The two configurations must be identical, and this requirement is not easy to achieve; therefore one is always left with uncertainty as to its degree of fulfilment. For the reasons above, wide use is made of the *transfer impedance* for the comparison of shielded cables. This is an absolute per unit length parameter representing the ratio between the voltage appearing on the internal side of the shield when a given current is flowing on the external shield surface and the current itself, or viceversa. Shielded cables are the subject of Chapter 8, where the influence of constructive parameters on the shielding performance is discussed in some detail.

REFERENCES

[1] J. Butler. "Shielding effectiveness—Why don't we have a consensus industry standard?" *Proc. 1997 IEEE Int. Symp. Electromagnetic Compatibility*. Austin, TX, 18–22 Aug. 1997, pp. 29–35.

[2] IEEE Std-299–2006. "IEEE Standard method for measuring the effectiveness of electromagnetic shielding enclosures." *Institute of Electrical and Electronics Engineers (IEEE)*, Piscataway, NJ, Feb. 28, 2007.

[3] L. O. Hoeft and J. S. Hofstra. "Experimental and theoretical analysis of the magnetic field attenuation of enclosures." *IEEE Trans. Electromagn. Compat.*, vol. 30, no. 3, pp. 326–340, Aug. 1988.

[4] L. Klinkenbusch. "On the shielding effectiveness of enclosures." *IEEE Trans. Electromagn. Compat.*, vol. 47, no. 3, pp. 589–601, Aug. 2005.

[5] J. E. Bridges. "Proposed recommended practices for the measurement of shielding effectiveness high-performance shielding enclosures." *IEEE Trans. Electromagn. Compat.*, vol. 10, no. 1, pp. 82–94, Mar. 1968.

[6] S. Celozzi. "New figures of merit for the characterization of the performance of shielding enclosures." *IEEE Trans. Electromagn. Compat.*, vol. 46, no. 1, p. 142, Feb. 2004.

[7] R. Araneo and S. Celozzi. "Actual performance of shielding enclosures." *Proc. 2004 IEEE Int. Symp. Electromagnetic Compatibility*. S. Clara, CA, 9–13 Aug. 2004, pp. 539–544.

[8] M. Abramowitz and I. A. Stegun. *Handbook of Mathematical Functions*, 9th ed. NewYork: Dover, 1972.

[9] M. A. Morgan and F. M. Tesche. "Statistical analysis of critical load excitations induced on a random cable system by an incident driving field: Basic concepts and methodology." *AFWL Interaction Note 249*, Jul. 1975.

[10] W. R. Graham and C. T. C. Mo. "Probability distribution of CW induced currents on randomly oriented sub-resonant loops and wires." *AFWL Interaction Note 321*, May 1976.

[11] R. H. Price, H. T. Davis, and E. P. Wenaas. "Determination of the statistical distribution of electromagnetic-field amplitudes in complex cavities." *Phys. Rev. E*, vol. 48, no. 6, pp. 4716–4729, Dec. 1993.

[12] C. F. Bunting and S.-P. Yu. "Field penetration in a rectangular box using numerical techniques: An effort to obtain statistical shielding effectiveness." *IEEE Trans. Electromagn. Compat.*, vol. 46, no. 2, pp. 160–168, May 2004.

[13] D. A. Hill, M. T. Ma, A. R. Ondrejka, B. F. Riddle, M. L. Crawford, and R. T. Johnk. "Aperture excitation of electrically large, lossy cavities." *IEEE Trans. Electromagn. Compat.*, vol. 36, no. 3, pp. 169–178, Aug. 1994.

[14] T. H. Lehman. "A statistical theory of electromagnetic fields in complex cavities." *AFWL Interaction Note 494*, May 1993.

[15] R. H. St. John and R. Holland. "Field-component statistics of externally illuminated overmoded cavities." *IEEE Trans. Electromagn. Compat.*, vol. 42, no. 2, pp. 125–134, May 2000.

[16] A. H. Panaretos, C. A. Balanis, and C. R. Birtcher. "Shielding effectiveness and statistical analysis of cylindrical scale fuselage model." *IEEE Trans. Electromagn. Compat.*, vol. 47, no. 2, pp. 361–366, May 2005.

[17] A. C. Marvin, J. F. Dawson, S. Ward, L. Dawson, J. Clegg, and A. Weissenfeld. "A proposed new definition and measurement of the shielding effect of equipment enclosures." *IEEE Trans. Electromagn. Compat.*, vol. 46, no. 3, pp. 459–468, Aug. 2004.

CHAPTER FOUR

Shielding Effectiveness of Stratified Media

Electromagnetic plane waves are the simplest solution of the time-harmonic Maxwell equations in a homogeneous and source-free spatial region. The study of their properties is useful for better understanding the behavior of more complex fields. For instance, the far field radiated by an arbitrary source has, locally and sufficiently far from the source, the characteristics of a plane wave (this usually allows an EM field impinging on a given structure to be approximated as a plane wave). In addition the exact field produced by any source can be expressed in terms of a continuous superposition of elemental plane-wave components (plane-wave spectrum representation). On the other hand, stratified media (in planar, cylindrical, and spherical configurations) are the simplest example of inhomogeneous media and are often considered as models of many shields.

4.1 ELECTROMAGNETIC PLANE WAVES: DEFINITIONS AND PROPERTIES

A plane wave (with a time-harmonic behavior $e^{j\omega t}$, suppressed for the sake of conciseness) is a frequency-domain EM field mathematically expressed as

$$\begin{aligned}\mathbf{E}(x,y,z) &= \mathbf{E}_0 e^{-j(k_x x + k_y y + k_z z)} = \mathbf{E}_0 e^{-j\mathbf{k}\cdot\mathbf{r}},\\ \mathbf{H}(x,y,z) &= \mathbf{H}_0 e^{-j(k_x x + k_y y + k_z z)} = \mathbf{H}_0 e^{-j\mathbf{k}\cdot\mathbf{r}},\end{aligned} \tag{4.1}$$

where $\mathbf{E}_0$ and $\mathbf{H}_0$ are constant complex vectors, and k_x, k_y, and k_z are complex scalars that define the wavevector $\mathbf{k} = k_x\mathbf{u}_x + k_y\mathbf{u}_y + k_z\mathbf{u}_z$. The vectors $\mathbf{E}_0$ and $\mathbf{H}_0$ define the

Electromagnetic Shielding by Salvatore Celozzi, Rodolfo Araneo and Giampiero Lovat

polarization of the plane wave, namely the temporal evolution of the vector direction in the plane on which the vector lies. In particular, according to the locus described by the tip of the vector, polarization can be linear, circular, or, in the most general case, elliptic. For circular and elliptic polarizations, the sense of rotation for increasing time can be clockwise or counterclockwise with respect to a fixed direction. The real and (the opposite of) the imaginary part of the wavevector **k** are the so-called phase vector $\boldsymbol{\beta}$ and attenuation vector $\boldsymbol{\alpha}$, respectively, so that $\mathbf{k} = \boldsymbol{\beta} - j\boldsymbol{\alpha}$ (note that $\boldsymbol{\beta}$ and $\boldsymbol{\alpha}$ are real vectors). Alternatively, the propagation vector $\boldsymbol{\gamma} = \boldsymbol{\alpha} + j\boldsymbol{\beta}$ can be used (i.e., $\mathbf{k} = -j\boldsymbol{\gamma}$). In any case, the magnitude of $\boldsymbol{\beta}$ gives the phase shift per unit length along the direction of $\boldsymbol{\beta}$, while the magnitude of $\boldsymbol{\alpha}$ determines the rate of attenuation along the direction of $\boldsymbol{\alpha}$. From Maxwell's equations it follows that the wavevector must satisfy the "separation equation"

$$\mathbf{k} \cdot \mathbf{k} = k_x^2 + k_y^2 + k_z^2 = k^2, \tag{4.2}$$

where k is the *wavenumber* of the medium defined as

$$k = \omega\sqrt{\mu\left(\varepsilon - j\frac{\sigma}{\omega}\right)} = \omega\sqrt{\mu\varepsilon_c} \tag{4.3}$$

(the principal branch of the square root is chosen so that the imaginary part of k is nonpositive). Recall that in the frequency domain the complex electric permittivity ε_c is introduced to account for conductive losses, meaning $\varepsilon_c = \varepsilon - j\sigma/\omega$. From Maxwell'sequations, it follows that the wavevector is related to the electric and magnetic field vectors through

$$\begin{aligned} \mathbf{k} \times \mathbf{E} &= \omega\mu\mathbf{H}, \\ \mathbf{k} \times \mathbf{H} &= \omega\varepsilon_c\mathbf{E}. \end{aligned} \tag{4.4}$$

Equations (4.4) imply that $\mathbf{E} \cdot \mathbf{H} = 0$, $\mathbf{k} \cdot \mathbf{E} = 0$, and $\mathbf{k} \cdot \mathbf{H} = 0$. From (4.4) it also follows that $\mathbf{E} \cdot \mathbf{E} = \eta^2 \mathbf{H} \cdot \mathbf{H}$, where η is the *intrinsic impedance* of the medium defined as

$$\eta = \sqrt{\frac{\mu}{\varepsilon_c}} = \sqrt{\frac{\mu}{\varepsilon - j\frac{\sigma}{\omega}}} = \sqrt{\frac{j\omega\mu}{\sigma + j\omega\varepsilon}} \tag{4.5}$$

(in this case the principal branch of the square root is chosen so that the real part of η is nonnegative [2]).

A characterization of a plane wave can be obtained by introducing the direction angles (ϕ, θ). Such direction angles (which are in general complex) are defined as

$$\begin{aligned} k_x &= k\cos\phi\sin\theta, \\ k_y &= k\sin\phi\sin\theta, \\ k_z &= k\cos\theta\,. \end{aligned} \tag{4.6}$$

Uniform (or *homogeneous*) plane waves are an important subset of plane waves. A uniform plane wave is defined as a plane wave for which both the direction angles (ϕ, θ) are real. Therefore, for a uniform plane wave, the wavevector can be written as $\mathbf{k} = k\,\mathbf{u}_r$, where $\mathbf{u}_r$ is a real unit vector given by

$$\mathbf{u}_r = \cos\phi \sin\theta \mathbf{u}_x + \sin\phi \sin\theta \mathbf{u}_y + \cos\theta \mathbf{u}_z. \tag{4.7}$$

For a uniform plane wave, the phase and attenuation vectors point in the same direction of the unit vector $\mathbf{u}_r$. It can easily be shown that inside an isotropic medium this is also the direction of the time-averaged power-flow vector (i.e., the real part of the Poynting vector $\mathbf{\Pi}$). Therefore, for a uniform plane wave, all the vectors $\boldsymbol{\beta}$, $\boldsymbol{\alpha}$, and $\mathbf{\Pi}$ point in the same direction: such a direction can be uniquely defined as the *direction of propagation* of the uniform plane wave. Since $\mathbf{u}_r \cdot \mathbf{E} = \mathbf{u}_r \cdot \mathbf{H} = 0$ (i.e., $E_r = H_r = 0$), a uniform plane wave is said to be transverse electromagnetic (or TEM) to the direction of propagation $\mathbf{u}_r$; furthermore we have $|\mathbf{E}| = |\eta||\mathbf{H}|$.

It is worth noting that for lossless media the wavenumber k is real, so the attenuation vector is zero. Moreover it is important to note that since for non-uniform plane waves the direction angles are complex, their attenuation vector is never zero and cannot be parallel to the phase vector, so a more involved relation exists among the $\boldsymbol{\beta}$, $\boldsymbol{\alpha}$, and $\mathbf{\Pi}$ vectors. Therefore the direction of propagation cannot be uniquely defined for non-uniform plane waves.

As is well known [1], an EM field in a source-free homogeneous region can be represented as the sum of two types of fields, a field that is transverse magnetic to an arbitrary direction x (TM_x, for which $H_x = 0$) and a field that is transverse electric to x (TE_x, for which $E_x = 0$). A general plane wave can thus be written as the sum of a TM_x plane wave and a TE_x plane wave. By defining the plane of incidence as the plane containing the phase vector $\boldsymbol{\beta}$ and the unit vector $\mathbf{u}_x$, the TM_x plane wave has the electric field lying in the plane of incidence (and it is also called E wave, vertically polarized wave, or p-polarized wave), while the TE_x plane wave has the magnetic field lying in the plane of incidence (and it is also called H wave, horizontally polarized wave, or s-polarized wave). Actual sources generally give rise to EM fields that are not uniform TEM (or TE, TM) plane waves. However, as already mentioned above, at distances far enough from the sources (i.e., for distances much larger than the longest significant wavelength associated with the frequency spectrum of the source and than its largest dimension), the assumption of a uniform plane wave is often a reasonable approximation.

In the following, we will first consider the problem of plane-wave incidence on a planar shield. The convenient formalism of the so-called equivalent transmission line will next be introduced. Such a formalism has the advantage of automatically including the boundary conditions for the fields. Last, the analysis of shielding in the presence of curved surfaces and near-field sources will be discussed.

4.2 UNIFORM PLANE WAVES INCIDENT ON A PLANAR SHIELD

The simplest shielding configuration is that depicted in Figure 4.1 where a uniform plane wave is impinging with an angle θ^{inc} on a single planar shield of infinite extension in the transverse y and z directions (i.e., whose dimensions are large enough to neglect edge effects) with a finite thickness d in the x direction. Although this configuration is almost not realizable and far from usual actual conditions, some basic ideas on shielding are based on this problem. For this reason it has been kept as a starting point of any further and deeper analysis.

The most useful way to solve the problem is based on the analogy existing between uniform plane-wave propagation in a stratified medium and the propagation of voltages and currents in a uniform transmission line. In the next subsection we briefly summarize the key steps in proving such analogy.

4.2.1 Transmission-Line Approach

We thus start with a plane wave propagating in a medium that is linear, homogeneous, stationary, isotropic, and possibly lossy and dispersive. Linearity and stationarity of the constitutive relations characterizing the EM behavior of the shield material guarantee that the boundary value problem may be conveniently posed in the frequency domain, by introducing the magnetic permeability μ and electric permittivity ε_c that, under the assumptions above, are complex frequency-dependent scalar quantities, meaning $\mu = \mu' - j\mu''$ (with $\mu' > 0$ and $\mu'' > 0$) and $\varepsilon_c = \varepsilon_c' - j\varepsilon_c''$ (with $\varepsilon_c' > 0$ and $\varepsilon_c'' > 0$). The possible occurrence of negative values of ε_c' and/or μ' will be discussed in Chapter 12. It should be noted that the hypothesis of linearity is often fulfilled in consideration of the reduced values of the incident high-frequency fields, while the stationarity assumption is often reasonable because of the different order of magnitude existing between the time scales of EM phenomena and typical changes (e.g., for thermal or corrosion issues) occurring in the materials characteristics.

As was mentioned above, a general plane wave can be written as the sum of a TM_x and a TE_x plane wave. The TM_x propagation in the ith region of a stratified medium

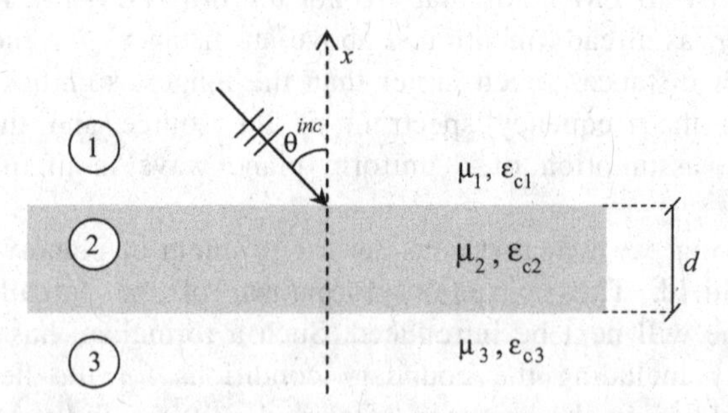

FIGURE 4.1 Uniform plane wave impinging on a shield of finite thickness d and of infinite transverse dimensions.

characterized by its relative permittivity ε_{cri} and permeability μ_{ri} is considered first. In such a case the magnetic field of the plane wave can be expressed as

$$\mathbf{H}^{\text{TM}}(x, y, z) = (\mathbf{u}_y H_y^{\text{TM}} + \mathbf{u}_z H_z^{\text{TM}}) e^{-j(\pm k_{xi}x + k_{yi}y + k_{zi}z)}, \tag{4.8}$$

where $k_{xi}^2 + k_{yi}^2 + k_{zi}^2 = k_i^2 = \omega^2 \mu_i \varepsilon_{ci} = \omega^2 \mu_0 \varepsilon_0 \mu_{ri} \varepsilon_{cri} = k_0^2 \mu_{ri} \varepsilon_{cri}$. In equation (4.8) the plus sign is chosen for plane waves propagating or decaying in the positive x direction, while the negative sign is for propagation or decay in the minus x direction.

From Maxwell's equations, we can easily derive the expression of the electric field. In particular, for the components of the electric field normal to the x direction,

$$\begin{aligned} E_y(x, y, z) &= \frac{1}{j\omega\varepsilon_{ci}} \mathbf{u}_y \cdot \nabla \times \mathbf{H}(x, y, z) = \pm \frac{k_{xi}}{\omega\varepsilon_{ci}} H_z(x, y, z), \\ E_z(x, y, z) &= \frac{1}{j\omega\varepsilon_{ci}} \mathbf{u}_z \cdot \nabla \times \mathbf{H}(x, y, z) = \mp \frac{k_{xi}}{\omega\varepsilon_{ci}} H_y(x, y, z). \end{aligned} \tag{4.9}$$

These expressions can be recast in the form

$$\mathbf{E}_\tau^{\text{TM}}(x, y, z) = \mp Z_i^{\text{TM}} [\mathbf{u}_x \times \mathbf{H}_\tau^{\text{TM}}(x, y, z)], \tag{4.10}$$

where the subscript τ indicates the vector component transverse to the x direction and the TM$_x$ characteristic impedance Z_i^{TM} is defined as

$$Z_i^{\text{TM}} = \frac{k_{xi}}{\omega\varepsilon_{ci}}. \tag{4.11}$$

With similar steps it can be shown that the transverse components of the TE$_x$ fields are related by

$$\mathbf{E}_\tau^{\text{TE}}(x, y, z) = \mp Z_i^{\text{TE}} [\mathbf{u}_x \times \mathbf{H}_\tau^{\text{TE}}(x, y, z)], \tag{4.12}$$

where the TE$_x$ characteristic impedance Z_i^{TE} is defined as

$$Z_i^{\text{TE}} = \frac{\omega\mu_i}{k_{xi}}. \tag{4.13}$$

It can immediately be seen that the transverse electric and magnetic fields in (4.10) and (4.12) correspond to the voltage and current, respectively, of an equivalent transmission line (TL) with propagation constant k_{xi} and characteristic impedance $Z_i^{\text{TM/TE}}$. In particular, the correspondence is such that

$$\begin{aligned} \mathbf{E}_\tau^{\text{TM/TE}}(x, y, z) &= V^{\text{TM/TE}}(x) e^{-j(k_{yi}y + k_{zi}z)} \mathbf{u}_E, \\ \mathbf{H}_\tau^{\text{TM/TE}}(x, y, z) &= I^{\text{TM/TE}}(x) e^{-j(k_{yi}y + k_{zi}z)} \mathbf{u}_H \end{aligned} \tag{4.14}$$

(where the relation $\mathbf{u}_E \times \mathbf{u}_H = \pm\mathbf{u}_x$ occurs among the unit vectors for propagation in the $\pm x$ direction). In addition, starting from Maxwell's equations, it can be shown that the functions $V^{\mathrm{TM/TE}}$ and $I^{\mathrm{TM/TE}}$ are solutions of TL-like equations. When dealing with planar stratified media (i.e., in which material discontinuities occur only along the x axis) the considerations above are still valid inside each layer.

In the case of a discontinuity between two half-spaces (region 1 and 2), the fulfilment of the boundary conditions at the interface (i.e., continuity of the transverse component of the fields) implies the presence of a reflected wave (in the same half-space where the incident plane wave originates) and a transmitted wave (in the other half-space). The continuity conditions also imply that the transverse dependence of the incident, reflected, and transmitted waves must be the same (i.e., $k_{y1} = k_{y2}$ and $k_{z1} = k_{z2}$). The fact that the transverse wavenumbers are the same in all regions leads to the well-known *law of reflection*, by which the direction angles of the incident and reflected waves are equal, (i.e., $\theta^{\mathrm{inc}} = \theta^{\mathrm{r}}$) and also to *Snell's law*, by which the direction angles inside each of the regions are related to each other:

$$k_1 \sin\theta^{\mathrm{inc}} = k_2 \sin\theta^{\mathrm{t}}. \tag{4.15}$$

In addition, from (4.14), the continuity of the transverse components of the field at the material interfaces implies the continuity of voltage and current at the junction between the two semi-infinite TLs. In general, when a number of layers is considered, voltages and currents are continuous at each material interface. In Figure 4.2 the analogy between a plane-wave transmission problem and propagation along the equivalent TL is sketched. In the figure, the incident, reflected, and transmitted wavevectors are assumed to be real (as the corresponding direction angles, i.e., uniform plane waves are assumed), but the analysis is general and valid also for non-uniform plane waves.

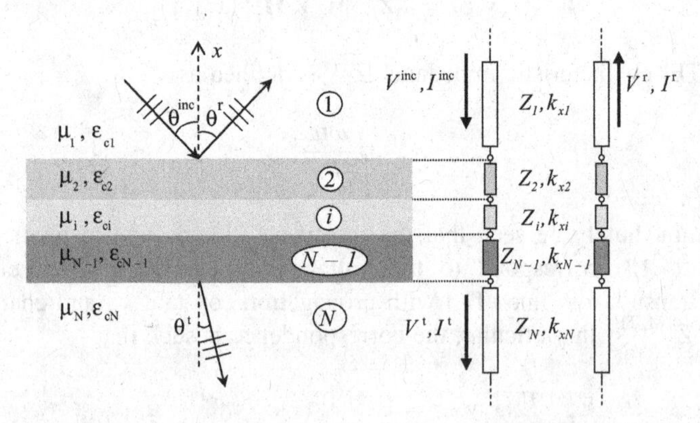

FIGURE 4.2 Uniform plane wave impinging on a planar multilayer shield and equivalent TL model.

It is possible to express the TM_x and TE_x characteristic impedances of the ith layer of a stratified medium in terms of the intrinsic impedance of the medium. In particular, this results in

$$Z_i^{\mathrm{TM}} = \eta_i \cos\theta_i \quad \text{and} \quad Z_i^{\mathrm{TE}} = \frac{\eta_i}{\sin\theta_i}, \tag{4.16}$$

or

$$Z_i^{\mathrm{TM}} = \eta_i \sqrt{1 - \left(\frac{k_1}{k_i}\sin\theta_1\right)^2} \quad \text{and} \quad Z_i^{\mathrm{TE}} = \frac{\eta_i}{\sqrt{1 - \left(\frac{k_1}{k_i}\sin\theta_1\right)^2}}, \tag{4.17}$$

where θ_1 is the direction angle of the incident wave in the first region (i.e., the incidence angle).

One might ask why, among all the possible choices, the TM_x-TE_x decomposition has been adopted. As stated above, the functions $V^{\mathrm{TM/TE}}$ and $I^{\mathrm{TM/TE}}$ behave as voltage and current on different TLs, with characteristic impedances $Z_i^{\mathrm{TM/TE}}$. Therefore any plane-wave reflection and transmission problem reduces to a TL problem, giving the exact solution that satisfies all the boundary conditions. One consequence is that TM_x and TE_x plane waves *do not couple* at a boundary. Therefore, if an incident plane wave is expressed as a combination of both TM_x and TE_x waves, the two problems are solved separately, and the solutions are eventually summed to obtain the *total* reflected or transmitted field.

4.2.2 The Single Planar Shield

In general, the two media surrounding the single planar shield of thickness d can be different. In what follows we assume that each material occupying the region i ($i = 1, 2, 3$) is characterized by its relative permittivity $\varepsilon_{\mathrm{c}ri}$ and permeability $\mu_{\mathrm{r}i}$, although in practical configurations media 1 and 3 are equal (usually air, i.e., $\varepsilon_{\mathrm{cr}1} = \varepsilon_{\mathrm{cr}3} = \mu_{\mathrm{r}1} = \mu_{\mathrm{r}3} = 1$). In the latter, for uniform plane-wave incidence, the electric and magnetic shielding effectivenesses SE_E and SE_H coincide. Both the incident and the transmitted waves in medium 3 are in fact uniform plane waves, for which $|\mathbf{E}| = |\eta||\mathbf{H}|$ (so that the ratio of the magnitudes of the electric and magnetic fields is the same in the presence and in the absence of the screen). In particular, with reference to Figure 4.1, the SE of the screen (in dB) is defined as

$$\mathrm{SE}_E = \mathrm{SE}_H = \mathrm{SE} = 20 \log \frac{|\mathbf{E}_3^0|}{|\mathbf{E}_3^{\mathrm{S}}|}, \tag{4.18}$$

where $\mathbf{E}_3$ indicates the electric field in the region 3, while the superscripts 0 and S indicate the absence and the presence of the shield, respectively.

For TE_x-polarized plane waves, it is immediate to see that the ratio between the amplitudes of the total electric fields occurring in (4.18) coincides with the ratio of the amplitudes of the relevant tangential components (no normal component of the electric field is present in this TE_x case). On the other hand, for TM_x-polarized waves, the amplitude of the normal component of the electric field is proportional to the amplitude of the transverse component through the factor $\tan\theta_3$, where θ_3 indicates the angle of transmission in medium 3. However, such an angle is independent of the presence of the screen, so also in this TM_x case we have

$$\frac{|\mathbf{E}_3^0|}{|\mathbf{E}_3^S|} = \frac{|\mathbf{E}_{3\tau}^0(1+\tan^2\theta_3)|}{|\mathbf{E}_{3\tau}^S(1+\tan^2\theta_3)|} = \frac{|\mathbf{E}_{3\tau}^0|}{|\mathbf{E}_{3\tau}^S|} \tag{4.19}$$

meaning the ratio of the amplitudes of the total electric fields is equal to the ratio of the amplitudes of the relevant transverse components.

Now, by means of a straightforward TL analysis based on the analogy illustrated in the previous subsection, it is simple to show that for a TM_x or TE_x uniform plane wave impinging on the shield with an angle θ^{inc}, the result is

$$\frac{|\mathbf{E}_3^S|}{|\mathbf{E}_3^0|} = \left|\frac{4Z_2Z_3}{(Z_1+Z_2)(Z_2+Z_3)e^{+jk_{x2}d} + (Z_1-Z_2)(Z_2-Z_3)e^{-jk_{x2}d}}\right|, \tag{4.20}$$

where $k_{xi} = k_0\sqrt{\mu_{ri}\varepsilon_{cri} - \mu_{r1}\varepsilon_{cr1}\sin^2\theta^{\mathrm{inc}}}$ and Z_i $(i = 1, 2, 3)$ are the TM_x or TE_x characteristic impedances.

When region 1 and 3 are air, by letting $Z_1 = Z_3 = Z_0$, $Z_2 = Z_S$, and $k_{x2} = k_{xS}$, we reduce (4.20) to

$$\frac{|\mathbf{E}_3^S|}{|\mathbf{E}_3^0|} = \left|\frac{4Z_0Z_S}{(Z_0+Z_S)^2}e^{-jk_{xS}d}\frac{(Z_0+Z_S)^2}{(Z_0+Z_S)^2-(Z_0-Z_S)^2e^{-j2k_{xS}d}}\right|. \tag{4.21}$$

Thus the SE is readily obtained as

$$\mathrm{SE} = -20\log\left|\frac{4Z_0Z_S}{(Z_0+Z_S)^2}e^{-jk_{xS}d}\frac{(Z_0+Z_S)^2}{(Z_0+Z_S)^2-(Z_0-Z_S)^2e^{-j2k_{xS}d}}\right| = R + A + M, \tag{4.22}$$

where

$$R = 20\log\left|\frac{(Z_0+Z_S)^2}{4Z_0Z_S}\right| = 20\log\left|\frac{(\zeta+1)^2}{4\zeta}\right|, \tag{4.23a}$$

$$A = 20\log\left|e^{jk_{xS}d}\right| = 8.686\,\alpha_{xS}d, \tag{4.23b}$$

$$M = 20\log\left|\frac{(Z_0+Z_S)^2-(Z_0-Z_S)^2e^{-j2k_{xS}d}}{(Z_0+Z_S)^2}\right| = 20\log\left|1-\frac{(\zeta-1)^2}{(\zeta+1)^2}e^{-j2k_{xS}d}\right|, \tag{4.23c}$$

and $\zeta = Z_0/Z_S$.

The first term, R, in equation (4.22) depends only on the free-space impedance Z_0 and the impedance of the shield medium Z_S, and it accounts for the first field reflection at the two shield interfaces that is due to the mismatch between the two impedances at both the interfaces. R is called the *reflection-loss* term, and it is always positive or null.

The second term, A, is a function of the shield characteristics only, and it accounts for the attenuation that a plane wave undergoes in traveling through an electrical depth equal to $k_{xS}d/k_0$ in the shield material. A is called the *absorption-loss* term, and it is always positive.

The last term, M, is associated with the wave that undergoes multiple reflections and consequent attenuation before passing through the shield. M is called the *multiple-reflection-loss* term, and it may be positive, null, or negative. M is often negligible with respect to the other two terms, especially in the high-frequency range.

To illustrate how the TL approach behaves, the typical values of the intrinsic impedance $\eta = Z^{TE}(k_x = k) = Z^{TM}(k_x = k)$ of shielding materials can be compared with the free-space impedance. Three typical shielding materials are selected: a copper casting alloy having relative permeability $\mu_r = 1$ and conductivity $\sigma = 11.8 \cdot 10^6$ S/m, a Duranickel with $\mu_r = 10.58$ (considered independent of frequency) and $\sigma = 2.3 \cdot 10^6$ S/m, and a commercial stainless steel having $\mu_r = 95$ and $\sigma = 1.3 \cdot 10^6$ S/m. The behavior of the intrinsic impedance of these media is plotted in Figure 4.3 in relation to their frequency. It is clearly evident that the intrinsic impedance of these shielding materials is much smaller than the free space impedance which is $\eta_0 \simeq 377\ \Omega$. In Figure 4.4, the frequency-dependences of the reflection loss R and of the absorption loss A terms are reported for the same materials as in Figure 4.3, for a planar shield that is 0.25 mm thick under normal (TEM_x) plane-wave incidence. In Figure 4.5, the trend of the multiple-reflection loss term M is also shown; it can be observed that M becomes negligible in comparison with the other two terms

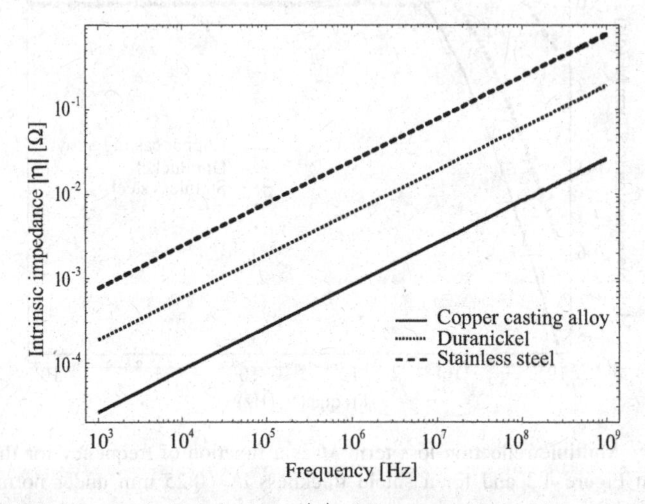

FIGURE 4.3 Intrinsic impedance $|\eta|$ of three typical shielding materials.

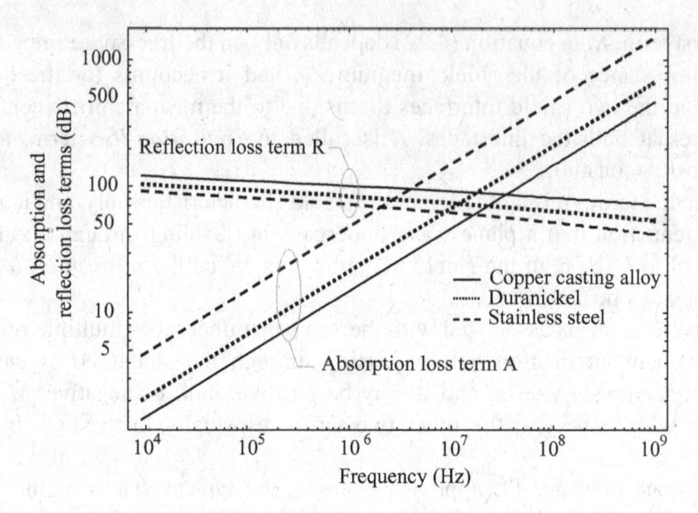

FIGURE 4.4 Reflection-loss R and absorption-loss A terms as functions of frequency for the materials considered in Figure 4.3 and for a shield thickness $d = 0.25$ mm under normal (TEM_x) plane-wave incidence.

above a frequency that depends on the material characteristics and on the shield thickness and that is generally lower than 1 MHz for typical shielding materials.

It is obvious that shielding effectiveness levels in excess of 120 dB are generally not achievable and also not measurable without extraordinary efforts. The preceding figures represent the trends of R and A, and they should be interpreted in the sense that above certain frequencies such quantities exceed any reasonable threshold of shielding level.

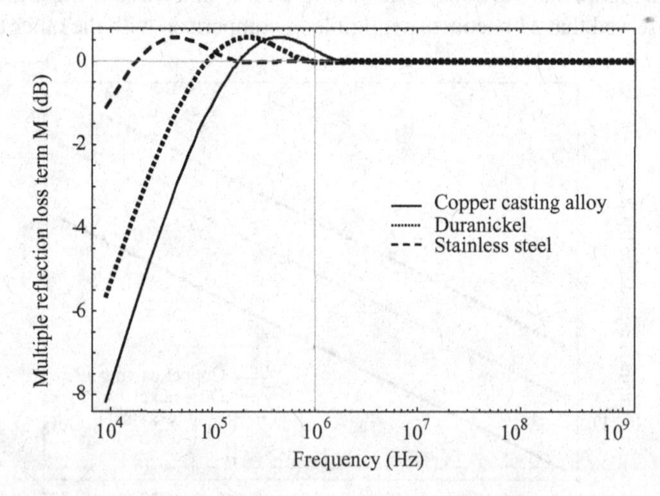

FIGURE 4.5 Multiple-reflection-loss term M as a function of frequency for the materials considered in Figure 4.3 and for a shield thickness $d = 0.25$ mm under normal (TEM_x) plane-wave incidence.

For the common case of sufficiently thick metals, the contribution of the multiple-reflection-loss term M can usually be neglected because of the high penetration loss inside the shielding material. In fact, we can assume $|\zeta| \gg 1$, and rewrite (4.23c) as

$$M = 20\log\left|1 - \frac{(\zeta-1)^2}{(\zeta+1)^2}e^{-j2k_{x\text{s}}d}\right| \cong 20\log\left|1 - e^{-2d(1+j)/\delta}\right|, \tag{4.24}$$

where the frequency-dependent skin depth δ, defined as

$$\delta = \sqrt{\frac{2}{\omega\mu_0\mu_r\sigma}} \tag{4.25}$$

has been introduced. From (4.25) it is evident that M is almost zero when $d/\delta \gg 1$.

It should be noted that when lossy media are considered in the problem of plane-wave transmission, if $\theta^{\text{inc}} \neq 0$ the transmitted plane wave is necessarily non-uniform. Equation (4.15) is still valid, but the direction angle θ^{t} assumes complex values. This is also what happens in the well-known case of *total reflection* at the interface between two lossless media. In the latter case, medium 1 (from which the incident wave comes) is denser than medium 2 (i.e., $k_1 > k_2$). The incidence angle θ^{inc} is greater than the *critical angle* $\theta_c = \sin^{-1}\sqrt{k_2/k_1}$, and it can immediately be seen that the wavenumber k_{x2} in medium 2 is purely imaginary. This gives rises to a non-uniform transmitted wave whose fields exponentially decay in medium 2 and whose power flow is in the transverse direction only. The case where three-layered regions are considered is still different. For instance, consider a plane wave propagating in medium 1 with $k_1 > k_0$ and impinging on a planar air gap (region 2) that separates two regions filled with medium 1 (regions 1 and 3). If the incidence angle in region 1 is greater than the critical angle θ_c, the plane waves in the air gap will be non-uniform and, in particular, exponentially decaying (evanescent or reactive fields). However, although individually the backward and forward waves are evanescent, their sum is a standing wave with a power flow in the x direction different from zero. A uniform transmitted wave in fact exists in region 3 (*electromagnetic tunneling*).

Now consider the problem of oblique incidence (for a uniform plane wave) at the interface with an half-space made of a lossy material. Again, the transmitted wave is a non-uniform plane wave. However, it can easily be shown that the attenuation vector $\boldsymbol{\alpha}$ is directed as $-\mathbf{u}_x$, while, if sufficiently high values of the conductivity σ are considered, the angle θ_β between $-\mathbf{u}_x$ and the phase vector $\boldsymbol{\beta}$ is very small, namely $\theta_\beta \simeq 0$. Therefore the transmitted wave can be approximately considered as a uniform plane wave propagating in a direction that is almost normal to the interface, independently of the angle of incidence θ^{inc}. This is also what usually happens in metallic shields. The propagation inside the shield can be approximately considered in a direction that is almost normal to the shield plane in the whole frequency spectrum of interest for shielding applications [4]. Moreover the EM field transmitted through the shield and propagating in region 3 (always considered as having the same electrical properties as region 1) will have the same direction of propagation it would have in the absence of the shield itself, which is that of the incident field. This will not be true in dealing with anisotropic materials, that is when the permittivity or the permeability of the shield material is expressed by means of a dyadic.

4.2.3 Multiple (or Laminated) Shields

Multiple or laminated shields are obtained by a stratification of two or more sheets of different materials. They are used because of the advantages of shielding performance or for other technical reasons occurring, for instance, when a shielded component is placed into a metallic enclosure. Between adjacent sheets a dielectric (air) gap may or may be not present; such a gap can considerably affect the overall performance of the shield. The simplest configuration consisting of a pair of planar shields separated by an air gap is usually called a double shield [6] in the shielding community, as will be discussed later.

The equivalent circuit of a multiple shield is that already reported in Figure 4.2. The analysis of that configuration is straightforward and can be simply carried out by means of the TL analogy. However, it may be useful to make explicit the equivalence between each shielding layer and a two-port network for which

$$\begin{bmatrix} V_1 \\ I_1 \end{bmatrix} = \begin{bmatrix} A & B \\ C & D \end{bmatrix} \begin{bmatrix} V_2 \\ I_2 \end{bmatrix} = \underline{\mathbf{T}} \begin{bmatrix} V_2 \\ I_2 \end{bmatrix}. \tag{4.26}$$

where $\underline{\mathbf{T}}$ is the transmission matrix. In particular, for a stratified planar structure made of N layers it results

$$\begin{bmatrix} V_{i-1} \\ I_{i-1} \end{bmatrix} = \begin{bmatrix} \cos(k_{xi}d_i) & -jZ_i \sin(k_{xi}d_i) \\ -\frac{j}{Z_i}\sin(k_{xi}d_i) & \cos(k_{xi}d_i) \end{bmatrix} \begin{bmatrix} V_i \\ I_i \end{bmatrix} = \underline{\mathbf{T}}_i \begin{bmatrix} V_i \\ I_i \end{bmatrix} \tag{4.27}$$

where d_i is the thickness of the i-th layer, while V_i and I_i are the equivalent voltage and current (corresponding to the transverse electric and magnetic field, respectively) at the interface between the i-th and the $i+1$-th layer. Thus, the input–output relationship reads

$$\begin{bmatrix} V_1 \\ I_1 \end{bmatrix} = \prod_{i=2,3\ldots N-1} \underline{\mathbf{T}}_i \begin{bmatrix} V_{N-1} \\ I_{N-1} \end{bmatrix} \tag{4.28}$$

to be completed with the two boundary conditions at the first and last interfaces. Using the input–output relationship of the transmission coefficient, and considering the practical case for which region 1 and N are air (i.e., $Z_1 = Z_N = Z_0$), we can express the SE of a multilayer shield as

$$\mathrm{SE} = 20 \log \left| \frac{1}{p} \prod_{n=2}^{N-1} e^{jk_{xn}d_n} \left(1 - q_n e^{-j2k_{xn}d_n}\right) \right| \tag{4.29}$$

where

$$p = \frac{2Z_0 \prod_{i=2}^{N-1} 2Z_i}{(Z_0 + Z_2)(Z_{N-1} + Z_0) \prod_{i=2}^{N-2} (Z_i + Z_{i+1})}, \tag{4.30a}$$

$$q_i = \frac{(Z_i - Z_{i-1})[Z_i - Z(d_i)]}{(Z_i + Z_{i-1})[Z_i + Z(d_i)]}, \tag{4.30b}$$

$$Z(d_{i-1}) = Z_i \frac{Z(d_i)\cos(k_{xi}d_i) + jZ_i \sin(k_{xi}d_i)}{Z_{i+1}\cos(k_{xi}d_i) + jZ(d_i)\sin(k_{xi}d_i)}, \tag{4.30c}$$

$$Z(d_{N-1}) = Z_0.$$

In the case of a planar multilayered shield, the TL formalism allows one to easily determine under which conditions maximum transmission (i.e., absence of reflection) occurs. For instance, in considering a planar shield (region 2) placed between two half-spaces filled with different materials (regions 1 and 3), it is simple to show that when the parameters of the media are chosen so that $Z_2 = \sqrt{Z_1 Z_3}$, the EM energy is totally transmitted from one half-space to the other; the role of the second layer is right that of a quarter-wave transformer matching two lengths of TLs [2]. In addition to the classical work of Schulz et al. [3], some considerations about shielding of plânar multilayer structures are presented in [4] and [5].

The explicit SE expression for a laminated shield of two different materials ($N = 4$) can be found in [3]. In [3] the SE expression for a double shield is also reported; the double shield is a practical case of considerable importance constituted by two sheets separated by an air gap ($N = 5$, $Z_1 = Z_3 = Z_N = Z_0$). When the two metallic sheets consist of the same material (i.e., $Z_2 = Z_4$) and of the same thickness (i.e., $d_2 = d_4$), it can be shown that the absorption- and reflection-loss terms are exactly twice those of a single sheet, while the multiple-reflection-loss term is more than twice that of a single sheet; because of the two new interior interfaces between air and the metallic material, the multiple-reflection shielding mechanism is exalted. Figure 4.6 shows the SE of double shields of copper casting alloy ($d_2 = d_4 = \Delta = 0.25$ mm) with $d_3 = \Delta$ and $d_3 = 10\Delta$, compared with the SE of a single layer shield with a double thickness ($d_2 = 2\Delta = 0.5$ mm). It can be seen that the larger the air gap, the more effective is the screen; this is true when the shield is electrically thin. When the shield is electrically thick, the double shield is considerably less effective than the single shield with a double thickness, except at the shielding inter-space resonances. More details on the subject can be found in [3].

4.3 PLANE WAVES NORMALLY INCIDENT ON CYLINDRICAL SHIELDING SURFACES

The EM problem is depicted schematically in Figure 4.7, where a uniform plane wave impinges on an infinitely long cylindrical shield, normally with respect to its

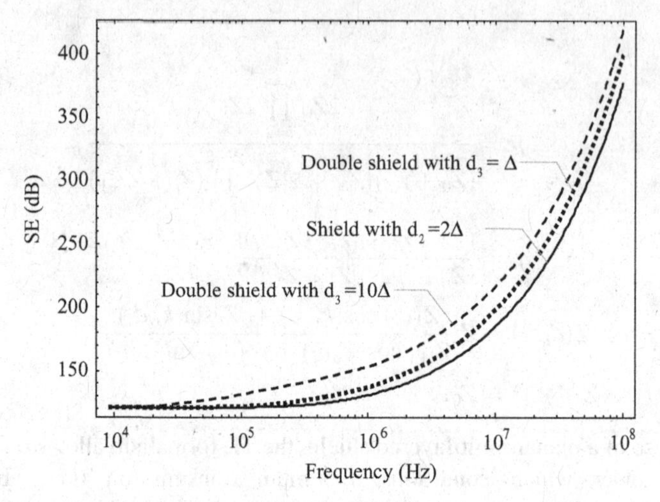

FIGURE 4.6 Comparison between the SE of double shields and a single-layer shield.

axis (i.e., the z axis). The cylindrical shield is characterized by its inner and outer radius ρ_1 and ρ_2, respectively, and by its constitutive parameters ε_c and μ (in particular, the material is assumed linear, stationary, isotropic, and homogeneous).

The most convenient decomposition for the incident uniform plane wave is in the linearly polarized TM_z and TE_z uniform plane waves. In the case of normal incidence (i.e., $k_z^{inc} = 0$), the TM_z and TE_z waves in fact do not couple, although in the more general case of oblique incidence (i.e., $k_z^{inc} \neq 0$) they do. The three regions corresponding to the interior of the shield, to the shield itself, and to the outer space are denoted as region 1, 2, and 3, respectively:

$$\begin{aligned} &\rho < \rho_1 && \text{(Region 1)}, \\ &\rho_1 \leq \rho < \rho_2 && \text{(Region 2)}, \\ &\rho_2 \leq \rho && \text{(Region 3)}. \end{aligned}$$

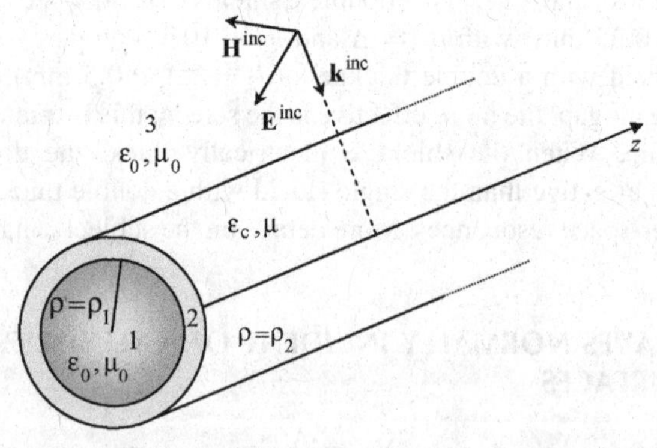

FIGURE 4.7 Uniform plane wave incident on an infinitely long cylindrical shield, normally with respect to its axis.

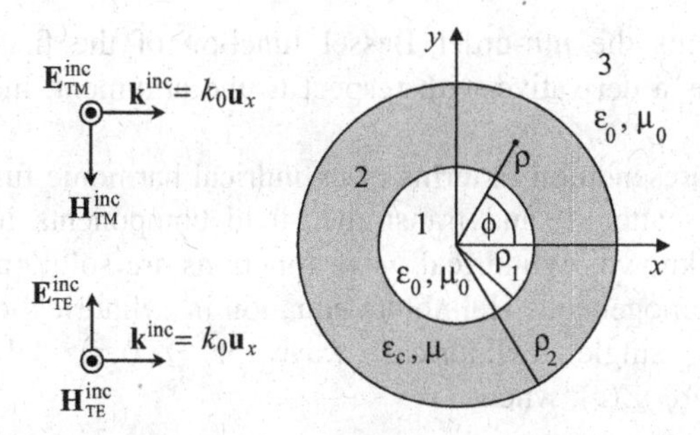

FIGURE 4.8 TM_z and TE_z uniform plane waves normally incident on an infinitely long cylindrical shield.

Since both the incident field and the structure are invariant along z, the problem is essentially a two-dimensional (2D) problem. Let us first consider a uniform TM_z-polarized plane wave normally incident on the shield, as illustrated in Figure 4.8.

Without loss of generality, we can assume that the wave propagates along the x axis of a Cartesian coordinate system having the z axis along the cylindrical shield axis. The electric and magnetic fields are expressed as

$$\mathbf{E}^{\mathrm{inc}}(x,y,z) = E_z^{\mathrm{inc}}(x)\mathbf{u}_z = E_0^{\mathrm{inc}} e^{-jk_0x}\mathbf{u}_z,$$
$$\mathbf{H}^{\mathrm{inc}}(x,y,z) = H_y^{\mathrm{inc}}(x)\mathbf{u}_y = H_0^{\mathrm{inc}} e^{-jk_0x}\mathbf{u}_y = -\frac{E_0^{\mathrm{inc}}}{\eta_0} e^{-jk_0x}\mathbf{u}_y, \tag{4.31}$$

or, in cylindrical coordinates, as

$$\mathbf{E}^{\mathrm{inc}}(\rho,\phi,z) = E_z^{\mathrm{inc}}(\rho,\phi)\mathbf{u}_z = E_0^{\mathrm{inc}} e^{-jk_0\rho\cos\phi}\mathbf{u}_z,$$
$$\mathbf{H}^{\mathrm{inc}}(\rho,\phi,z) = H_\rho^{\mathrm{inc}}(\rho,\phi)\mathbf{u}_\rho + H_\phi^{\mathrm{inc}}(\rho,\phi)\mathbf{u}_\phi = -\frac{E_0^{\mathrm{inc}}}{\eta_0}(\mathbf{u}_\rho\sin\phi + \mathbf{u}_\phi\cos\phi)e^{-jk_0\rho\cos\phi}. \tag{4.32}$$

Because of the cylindrical symmetry, the problem may be conveniently studied representing the incident fields in cylindrical harmonic functions through a Fourier series expansion of the fields in (4.32). In particular, the components of the fields tangential to the cylindrical surfaces can be expressed as

$$E_z^{\mathrm{inc}}(\rho,\phi) = E_0^{\mathrm{inc}} \sum_{n=-\infty}^{\infty} j^{-n} J_n(k_0\rho)\, e^{jn\phi},$$
$$H_\phi^{\mathrm{inc}}(\rho,\phi) = \frac{1}{j\omega\mu_0}\frac{\partial E_z^{\mathrm{inc}}}{\partial\rho} = -j\frac{E_0^{\mathrm{inc}}}{\eta_0} \sum_{n=-\infty}^{\infty} j^{-n} J_n'(k_0\rho)\, e^{jn\phi}, \tag{4.33}$$

where $J_n(\cdot)$ represents the nth-order Bessel function of the first kind and the prime symbol means a derivative with respect to the argument, meaning $J_n'(\xi) = \mathrm{d}J_n(\xi)/\mathrm{d}\xi$.

An analogous representation in terms of cylindrical harmonic functions may be considered for the scattered- and transmitted-field components in all the three regions. As is well known, cylindrical wave functions are solutions of the three-dimensional (3D) homogeneous Helmholtz equation in cylindrical coordinates [2]. In particular, the single cylindrical wave $\Psi^{\mathrm{C}}(\rho, \phi, z)$ takes the form $\Psi^{\mathrm{C}}(\rho, \phi, z) = \mathrm{P}(\rho)\Phi(\phi)Z(z)$, where

$$\begin{aligned} Z(z) &= a_z \cos(k_z z) + b_z \sin(k_z z), \\ \Phi(\phi) &= a_\phi \cos(k_\phi \phi) + b_\phi \sin(k_\phi \phi), \\ \mathrm{P}(\rho) &= a_\rho C_{k_\phi}^{(1)}(k\rho) + b_\rho C_{k_\phi}^{(2)}(k\rho) \,. \end{aligned} \tag{4.34}$$

The functions $C_\nu^{(1)}$ and $C_\nu^{(2)}$ in (4.34) are any two independent cylindrical Bessel functions chosen among J_ν, Y_ν, $H_\nu^{(1)}$, and $H_\nu^{(2)}$. In bounded regions, oscillatory Bessel functions of the first and second kind (J_ν and Y_ν, respectively) are usually adopted to represent standing waves, whereas in unbounded regions, the Hankel functions of the first and second kind ($H_\nu^{(1)}$ and $H_\nu^{(2)}$) are used to represent inward and outward propagating waves, respectively. For z-independent problems (i.e., $k_z = 0$), it is assumed $Z(z) = 1$ and the cylindrical wave is simply $\Psi^{\mathrm{C}}(\rho, \phi) = \mathrm{P}(\rho)\Phi(\phi)$.

For the considered problem of Figure 4.8, periodicity in ϕ enforces k_ϕ to be an integer, meaning $k_\phi = n$. The reflected field in region 3, being an outward wave, requires the representation of its radial dependence through Hankel functions of the second kind $H_n^{(2)}$. The transmitted field inside the shield (region 2) will be instead represented in terms of both Bessel functions of the first and second kind J_n and Y_n, whereas the transmitted field inside the region 1 will be expressed in terms of the only J_n functions (since all the other functions are singular at the origin). It is immediate to see that it is not possible to satisfy the boundary conditions (i.e., continuity of the components of the fields tangential to the shield-air interfaces) with a single cylindrical wave function (each cylindrical wave function in fact cannot match the more complicated dependence on the spatial coordinates of the fields in (4.33)). However, since the set of cylindrical functions is complete, an infinite linear combination of Ψ^{C} functions with unknown coefficients can be used to represent the reflected and transmitted fields in regions 1, 2, and 3. After recognizing that the series of sinusoidal functions is a Fourier series, one can rewrite the components of the fields tangential to the cylindrical surfaces as

$$\begin{aligned} E_z^{(1)}(\rho, \phi) &= \sum_{n=-\infty}^{\infty} a_n j^{-n} J_n(k_0\rho)\, e^{jn\phi}, \\ H_\phi^{(1)}(\rho, \phi) &= \frac{1}{j\omega\mu_0}\frac{\partial E_z^{(1)}}{\partial \rho} = -j\frac{1}{\eta_0}\sum_{n=-\infty}^{\infty} a_n j^{-n} J_n'(k_0\rho)\, e^{jn\phi}, \end{aligned} \tag{4.35a}$$

$$E_z^{(2)}(\rho,\phi) = \sum_{n=-\infty}^{\infty} j^{-n}[b_n J_n(k\rho) + c_n Y_n(k\rho)]e^{jn\phi},$$
$$H_\phi^{(2)}(\rho,\phi) = -j\frac{1}{\eta}\sum_{n=-\infty}^{\infty} j^{-n}[b_n J_n'(k\rho) + c_n Y_n'(k\rho)]e^{jn\phi}, \tag{4.35b}$$

$$E_z^{(3)}(\rho,\phi) = \sum_{n=-\infty}^{\infty} j^{-n}[E_0^{\mathrm{inc}} J_n(k_0\rho) + d_n H_n^{(2)}(k_0\rho)]e^{jn\phi},$$
$$H_\phi^{(3)}(\rho,\phi) = -j\frac{1}{\eta_0}\sum_{n=-\infty}^{\infty} j^{-n}[E_0^{\mathrm{inc}} J_n'(k_0\rho) + d_n H_n^{(2)'}(k_0\rho)]e^{jn\phi}, \tag{4.35c}$$

where the prime symbol means again a derivative with respect to the argument $C_n'(\xi) = \mathrm{d}C_n(\xi)/\mathrm{d}\xi$. The unknown coefficients a_n, b_n, c_n, d_n, can be determined by enforcing the continuity of the tangential components of the electric and magnetic field at the shield-air interfaces (i.e., $\rho = \rho_1$ and $\rho = \rho_2$). By using the orthogonality relation among exponential functions, the following linear system is obtained:

$$\begin{bmatrix} J_n(k_0\rho_1) & -J_n(k\rho_1) & -Y_n(k\rho_1) & 0 \\ \frac{1}{\eta_0}J_n'(k_0\rho_1) & -\frac{1}{\eta}J_n'(k\rho_1) & -\frac{1}{\eta}Y_n'(k\rho_1) & 0 \\ 0 & J_n(k\rho_2) & Y_n(k\rho_2) & -H_n^{(2)}(k_0\rho_2) \\ 0 & \frac{1}{\eta}J_n'(k\rho_2) & \frac{1}{\eta}Y_n'(k\rho_2) & -\frac{1}{\eta_0}H_n^{(2)'}(k_0\rho_2) \end{bmatrix} \begin{bmatrix} a_n \\ b_n \\ c_n \\ d_n \end{bmatrix} = \begin{bmatrix} 0 \\ 0 \\ E_0^{\mathrm{inc}} J_n(k_0\rho_2) \\ \frac{E_0^{\mathrm{inc}}}{\eta_0} J_n'(k_0\rho_2) \end{bmatrix}. \tag{4.36}$$

Once the system (4.36) is solved for each n, (4.35) gives the fields everywhere in space. However, in order to obtain the value of the *electric* SE on the cylinder axis, $\rho = 0$, only one coefficient is required (though this is not always the point of maximum field). For $\rho = 0$, all the functions in (4.35a) but the one corresponding to the coefficient a_0 are in fact identically zero. Therefore the total electric field on the cylinder axis in the presence of the shield E_z^{S} is $E_z^{\mathrm{S}}(0,\phi) = a_0$; on the other hand, the electric field in the absence of the screen E_z^0 is simply the incident field, meaning $E_z^0(0,\phi) = E_0^{\mathrm{inc}}$. The electric SE on the cylinder axis is thus

$$\mathrm{SE}_{E(\mathrm{dB})} = 20\log\frac{|\mathbf{E}^0|}{|\mathbf{E}^{\mathrm{S}}|} = 20\log\left|\frac{E_z^0}{E_z^{\mathrm{S}}}\right| = 20\log\left|\frac{E_0^{\mathrm{inc}}}{a_0}\right|, \tag{4.37}$$

where from (4.36),

$$a_0 = \frac{\dfrac{4jE_0^{\mathrm{inc}}}{(\pi k_0\eta_0)^2\mu_r\rho_1\rho_2}}{\begin{vmatrix} J_0(k_0\rho_1) & -J_0(k\rho_1) & -Y_0(k\rho_1) & 0 \\ \frac{1}{\eta_0}J_0'(k_0\rho_1) & -\frac{1}{\eta}J_0'(k\rho_1) & -\frac{1}{\eta}Y_0'(k\rho_1) & 0 \\ 0 & J_0(k\rho_2) & Y_0(k\rho_2) & -H_0^{(2)}(k_0\rho_2) \\ 0 & \frac{1}{\eta}J_0'(k\rho_2) & \frac{1}{\eta}Y_0'(k\rho_2) & -\frac{1}{\eta_0}H_0^{(2)'}(k_0\rho_2) \end{vmatrix}}. \tag{4.38}$$

In order to obtain the *magnetic* SE, also the radial component of the magnetic field is needed. In particular, it simply results

$$H_\rho^{(1)}(\rho,\phi) = \frac{1}{j\omega\mu_0}\frac{1}{\rho}\frac{\partial E_z^{(1)}}{\partial\phi} = -\frac{1}{\eta_0}\sum_{n=-\infty}^{\infty} n a_n j^{-n}\frac{J_n(k_0\rho)}{k_0\rho}e^{jn\phi}. \tag{4.39}$$

From (4.35a) and (4.36) and by using the properties of Bessel functions, it can thus be seen that $|\mathbf{H}^{(1)}(0,\phi)| = |a_1|/\eta_0$ (in particular, from $C_{-n}(\xi) = (-1)^n C_n(\xi)$, it follows that $a_{-n} = a_n$). Therefore, by taking into account that $|\mathbf{H}^{\text{inc}}(0,\phi)| = |E_0^{\text{inc}}/\eta_0|$, one can obtain the magnetic SE as

$$\text{SE}_{H(\text{dB})} = 20\log\frac{|\mathbf{H}^0|}{|\mathbf{H}^S|} = 20\log\left|\frac{E_0^{\text{inc}}}{a_1}\right|. \tag{4.40}$$

It should be noted that in this cylindrical configuration, contrary to what happens for uniform plane waves incident on a planar shield, the electric and magnetic SE are in general dramatically different.

The case of TE_z normal incidence can be studied in a similar way starting from

$$\begin{aligned}\mathbf{H}^{\text{inc}}(x,y,z) &= H_z^{\text{inc}}(x)\mathbf{u}_z = H_0^{\text{inc}}e^{-jk_0x}\mathbf{u}_z,\\ \mathbf{E}^{\text{inc}}(x,y,z) &= E_y^{\text{inc}}(x)\mathbf{u}_y = E_0^{\text{inc}}e^{-jk_0x}\mathbf{u}_y = \eta_0 H_0^{\text{inc}}e^{-jk_0x}\mathbf{u}_y.\end{aligned} \tag{4.41}$$

In a still simpler way one can apply duality to obtain

$$\text{SE}_{E(\text{dB})} = 20\log\frac{|\mathbf{E}^0|}{|\mathbf{E}^S|} = 20\log\left|\frac{H_0^{\text{inc}}}{A_1}\right| \tag{4.42}$$

and

$$\text{SE}_{H(\text{dB})} = 20\log\frac{|\mathbf{H}^0|}{|\mathbf{H}^S|} = 20\log\left|\frac{H_z^0}{H_z^S}\right| = 20\log\left|\frac{H_0^{\text{inc}}}{A_0}\right|, \tag{4.43}$$

where the unknown coefficients can be obtained by solving the linear system

$$\begin{bmatrix} J_n(k_0\rho_1) & -J_n(k\rho_1) & -Y_n(k\rho_1) & 0\\ \eta_0 J_n'(k_0\rho_1) & -\eta J_n'(k\rho_1) & -\eta Y_n'(k\rho_1) & 0\\ 0 & J_n(k\rho_2) & Y_n(k\rho_2) & -H_n^{(2)}(k_0\rho_2)\\ 0 & \eta J_n'(k\rho_2) & \eta Y_n'(k\rho_2) & -\eta_0 H_n^{(2)'}(k_0\rho_2)\end{bmatrix}\begin{bmatrix}A_n\\B_n\\C_n\\D_n\end{bmatrix} = \begin{bmatrix}0\\0\\H_0^{\text{inc}}J_n(k_0\rho_2)\\ \eta_0 H_0^{\text{inc}}J_n'(k_0\rho_2)\end{bmatrix}. \tag{4.44}$$

Figure 4.9 shows the SE versus frequency of a glass cylindrical surface with $\varepsilon_r = 3.8$, $\rho_1 = 10$ cm, and $\rho_2 = 10.1$ cm. It is possible to note that for both the TM_z

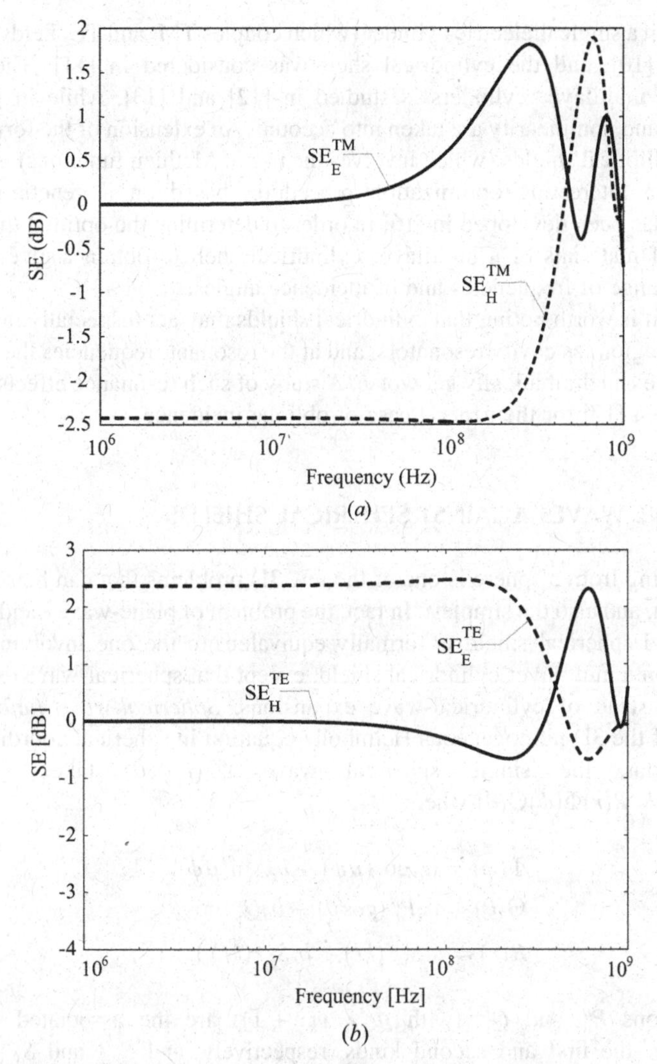

FIGURE 4.9 SE of a glass cylindrical surface with $\varepsilon_r = 3.8$, $\rho_1 = 10$ cm, and $\rho_2 = 10.1$ cm under TM_z and TE_z normal incidence.

and TE_z normal incidence, the electric and magnetic SE are very different. In addition negative values of SE can be observed.

The problem of EM plane-wave transmission through a hollow metallic cylinder has been studied in [6], where the shield thickness was assumed electrically small, and next, for arbitrary thickness values, in [7].

A numerical study of the normal plane-wave incidence on a circular cylindrical shield consisting of a single dielectric layer is reported in [8], and results for multilayer circular cylindrical structures appear in [9]. The oblique plane-wave incidence on a single dielectric cylinder (which couples TM_z and TE_z fields) was first studied in [10], and the cylindrical shell was considered in [11]. The case of anisotropic multilayer cylinders is studied in [12] and [13], while in [14] both anisotropy and nonlinearity are taken into account. An extension of the formalism to deal with elliptical shields (which involves the use of Mathieu functions) is reported in [15]. An interesting optimization procedure, based on a genetic-algorithm approach, has been developed in [16] in order to determine the optimal thicknesses and type of materials of a multilayer cylindrical shell to obtain a given SE in a particular range of frequencies and of incidence angles.

Finally, it is worth noting that cylindrical shields may act (especially in the high-frequency region) as cavity resonators, and at the resonant frequencies the shielding performance can dramatically get worse. A study of such resonance effects has been carried out in [17] for the general case of oblique incidence.

4.4 PLANE WAVES AGAINST SPHERICAL SHIELDS

The scattering from a sphere is one of the few 3D problems that can be solved in a closed form, and also the simplest. In fact, the problem of plane-wave incidence on a multilayered spherical shield is formally equivalent to the one involving normal incidence on a multilayer cylindrical shield, except that spherical-wave expansions are used instead of cylindrical-wave expansions. *Spherical-wave functions* are solutions of the 3D homogeneous Helmholtz equation in spherical coordinates [2]. In particular, the single spherical wave $\Psi^S(r,\phi,\theta)$ takes the form $\Psi^S(r,\phi,\theta) = R(r)\Phi(\phi)\Theta(\theta)$ where

$$\begin{aligned}
\Phi(\phi) &= a_\phi \cos(\mu\phi) + b_\phi \sin(\mu\phi), \\
\Theta(\theta) &= a_\theta P_\nu^\mu(\cos\theta) + b_\theta Q_\nu^\mu(\cos\theta), \\
R(r) &= a_r S_\nu^{(1)}(kr) + b_r S_\nu^{(2)}(kr)\ .
\end{aligned} \tag{4.45}$$

The functions P_ν^μ and Q_ν^μ (with $\mu^2 = \nu(\nu+1)$) are the associated Legendre functions of the first and second kinds, respectively, and $S_\nu^{(1)}$ and $S_\nu^{(2)}$ are two independent *spherical* Bessel functions chosen among j_ν, y_ν, $h_\nu^{(1)}$, and $h_\nu^{(2)}$. It should be noted that in problems containing the z axis, the b_θ coefficient must be considered zero in order to avoid singularities; moreover ν and μ are restricted to be integers ($\nu = n$ and $\mu = m$). The product of the functions Φ and Θ are usually called *spherical harmonics* ($Y_{mn}(\phi,\theta) = \Phi(\phi)\Theta(\theta)$). In problems involving spherical waves, the TE-TM decomposition of a given EM field in spherical coordinates is usually performed with respect to the r direction.

Several papers in the literature analyze scattering from a multilayered sphere as shown in Figure 4.10*a*. Aden and Kerker were first to investigate the incidence of a uniform plane wave on a concentric spherical shell [18], and shortly after, in [19], approximate results were obtained in the limit of very thin shells. Multilayered spherical shields were studied in [20–22], and the possible anisotropy of the

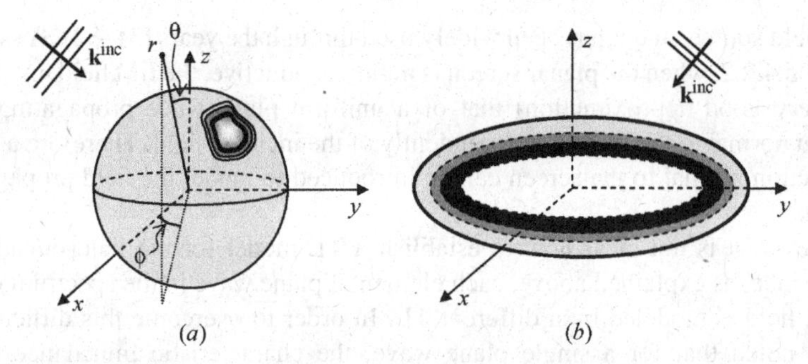

FIGURE 4.10 Plane-wave incidence on a multilayer spherical shield (*a*) and on a multilayer spheroidal shield (*b*).

materials was taken into account in [23] for a single sphere and in [24] for a multilayered spherical structure. Resonance effects in a spherical shell were studied in [25], and, more recently, in [26] the use of lossy dielectrics was suggested to mitigate such undesired effects, determining (for a double-layered spherical shield) an approximate value for the optimum conductivity of a metal layer that minimizes the Q factor at the fundamental resonant frequency of the structure.

Other 3D shielding problems solvable in closed form involve shields of ellipsoidal shape, as shown in Figure 4.10*b*. The mathematical formalism requires the use of spheroidal wavefunctions [27], and analyses of different configurations (involving a single spheroid, a spheroidal shell, and a multilayer spheroid) can be found in [28–30].

4.5 LIMITS TO THE EXTENSION OF THE TL ANALOGY TO NEAR-FIELD SOURCES

Since the analytical solution of the field equations is complicated, even for simple shield configurations, the TL approach has been widely used for evaluating the performances of shields also against near-field sources, because of its inherent formal simplicity. However, impinging fields are different from plane waves, and so require that the TL analogy be carefully used to avoid incorrect results.

Generally, when a planar screen is placed in the near-field region of a finite source, the incident EM field produced by the source can be represented by an integral superposition of plane waves having all the possible directions of propagation and polarizations. In principle, one could apply the TL approach to each of these elemental plane waves, and then reconstruct the field beyond the screen by summing up the contributions of all the transmitted elemental plane waves. However, the TL parameters would be different for different plane waves and hence a unique TL cannot be defined for the total field.

Despite these considerations, an approximate TL approach does exist for the evaluation of the SE of an infinite highly conductive planar screen in the presence of a near-field source, and it has been widely used through the years [31–32]. As shown in Section 4.2.2, when the planar screen is highly conductive, the EM field inside it is (with very good approximation) that of a uniform plane wave propagating in a direction normal to the screen, independently of the incident field. Therefore a TL in the direction normal to the screen can be introduced to model the field propagation inside it.

However, it is not clear how to establish a TL model for the field outside the screen, since, as explained above, each elemental plane wave in the spectrum of the incident field is modeled by a different TL. In order to overcome this difficulty, it may be noted that for a single plane wave, the characteristic impedance of its equivalent TL coincides with its *wave impedance* in the TL direction. According to [33], the wave impedance along an arbitrary direction **u** is defined as the ratio of the transverse (with respect to **u**) electric-field component to the mutually perpendicular transverse magnetic-field component. Therefore the idea is to use the wave impedances of the incident field at the screen interfaces as the characteristic impedances of the TLs modeling the field outside the screen.

It should be noted that the wave impedances of the incident field actually depend on the observation point on the screen interface. For instance, let us consider an infinitesimal electric or magnetic dipole, which can be used as a first approximation to model finite sources of small dimensions. With respect to the arbitrarily oriented reference system (x', y', z'), the electric- and magnetic-field components produced by an infinitesimal electric dipole (ED) of moment $Il\,\mathbf{u}_{z'}$ can be expressed in spherical coordinates as

$$
\begin{aligned}
E_r(r',\theta') &= \eta_0 \frac{Il}{2\pi} \cos\theta' \left[\frac{1}{r'^2} + \frac{1}{jk(r')^3} \right] e^{-jkr'}, \\
E_\theta(r',\theta') &= \eta_0 \frac{Il}{4\pi} \sin\theta' \left[\frac{jk}{r'} + \frac{1}{(r')^2} + \frac{1}{jk(r')^3} \right] e^{-jkr'}, \\
H_\phi(r',\theta') &= \frac{Il}{4\pi} \sin\theta' \left[\frac{jk}{r'} + \frac{1}{(r')^2} \right] e^{-jkr'},
\end{aligned}
\tag{4.46}
$$

while an infinitesimal magnetic dipole (MD) of moment $IS\,\mathbf{u}_{z'}$ produces a field

$$
\begin{aligned}
E_\phi(r',\theta') &= \frac{j\omega\mu_0 IS}{4\pi} \sin\theta' \left[\frac{jk}{r'} + \frac{1}{(r')^2} \right] e^{-jkr'}, \\
H_r(r',\theta') &= \frac{j\omega\mu_0 IS}{2\pi\eta_0} \cos\theta' \left[\frac{1}{(r')^2} + \frac{1}{jk(r')^3} \right] e^{-jkr'}, \\
H_\theta(r',\theta') &= \frac{j\omega\mu_0 IS}{4\pi\eta_0} \sin\theta' \left[\frac{jk}{r'} + \frac{1}{(r')^2} + \frac{1}{jk(r')^3} \right] e^{-jkr'}.
\end{aligned}
\tag{4.47}
$$

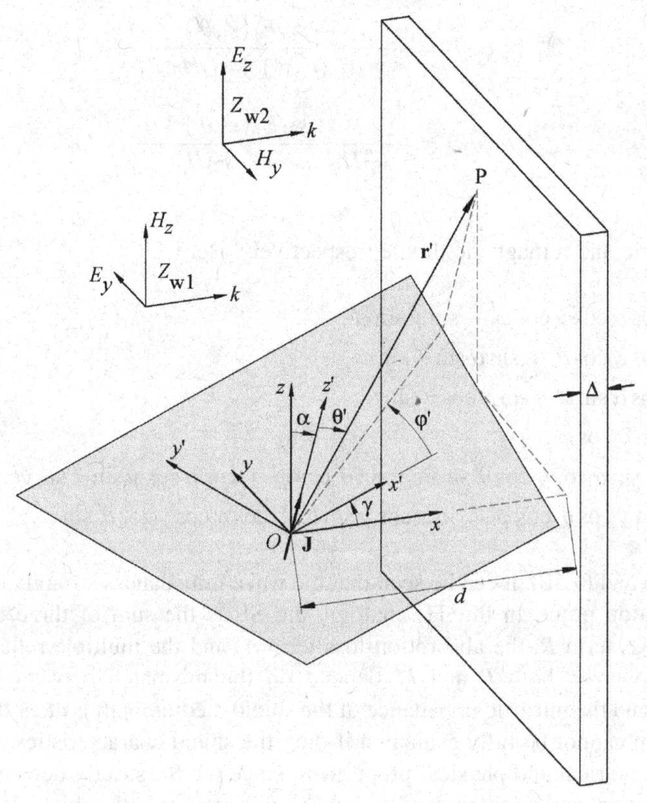

FIGURE 4.11 Infinitesimal dipole source arbitrarily oriented illuminating an infinite planar shield.

With respect to the reference system (x, y, z) shown in Figure 4.11, two wave impedances Z_{w1} and Z_{w2} can be defined as

$$Z_{w1} = +\frac{E_y^{inc}}{H_z^{inc}}, \tag{4.48a}$$

$$Z_{w2} = -\frac{E_z^{inc}}{H_y^{inc}}. \tag{4.48b}$$

It is possible to move from the unprimed reference system to the primed one by a rotation around the z axis of an angle γ and then around the new rotated x axis of an angle α. The wave impedances are then given by

$$\begin{aligned} Z_{w1}^{ED}(r, \phi, \theta) &= +\frac{s_{22}E_r(r', \theta') + s_{23}E_\theta(r', \theta')}{s_{21}H_\varphi(r', \theta')}, \\ Z_{w2}^{ED}(r, \phi, \theta) &= -\frac{s_{12}E_r(r', \theta') + s_{13}E_\theta(r', \theta')}{s_{11}H_\varphi(r', \theta')}, \end{aligned} \tag{4.49a}$$

$$Z_{w1}^{MD}(r,\phi,\theta) = +\frac{s_{11}E_\phi(r',\theta')}{s_{12}H_r(r',\theta') + s_{13}H_\theta(r',\theta')},$$
$$Z_{w2}^{MD}(r,\phi,\theta) = -\frac{s_{21}E_\phi(r',\theta')}{s_{22}H_r(r',\theta') + s_{23}H_\theta(r',\theta')}, \tag{4.49b}$$

for an electric and a magnetic dipole, respectively. Here

$$\begin{aligned}
s_{11} &= \cos\alpha\cos\gamma\cos\varphi' - \sin\gamma\sin\varphi',\\
s_{12} &= \cos\alpha\cos\theta' + \sin\alpha\sin\theta'\sin\varphi',\\
s_{13} &= \cos\alpha\sin\theta - \sin\alpha\cos\theta'\sin\varphi',\\
s_{21} &= \sin\alpha\cos\varphi',\\
s_{22} &= -\sin\alpha\cos\gamma\cos\theta' + \sin\gamma\sin\theta'\cos\varphi' + \cos\alpha\cos\gamma\sin\theta'\sin\varphi',\\
s_{23} &= \sin\gamma\cos\theta'\cos\varphi + \sin\alpha\cos\gamma\sin\theta' + \cos\alpha\cos\gamma\cos\theta'\sin\varphi'\ .
\end{aligned} \tag{4.50}$$

From (4.49) and (4.50), it can be seen that the wave impedances strongly depend on the observation point. In the TL analogy, the SE is the sum of three terms: the reflection-loss term R, the absorption-loss term A, and the multiple-reflection-loss term M. However, both R and M depend on the mismatch between the wave impedance and the intrinsic impedance of the shield medium η. It is clear that the SE of the screen cannot be fully evaluated if only the shield characteristics are known (i.e., its geometrical and physical properties), since the SE strictly depends also on the size, location, and orientation of the source with respect to the shield. It should be noted that the source identification is a necessary prerequisite to any shielding analysis. A first fundamental step in the source-identification process when dealing with near fields is the distinction between high- and low-impedance field sources. High-impedance sources produce a near field dominated by the electric field **E** (i.e., $|Z_{w1,2}| \gg \eta_0$), while low-impedance field sources are characterized by a near field dominated by the magnetic field **H** (i.e., $|Z_{w1,2}| \ll \eta_0$).

In practice, when the TL analogy is applied to evaluate the SE of an infinite planar conducting screen, the impedances of the impinging field are computed along the axis drawn from the source orthogonally to the shield (x axis in Figure 4.11). Two significant configurations are those in which a dipole is normal to the shield (i.e., $\gamma = 0$ and $\alpha = \pi/2$) or parallel to the shield (i.e., $\gamma = 0$ and $\alpha = 0$). From equations (4.46) through (4.50), by assuming $\theta' = 0$ for orthogonal dipoles ($\perp$) and $\theta' = \pi/2$ for parallel dipoles ($\|$), respectively, the following expressions for the wave impedances can be obtained:

$$Z_{w2}^{ED\perp}(x,\omega) = \eta_0\frac{\xi^2+\xi+1}{\xi^2+\xi}, \tag{4.51a}$$

$$Z_{w1,2}^{MD\perp}(x,\omega) = \eta_0\frac{\xi^2+\xi}{\xi^2+3\xi+3}, \tag{4.51b}$$

$$Z_{w1,2}^{ED\parallel}(x,\omega) = \eta_0 \frac{\xi^2 + 3\xi + 3}{\xi^2 + \xi}, \tag{4.51c}$$

$$Z_{w1}^{MD\parallel}(x,\omega) = \eta_0 \frac{\xi^2 + \xi}{\xi^2 + \xi + 1}, \tag{4.51d}$$

where $\xi = jk_0x$. It should be noted that the wave impedances are both defined only when the electric dipole is parallel to the shield or the magnetic dipole is perpendicular to the screen.

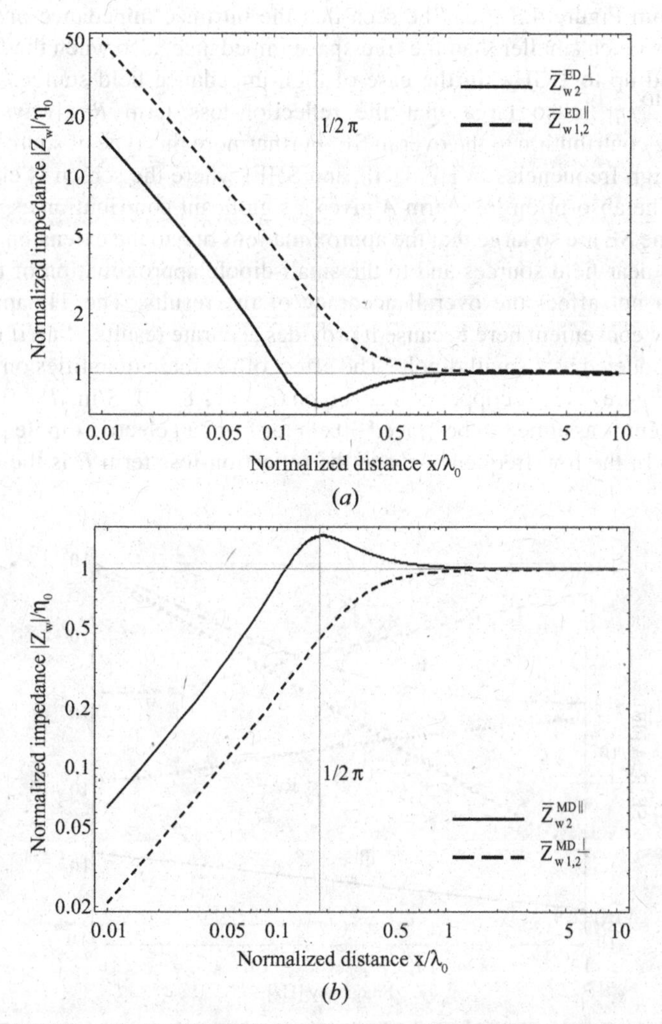

FIGURE 4.12 Magnitude of the normalized wave impedances as functions of the normalized distance $\bar{x} = x/\lambda_0$. Electric dipole (*a*) and magnetic dipole (*b*).

The behavior of the wave impedances is illustrated in Figure 4.12, where the normalized wave impedances $|\bar{Z}_{w1,2}^{ED/MD}| = |Z_{w1,2}^{ED/MD}|/\eta_0$ are reported as functions of the normalized distance $\bar{x} = x/\lambda_0$. For normalized distances smaller than $1/(2\pi)$ (i.e., in the near-field region), the wave impedances of the field produced by the electric dipole are larger than the free-space impedance η_0, while those corresponding to the magnetic dipole are smaller. It can be concluded that the electric dipole is a high-impedance field source, while the magnetic dipole is a low-impedance field source.

To illustrate how the TL approach behaves when it is extended to study near-field sources, the typical values of the intrinsic impedance of shielding materials are compared with the wave impedances of the impinging field produced by the source. From Figure 4.3 it can be seen that the intrinsic impedance of shielding materials is much smaller than the free-space impedance, also when the frequency is increased up to 1 GHz. In the case of high-impedance field sources, the ratio $\zeta = |Z_{w1,2}^{ED/MD}/\eta|$ is so large that the reflection-loss term R always gives a remarkable contribution to the overall SE. Furthermore this type of sources usually works at high frequencies (VHF, UHF, and SHF) where the screen is electrically thick and the absorption-loss term A gives a significant contribution as well. The values of the SE are so large that the approximations due to the extension of the TL analogy to near-field sources and to the small-dipole approximation of the actual sources do not affect the overall accuracy of the results. The TL approach is particularly convenient here because it provides accurate results, even if the actual source is modeled as a small dipole. The effect of the main quantities on the SE is shown in Figure 4.13. A copper planar screen ($\sigma = 11.8 \cdot 10^6$ S/m, $d = 0.5$ mm) of infinite extent is assumed to be placed 30 cm far from an electric dipole parallel to the shield. In the low-frequency range the reflection-loss term R is the dominant

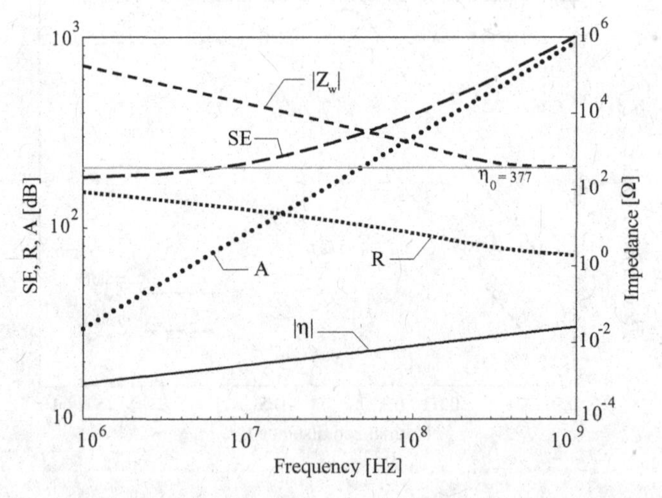

FIGURE 4.13 Magnitude of the contributions to the SE of an infinite copper planar screen of thickness $d = 0.5$ mm illuminated by a parallel electric dipole as a function of frequency.

contribution to the SE, due to the several orders of mismatch between the incident-wave impedances $|Z_{w1,2}^{ED}|$ and the intrinsic characteristic impedance of copper $|\eta|$ (the multiple-reflection-loss term M is practically zero). As the frequency is increased, the screen becomes more and more electrically distant from the source; the values $|Z_{w1,2}^{ED}|$ of the wave impedances decrease toward that of the free-space impedance of a normally incident plane wave η_0, the characteristic impedance $|\eta|$ of copper increases, and the contribution of R to the SE reduces while the contribution of A increases. The SE always assumes large values, varying from 150 to 1000 dB. In this range of values, the errors introduced by the TL analogy are negligible.

On the other hand, in the case of low-impedance field sources, the TL analogy needs to be applied with several attentions to avoid incorrect results. Low-impedance field sources (e.g., loops or straight conductors carrying an electric current) are mainly source of magnetic field in the low-frequency range (e.g., ULF and VLF). At low frequencies the conducting shield is electrically small ($d \ll \lambda_0$), and the contribution of the absorption-loss term A is almost negligible. For the materials shown in Figure 4.3, the magnitude of the wave impedances is from one to three orders of magnitude larger than that of their intrinsic impedance $|\eta|$: the reflection-loss term is the dominant contribution to the SE, although its values are much smaller than those attained in the case of high-impedance field sources. Moreover the multiple-reflection-loss term M can assume negative values, further reducing the SE values. This behavior is illustrated in Figure 4.14, where a copper planar screen ($\sigma = 11.8 \cdot 10^6$ S/m, $d = 0.5$ mm) of infinite extent is assumed to be placed 30 cm far from a magnetic dipole parallel to the screen. Below 200 Hz, the SE is practically zero and the R and M terms are almost equal but opposite in sign. From 200 up to 200 kHz,

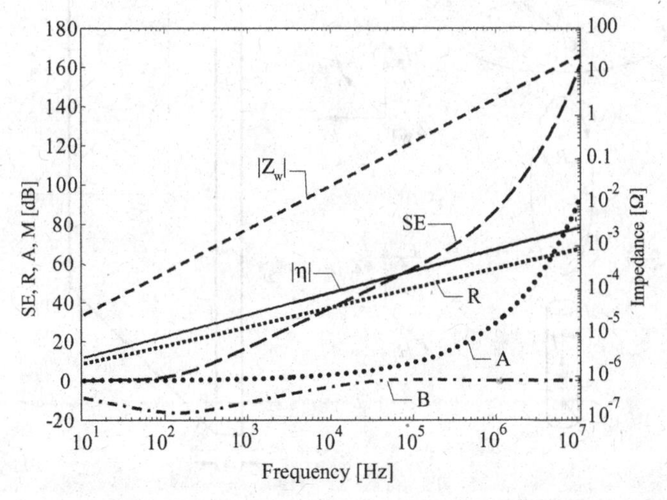

FIGURE 4.14 Magnitude of the contributions to the SE of an infinite copper planar screen illuminated by a parallel magnetic dipole as a function of frequency.

the main contribution to the SE is mainly due to the reflection-loss term R, since the mismatch between the wave impedances and the copper intrinsic impedance increases. Above 200 kHz, the absorption-loss term A starts to contribute to the overall SE while the contribution of the M term drops to zero.

A comparison of Figures 4.13 and 4.14 shows that a difference of about one order of magnitude exists between the SE of the screen when it is illuminated by the high-impedance field source and by the low-impedance field source. From a physical point of view, the main difference is the role played by the eddy-current cancellation mechanism (see Appendix B), which is practically negligible at very low frequencies, and the different weight of the reflection- and multiple-reflection-loss terms. For low-impedance field sources these two terms are the dominant contributions to the SE. Hence any small error in the computation of these terms deeply affects the accuracy of the final results provided by the TL analogy. For this reason it is important to compute the wave impedances of the field produced by actual sources as accurately as possible.

In Appendix B some canonical problems of VLF magnetic shielding are revised and the analytical solutions are provided. These problems deal with typical sources of practical interest, as illustrated in Figure 4.15: the current loop placed in a plane normal to the shield, the current loop placed in a plane parallel

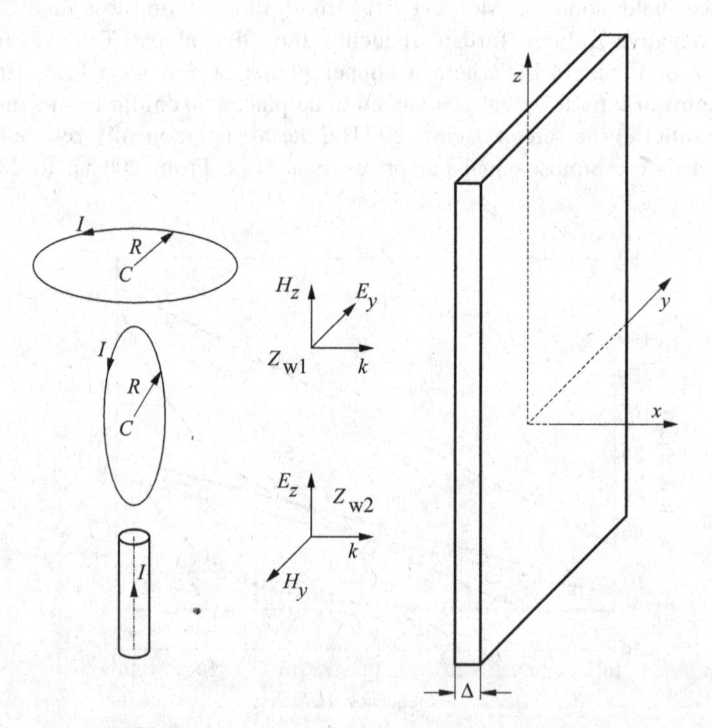

FIGURE 4.15 Basic low-frequency field sources.

to the shield, the straight wire conductor parallel to the shield. Comparisons between the results obtained through exact formulations and the approximate TL analogy are also reported, showing that the latter can provide very accurate results.

For these sources it is possible to compute the correct wave impedances along the normal to the shield (x axis) as

$$Z_{\mathrm{w1}}^{\mathrm{loop}\perp} = -j\omega\mu_0 \frac{\int_0^\infty [\int_0^\infty [I_1(\beta R)/\beta] e^{-\sqrt{\alpha^2+\beta^2}x} \mathrm{d}\alpha] \mathrm{d}\beta}{\int_0^\infty [\int_0^\infty [\beta I_1(\beta R)/\sqrt{\alpha^2+\beta^2}] e^{-\sqrt{\alpha^2+\beta^2}x} \mathrm{d}\alpha] \mathrm{d}\beta}, \tag{4.52a}$$

$$Z_{\mathrm{w1,2}}^{\mathrm{loop}\parallel} = -j\omega\mu_0 \frac{\int_0^\infty \left(\nu/\sqrt{\nu^2-k_0^2}\right) J_1(\nu R) e^{-\sqrt{\nu^2-k_0^2}x} \mathrm{d}\nu}{\int_0^\infty \nu J_1(\nu R) e^{-\sqrt{\nu^2-k_0^2}x} \mathrm{d}\nu}, \tag{4.52b}$$

$$Z_{\mathrm{w2}}^{\mathrm{wire}\parallel} = \eta_0 \frac{K_0(jk_0 x)}{K_1(jk_0 x)}, \tag{4.52c}$$

where $I_1(\cdot)$ is the first-order modified Bessel functions of the first kind while $K_0(\cdot)$ and $K_1(\cdot)$ are the zero- and first-order modified Bessel functions of the second kind, respectively.

In Figure 4.16, a comparison among these wave impedances (radius of the loop $R = 30\,\mathrm{cm}$) and the wave impedances of an infinitesimal magnetic dipole for a shield-to-source distance of 30 cm is shown. Even if the sources are all characterized by very low values of the wave impedances, some differences are noticeable. Hence, when a real source is replaced with a small magnetic dipole, the wave impedances

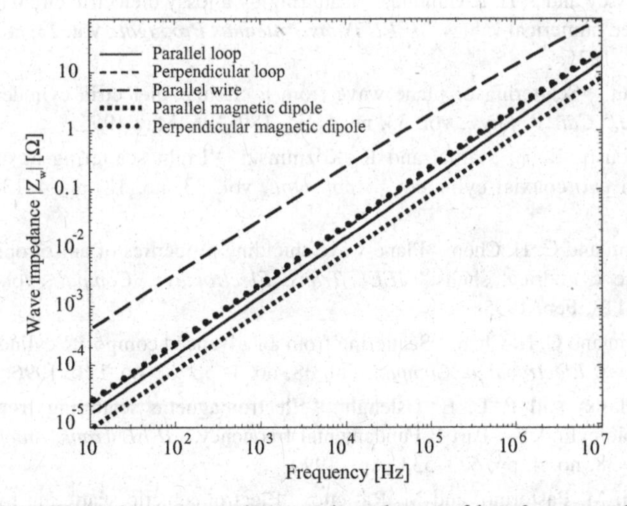

FIGURE 4.16 Comparison among the wave impedances of low-frequency field sources.

must be accurately evaluated since small differences can provide inaccurate results when the TL analogy is adopted.

In the low-frequency range, where the approach is more critical, the TL analogy has been widely tested with high-permeability and high-conductivity materials [1, 31, 32, 37], showing results in reasonable agreement with those obtained through measurements and those calculated with exact formulations [33–37].

REFERENCES

[1] R. F. Harrington. *Time-Harmonic Electromagnetic Fields*, 2nd ed. Piscataway, NJ: Wiley-IEEE Press, 2001.

[2] J. A. Stratton. *Electromagnetic Theory.* Piscataway, NJ: Wiley-IEEE Press, 2007.

[3] R. B. Schulz, V. C. Plantz, and D. R. Brush. "Shielding theory and practice." *IEEE Trans. Electromagn. Compat.*, vol. 30, no. 3, pp. 187–201, Aug. 1988.

[4] C. A. Klein. "Microwave shielding effectiveness of EC-coated dielectric slabs." *IEEE Trans. Microwave Theory Tech.*, vol. 38, no. 3, pp. 321–324, Mar. 1990.

[5] H. Cory, S. Shiran, and M. Heilper. "An iterative method for calculating the shielding effectiveness and light transmittance of multilayered media." *IEEE Trans. Electromagn. Compat.*, vol. 35, no. 4, pp. 451–456, Nov. 1993.

[6] D. Schieber. "Shielding performance of metallic cylinders." *IEEE Trans. Electromagn. Compat.*, vol. 15, pp. 12–16, Feb. 1973.

[7] T. K. Wu and L. L. Tsai. "Shielding properties of thick conducting cylindrical shells." *IEEE Trans. Electromagn. Compat.*, vol. 16, no. 4, pp. 201–204, Nov. 1974.

[8] A. K. Datta and S. C. Som. "Numerical study of the scattered electromagnetic field inside a hollow dielectric cylinder. 1: Scattering of a single beam." *Appl. Opt.*, vol. 14, no. 7, pp. 1516–1523, Jul. 1975.

[9] H. E. Bussey and J. H. Richmond. "Scattering by a lossy dielectric circular cylindrical multilayer, numerical values." *IEEE Trans. Antennas Propagat.*, vol. 23, no. 5, pp. 723–725, Sep. 1975.

[10] J. R. Wait. "Scattering of plane wave from a circular dielectric cylinder at oblique incidence." *Can. J. Phys.*, vol. 33, no. 5, pp. 189–195, May 1955.

[11] H. A. Yousif, R. E. Mattis, and K. Kozminski. "Light scattering at oblique incidence on two coaxial cylinders." *Appl. Opt.*, vol. 33, no. 18, pp. 4013–4024, Jun. 1994.

[12] C.-N. Chiu and C. H. Chen. "Plane-wave shielding properties of anisotropic laminated composite cylindrical shells." *IEEE Trans. Electromagn. Compat.*, vol. 37, no. 1, pp. 109–113, Feb. 1995.

[13] C.-N. Chiu and C. H. Chen. "Scattering from an advanced composite cylindrical shell." *IEEE Trans. Electromagn. Compat.*, vol. 38, no. 1, pp. 62–66, Feb. 1996.

[14] M. A. Hasan and P. L. E. Uslenghi. "Electromagnetic scattering from nonlinear anisotropic cylinders—Part I: Fundamental frequency." *IEEE Trans. Antennas Propagat.*, vol. 38, no. 4, pp. 523–533, Apr. 1990.

[15] S. Caorsi, M. Pastorino, and M. Raffetto. "Electromagnetic scattering by multilayer elliptic cylinder under transverse-magnetic illumination: Series solution in terms of

Mathieu functions." *IEEE Trans. Antennas Propagat.*, vol. 45, no. 6, pp. 926–935, Jun. 1997.

[16] M. H. Öktem and B. Saka. "Design of multilayer cylindrical shields using a genetic algorithm." *IEEE Trans. Electromagn. Compat.*, vol. 43, no. 2, pp. 170–176, May 2001.

[17] L. G. Guimarães and J. P. R. Furtado de Mendonça. "Analysis of the resonant scattering of light by cylinders at oblique incidence." *Appl. Opt.* vol. 36, no. 30, pp. 8010–8019, Oct. 1997.

[18] A. L. Aden and M. Kerker. "Scattering of electromagnetic waves from two concentric spheres." *J. Appl. Phys.*, vol. 22, no. 10, pp. 1242–1246, Oct. 1951.

[19] M. G. Andreasen. "Back-scattering cross section of thin, dielectric, spherical shell." *IRE Trans. Antennas Propagat.*, vol. 33, no. 5, pp. 267–270, Jul. 1957.

[20] J. R. Wait. "Electromagnetic scattering from a radially inhomogeneous sphere." *Appl. Sci. Res. Sect. B*, vol. 10, no. 5–6, pp. 441–449, 1963.

[21] R. Bhandari. "Scattering coefficients for a multilayered sphere: Analytic expressions and algorithms." *Appl. Opt.*, vol. 24, no. 13, pp. 1960–1967, Jul. 1985.

[22] Z. S. Wu and Y. P. Y. Wang. "Electromagnetic scattering for multilayered sphere: Recursive algorithms." *Radio Sci.*, vol. 26, pp. 1393–1401, 1991.

[23] J. C. Monzon. "Three-dimensional field expansion in the most general rotationally symmetric anisotropic material: Application to scattering by a sphere." *IEEE Trans. Antennas Propagat.*, vol. 37, no. 6, pp. 728–735, Jun. 1989.

[24] D. K. Cohoon. "An exact solution of Mie type for scattering by a multilayer anisotropic sphere." *J. Electromag. Waves Appl.*, vol. 3, no. 5, pp. 421–448, 1989.

[25] R. L. Hightower and C. B. Richardson. "Resonant Mie scattering from a layered sphere." *Appl. Opt.*, vol. 27, no. 23, pp. 4850–4855, Dec. 1988.

[26] T. Yamane, A. Nishikata, and Y. Shimizu. "Resonance suppression of a spherical electromagnetic shielding enclosure by using conductive dielectrics." *IEEE Trans. Electromagn. Compat.*, vol. 42, no. 4, pp. 441–448, Nov. 2000.

[27] L.-W. Li, X.-K. Kang, and M.-S. Leong. *Spheroidal Wave Functions in Electromagnetic Theory.* New York: Wiley, 2002.

[28] S. Asano and G. Yamamoto. "Light scattering by a spheroidal particle." *Appl. Opt.*, vol. 14, no. 1, pp. 29–49, Jan. 1975.

[29] V. G. Farafonov, N. V. Voshchinnikov, and V. V. Somsikov. "Light scattered by a core-mantled spheroidal particle." *Appl. Opt.*, vol. 35, n. 27, pp. 5412–5426, Sep. 1996.

[30] I. Gurwich, M. Kleiman, N. Shiloah, and A. Cohen. "Scattering of electromagnetic radiation by multilayered spheroidal particles: Recursive procedure." *Appl. Opt.*, vol. 39, no. 3, pp. 470–477, Jan. 2000.

[31] R. B. Schulz. "ELF and VLF shielding effectiveness of high-permeability materials." *IEEE Trans. Electromagn. Compat.*, vol. 10, no. 1, pp. 95–100, Mar. 1968.

[32] R. B. Schulz, V. C. Plantz, and D. R. Brush. "Low-frequency shielding resonance." *IEEE Trans. Electromagn. Compat.*, vol. 10, no. 1, pp. 7–15, Mar. 1968.

[33] J. R. Moser. "Low-frequency shielding of circular loop electromagnetic field sources." *IEEE Trans. Electromagn. Compat.*, vol. 9, no. 1, pp. 6–18, Jan. 1967.

[34] A. Nishikata and A. Sugiura. "Analysis for electromagnetic leakage through a plane shield with an arbitrarily-oriented dipole source." *IEEE Trans. Electromagn. Compat.*, vol. 34, no. 8, pp. 284–291, Aug. 1992.

[35] Y. Du, T. C. Cheng, and A. S. Farag. "Principles of power frequency magnetic field shielding with flat sheets in a source of long conductors." *IEEE Trans. Electromagn. Compat.*, vol. 38, no. 8, pp. 450–459, Aug. 1996.

[36] R. Araneo and S. Celozzi. "Exact solution of low-frequency coplanar loops shielding configuration." *IEE Proc. Sci., Meas. Tech.*, vol. 149, no. 1, pp. 37–44, Jan. 2002.

[37] R. G. Olsen, M. Istenic, and P. Zunko. "On simple methods for calculating ELF shielding of infinite planar shields." *IEEE Trans. Electromagn. Compat.*, vol. 45, no. 3, pp. 538–547, Aug. 2003.

CHAPTER FIVE

Numerical Methods for Shielding Analyses

The numerical approximation of Maxwell's equations is known as *computational electromagnetics* (CEM). Computational electromagnetics is the scientific discipline that studies electric and magnetic fields and their interaction (electromagnetics) using intrinsically and routinely a digital computer to obtain numerical results (computational). Maxwell's equations are the starting point for the study of any electromagnetic (EM) problem. Nevertheless, their actual solution is generally complex, and for realistic problems a *numerical modeling* is usually required. CEM may be broadly defined as the branch of electromagnetics that develops and solves a numerical model of physical EM phenomena, introducing approximations that are an intrinsic part of any computer model. The main issues involved in developing a computer model (i.e., classification of model types, steps involved in developing the model, desirable attributes of the model, and role of approximations) are well described in [1]. What is worth pointing out here is that numerical modeling is an activity distinct from computation [2].

The adoption of any numerical technique always requires one to be familiar with its basic formulation in order to apply it correctly and be aware of its accuracy, efficiency, and utility with respect to the specific problem under analysis. The widespread adoption of numerical modeling, boosted by the availability of powerful commercial codes that are claimed to be general purpose, has frequently led to the application of numerical techniques by unfamiliar users to problems for which they are not designed, resulting in inefficient simulations and/or inaccurate results.

A classification of the numerical methods is possible based on several criteria. The formulation of any engineering model starts expressing the physical laws used

Electromagnetic Shielding by Salvatore Celozzi, Rodolfo Araneo and Giampiero Lovat

in describing the observed phenomena in a functional form of the type $L[\phi] = g$, where $L[\cdot]$ is a functional operator, ϕ is the unknown field function of the problem, and g is a known source function.

A first characteristic to be used in the classification can be drawn from looking at the *domain* (time or frequency) in which the operator, the field, and the source functions are defined. This allows for a distinction to be made between the *time-domain* (TD) methods and the *frequency-domain* (FD) methods. Several reasons can be provided for modeling in one such domain rather than in the other. In general, the TD formulation is suitable for studying transient phenomena, when broadband information is sought or when problems involve nonlinear or time-varying media, while the FD formulation is straightforward for studying the steady-state response, when a single-frequency or a narrow-band response is sought, or when dealing with high-resonant structures. In applying this reasoning, it is important to keep in mind that no one method is capable of solving every problem. Identifying and developing an approach that is best suited to a particular problem always involves trade-offs among a variety of considerations and choices, such as the geometry of the problem, the involved media, the requested output information, the efficiency and accuracy of the method.

Another type of broad classification is possible about the nature of the functional operator $L(\cdot)$. The functional operator may be expressed in a differential or in an integral form, by way of *differential-equation* (DE) or *integral-equations* (IE) methods, respectively. Since the DE methods call for physical laws to be enforced at all points in space, they allow one to easily deal with complex materials with fine features and irregular shapes, such as anisotropic or inhomogeneous materials, but they require the discretization of the entire computational space and the enforcement of an artificial numerical boundary to solve open-region problems. In contrast, the IE methods call for physical laws to be enforced only at significant surfaces of the scatterers involved in the problem, and hence IE methods are used to solve a certain type of problems with fewer unknowns. Furthermore they allow open-boundary problems to be treated rigorously. However, the equations are generally more complex than those of the DE methods and cannot be easily applied to complex media. Besides the broad classes of DE and IE methods, there are methods based on operators of a particular type, such as *optical methods* based on ray tracing, physical optics (PO), and the geometric theory of diffraction (GTD). However, they cannot be considered *full-wave* methods (i.e., methods that approximate Maxwell's equations without any initial physical approximation), since they are based on optical approximations.

Describing the whole field of CEM is a daunting task, and beyond the scope of the present book. This chapter provides a short but comprehensive survey of numerical methods that are usually applied in the field of the shielding theory. The purpose is to enable the reader to gain a perspective on the variety of techniques presently available or being developed. The overview nevertheless will be selective and focused on the basic concepts, features, and proposed improvements relevant for an application of the methods used to analyze shielding problems.

5.1 FINITE-ELEMENT METHOD

The finite-element method (FEM) [3–11] is a numerical technique capable of obtaining approximate solutions to a wide variety of engineering problems. Although the FEM is conceptually more complex and difficult to program than other well-developed methods, it has matchless power for handling problems involving complex geometries and inhomogeneous media. The distinguished features of the method are its systematic generality, diversity and flexibility all of which make it possible to solve a broad field of problems in different engineering disciplines involving equilibrium, eigenvalue, and propagation problems, for example, in civil, mechanical, aerospace, and electrical-electronic engineering, and to construct general-purpose computer programs for solving coupled multidiscipline problems involving, for example, thermomechanic or thermoelectric properties.

Roughly, the FEM is a numerical technique used to solve the boundary-value problems arising in the mathematical modeling of physical systems. These systems are generally described by a governing equation of the form

$$L[\phi] = g, \tag{5.1}$$

which has to be solved for the unknown function ϕ in a domain Ω, together with the boundary conditions $B[\phi] = 0$ on the boundary Γ that encloses the domain. In (5.1), $L[\cdot]$ is the operator (differential or integral) that describes the problem while g is the known source term; of course, both ϕ and g can be either scalar or vector functions.

In electromagnetics the governing equation ranges from simple Poisson equations to more complicated scalar and vector wave equations; the boundary conditions range from simple Dirichlet or Neumann conditions to more sophisticated impedance, radiation, and higher order conditions. The basic idea of the finite-element analysis is to search an approximate solution of the whole problem by envisioning the solution region Ω as built up of many small interconnected subregions. The solutions are therefore expressed individually before they are put together to approximate the entire problem. An example of the possible original problem is sketched in Figure 5.1*a*, and its subregion triangular model is shown in Figure 5.1*b*. In essence, a complex problem reduces to considering a series of greatly simplified problems. Since the elements can be put together in a variety of ways, they can be used to represent exceedingly complex shapes. An approximate solution of the boundary-value problem can be searched by using the Rayleigh-Ritz variational approach or the weighted residual approach.

A place to start is with a review of the work of Mikhlin [12]. If L is a real self-adjoint operator, the solution of (5.1) makes stationary the functional

$$I[\phi] = \frac{1}{2}\langle L\phi, \phi\rangle - \frac{1}{2}\langle \phi, g\rangle - \frac{1}{2}\langle g, \phi\rangle, \tag{5.2}$$

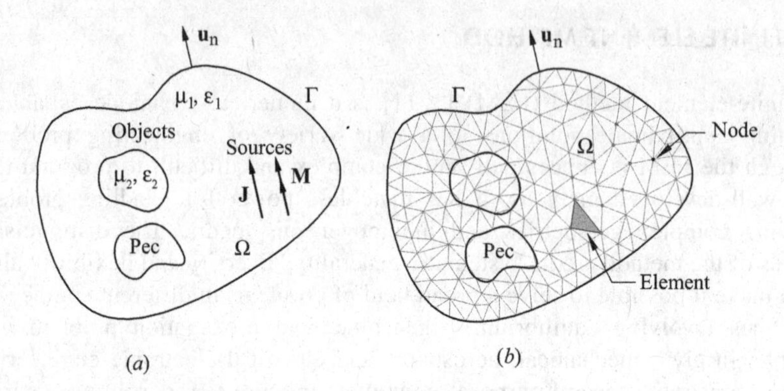

FIGURE 5.1 Electromagnetic sources in an inhomogeneous region Ω (a) and two triangular cell model (b).

where the brackets denote the inner product defined as

$$\langle \phi, \psi \rangle = \int_{\Omega} \phi \psi^* \, d\Omega. \tag{5.3}$$

If the operator $L[\cdot]$ is complex, the same functional can be retained, but the inner product must be modified not considering the complex conjugate. The Rayleigh-Ritz method yields an approximate solution to the variational problem (5.2) directly, that is, without recourse to the associated differential equation (5.1). From an appropriately selected set of linearly independent real expansion (or basis) functions u_j, an approximate solution $\tilde{\phi}$, satisfying the prescribed boundary conditions, is constructed in the form of a finite series

$$\tilde{\phi} = \sum_{j=1}^{N} x_j u_j, \tag{5.4}$$

where the coefficients x_j can be complex. The procedure is said to converge to the exact solution if $\tilde{\phi} \to \phi$ as N tends to infinity. Introducing (5.4) into (5.2), and minimizing the functional by forcing its partial derivatives with respect to the coefficients x_i to vanish, leads to

$$\sum_{j=1}^{N} \langle L u_j, u_i \rangle x_j = \langle g, u_i \rangle, \qquad i = 1, 2, \ldots, N, \tag{5.5}$$

which can be cast in a matrix form as $\underline{\mathbf{A}} \cdot \mathbf{x} = \mathbf{b}$.

When the functional does not exist, it is possible to apply one of the techniques referred to as the weighted residual methods. Use of the weighted residual is more general and of wider application because this technique is not limited to the class of variational problems. As the name indicates, the result sought is an approximate solution in the same form as in the Rayleigh-Ritz method, obtained by weighting the residual of the equation (5.1). Substitution of (5.4) in the functional equation (5.1) results in a nonzero residual:

$$R = L\tilde{\phi} - g \neq 0. \tag{5.6}$$

Since the best approximation will reduce the residual R to the least value at all points of the domain Ω, the weighted residual methods enforce to zero the integral conditions

$$\langle w_i, R \rangle = \int_{\Omega} w_i R \, \mathrm{d}\Omega = 0, \quad i = 1, \ldots, N, \tag{5.7}$$

where $\{w_i\}_N$ are an appropriately selected set of weighting (testing) functions. By enforcing the inner product to zero for each function w_i, the final system is readily obtained as

$$\sum_{j=1}^{N} \langle w_i, Lu_j \rangle x_j = \langle w_i, g \rangle. \tag{5.8}$$

Although several weighted residual methods result from different choices of the testing functions (point-matching method, subdomain-collocation method, least-squares method), the most common practice is to set the weighting functions equal to the expansion functions u_j. This particular choice is called *Galerkin's method*, which reduces to the Rayleigh-Ritz method if the operator L is self-adjoint.

The solution of any continuum problem by means of the FEM always involves an orderly step-by-step process. The first step consists in subdividing the solution domain Ω into nonoverlapping subregions, called *elements*. A variety of element shapes are available and different element shapes may be used in the same problem, both in two and three dimensions. Nevertheless, the most common elements are triangles and tetrahedra: they are in fact the simplest shape into which a 2D or 3D region can be broken and are well-suited to automatic mesh generation, for which efficient algorithms have been developed. Conforming elements with curved sides have been proposed too, which reduce errors in modeling the region of interest, especially when it has curved boundaries, but with an additional computational cost.

The next step is the choice of the real-valued independent expansion functions u_j. They are generally subdomain functions defined over a finite support comprising a few number of adjacent elements where they approximate the variations of the unknown function.

Once the finite-element model has been established, the third step requires one to derive the governing matrix equations over a single element. Galerkin's method or the variational Rayleigh-Ritz method is generally used to produce weak forms of the governing equations.

To find the solution over the domain Ω, all the elements must be assembled in a global matrix. The basis for the assembly procedure stems from the fact that all the elements sharing the support of a basis function contribute to the weight of the basis function. The step is usually done through a connectivity matrix that maps the local numbers of the simplices of the mesh to the numbers corresponding to the whole mesh.

Before being ready for solution, the system equations must be modified to account for the boundary conditions of the problem. The boundary conditions can be either of the essential or of the natural type [6]. The essential condition must be explicitly built into the finite-element model in order to be satisfied by the solution, even in the limit of an infinite number N of expansion functions; the natural condition is satisfied by the model but only in an approximate weak sense, it being exactly satisfied only in the limit case.

The last step requires one to solve the matrix system in order to obtain the unknown coefficients of the expansion functions. The nature of the resulting set of equations depends on the type of problem that is being solved: in particular, the algebraic system is linear for a steady-state problem, whereas it is nonlinear for an eigenvalue problem. Moreover a linear or nonlinear matrix differential equation must be solved if the problem is unsteady (e.g., when the coefficients are time dependent). Apart from the nature of the resulting matrix equations, the common characteristic of them all that distinguishes finite-element analysis is that the matrices are deeply sparse. This feature makes possible the use of special-purpose methods that exploit the matrix sparsity to optimize the storage and the computational cost.

The FEM was first applied (in electromagnetics) in electrostatic problems and for the computation of homogeneous waveguide modes in the frequency domain, where the primary unknowns are scalar in nature (the electric voltage and the longitudinal components of electric or magnetic field, respectively). Node-based elements were developed and particular polynomial expressions (the Lagrangian functions) were adopted to approximate the behavior of the field over the element, since they are easy both to be integrated and to be differentiated. The scalar field ϕ can be approximated in the meshed region Ω by the weighted sum

$$\phi(x, y, z) = \sum_{j=1}^{M} \alpha_j(x, y, z)\, \Phi_j, \tag{5.9}$$

where Φ_j are the unknown values of the variable ϕ at the M nodes of the mesh and α_j are linearly independent scalar basis functions. The expansion function α_j is a subdomain function, meaning it is nonzero only on the elements around the node j, whose space regions define the support of the function. The number of regularly spaced nodes in the single element depends on the order k of the polynomial functions, thus giving rise to first-order ($k = 1$) and higher order ($k > 1$) elements. Two-dimensional basis functions of first and second order are shown in Figure 5.2 for a triangular element. To achieve higher accuracy in the finite-element solution, it

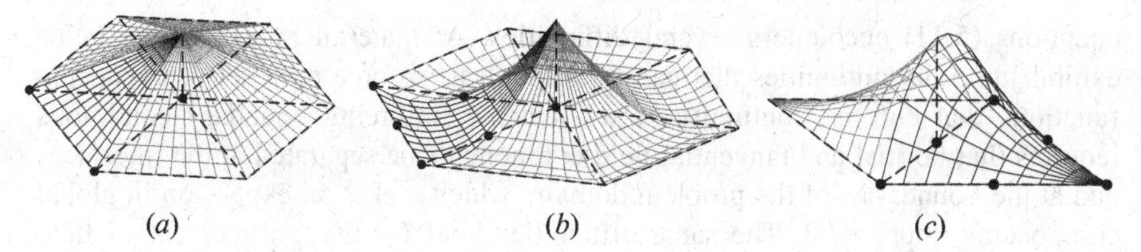

FIGURE 5.2 Subdomain triangular basis functions of the first (*a*) and second (*b*, *c*) order.

is possible to resort to finer subdivisions with lower order elements, at the price of increasing computing time and memory due to the larger number of unknowns, or to resort to higher order interpolation functions, at the price of more complex formulations. A huge amount of literature exists on the systematic construction of Lagrangian and general hierarchical scalar functions, which the interested reader is referred to [6–11].

The scalar formulation was extended in the frequency domain to inhomogeneous waveguides and to three-dimensional (3D) problems. Although there are a number of possible equations describing a general 3D EM scattering problem, without any loss of generality it is possible to limit the attention to the use of the vector wave equations in the double-curl form for the electric and magnetic field as

$$\nabla \times \left(\frac{1}{\mu_r}\nabla \times \mathbf{E}\right) - k_0^2\varepsilon_r\mathbf{E} = -jk_0\eta_0\mathbf{J}, \tag{5.10a}$$

$$\nabla \times \left(\frac{1}{\varepsilon_r}\nabla \times \mathbf{H}\right) - k_0^2\mu_r\mathbf{H} = \frac{1}{\varepsilon_r}\nabla \times \mathbf{J}. \tag{5.10b}$$

Scalar representation is employed so that each component of the vector field quantities can be individually approximated on the finite-element scheme. By applying the Galerkin method, we can derive the weak forms as

$$\iiint_\Omega \left[\frac{1}{\mu_r}(\nabla \times \mathbf{E}) \times \nabla\alpha_i - k_0^2\varepsilon_r\mathbf{E}\alpha_i + jk_0\eta_0\mathbf{J}\alpha_i\right] d\Omega = -\oint\!\!\!\oint_\Gamma \left(\frac{1}{\mu_r}\mathbf{u}_n \nabla \times \mathbf{E}\alpha_i\right) d\Gamma, \tag{5.11a}$$

$$\iiint_\Omega \left[\frac{1}{\varepsilon_r}(\nabla \times \mathbf{H}) \times \nabla\alpha_i - k_0^2\mu_r\mathbf{H}\alpha_i - \frac{1}{\varepsilon_r}\nabla \times \mathbf{J}\alpha_i\right] d\Omega = -\oint\!\!\!\oint_\Gamma \left(\frac{1}{\varepsilon_r}\mathbf{u}_n \nabla \times \mathbf{H}\alpha_i\right) d\Gamma, \tag{5.11b}$$

where α_i is a scalar basis function and $\mathbf{u}_n$ is the outward-pointing unit vector normal to the closed boundary Γ of the volume Ω. It can be noted that the final matrix system has dimension $3N$, comprising complex 3×3 submatrices of dimension N, since three unknowns are placed at each node of the mesh. The application of equations (5.11) encounters several difficulties. At material interfaces the fields exhibit jump discontinuities that can be difficult to enforce on a set of expansion functions that enforce continuity between cells. Enforcing boundary conditions requires that normal and tangential field components be separated at the interfaces and at the boundaries of the problem domain, which makes an expansion in global components impractical. The same difficulties hold for the enforcement of field singular behaviors at edges and corners. Furthermore the Lagrangian discretization of the wave equations produces spurious functions in the eigenspectrum of the curl operator (known as *spurious modes* or vector parasites).

A body of literature exists on the issue of spurious modes, which at first were a source of confusion in the EM community. Spurious modes are numerical solutions of the vector wave equation, either in the double-curl or in the Helmholtz form, that have no correspondence to physical reality. Largely influenced by the first works on the subject [13,14], the early thinking attributed this problem to a deficiency in enforcing the solenoidal nature of the field in the approximation process. Consequently modified functionals and weak forms were proposed, associated with the so-called penalty method [15], to constrain the magnetic and electric flux vectors to be truly solenoidal. Successively, a body of literature has shown that the true cause of spurious modes is the incorrect approximation of the null-space of the curl operator [16–18]. The curl-curl operator admits eigenfunctions that have a solenoidal flux $\nabla \cdot \mathbf{D} = 0$ and that correctly represent a time-varying field in a source-free region, and irrotational eigenfunctions of the form $\mathbf{E} = -\nabla\Phi$ that form the null-space of the curl operator corresponding to the eigenvalue $k_0 = 0$ and thus represent a static field. Unless the basis functions are orthogonal to all the eigenfunctions in the null-space, the resulting matrix representing the operator will have some eigenvectors approximating those functions. The difficulty lies in the fact that Lagrangian basis functions cannot adequately represent the null-space eigenfunctions (since the approximation is very poor), the zero eigenvalues are approximated with large numbers, and the eigenvectors corresponding to the poorly approximated eigenfunctions form the spurious modes. The new philosophy that originated from these observations was to represent the null-space eigenfunctions as accurately as possible, as opposed to trying to suppress these eigenfunctions by making it impossible for the basis-function set to model the irrotational solutions. Provided that the basis functions are able to approximate the null-space of the curl operator, the eigenvalue $k_0 = 0$ will be computed exactly with its associated eigenfunctions, and it will be necessary only to ignore these static solutions.

The spurious-mode problem was fully solved in the early 1980s [19–22] when a new family of mixed-order subsectional vector basis functions was introduced providing a spurious eigenvalue free discretization of the curl-curl operator. The important properties of these functions are that they impose condition of tangential continuity for the unknown vector between adjacent cells, without enforcing any constraint on the normal component and that they exhibit identically zero divergence. For these reasons they are referred to as *curl-conforming tangential vector elements* or often *edge elements* because in the linear formulation (only) the

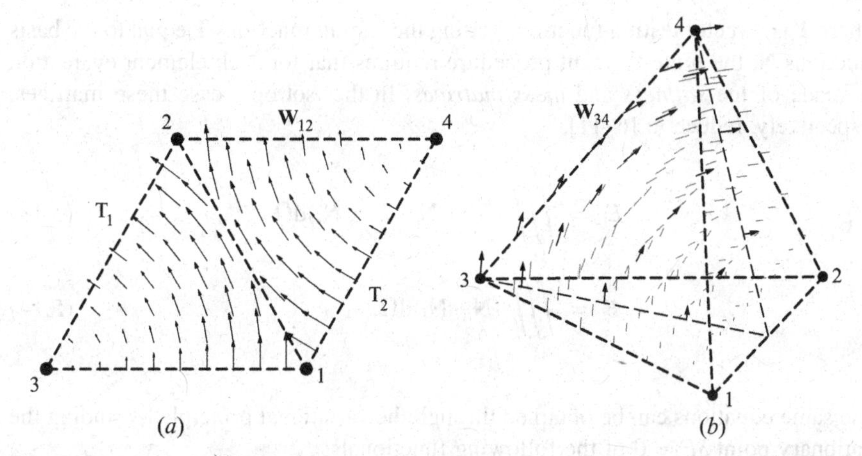

FIGURE 5.3 First-order two-dimensional triangular vector basis function (*a*) and three-dimensional-tetrahedral-vector basis function (*b*).

bases are associated with edges. The success of these vector functions in eliminating spurious modes is due to their providing a consistent representation of the field and the associated scalar potential over an arbitrary mesh of elements, and not to their solenoidal nature. A two-dimensional and a three-dimensional vector basis function of the first-order are shown in Figure 5.3. It has been shown in [23] that the edge elements are one of the several manifestations in electromagnetics of the differential *p*-forms described by Whitney [24] that, more generally, assign degrees of freedom to simplices of a given mesh. The 0-form is the classical linear polynomial interpolant associated with the nodes of the mesh (potential function); the 1-form is the edge element associated with an edge that ensures the continuity of the vector components tangential to facets containing the edge (electric and magnetic fields); the 2-form is the facet-element that maintains the continuity of the normal component across a facet (currents and fluxes); the 3-form is a constant on tetrahedron (divergence of currents and fluxes).

In the framework of the vector elements, the Helmholtz equations can be converted into a weak form through the weighted residual method, thus obtaining

$$\iiint\limits_{\Omega}\left(\frac{1}{\mu_r}\nabla\times\mathbf{T}\cdot\nabla\times\mathbf{E}-k_0^2\varepsilon_r\mathbf{T}\cdot\mathbf{E}+jk_0\eta_0\mathbf{T}\cdot\mathbf{J}\right)d\Omega = -\oiint\limits_{\Gamma}\left(\frac{1}{\mu_r}\mathbf{T}\cdot\mathbf{u}_n\times\nabla\times\mathbf{E}\right)d\Gamma \tag{5.12a}$$

$$\iiint\limits_{\Omega}\left[\frac{1}{\varepsilon_r}\nabla\times\mathbf{T}\cdot\nabla\times\mathbf{H}-k_0^2\mu_r\mathbf{T}\cdot\mathbf{H}-\mathbf{T}\cdot\nabla\times\left(\frac{1}{\varepsilon_r}\mathbf{J}\right)\right]d\Omega = -\oiint\limits_{\Gamma}\left(\frac{1}{\varepsilon_r}\mathbf{T}\cdot\mathbf{u}_n\times\nabla\times\mathbf{H}\right)d\Gamma \tag{5.12b}$$

where **T** is a vector testing function. Setting the testing functions **T** equal to the basis functions **N**, the finite-element procedure requires that for each element evaluation be made of the *stiffness* and *mass matrices*. In the isotropic case these matrices, respectively, reduce to [6–11],

$$E_{ij} = \iiint_{\Omega} (\nabla \times \mathbf{N}_i \cdot \nabla \times \mathbf{N}_j) \mathrm{d}\Omega, \tag{5.13}$$

$$F_{ij} = \iiint_{\Omega} (\mathbf{N}_i \cdot \mathbf{N}_j) \mathrm{d}\Omega. \tag{5.14}$$

The same equations can be obtained through the variational principle by finding the stationary point $\delta F = 0$ of the following functionals:

$$F[\mathbf{E}] = \frac{1}{2} \iiint_{\Omega} \left(\frac{1}{\mu_{\mathrm{r}}} \nabla \times \mathbf{E} \cdot \nabla \times \mathbf{E} - k_0^2 \varepsilon_{\mathrm{r}} \mathbf{E} \cdot \mathbf{E} \right) \mathrm{d}\Omega$$
$$+ \iiint_{\Omega} (jk_0 \eta_0 \mathbf{E} \cdot \mathbf{J}) \mathrm{d}\Omega + \oint\!\!\oint_{\Gamma} \left(\frac{1}{\mu_{\mathrm{r}}} \mathbf{E} \cdot \mathbf{u}_n \times \nabla \times \mathbf{E} \right) \mathrm{d}\Gamma, \tag{5.15}$$

$$F[\mathbf{H}] = \frac{1}{2} \iiint_{\Omega} \left(\frac{1}{\varepsilon_{\mathrm{r}}} \nabla \times \mathbf{H} \cdot \nabla \times \mathbf{H} - k_0^2 \mu_{\mathrm{r}} \mathbf{H} \cdot \mathbf{H} \right) \mathrm{d}\Omega$$
$$+ \iiint_{\Omega} \left[\mathbf{H} \cdot \nabla \times \left(\frac{\mathbf{J}}{\varepsilon_{\mathrm{r}}} \right) \right] d\Omega + \oint\!\!\oint_{\Gamma} \left(\frac{1}{\varepsilon_{\mathrm{r}}} \mathbf{H} \cdot \mathbf{u}_n \times \nabla \times \mathbf{H} \right) \mathrm{d}\Gamma. \tag{5.16}$$

With the use of edge elements (first-order continuous-tangent/linear-normal vector elements), each edge is associated with a vector function that straddles all the elements sharing the common edge. Hence, over each element there are six vector functions, one per edge, and the elemental matrices are of dimension six. The important feature of the vector finite-element formulation is that the unknowns are generally associated with the Whitney p-forms (edges, faces, and cells) and not with only the node as in the nodal formulation.

A potential disadvantage of the FEM, common to all those methods based on a discrete form of partial differential equations (PDEs), arises when dealing with the solution of open-region scattering problems, known as open-domain or open-boundary problems. Since any PDE method is a finite-domain method by its own nature (i.e., only a finite number of elements are used to discretize a bounded domain), an artificial outer boundary must be introduced to bound the region of interest that is subdivided into meshes. To avoid the error caused by a simple truncation of the mesh, special techniques must be applied to represent the exact asymptotic behavior of the fields in the infinite exterior domain (described by the regularity condition in the static case and the radiation condition in the dynamic one). A very huge amount of literature exists on the finite-element open-boundary techniques, both for static fields and traveling waves [25–27].

As concerns statics, among all the available techniques, the ballooning technique [28] and the infinite-element method [29] have received more attention. In the ballooning technique, the interior region is surrounded by an annular region constituted of a mesh of super-finite elements. With a recursive process, new annulus are added, with the inner nodes overlapping the outer nodes of the previous annulus. This way the radius of the whole outer boundary increases in a geometric progression and the system rapidly converges. The infinite-element method attempts to represent the specific field decay in the exterior region by modifying the interpolating functions or by mapping the infinite element onto a finite region. In the first approach, the field variation in the infinite elements is represented by means of a decay interpolation function that properly makes the solution in the exterior region decay as a function of the radial distance. In the second approach, a singular mapping function is used to map the finite domain of a regular element into an infinite domain. The mapping function allows the classical regular shape functions to be used on the infinite elements to describe the asymptotic behavior of the field.

As regards the dynamic problems, appropriate radiation conditions are usually enforced on the outer boundary in order to absorb the outgoing waves with minimum nonphysical reflection, so as to represent a fictitious exterior region of infinite extent. Basically two types of conditions can be used on the artificial outer boundary in order to truncate the mesh: global and local conditions.

Global boundary conditions are exact within the numerical approximation, but they result in a full matrix, at least insofar as the boundary nodes are concerned. They mainly include the combined use of various types of boundary-integral equations with the finite-element formulation, arising in hybrid methods that retain the most efficient characteristics of both finite methods and integral-equation methods. Integral methods (considered in the next sections) can in fact handle unbounded problems very effectively, but they become computationally intensive when complex inhomogeneous media are present. In contrast, FEM easily handles nonhomogeneous media, requiring less computational effort for its sparse and banded matrix, thus being more suitable for boundary-value problems. Hence the basic recommended technique is to use FEM to treat the bounded and inhomogeneous regions and to use a coupled integral-equation method to treat the unbounded homogeneous region. Hybrid methods worthy of note are the hybrid finite-element method and method of moments (FEM/MoM) [30] and various formulations of the hybrid finite-element method and boundary-integral method (FEM/BEM) [31]. By means of these techniques, a closed (bounded) artificial surface separates an interior possibly nonhomogeneous region (where the differential equation is solved by means of the FEM) from an unbounded homogeneous exterior region where the EM problem is described by an integral equation (obtained from the Helmholtz equation through the use of a suitable Green's function). These hybrid methods can be effective in obtaining an exact solution for the field at infinity, but they have their deficiencies, mainly in possible internal resonances, especially at high frequencies, and in populated system matrices that corrupt the highly sparse and banded nature of the finite-element system matrices, which is the distinctive advantage of the FEM.

To retain the sparsity of the FEM matrices, a class of *local* absorbing boundary conditions (ABCs) has been developed with more attractive numerical features. However, because they are a truncated form of exact asymptotic expansions, the local boundary conditions are inexact and errors can be introduced in the solution by the presence at the outer boundary of reflected waves that are not totally absorbed. The most popular ABC was derived by Bayliss, Gunzburger, and Turkel (BGT) [32].

As an alternative to ABCs, Berenger [33,34] has introduced the concept of a perfectly matched layer (PML), and this has led to new fervor in CEM as well as in all the fields of computational physics. The innovation in this approach is that instead of enforcing an approximate mathematical absorbing condition on the outer boundary, the computational domain is surrounded with a fictitious medium, having certain constitutive parameters, that can absorb (without any reflection) EM plane waves of arbitrary polarization and angle of incidence at all frequencies. This medium is consequently referred to as perfectly matched to the medium of the interior computational domain. It was shown that in the absorbing layers it is sufficient to "split" each vector-field component into two orthogonal components in order to "split" the Maxwell equations into two sets of (unphysical) first-order PDEs. By properly choosing the constitutive and loss parameters of the medium, a perfectly matched planar interface is derived. Within the PML medium, the transmitted wave propagates with the same speed and direction as the impinging wave while simultaneously undergoing exponential decay along the normal to the PML interface. Thus the PML region can be terminated with a perfect electric conductor (PEC) and still the reflections from this PEC that re-enter the interior computational space are negligible.

Following the first Berenger works, a number of papers published on the topic proposed several modifications and improvements to enhance the PML performance. The original split-field formulation gave rise to more effective interpretations: the PML concept was restated in a stretched coordinate form and was shown to behave as a uniaxial anisotropic medium (UPML) characterized by both magnetic permeability and electric permittivity tensors; this also led to an attempt to realize a physical PML based on a Maxwellian formulation.

In recent years the finite-element formulation has been extended in the time domain [35,36] where it has received much attention because of the peculiar potential to simulate transient phenomena and perform broadband characterizations. Its discussion is beyond the scope of this chapter; for details, the reader is referred to the appropriate literature (e.g., [8,9,11]), with the caveat that the method has not been much applied in shielding applications. It is sufficient to remark that by expanding the electric field with vector basis functions whose unknown coefficients are now time dependent as

$$\mathbf{E}(\mathbf{r},t) = \sum_{j=1}^{N} u_j(t)\mathbf{N}_j(\mathbf{r}), \tag{5.17}$$

the curl-curl equation for the electric field can be recast in the following weak form through the classical Galerkin or variational method:

$$\underline{\underline{\mathbf{E}}} \cdot \frac{\mathrm{d}^2\mathbf{u}}{\mathrm{d}t^2} + \underline{\underline{\mathbf{F}}} \cdot \frac{\mathrm{d}\mathbf{u}}{\mathrm{d}t} + \underline{\underline{\mathbf{S}}} \cdot \mathbf{u} + \mathbf{f} = \mathbf{0}. \tag{5.18}$$

The ordinary differential equation (5.18) in the time domain can be integrated by employing several time difference schemes; among them, the Newmark method is generally preferred.

5.2 METHOD OF MOMENTS

The method of moments (MoM) [37–41] has been shown to be a very powerful technique for solving EM problems involving radiation and scattering from arbitrarily shaped objects. The method is originally based on the idea of taking a linear functional equation for the unknown (in the frequency or time domain) and representing it by a linear matrix equation. The idea is quite old, and can be traced back to Galerkin who developed it around 1915, but it did not become popular before the advent of high-speed computers because of the tedious computation of the matrix required for its use. The method requires two steps: the development of an appropriate integral equation describing the EM problem and the application of the matrix-method procedure [42] to reduce the functional equation to a matrix one.

In solving boundary-value problems constituted by the wave equation plus boundary conditions and radiation condition at infinity, it is possible to carry out a surface functional equation. This way the 3D problem is reduced to a 2D one in which the unknown is an appropriate equivalent surface current, electric or magnetic. The introduction of any special spatial coordinate grid is unnecessary, and the only condition on the desired unknown function is that it satisfies the integral equation. In scattering problems from perfectly conducting bodies as the one depicted in Figure 5.4, the electric-field integral equation (EFIE) and the magnetic-field integral equation (MFIE) are stated as

$$\mathbf{u}_n \times \mathbf{E}^{\mathrm{s}}(\mathbf{r})|_S = -\mathbf{u}_n \times \mathbf{E}^{\mathrm{inc}}(\mathbf{r})|_S, \tag{5.19a}$$

$$\mathbf{u}_n \times \mathbf{H}^{\mathrm{s}}(\mathbf{r}) - \mathbf{J}(\mathbf{r})|_S = -\mathbf{u}_n \times \mathbf{H}^{\mathrm{inc}}(\mathbf{r})|_S, \tag{5.19b}$$

where S is the boundary of the conductive object and $\mathbf{u}_n$ is the unit vector normal to S pointing outward from the body. In (5.19) the superscripts 'inc' and 'S' indicate the incident and the scattered fields, respectively. The incident field is that produced by the actual sources in the absence of any object, whereas the scattered field is that due to the equivalent sources radiating in free space [37–41]. The EFIE is the most popular integral equation because it can be applied to objects of arbitrary shape, unlike the MFIE whose validity is limited only to closed surfaces.

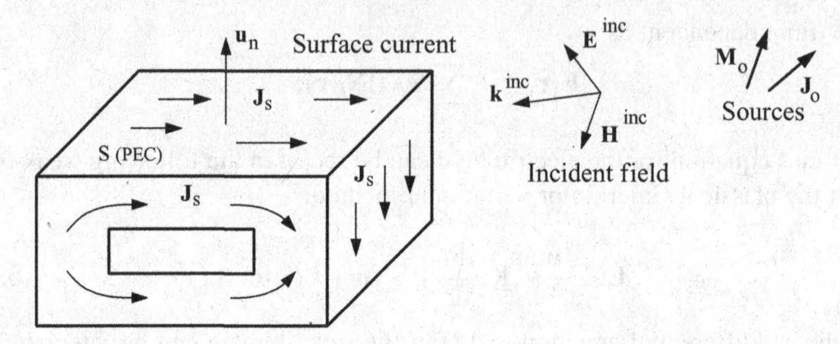

FIGURE 5.4 Electromagnetic sources radiating in presence of a PEC surface.

As regards the EFIE in (5.19a), the mixed-potential representation is usually adopted, according to which

$$\begin{aligned}
\mathbf{E}^{s}(\mathbf{r}) &= -j\omega\mu\mathbf{A}(\mathbf{r}) - \nabla\Phi(\mathbf{r}),\\
\mathbf{A}(\mathbf{r}) &= \iint_S G(\mathbf{r},\mathbf{r}')\mathbf{J}_S(\mathbf{r}')\mathrm{d}S',\\
\Phi(\mathbf{r}) &= \frac{1}{\varepsilon}\iint_S G(\mathbf{r},\mathbf{r}')\rho_{eS}(\mathbf{r}')\mathrm{d}S',\\
\nabla_S\cdot\mathbf{J}_S(\mathbf{r}) &= -j\omega\rho_{eS}(\mathbf{r}),
\end{aligned} \tag{5.20}$$

where $\mathbf{r}$ is the observation point, $\mathbf{r}'$ the source point, $G(\mathbf{r},\mathbf{r}')$ is the free-space scalar Green function defined in (1.40), and $\mathbf{J}_S$ is the unknown surface current density. By inserting (5.20) in (5.19a), we can obtain a Fredholm integral equation of the first kind in the original form derived by Maue [43] as

$$\begin{aligned}
&jk\eta\iint_S [G(\mathbf{r},\mathbf{r}')\mathbf{J}_S(\mathbf{r}')\times\mathbf{u}_n]\mathrm{d}S'\\
&\quad+\frac{\eta}{jk}\iint_S \{[\nabla' G(\mathbf{r},\mathbf{r}')\times\mathbf{u}_n]\cdot\mathbf{J}_S(\mathbf{r}')\,\mathrm{d}S' = \mathbf{E}^{\mathrm{inc}}(\mathbf{r})\times\mathbf{u}_n, \quad \mathbf{r}\in S,
\end{aligned} \tag{5.21}$$

where the integration spans the complete body surface S. Other forms of EFIEs are possible [37–41], but the mixed-potential form has gained success because it facilitates the explicit transfer of one derivative to the basis functions for $\mathbf{J}_S$ and another to the testing functions. This has proved to be very advantageous when dealing with arbitrarily shaped objects.

As concerns the MFIE in (5.19b), for observation points on the surface of the scatterer, it involves an improper integral that must be splitted into two definite parts

[39,44], leading to a Fredholm integral equation of the second kind, that is,

$$\frac{4\pi - \Omega_0(\mathbf{r})}{4\pi}\mathbf{J}_S(\mathbf{r}) - PV\iint_S \{[\nabla' G(\mathbf{r},\mathbf{r}') \times \mathbf{J}_S(\mathbf{r}')] \times \mathbf{u}_n\}\, dS' = \mathbf{u}_n \times \mathbf{H}^{\text{inc}}(\mathbf{r}), \mathbf{r} \in S, \tag{5.22}$$

where the surface integral must be performed in the Cauchy principal value sense and $\Omega_0(\mathbf{r})$ stands for the solid angle external to the scatterer surface. For the case of planar surfaces, the equation reduces to that originally proposed by Maue.

The second step of the method of moments is the reduction of the linear field functional equation to a matrix equation. As in the FEM, this may be accomplished by applying the Rayleigh-Ritz variational method or Galerkin's method [45,46]. The mathematical details are the same as those described for the finite-element formulation. So the functional equation can be cast in the following concise form:

$$L[\mathbf{f}] = \mathbf{g}, \tag{5.23}$$

where L is the linear integro-differential operator, $\mathbf{g}$ is the excitation vector function (e.g., the incident field), and $\mathbf{f}$ is the response vector function (e.g., the unknown surface current). The unknown function is expanded in a series of basis functions as

$$\mathbf{f} = \sum_{n=1}^{N} F_n \mathbf{f}_n, \tag{5.24}$$

where F_n are unknown constants (also called *moments*). When the number N of basis functions tends to infinity (so that $\mathbf{f}_n$ form a complete set of basis functions), the expansion (5.24) converges to the exact solution. For approximate solutions, the summation is over a finite number N. In order to determine the unknown coefficients F_n, a matrix equation has to be obtained. In particular, a set of weighting functions $\mathbf{w}_n = \mathbf{f}_n$ is introduced. Then the inner product of (5.23) with each $\mathbf{w}_n$ is taken, thus obtaining

$$\sum_{n=1}^{N} F_n \langle \mathbf{w}_m, L(\mathbf{f}_n)\rangle = \langle \mathbf{w}_m, \mathbf{g}\rangle \qquad \text{for } m = 1, 2, 3, \ldots, N. \tag{5.25}$$

The set of equations (5.25) can be written in a matrix form as

$$\underline{\mathbf{Z}} \cdot \mathbf{F} = \mathbf{G}, \tag{5.26}$$

where $\underline{\mathbf{Z}}$ is known as the impedance matrix. One of the main task of the MoM discretization lies in the choice of the basis functions, which must be linearly independent. Although a huge literature exists on this subject, the commonly used basis functions are of two types: entire-domain basis functions (which exist over the full domain) and subdomain basis functions (which exist only on one of the N nonoverlapping patches into which the computational domain is discretized). The subdomain basis functions are the most used, especially when developing general-purpose codes for the modeling of arbitrarily shaped objects, since they do not require any prior knowledge of the nature of the unknown function to be represented.

Although both scalar and vector basis functions are available, vector basis functions are generally preferred because they allow the required boundary and continuity conditions to be enforced in a rather simple way. Among the most popular ones are the linear triangular rooftop Rao–Wilton–Glisson (RWG) functions [47] that were originally proposed for use with the EFIE and triangular patch modeling (the most appropriate for modeling complex surfaces). The RWG function is defined on two adjacent triangular patches as shown in Figure 5.5. On the common edge of the two patches, the normal component of the basis function is equal on both patches, so no nonphysical line charge appears at the element's boundary. The direction of the vector basis function on a triangle is radial with respect to its vertex, and the intensity is a linear function of the edge–vertex distance. On each triangle there are three basis functions associated with it (one for each edge), except for patches at the boundary of the surface (no basis function is associated with such

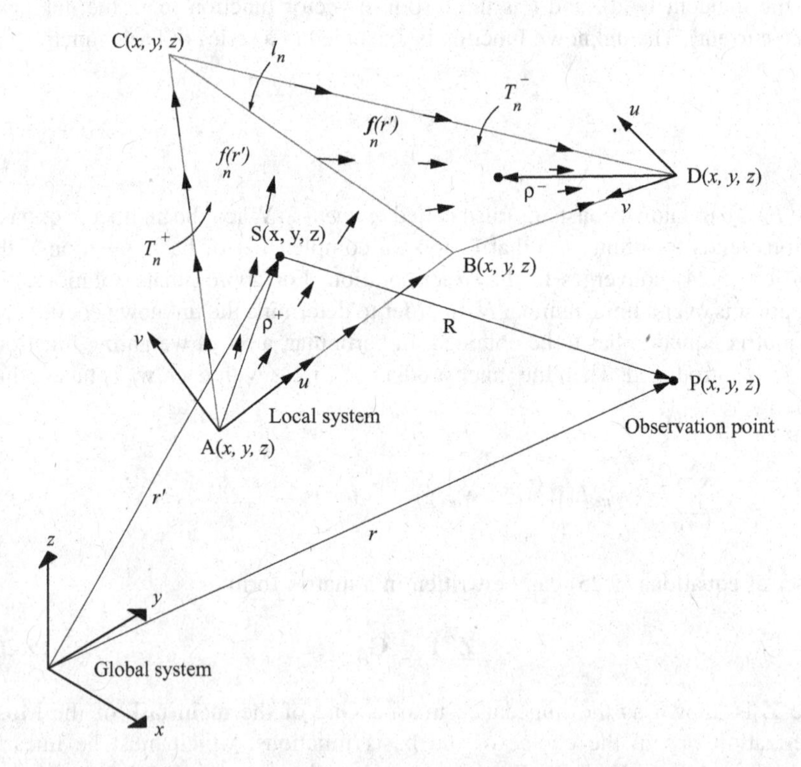

FIGURE 5.5 Triangular rooftop vector RWG basis function.

edges). According to Figure 5.5 the RWG basis function associated with the edge n between the triangles T_n^- and T_n^+ is thus defined as

$$\mathbf{f}_n(\mathbf{r}') = \begin{cases} \frac{\boldsymbol{\rho}^-}{h^-}, & \mathbf{r}' \in T_n^-, \\ \frac{\boldsymbol{\rho}^+}{h^+}, & \mathbf{r}' \in T_n^+, \\ 0, & \text{elsewhere}, \end{cases} \tag{5.27}$$

where $h^{\pm}$ are the heights of the triangles $T_n^{\pm}$ with respect to the edge n.

In summary, the RWG function is characterized by the presence of a constant normal component across the common edge between two triangle patches, zero normal components along the other four edges, and linear tangential components along all the edges. Since RWG functions are divergence conforming (i.e., they have continuous normal components across element boundaries), these functions do not permit any fictitious charge accumulation at the edge between two cells. Thus the nonphysical charge accumulation that was identified as a source of difficulty in the first EFIE formulations is eliminated. The RWG functions are the lowest order member (continuous normal-linear tangent CN/LT) of a general family of mixed-order divergence-conforming functions [19,48]. Although applications of higher order basis functions exist [37–41], RWG functions are widely used in all the integral formulations because of their inherent simplicity.

When the weighting functions are used, the inner product is typically defined as

$$\langle \mathbf{w}, \mathbf{g} \rangle = \iint_S \mathbf{w}^* \cdot \mathbf{g} \, \mathrm{d}S. \tag{5.28}$$

The calculation of the MoM matrix elements consequently requires the computation of two surface integrals, one over the source patch and one over the observation patch. Simpler techniques may be applied to reduce the testing-procedure cost such as point-matching techniques or testing razor-blade functions [38–41]. However, the simplifications in testing an integral equation are often paid for with a lost of symmetry in the impedance matrix $\underline{\mathbf{Z}}$.

Integral formulations require the computation of potential integrals involving the Green function and its gradient. In this connection the most severe problem is the singular behavior of the integral kernel, which can be hardly handled with classical Gaussian-quadrature rules. Much research has been focused on the analytical computation of free-space static and dynamic potential integrals that arise in EFIE and MFIE formulations with triangular basis functions [49]. Usually the singular part of the integrands, involving the free-space scalar Green function or its gradient, is extracted and its contribution is then evaluated with the aid of analytical procedures.

The MoM solutions of EFIE and MFIE are among the most successful numerical methods for the solution of EM radiation and scattering problems. Most important are EFIE and MFIE's robustness and insensitivity against dispersion errors. However, in their simplest form, they suffer some major problems that must be adequately treated.

The EFIE and MFIE fail in the solution of EM scattering from bounded objects at frequencies corresponding to the resonant frequencies of the cavities formed by hollow conductors with the same shape of the scatterers. In being only based on the tangential electric or magnetic field on the surface of the scatterer, the EFIE and MFIE formulations do not contain enough information to distinguish the exterior problem from the interior one. In correspondence of the resonant frequencies of the interior problem, the eigenvalue of the integral equation becomes zero, and the resulting homogeneous equation admits the interior resonant current as an eigenfunction. This contaminates the physical surface current as a spurious solution, causing the boundary conditions to be no longer satisfied. To circumvent internal resonance difficulties, several remedies have been suggested. The most effective one is the use of a linear combination of the EFIE and MFIE, which is widely known as the combined-field integral equation (CFIE) [50]. The resulting equation can thus be written as

$$\begin{aligned}
&\alpha\left\{jk\eta\iint_S [G(\mathbf{r},\mathbf{r}')\mathbf{J}_S(\mathbf{r}')]\times\mathbf{u}_n]\mathrm{d}S' + \frac{\eta}{jk}\iint_S [\nabla' G(\mathbf{r},\mathbf{r}')\times\mathbf{u}_n]\nabla\cdot\mathbf{J}_S(\mathbf{r}')\mathrm{d}S'\right\}\\
&+(1-\alpha)\eta\left\{\frac{4\pi-\Omega_0(\mathbf{r})}{4\pi}\mathbf{J}_S(\mathbf{r}) - PV\iint_S [\nabla' G(\mathbf{r},\mathbf{r}')\times\mathbf{J}_S(\mathbf{r}')]\times\mathbf{u}_n\mathrm{d}S'\right\}\\
&=\alpha\mathbf{E}^{\mathrm{inc}}(\mathbf{r})\times\mathbf{u}_n + (1-\alpha)\eta\mathbf{u}_n\times\mathbf{H}^{\mathrm{inc}}(\mathbf{r}),\qquad \mathbf{r}\in S,
\end{aligned}\tag{5.29}$$

where the parameter α is a real number in the range $0 < \alpha < 1$ that is used to adjust the relative weights of the EFIE and MFIE. Several studies on the optimum choice of α have found that a value of 0.2 is adequate. This way the method yields always a unique solution but at the expense of an additional computational overhead. Another technique, known as the combined-source integral equation (CSIE) [51], considers combined electric and magnetic sources on the surface of the conducting body and enforces the tangential electric-field boundary condition. The final equation is

$$\begin{aligned}
&jk\eta\iint_S [G(\mathbf{r},\mathbf{r}')\mathbf{J}_S(\mathbf{r}')\times\mathbf{u}_n]\mathrm{d}S' + \frac{\eta}{jk}\iint_S \{[\nabla' G(\mathbf{r},\mathbf{r}')\times\mathbf{u}_n]\nabla\cdot\mathbf{J}_S(\mathbf{r}')\}\mathrm{d}S'\\
&+\frac{4\pi-\Omega_0(\mathbf{r})}{4\pi}\mathbf{M}_S(\mathbf{r}) - PV\iint_S \{[\nabla' G(\mathbf{r},\mathbf{r}')\times\mathbf{M}_S(\mathbf{r}')]\times\mathbf{u}_n\}\,\mathrm{d}S' = \mathbf{E}^{\mathrm{inc}}(\mathbf{r})\times\mathbf{u}_n,\\
&\mathbf{r}\in S,
\end{aligned}\tag{5.30}$$

where $\mathbf{M}_S = \eta(1-\alpha)(\mathbf{u}_n\times\mathbf{J}_S)/\alpha$ is the equivalent surface magnetic current, and again $0 < \alpha < 1$. This approach has the same degree of complexity and computational effort as the CFIE, but the computed current is not the true one induced on the surface. Many other techniques have been proposed through the years (the combined-Helmholtz integral-equation formulation, the extended boundary-condition method, the parasitic-body technique, the correction-factor technique, and methods based on iterative techniques and singular-value decompositions) and can be applied more or less effectively to overcome the internal resonance problem.

It is worth noting that this kind of problem is usually encountered in the shielding practice in the study of enclosures with small apertures near the resonant frequencies of the cavity. It has been observed that when RWG basis functions are used in conjunction with EFIE, very fine discretization is needed to obtain convergence in the results. The problem is more evident when internal sources radiate inside the enclosure; in this case discretizations up to $\lambda/100$ may be needed for the radiated fields near the resonant frequencies to be accurately computed.

Other difficulties arise with integral formulations when they are used in the low-frequency limit, namely for electrically small conducting scatterers. The classical EFIE with RWG basis functions is in fact ill posed at the dc limit: the problem arises in the impedance matrix $\underline{\mathbf{Z}}$ whose condition number approaches infinity as the frequency tends to zero. Since the low-frequency limit is equivalent to considering small-sized elements, a deteriorated condition number is obtained in the attempt to obtain better accuracy when dealing with complex geometries that require fine mesh refinements with smaller elements. By applying the Galerkin weighted residual method, it is easy to reduce (5.21) in the dc limit as

$$\iint_S \left\{ \nabla \cdot \mathbf{J}_S(\mathbf{r}) \iint_S \left[\frac{1}{|\mathbf{r}-\mathbf{r}'|} \nabla \cdot \mathbf{J}_S(\mathbf{r}') \right] \mathrm{d}S' \right\} \mathrm{d}S = 2, \tag{5.31}$$

where it is evident that any solenoidal vector field can be a solution, resulting in a singular impedance matrix $\underline{\mathbf{Z}}$. The instability is consequence of the decoupling of the electric field and charge density from the magnetic field and current density in the static limit in nonconductive media. This decoupling of fields manifests in the current that undergoes a natural Helmholtz decomposition. It separates itself into a solenoidal component that produces only a magnetic field and a complementary irrotational component that vanishes as the frequency goes to zero, producing through its divergence a finite electric charge $\rho_e = \lim_{\omega \to 0}(\nabla \cdot \mathbf{J}_{\mathrm{irr}}/j\omega)$, which is the source of the electric field. Near the dc limit the contribution of the magnetic vector potential $\mathbf{A}$ is lost, and the remaining information from the electric scalar potential term $\nabla\Phi$ is not sufficient to determine the predominant solenoidal current. To eliminate this instability, a loop-star formulation has been proposed [52,53]. The rationale behind this formulation is to separate the solenoidal part of the current (described by the loop-type basis functions $\mathbf{f}_i^L$) from the remaining nonsolenoidal part (which is described by the star-type $\mathbf{f}_i^S$ basis functions). These functions are easy to generate by rearranging the original RWG basis functions: for simply connected structures, a loop basis function is associated with each interior vertex, while a star basis function is always associated with each triangle. Since the RWG basis functions are divergence conforming (ensuring continuity of the normal component of the current across all the edges and not of the parallel component), any resulting basis function cannot be curl conforming (the curl of the current exhibits line delta singularities along the edges). Therefore the Helmholtz splitting can be enforced only in a weak sense, meaning the loop basis functions are solenoidal vector functions ($\nabla \cdot \mathbf{f}_i^L = 0$), whereas the star basis functions are irrotational only approximately ($\nabla \times \mathbf{f}_i^S \simeq \mathbf{0}$). The loop-star decomposition dramatically improves the condition number of the EFIE matrix in the low-frequency range. Using loop-star basis functions in (5.33) results in

$$\iint\limits_S \left[\nabla \cdot \mathbf{f}_m^S \sum_n J_n^S \iint\limits_S \left(\frac{1}{|\mathbf{r}-\mathbf{r}'|} \nabla \cdot \mathbf{f}_n^S \right) \mathrm{d}S' \right] \mathrm{d}S = 0 \quad \text{for } m = 1,2,3,\ldots,N, \tag{5.32}$$

which gives the unique solution $J_n^S = 0$, since the span of the star basis functions does not include any solenoidal vector field. To solve the EFIE low-frequency instability, other strategies have been proposed such as the loop-tree and loop-cotree decompositions. They are based on the same rationale of the loop-star decomposition and are very similar to the tree-cotree decomposition in the vector finite-element methods.

As concerns the MFIE, it has been shown that this integral equation is free of any breakdown problem and can be solved at an arbitrary low frequency by the conventional MoM with the RWG basis. Being based on the magnetic field (which is sufficient to determine the solenoidal current component), the resulting MoM matrix is always well conditioned, no matter how low the frequency is and is solvable if the surface is closed.

A last difficulty arises in the solution of large-scale EM problems. The MoM solution of integral equations in fact requires to solve a fully populated matrix equation whose numerical complexity is $O(N^2)$, where N denotes the number of unknowns. The computational time and storage required by the numerical procedure is too cumbersome for handling electrically large problems, so as to make the solution of the problem impractical. Great effort has been devoted to reduce the computational complexity of conventional MoM techniques, and many fast algorithms have been proposed [54]. Among them, the fast multipole method (FMM) [55–57] has found wide applications, due to its accuracy, flexibility, and great efficiency, especially if integrated with parallel computing technology. The FMM provides an efficient mechanism for the numerical convolution of the free-space Green function with the source and test distribution, reducing the computational complexity of the solution of the matrix system (5.26) to $O(N^{3/2})$ in the simple single-stage form. In its multilevel version (MLFMM), the complexity is further reduced to $O(N \log N)$.

The basic principle behind the FMM consists in decomposing the computation of the matrix-vector products required by the iterative solver into two parts: one involving the interaction between nearby sources and the other involving the interaction between well separated ones:

$$\underline{\mathbf{Z}}^{\mathrm{NEAR}} \cdot \mathbf{x} + \underline{\mathbf{Z}}^{\mathrm{FAR}} \cdot \mathbf{x} = \mathbf{G}. \tag{5.33}$$

The idea is to perform the matrix-vector product for far-field terms indirectly, without computing and storing the relevant matrix elements. Furthermore the near-field matrix $\underline{\mathbf{Z}}^{\text{NEAR}}$ presents a sparse nature, leading to a further reduction of the computational cost. The foundation of the whole computation is the subdivision of the N basis functions into M localized groups, each containing G_m functions, and the identification for each group m of the bordering B_m groups that are in the near-field region. By Gegenbauer's addition theorem for spherical harmonics and a plane-wave expansion in the k-space, the 3D free-space Green function can be expressed approximately as

$$\frac{e^{jk|\mathbf{R}+\mathbf{d}|}}{|\mathbf{R}-\mathbf{d}|} \simeq -\frac{jk}{4\pi} \oiint e^{-j\mathbf{k}\cdot\mathbf{d}} T_L(kR, \mathbf{u}_k \cdot \mathbf{u}_R) \mathrm{d}\mathbf{u}_k, \tag{5.34}$$

where

$$T_L(kR, \mathbf{u}_k \cdot \mathbf{u}_R) = \sum_{l=0}^{L} (-j)^l (2l+1) h_l^{(1)}(kR) P_l(\mathbf{u}_k \cdot \mathbf{u}_R) \tag{5.35}$$

is the so-called translation operator. In (5.34) the integration on the unit wave vector $\mathbf{u}_k$ is over the Ewald sphere, whereas in (5.35) $h_l^{(1)}(\cdot)$ is the lth spherical Hankel function of the first kind and $P_l(\cdot)$ is the lth order Legendre polynomial. The expression is exact if the summation is performed over an infinite number of terms ($L = \infty$). The FMM can be applied to all the integral equations. In the following discussion, reference will be made to the EFIE that can be written as

$$\mathbf{u}_n \times \left[jk\eta \iint_S \underline{\mathbf{G}}_e(\mathbf{r}, \mathbf{r}') \cdot \mathbf{J}_S(\mathbf{r}') \, \mathrm{d}S' \right] = \mathbf{u}_n \times \mathbf{E}^{\text{inc}}(\mathbf{r}), \quad \mathbf{r} \in S, \tag{5.36}$$

where $\underline{\mathbf{G}}_e(\mathbf{r}, \mathbf{r}')$ is the electric dyadic Green function introduced in (1.42). The expression of the generic impedance matrix term Z_{ij} is

$$Z_{ij} = \iint_S \mathbf{t}_i(\mathbf{r}') \left[\iint_S \underline{\mathbf{G}}_e(\mathbf{r}, \mathbf{r}') \cdot \mathbf{j}_j(\mathbf{r}') \mathrm{d}S' \right] \mathrm{d}S, \tag{5.37}$$

where $\mathbf{j}_i$ and $\mathbf{t}_i$ are the basis and testing vector functions, respectively. With the aid of the expansion above, the far-field matrix can be expressed in the following form:

$$Z_{ij}^{\text{FAR}} = -\frac{jk}{4\pi} \oiint \mathbf{V}_{mi}(\mathbf{u}_k) \cdot \mathbf{V}_{m'j}^*(\mathbf{u}_k) T_L(kr_{mm'}, \mathbf{u}_k \cdot \mathbf{u}_{r_{mm'}}) \, \mathrm{d}\mathbf{u}_k. \tag{5.38}$$

The notation mi stands for the ith testing function belonging to the observation group m, while $m'j$ for the jth basis function belonging to the source group m'. In the expression above $\mathbf{V}_{mi}(\mathbf{u}_k)$ and $\mathbf{V}_{m'j}(\mathbf{u}_k)$ are the receiving and radiation patterns of the testing and basis functions, respectively, given by

$$\begin{aligned}\mathbf{V}_{mi}(\mathbf{u}_k) &= \iint_S e^{-jk\,\mathbf{u}_k\cdot\mathbf{r}_{mi}}(\underline{\mathbf{I}} - \mathbf{u}_k\mathbf{u}_k)\cdot\mathbf{t}_i(\mathbf{r}')\mathrm{d}S',\\ \mathbf{V}_{m'j}(\mathbf{u}_k) &= \iint_S e^{-jk\,\mathbf{u}_k\cdot\mathbf{r}_{m'j}}(\underline{\mathbf{I}} - \mathbf{u}_k\mathbf{u}_k)\cdot\mathbf{j}_j(\mathbf{r}')\mathrm{d}S'.\end{aligned} \tag{5.39}$$

Finally the matrix-vector product can be performed as

$$\begin{aligned}\sum_{j=1}^{N} Z_{ij}x_j = &\sum_{m'\in B_m}\sum_{j\in G_{m'}} Z_{ij}^{\text{NEAR}}x_j\\ &-\frac{jk}{4\pi}\oiint\left[\underbrace{\mathbf{V}_{mi}(\mathbf{u}_k)}_{\text{DISAGGREGATION}}\cdot\sum_{m'\notin B_m}\underbrace{T_L(kr_{mm'},\mathbf{u}_k\cdot\mathbf{u}_{r_{mm'}})}_{\text{TRANSLATION}}\underbrace{\sum_{j\in G_{m'}}\mathbf{V}^*_{m'j}(\mathbf{u}_k)x_j}_{\text{AGGREGATION}}\right]\mathrm{d}\mathbf{u}_k.\end{aligned} \tag{5.40}$$

The product for the far-field terms requires three fundamental sweeps in the transformed k-space as shown in Figure 5.6: aggregations of the functions belonging to the source group, translation of the overall effect of the source group on the observation group, and disaggregation to compute the convolution with the specific testing function belonging to the observation group. For the full details of the procedure together with its numerical aspects (i.e., number of multipoles and number of directions for accurate integration over $\mathbf{u}_k$), the interested reader is referred to the specific literature [54–57].

The FMM can be further generalized into a multilevel version (MLFMM) that employs a recursive subdivision of the spatial domain and a corresponding hierarchical tree structure. The sources are initially subdivided into groups; then each group is recursively divided into smaller and smaller groups so as to establish a tree structure made of *fathers* (coarse levels) and *children* (fine levels). The MLFMM decomposes the matrix-vector product into two sweeps. For L levels of

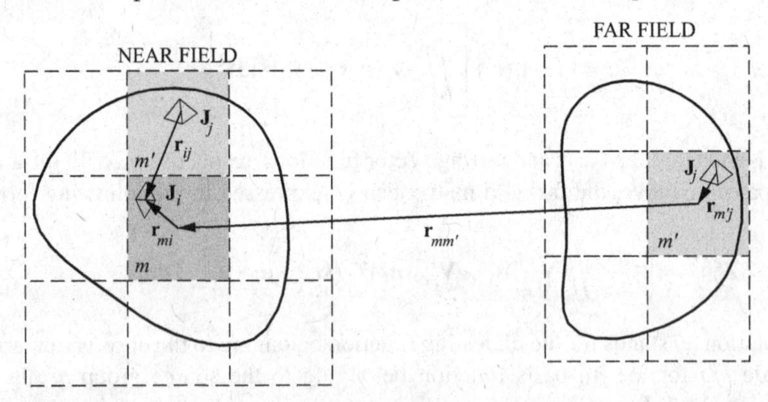

FIGURE 5.6 Domain decomposition according to the fast multipole method.

decomposition (with L referring to the finest level) the first sweep consists in constructing outer multipole expansions for each group from level L to level 2. Progressing from the finest level to the coarsest one, the groups become larger and larger, and the number of plane-wave directions needed to represent the radiation patterns in the k-space should increase. Thus the outgoing expansions of each group from level $L-1$ to 2 are computed from the expansions of their children groups using interpolation and shifting. The second sweep consists in constructing local incoming multipole expansions contributed from well-separated groups from level 2 to L. At the coarsest level 2, the local multipole expansions are constructed by translating the outgoing-wave expansions from well-separated groups at the same level. At the other levels (from 3 to L), the local expansions include the contribution of incoming waves received by parent groups using anterpolation (down-interpolation) with shifting and incoming waves from all the well-separated groups at the same level whose parent groups are not well separated. For L levels of decomposition, the far-field matrix-vector product is subdivided into summations of $L-1$ products among aggregation, translation, and disaggregation sparse matrices, plus a full matrix containing the interaction between near elements.

The renewed vigor pursued in the development of efficient transient simulators in recent years has led to a growing application of integral-equation methods in the time domain. For analyzing surface scattering phenomena, integral-equation techniques show unquestionable advantages over PDE methods: they require discretization of the scatterer surface as opposed to the volume surrounding the scatterer and do not call for ABCs, but automatically enforce the radiation condition through the use of the Green function. Notwithstanding their intrinsic qualities, historically two principal hurdles has prevented their use in the time domain. One is that many traditionally applied marching-on-in-time (MOT) schemes have been shown to be prone to late-time instabilities, and another, that their computational cost scales unfavourably with problem size, thus making the analysis of electrically large structures practically impossible with currently available resources. The time schemes are nearly always implicit, requiring the solution of a matrix equation at each time step. However, recent developments (e.g., stabilized MOT schemes and the multilevel plane-wave time-domain algorithm, PWTD) have cast new attention on the topic. The most representative and widely used time-domain integral equations (TDIE) remain the EFIE and MFIE. For problems in vacuum they can be recast in time domain as

$$\mathbf{u}_n \times \mathbf{E}^{\mathrm{inc}}(\mathbf{r},t) = \mathbf{u}_n \times \left\{ \frac{\mu_0}{4\pi}\frac{\partial}{\partial t}\left[\iint_S \frac{\mathbf{J}_S(\mathbf{r}',\tau)}{R}\mathrm{d}S'\right] - \frac{\nabla}{4\pi\varepsilon_0}\iint_S \left[\int_{-\infty}^{t} \frac{\nabla'\cdot\mathbf{J}_S(\mathbf{r}',\tau)}{R}\mathrm{d}t'\right]\mathrm{d}S' \right\}, \qquad \mathbf{r}\in S, \tag{5.41a}$$

$$\mathbf{u}_n \times \mathbf{H}^{\mathrm{inc}}(\mathbf{r},t) = -\frac{1}{4\pi}\mathbf{u}_n \times \iint_S \left\{ \frac{1}{cR}\left[\frac{\partial}{\partial\tau}\mathbf{J}(\mathbf{r}',\tau)\right] + \frac{1}{R^2}\mathbf{J}(\mathbf{r}',\tau)\right\} \times \mathbf{u}_R \mathrm{d}S', \quad \mathbf{r}\in S, \tag{5.41b}$$

where $\tau = t - R/c$ is the retarded time and $R = |\mathbf{r} - \mathbf{r}'|$ is the distance between the source and the observation points. The unknown surface current $\mathbf{J}_S(\mathbf{r}, t)$ is then represented by using spatial and temporal basis functions $\mathbf{j}_n(\mathbf{r})$ and $T_j(t)$ such that

$$\mathbf{J}_S(\mathbf{r}, t) = \sum_{j=0}^{N_t} \sum_{n=1}^{N} J_{n,j}\, \mathbf{j}_n(\mathbf{r})\, T_j(t) \tag{5.42}$$

to obtain a matrix equation of the form:

$$\underline{\boldsymbol{\alpha}} \cdot \mathbf{I}_j = \mathbf{F}_j - \sum_{l=1}^{j} \underline{\boldsymbol{\beta}_l} \cdot \mathbf{I}_{j-l} \tag{5.43}$$

that must be solved at each time step. For more details, the interested reader is referred to the specific literature [58].

5.3 FINITE-DIFFERENCE TIME-DOMAIN METHOD

The finite-difference time-domain (FDTD) method is the most popular numerical technique for the solution of problems in electromagnetics [58–62]. This is mainly due to its inherent marking characteristics: the method is simple because it does not make use of linear algebra, it is accurate and robust, it is a systematic direct approach that does not require the solution of any linear system, it has wide modeling capabilities that allow one simulate generally complex materials (dispersive, nonlinear, anisotropic), it naturally treats impulse behavior as a time-domain method, and it is an extensively computer-based method that can be parallelized rather easily. Since its first appearance in 1966 when it was proposed by Yee [63], the FDTD related research activity has been continuously running and enhancements of the method have been presented over the years. The goal of the present survey, which is necessarily incomplete, is to explain the most important features of the FDTD method, in the light of its development up to the current state-of-the-art, highlighting some of the most successful applications.

The method is a marching-in-time procedure that simulates the continuous propagation of actual EM waves in a finite spatial region. Yee discretized in a simple and elegant way the differential form of the time-dependent Maxwell curl equations system for the lossless material case, originating a set of finite-difference equations that yields the present fields throughout the computational domain in terms of the past fields. These equations were based on volumetric-time sampling of the unknown electric field **E** and magnetic field **H** within the space containing and surrounding the structure of interest and over a finite period of time.

To carry out the finite-difference expressions for the complete Maxwell equations, it is necessary to introduce the following notation: a grid point of the space is denoted as $(i, j, k) = (i\Delta x, j\Delta y, k\Delta z)$ and any function of space and time evaluated at a discrete point in the grid and at a discrete point in time is denoted as $F(i\Delta x, j\Delta y, k\Delta z, n\Delta t) = F^n_{i,j,k}$, where $\Delta x, \Delta y, \Delta z$, and Δt are the lattice space

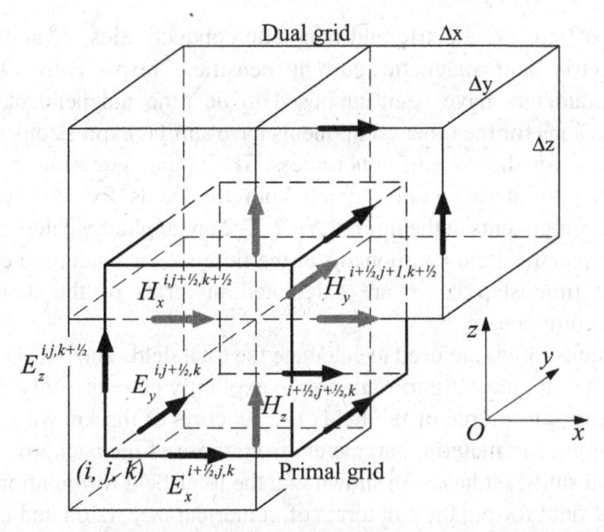

FIGURE 5.7 Spatial arrangement of field components in the Yee FDTD grid.

increments in the three coordinate directions and the time increment, respectively, and i, j, k, and n are integers. Yee used an electric-field grid that was spatially offset with respect to a magnetic-field grid: each magnetic (electric) field component is surrounded by four circulating electric (magnetic) field components. Furthermore the two grids are temporally interlaced, being offset of half the time increment. Figure 5.7 illustrates the two grids.

Using central-difference expressions for the space and time derivatives, and semi-implicit approximations for the non-time-collocated quantities [59], the following explicit expressions can easily be obtained for the x components of the total electric and magnetic fields:

$$E_x|_{i+1/2,j,k}^{n+1} = \left[\frac{1-(\sigma^{\mathrm{e}}_{i+1/2,j,k}\Delta t/2\varepsilon_{i+1/2,j,k})}{(1+\sigma^{\mathrm{e}}_{i+1/2,j,k}\Delta t/2\varepsilon_{i+1/2,j,k})}\right] E_x|_{i+\frac{1}{2},j,k}^{n} + \left[\frac{\Delta t/\varepsilon_{i+1/2,j,k}}{1+(\sigma^{\mathrm{e}}_{i+1/2,j,k}\Delta t/2\varepsilon_{i+1/2,j,k})}\right]$$
$$\cdot\left(\frac{H_z|_{i+1/2,j+1/2,k}^{n+1/2} - H_z|_{i+1/2,j-1/2,k}^{n+1/2}}{\Delta y} - \frac{H_y|_{i+1/2,j,k+1/2}^{n+1/2} - H_y|_{i+1/2,j,k-1/2}^{n+1/2}}{\Delta z} - J_S|_{i+1/2,j,k}^{n+1/2}\right), \tag{5.44a}$$

$$H_x|_{i,j+1/2,k+1/2}^{n+1/2} = \left[\frac{1-(\sigma^{\mathrm{m}}_{i,j+1/2,k+1/2}\Delta t/2\mu_{i,j+1/2,k+1/2})}{1+(\sigma^{\mathrm{m}}_{i,j+1/2,k+1/2}\Delta t/2\mu_{i,j+1/2,k+1/2})}\right] H_x|_{i+\frac{1}{2},j,k}^{n-1/2}$$
$$+\left[\frac{\Delta t/\mu_{i,j+1/2,k+1/2}}{1+(\sigma^{\mathrm{m}}_{i,j+1/2,k+1/2}\Delta t/2\mu_{i,j+1/2,k+1/2})}\right] \tag{5.44b}$$
$$\cdot\left(\frac{E_y|_{i,j+1/2,k+1}^{n} - E_y|_{i,j+1/2,k}^{n}}{\Delta z} - \frac{E_z|_{i,j+1,k+1/2}^{n} - E_z|_{i,j,k+1/2}^{n}}{\Delta y} - M_S|_{i,j+1/2,k+1/2}^{n}\right),$$

where σ^e and σ^m are the electric and magnetic conductivities, J_S and M_S are the impressed electric and magnetic current densities, respectively, and all the constitutive parameters have been supposed to be time independent. The time-stepping expressions for the other components have similar expressions [59] that are not reported here for the sake of conciseness. The update equations are used in a leapfrog scheme to incrementally march both the fields forward in time. The magnetic-field components at the time step $n + 1/2$ are explicitly calculated in terms of the adjacent electric-field components at the time step n. Then, the electric-field components at time step $n + 1$ are calculated in terms of the new computed magnetic-field components.

The expressions (5.44) are used to calculate the total field. The FDTD method can also be cast in a scattered-field formulation to explicitly compute only the scattered fields. In this case the source of the fields is a function of the known incident field and of the difference in material parameters from those of the background medium. Through several studies it has been shown that the total-field formulation is superior to the scattered-field formulation in terms of numerical dispersion and also because it can easily accommodate nonlinear media. However, the scattered-field formulation is used in the so-called total-field/scattered-field technique employed to simulate an incident-wave source condition [64]. In this approach the total fields are calculated only over an interior subsection of the computational domain, while the scattered fields are calculated in the remaining exterior portion. Consistency between the two schemes is preserved by specifying the incident field over the boundary between the two regions. In this way the absorbing boundary conditions are illuminated only with the scattered fields which are more readily absorbed.

The space-time-stepping algorithm persists in having great usefulness because of its soundness and robustness. Unlike the FEM that solves for the electric or magnetic field alone by means of the double-curl wave equation, the FDTD method solves for both electric and magnetic fields in space and time by means of coupled Maxwell curl equations. This way it is possible to model both electric and magnetic material properties in an accurate and rather simple way and also some peculiar field characteristics, such as singularities near corners and edges.

The original leapfrog algorithm is naturally second-order accurate in both space and time. Continuity of the tangential field components is naturally maintained across an interface between different materials parallel to one of the lattice coordinate axes, as in the FEM vector formulation. The particular structure of the space lattice (with a primal electric grid spatially interlaced with a dual magnetic grid) implicitly enforces the two Gauss laws. Hence a key point of the method is that unlike the FEM, it is not affected at all by spurious unphysical solutions, since all four Maxwell equations are discretized in a consistent way. The leapfrog time-stepping algorithm is fully explicit, thereby avoiding problems involved with matrix inversion as in the FEM and MoM, and is nondissipative, allowing numerical wave modes to propagate without any nonphysical artifact decay.

The FDTD method allows for the straightforward simulation of general complex media (dispersive, anisotropic, nonlinear, and time varying). Direct integration methods based on auxiliary differential equations, recursive convolution techniques, and the Z-transform method have been proposed for the analysis of Drude cold plasma, multi-term Debye and Lorentz dispersive dielectrics [65], ferrites [66], and nonlinear materials [67]. Furthermore surface-impedance boundary conditions (SIBCs) have been well implemented in the FDTD technique (Maloney-Smith, Beggs, Lee methods [59])—as in the FEM—for the analysis outside lossy dielectrics, conducting structures, thin conducting shields, without having to model their interiors. This avoids the need to resolve the decay of penetrating fields due to the skin effect, which is a huge reduction in computer burden.

Notwithstanding the above-mentioned advantages, several critical aspects of the FDTD method must be accounted for to understand its operation and its accuracy limits, especially when dealing with electrically large structures or with fine generally shaped details.

When using the explicit second-order accurate leapfrog scheme, the time increment Δt must be bounded to ensure numerical stability. The stability analysis is usually performed by standard von Neumann analysis [68]. In principle, the procedure performs a spatial Fourier transform along all the dimensions, thereby reducing the finite-difference scheme to a time recursion in terms of transformed quantities, and verifies that no Fourier components are exponentially growing with respect to time. Therefore instantaneous values of electric and magnetic fields $F_\alpha|_{i,j,k}^n$ distributed in space across the grid are Fourier-transformed into waves in the spectral domain to represent a spectrum of spatial sinusoidal modes, as $F_\alpha|_{i,j,k}^n = F_\alpha^n \exp\left[-j\left(k_x i\Delta x + k_y j\Delta y + k_z k\Delta z\right)\right]$. With the vector that contains all the six field transformed components $[E_x^n, E_y^n, E_z^n, H_x^n, H_y^n, H_z^n]$ denoted by $\mathbf{X}^n$, the time-marching recursive scheme can be cast in the following matrix form:

$$\mathbf{X}^{n+1} = \underline{\mathbf{\Lambda}} \cdot \mathbf{X}^n, \tag{5.45}$$

where $\underline{\mathbf{\Lambda}}$ is a 6×6 matrix depending on the particular space-time differential scheme that contains $k_x, k_y, k_z, \Delta x, \Delta y, \Delta z$, and Δt. By checking the eigenvalues of the matrix $\underline{\mathbf{\Lambda}}$, it is possible to determine the stability conditions of the system. If the magnitudes of all the eigenvalues are smaller than or equal to unity, the scheme is stable. Otherwise, if the magnitude of only one eigenvalue is larger than one, the scheme is unstable. The stability analysis for the uniform structured grid FDTD method in a homogeneous lossless medium leads to the well-known explicit Courant-Friedrich-Levy (CFL) condition for the maximum allowable time step:

$$\Delta t \leq \left[v \sqrt{\frac{1}{(\Delta x)^2} + \frac{1}{(\Delta y)^2} + \frac{1}{(\Delta z)^2}} \right]^{-1}, \tag{5.46}$$

where $v = 1/\sqrt{\mu\varepsilon}$ is the wave velocity in the medium. To derive the stability condition for graded meshes with inhomogeneous materials, the updating scheme is recast in a single recursive matrix equation $\mathbf{F}^{n+1} = \underline{\mathbf{A}} \cdot \mathbf{F}^n + \mathbf{S}^n$, where $\mathbf{F}^n$ and $\mathbf{S}^n$ are vectors containing all the field unknowns and the impressed sources of the computational space, respectively. From the maximum eigenvalue of the sparse system matrix $\underline{\mathbf{A}}$, it is possible to verify whether the scheme is stable. This computation does not represent a serious drawback, since the eigenvalue calculation of large sparse matrices is well developed. Other approaches are possible to derive the stability conditions, such as the energy method [69] and the use of stability polynomials [70].

The FDTD algorithm, as does any PDE method, causes dispersion of the simulated waves in the computational lattice. The phase velocity of the numerical wave modes can be different from the real velocity by an amount that depends on the wavelength, direction of propagation, and grid discretization. The numerical dispersion causes propagating waves to accumulate delay or phase errors that can lead to nonphysical results, such as broadening and ringing pulsed waveforms, imprecise cancellation of multiple-scattered waves, anisotropy, and pseudorefraction. The dispersion of the scheme is usually derived by the above-mentioned von Neumann analysis. If the fields are assumed to be monochromatic waves with angular frequency ω (i.e., $F_\alpha^n = F_\alpha \exp(j\omega\Delta t)$), (5.45) reduces to $e^{j\omega n\Delta t}\left(e^{j\omega\Delta t}\underline{\mathbf{I}} - \underline{\mathbf{A}}\right) \cdot \mathbf{X} = \mathbf{0}$, where $\mathbf{X} = [E_x, E_y, E_z, H_x, H_y, H_z]$. For a nontrivial solution, the determinant of the coefficient matrix should be zero, which leads, after some manipulations, to the dispersion relationship. For 3D homogeneous lossless media, the numerical dispersion relation results in

$$\left[\frac{1}{v\Delta t}\sin\left(\frac{\omega\Delta t}{2}\right)\right]^2 = \left[\frac{1}{\Delta x}\sin\left(\frac{k_x\Delta x}{2}\right)\right]^2 + \left[\frac{1}{\Delta y}\sin\left(\frac{k_y\Delta y}{2}\right)\right]^2 + \left[\frac{1}{\Delta z}\sin\left(\frac{k_z\Delta z}{2}\right)\right]^2. \tag{5.47}$$

It is easy to show that (5.47) becomes identical to the dispersion relation for a physical wave in the limit as Δx, Δy, Δz, and Δt approach zero. The key implication of the dispersion relation is that the velocity of the numerical waves depends on the direction of propagation (the space lattice represents an anisotropic medium). In order to mitigate the effects of dispersion, a good resolution in space must be employed, with the proviso that the choice of the spatial increments must represent a trade-off between accuracy and computational resources required by the scheme, dispersion, and numerical errors. Typically the resolution is set to properly sample the highest near-field spatial frequencies, using 10 to 20 samples per wavelength. However, when dealing with large structures or long time simulations (where the use of a large number of samples is not always practical because of the consequent high cost of computing), dispersion can cause significant errors to arise in the calculated fields. Over the years several methods have been introduced with the goal of reducing the numerical-dispersion error inherent in the original Yee algorithm. These methods have typically relied on higher order approximations of the spatial or temporal derivatives: second-/fourth-order accurate in time and fourth-order accurate in space (2, 4)/(4, 4) FDTD algorithms have been proposed. However, because of the problems associated with the increased spatial stencils, which require special boundary conditions over material discontinuities, these algorithms have not enjoyed widespread use. More sophisticated techniques have then been proposed, such as the pseudospectral time-domain (PSTD) method and techniques developed from the theory of wavelets, such as the multiresolution time-domain method (MRTD). For a review of these methods, the reader is referred to [71].

As for any PDE technique that relies on the discretization of the computational space, the FDTD method requires special boundary conditions to truncate the computational domain, since the tangential components of the fields along the outer boundary cannot be updated. As previously explained when dealing with the FEM, also in the FDTD method global and local boundary conditions have been developed and the PML technique is well established. The PML approach has been shown to provide significantly better accuracy and efficiency than other techniques so as to become the standard. Thus the previous sizable literature on radiation boundary conditions is obsolete by now. Anyway, research is still active concerning the development of ABCs that are local, in the sense that the field at any point on the boundary of the lattice is updated using the neighboring fields. Mei and Fang have proposed a superabsorption technique that was applied to many ABCs to improve their performance. More recently Ramahi has presented a grid-termination scheme based on two complementary boundary operators whose reflection coefficients are identically opposite and that can be concurrently used to cancel out the errors associated with reflections from the ABCs. Meanwhile exact grid-termination techniques are being developed. They are generally based on the Kirchhoff integral formula and have proved to yield accurate results. However, because they are nonlocal, they are computationally expensive. For more details on analytical boundary conditions, the reader is referred to the literature [58–62].

As previously explained, an alternative approach to an ABC was suggested by Berenger [33,34], who terminated the outer boundary of the space lattice with a PML lossy material that dampens the outgoing fields, matching any plane wave of arbitrary incidence, polarization, and frequency at the boundary. What explained for the FEM holds for the FDTD method as well. Since the first publication very fervent research activity has continued to make improvements and extensions of the PML technique. All the numerical aspects of the simulation of the PML with the FDTD method have been assessed: the best choice for the profile of conductivities, thickness, maximum theoretical reflection coefficient, the best time-stepping within the medium (exponential or central differencing), the accuracy in terms of the computed reflection coefficient and numerical errors, the stability of the scheme, and the dependence of the performance on spatial discretization.

The original field-splitting method, which can be considered just a mathematical convenience, was replaced by more sophisticated formulations by introducing complex coordinate stretching into Maxwell's equations and by showing that the PML is a particular passive lossy dielectric and magnetic medium with appropriate conductivity and Debye dispersion characteristics. The original PML was matched only to free space; successive works extended the PML formulation to allow for the truncation of space filled with lossy media, Lorentz and Debye dispersive materials, and anisotropic dielectrics. The original implementation of the PML only absorbed propagating waves; modified versions of the technique were presented to absorb also evanescent energy. A huge amount of literature exists on the subject of PML in the

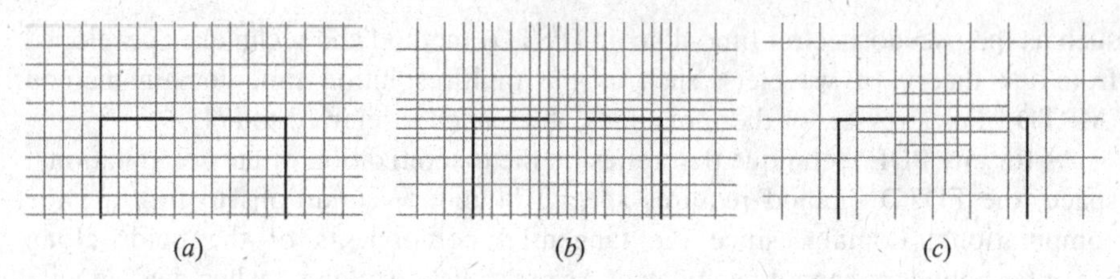

FIGURE 5.8 Structured mesh (*a*); graded mesh (*b*); mesh with subgrid (*c*).

FDTD method. The interested reader is referred to [58–62] and to the chapter "Advances in the Theory of Perfectly Matched Layers" in [54].

A fundamental issue related with any grid-based numerical technique is that the distance scale over which the physical processes or structural features must be resolved can range over several orders of magnitude. When using a uniform structured mesh (as that on which the FDTD method was originally formulated, see Figure 5.8*a*), this issue makes a problem arise. The use of a globally fine grid would lead to an excessively cumbersome simulation, in terms of both spatial samplings and consequently needed time steps. Furthermore such an accurate global simulation would be unnecessary since only few details can call for a fine resolving. To address this shortcoming, several methods have been proposed over the years. The concern ever-present to all the techniques that modify the original Yee algorithm is to preserve the numerical stability of the explicit scheme: any modification, even local, in the space discretization could lead to more or less fast instabilities in the time-stepping.

A first method consists in using a graded mesh, as shown in Figure 5.8*b*, with varying spatial increments along different coordinate directions, that permits finer discretizations in areas of rapid field fluctuations. It has been shown that these grids preserve the second-order accuracy of the original constant cell-size algorithm. However, attention must be paid on the shape factor of the rectangular cells to avoid to excessively stretch them, thus avoiding dispersion and numerical approximation errors. The method allows to reduce the required space samples, but not the number of time steps. In fact, the smallest distance existing in a mesh directly influences the width of the time step usable in the simulation through the Courant stability criterion.

To alleviate this drawback, other methods have been proposed based on the contour-path model. They obtain the updating equations from the integral form of Faraday's law rather than the differential form and result in modified equations only for cells where the thin structures are present. These techniques work well only for very fine subcellular features, such as thin narrow slots, wires, or joints in conducting screens. Moreover stability issues arise on the modified time-stepping schemes. A more general approach consists in using a grid finer than the rest of the problem space only in those subdomains where a more accurate modelling is required (Figure. 5.8*c*). Such a technique is called *subgridding*, and several schemes have been put forward over the years [72–75]. Away from the boundary between these two grids the standard FDTD update equations can be used.

The key issue common to all these techniques is the coupling of the coarse grid with the fine grid, where smaller space increments and time steps are present. The space increments in the base mesh must be an integer multiple N of those in the coarse grid, so that each time step in the base grid corresponds to N time steps in the submesh. To compute the fields on the boundaries, different spatial-interpolation schemes have been proposed and also time-interpolation or extrapolation schemes have been often used. In any subgridding technique, important numerical aspects arise, such as long-term stability, energy and divergence conservation, dispersion, and unphysical numerical reflections. To construct a consistent subgridding scheme, it is necessary to maintain some important properties of continuous Maxwell equations throughout the discretization process, as described in [75].

As originally formulated, the FDTD method makes use of structured mesh. This dictates that for generally shaped objects whose surface does not conform to the orthogonal lattice, the desired boundary conditions cannot be enforced directly on the physical boundary but rather on a staircased approximation of it. It has been observed that consequently, the solution does not converge to the correct one, no matter how fine the mesh is made to better resolve the boundaries. To overcome this problem, the contour-path concept was extended to obtain locally conformal models. In a standard Cartesian grid, the original time-stepping scheme is used everywhere except in the vicinity of the material boundaries, where special updating equations are employed to account for their curved shape. These equations are based on the standard integral form of Ampere's and Faraday's laws. Further generalizations have led to locally unstructured mesh and, eventually, to globally unstructured mesh. The development of a stable unstructured FDTD method is the subject of a research activity that is very extensive and still continuing. The developments proposed through the years have brought so many changes to Yee's original scheme that it is not possible anymore to strictly speak of the FDTD method but rather of new improved methods, such as the finite integration technique (FIT), the finite-volume time-domain method (FVTD), and the cell method. These approaches are well-developed self-consistent methods and not just simple generalizations of the FDTD technique. Nevertheless, they preserve several elements in common with FDTD when applied to structured meshes.

For a full understanding of the method, a last problem needs to be addressed that is related with its explicit nature. Because the FDTD method is based on a direct explicit algorithm, it must always satisfy the CFL condition that dictates that the maximum time-step size is limited by the minimum cell size. Therefore the FDTD method turns out to be inefficient when dealing with the class of problems where the cell size needed to resolve the fine-scale geometric details of the scatterer is much smaller than the shortest wavelength of a significant spectral component of the source. The small time step would create a significant increase in calculation time. To circumvent the CFL constraint, the explicit leapfrog time-stepping scheme could be replaced with the fully implicit Crank-Nicolson scheme, but it requires a system of linear algebraic equations to be solved for each time level. An efficient alternative that is being currently researched is to use an alternating-direction-implicit (ADI) time-stepping algorithm [76–79].

The basic concept of the ADI technique is to alternate an implicit approximation of the space's second derivatives with an explicit one. The original technique alternated the two approximations with respect to the three coordinate directions, requiring three implicit substep computations for each FDTD cycle, and it has never been found to be completely stable. A different, newly developed ADI technique alternates the two approximations with respect to the sequence of the terms on the right-hand side (RHS) of the Maxwell equations, thus requiring only two substeps. The calculation for a discrete time step is broken up into two computational half-time steps. For the first substep, the first partial derivative on the RHS is replaced with an implicit difference approximation of its unknown values at the present time, while the second partial derivative is replaced with an explicit finite-difference approximation of its known values at the past time. For the second substep, the opposite is done. For example, the x component of the electric field in a source-free isotropic medium is advanced as

$$\frac{E_x|_{i+1/2,j,k}^{n+1/2}-E_x|_{i+1/2,j,k}^{n}}{\Delta t/2} =\frac{1}{\varepsilon}\left[\frac{H_z|_{i+1/2,j+1/2,k}^{n+1/2}-H_z|_{i-1/2,j+1/2,k}^{n+1/2}}{\Delta y}-\frac{H_y|_{i+1/2,j,k+1/2}^{n}-H_y|_{i+1/2,j,k-1/2}^{n}}{\Delta z}\right], \tag{5.48a}$$

$$\frac{E_x|_{i+1/2,j,k}^{n+1}-E_x|_{i+1/2,j,k}^{n+1/2}}{\Delta t/2} =\frac{1}{\varepsilon}\left[\frac{H_z|_{i+1/2,j+1/2,k}^{n+1/2}-H_z|_{i-1/2,j+1/2,k}^{n+1/2}}{\Delta y}-\frac{H_y|_{i+1/2,j,k+1/2}^{n+1}-H_y|_{i+1/2,j,k-1/2}^{n+1}}{\Delta z}\right]. \tag{5.48b}$$

Similar equations hold for the other components of the electric and magnetic fields. It is important to observe that no time lag appears between them in the formulation, contrarily to the conventional leapfrog scheme. Through some manipulations, the ADI-FDTD equations can be simplified to enable efficient computation yielding an implicit tridiagonal scheme for the electric field and an explicit updating scheme for the magnetic field at each substep. Further simplification of the implicit scheme is possible by taking appropriate directional scans of each component by which several separate tridiagonal systems may be solved with a lower number of equations [80], or by exploiting appropriate sequences of ascending indexes [81]. Care must be taken in interpreting the field values at the half-time step, since these values are accurate only to first order. It is useful to use them just as intermediate values. It has been shown that the ADI-FDTD scheme is an $O(\Delta t^2)$ perturbation of the implicit Crank–Nicolson formulation and right the perturbation term allows factorization into the two-step procedure.

The ADI-FDTD scheme is second-order accurate in space and time, the same as the standard FDTD. It forms tridiagonal matrices whose solution techniques are very simple and efficient. Thus the method requires less memory, possesses faster solution times than other implicit methods, while retaining the simplicity that is inherent in the traditional FDTD technique. Furthermore the method is unconditionally stable regardless of the time step, since the matrix $\underline{\mathbf{A}}$ considered after (5.46) has always eigenvalues smaller or equal to unity. The scheme uses two substep-marching procedures, each having either numerical growth or dissipation; since the growth and dissipation exactly cancel out each other, the overall ADI-FDTD scheme is unconditionally stable. Hence the method removes the CFL stability constraint and allows for any choice of Δt, which is therefore only dictated by sampling and accuracy considerations. In fact, although the method is unconditionally stable, the accuracy of the numerical results gets worse when the time step increases, because of the numerical-dispersion error. The accuracy degrades quickly with increments of 10 to 20 in the time step beyond the Courant limit. In addition the method shows three major drawbacks. First, the left-hand and right-hand sides of the original updating equations of the scheme are not balanced with regard to the time steps, and certainly the unbalanced effects limit the accuracy of this scheme. Second, the method has a large numerical anisotropy error compared with Yee's original FDTD method. This is because the use of the ADI technique leads to asymmetric results, even for exactly symmetric computational setups and symmetric sources. Finally, in the ADI-FDTD scheme, three time steps are used to define the field components (one time step for each component) and two sub-iterations are required for field advancement. It is necessary to solve six tridiagonal matrices and six explicit updates for one full advancement cycle, and therefore the scheme has low efficiency. Because these are the main drawbacks to a wide application of the method, much research activity has focused on improved schemes for ADI techniques.

5.4 FINITE INTEGRATION TECHNIQUE

Since the publication of Maxwell's treatise, it has been standard practice to give a mathematical formulation of the electromagnetic theory in terms of differential formalism. However, the laws of EM phenomena, such as Faraday's and Ampere's laws, were originally formulated using global quantities, such as charges, currents, electric and magnetic fluxes, electromotive and magnetomotive forces, which, being integral quantities, are directly measurable. A discrete direct formulation of Maxwell's equations in their integral form is also possible, and suitable for numerical computation. This way the electromagnetism can be described in a finite form from the beginning and the differential formalism can be deduced as a consequence whenever it is necessary. The theoretical framework for solving Maxwell's equations in integral form was first described in [83], where an approach called the finite integration technique (FIT) was applied, resulting in a set of matrix equations (called Maxwell's grid equations, MGEs) as an analogue to the continuous ones. However, only recently the discrete formalism (along with the concepts of algebraic topology) has been widely used to study Maxwell's equations in integral form [84–88], leading to the development of a computational methodology (the cell method [89,90]) that extends the FIT to many physical theories such as electrodynamics, mechanics, and thermal conduction.

The starting point of the FIT is the use of a system of two computational grids, the primary grid G and the dual grid $\tilde{G}$. This requires two steps:

1. The restriction of the EM problem to a simply connected and bounded 3D space region.
2. The decomposition of the computational domain into a finite number of simplicial cells to build up the primary spatial grid G, defined by the set of primary elements $G = \{P, L, S, V\}$ consisting of points P, lines L, surfaces S, and volumes V, and the dual spatial grid $\tilde{G}$ defined by the set of dual elements $\tilde{G} = \{\tilde{P}, \tilde{L}, \tilde{S}, \tilde{V}\}$.

The method is a very general cell-based approach, since the generic primary cell can take an hexahedron, a tetrahedron, or other geometric forms that maintain the constraint that all the cells fit exactly to each other, meaning the intersection of two different cells is empty or a surface or a line. Once the primary grid G is built up, the dual grid $\tilde{G}$ is defined by taking the primary cell barycenters as boundary vertices and the dual lines as piece-lines made out of two separate straight lines connecting the dual nodes and the barycenter of the common primary surface (barycentric grids). In the simplest case of structured orthogonal Cartesian grids, these cell complexes reduce to the electric and magnetic grids introduced by Yee in the FDTD method. Between the elements of the primary and dual grids, the following bijective mapping holds: $P \leftrightarrow \tilde{V}, L \leftrightarrow \tilde{S}, S \leftrightarrow \tilde{L}, V \leftrightarrow \tilde{P}$. The indexes are usually chosen such that the primary element has the same index as the dual element. Each edge of the cells includes an orientation, and each polygonal facet is associated with a direction [84,85].

After the definition of the two grid complexes, the state variables of the FIT are introduced. They are the electric and magnetic grid voltages (V_i, $\tilde{F}_i$) and fluxes ($\tilde{\Psi}_i$, Φ_i), and the electric charge ($\tilde{Q}_i$) and the electric current ($\tilde{I}_i$), which are defined as the integrals of the electric- and magnetic-field vectors over elementary objects (line, surfaces, volumes) of the computational grids. The electric voltages and the magnetic fluxes are associated with the primary elements, while the magnetic voltages, the electric fluxes, and the electric charges and currents are associated with dual elements [84,85]. With these integral variables Maxwell's equations can be rewritten into a set of matrix-vector equations, which are referred to as Maxwell grid equations (MGEs):

$$G: \left\{ \underline{\mathbf{C}} \cdot \mathbf{V} = -\frac{\mathrm{d}}{\mathrm{d}t}\mathbf{\Phi};\ \underline{\mathbf{S}} \cdot \mathbf{\Phi} = \mathbf{0} \right\}, \quad \tilde{G}: \left\{ \underline{\tilde{\mathbf{C}}} \cdot \tilde{\mathbf{F}} = -\frac{\mathrm{d}}{\mathrm{d}t}\tilde{\mathbf{\Psi}} + \tilde{\mathbf{I}};\ \underline{\tilde{\mathbf{S}}} \cdot \tilde{\mathbf{\Psi}} = \tilde{\mathbf{Q}} \right\}. \tag{5.49}$$

The matrix $\underline{\mathbf{C}}$ represents the discrete curl-operator on the grid G. It contains only topological information on the incident relation of the cell edges within G and on their orientation. Therefore its entries C_{ij} are ± 1 if the edge L_j is contained in the boundary of facet A_i, and zero otherwise. In terms of algebraic topology, the discrete curl-operator is identical to the coboundary process operator δ [84] that generates a 2-cochain (i.e., degrees of freedom connected to 2D cell surfaces) from a 1-cochain (i.e., degrees of freedom allocated on 1D cell edges). Analogously, the matrix $\underline{\mathbf{S}}$ is the discrete div-operator of the primary grid G. Its entries S_{ij} are ± 1 if the facet A_j is contained in the boundary of the cell V_i, and zero otherwise. It corresponds to the coboundary operator applied to cochains of degree two yielding cochains of degree three (i.e., degrees of freedom connected to cell volumes). The same applies to the matrices $\underline{\tilde{\mathbf{C}}}$ and $\underline{\tilde{\mathbf{S}}}$ on the dual grid $\tilde{G}$.

Equations (5.49) are exact. The approximation of the method enters when the voltage and flux state variables associated with the two different cell complexes are related to each other. This is done through the *material matrices*, which are an analogue of the constitutive relations of continuous fields. Without considering permanent polarization vectors, the discrete constitutive relations are

$$\mathbf{\Psi} = \underline{\mathbf{M}}_{\varepsilon} \cdot \mathbf{V}, \quad \mathbf{I} = \underline{\mathbf{M}}_{\kappa} \cdot \mathbf{V}, \quad \mathbf{\Phi} = \underline{\mathbf{M}}_{\mu} \cdot \mathbf{F}. \tag{5.50}$$

The material matrices are defined as $\underline{\mathbf{M}}_{\varepsilon} = \underline{\tilde{\mathbf{D}}}_A \cdot \underline{\mathbf{D}}_{\varepsilon} \cdot \underline{\mathbf{D}}_L^{-1}$, $\underline{\mathbf{M}}_{\kappa} = \underline{\tilde{\mathbf{D}}}_A \cdot \underline{\mathbf{D}}_{\kappa} \cdot \underline{\mathbf{D}}_L^{-1}$, and $\underline{\mathbf{M}}_{\mu} = \underline{\mathbf{D}}_A \cdot \underline{\mathbf{D}}_{\mu} \cdot \underline{\tilde{\mathbf{D}}}_L^{-1}$, where the area matrices $\underline{\mathbf{D}}_A$ hold the areas of facets, the length matrices $\underline{\mathbf{D}}_L$ hold the lengths of edges, and the central matrices contain a proper volume average of the permittivity ε, conductivity κ, and permeability μ, respectively [86–90]. The material matrices can be dubbed discrete Hodge operators because they couple edge degrees of freedom (called discrete 1-forms) with dual facet degrees of freedom (called discrete 2-forms) [88]. The constitutive matrices must be symmetric and positive definite in order to ensure stability of the numerical scheme [86–88]. Furthermore they must ensure consistency of the numerical method, which is related to the consistency of the algebraic constitutive relations with the local constitutive relations $\mathbf{D} = \varepsilon\mathbf{E}$ and $\mathbf{B} = \mu\mathbf{H}$, respectively [90]. The fulfilment of the consistency condition [91], together with stability, ensures the convergence of the method when the grid is properly refined (fields uniform inside the cell) [90].

In the simplest case of Yee's dual-orthogonal grid system, the directions associated with the facet and with the dual edge penetrating this facet are identical; that is, the primary (dual) edges and dual (primary) facets intersect each other with angles of 90 degrees. The two cell complexes represent a so-called Delaunay-Voronoi grid doublet. For simple isotropic media the material matrices are diagonal (with $M_{\varepsilon,ii} = \varepsilon \tilde{A}_i / L_i$, $M_{\mu,ii} = \mu A_i / \tilde{L}_i$, $M_{\kappa,ii} = \kappa \tilde{A}_i / L_i$), $\underline{\mathbf{M}}_{\varepsilon}$ and $\underline{\mathbf{M}}_{\mu}$ are positive definite, and $\underline{\mathbf{M}}_{\kappa}$ is positive semidefinite [86]. Only the material matrices contain the metrical information on the discrete cell complexes. Hence, the complete set of equations can be subdivided into two different groups: metric-free equations (5.49), arising from grid topology, and metric-dependent equations (5.50).

The outstanding feature of the discrete representation lies in the fact that the set of matrix equations is a consistent discrete representation of the original field equations, meaning that it maintains all the analytical properties of the EM fields when moving from the continuous space $\mathbb{R}^3$ to the grid space doublet $G - \tilde{G}$. The numerical solution no longer relates to only a sequence of numbers but to vectors with exact algebraic properties enabling an independent cross-check of accuracy. By simple topological considerations, the following key properties can be easily proved [86]: $\underline{\mathbf{C}} = \underline{\tilde{\mathbf{C}}}^T$, $\underline{\mathbf{S}} \cdot \underline{\mathbf{C}} = \underline{\mathbf{0}}$, and $\underline{\tilde{\mathbf{S}}} \cdot \underline{\tilde{\mathbf{C}}} = \underline{\mathbf{0}}$. The first property is referred to as the generalized symmetry or duality property of the curl-operator. It is also possible to define a discrete grad-operator $\underline{\mathbf{G}}$ with the following properties: $\underline{\mathbf{G}} = -\underline{\tilde{\mathbf{S}}}^T$, $\underline{\tilde{\mathbf{G}}} = -\underline{\mathbf{S}}^T$, $\underline{\mathbf{C}} \cdot \underline{\mathbf{G}} = \underline{\mathbf{0}}$, and $\underline{\tilde{\mathbf{C}}} \cdot \underline{\tilde{\mathbf{G}}} = \underline{\mathbf{0}}$.

The above-mentioned topological properties, together with the symmetry and positive definiteness of the material matrices, are exploited to derive a number of important theorems for the discrete electromagnetism. Some important properties from a numerical point of view are reported below. For other theorems and properties, the reader is referred to the literature.

The eigenvalue equation for the electric field in a lossless media $\varepsilon^{-1}\nabla \times (\mu^{-1}\nabla \times \mathbf{E}) = \omega^2\mathbf{E}$ can be rewritten on the two cell complexes as $(\underline{\mathbf{M}}_\varepsilon^{-1} \cdot \underline{\tilde{\mathbf{C}}} \cdot \underline{\mathbf{M}}_\mu^{-1} \cdot \underline{\mathbf{C}}) \cdot \mathbf{V} = \omega^2\mathbf{V}$. Because of the duality property of the curl-operator, and because the material matrices are symmetric and positive definite, the system matrix of the algebraic eigenvalue problem can be transformed to a symmetric and positive semi-definite matrix $\underline{\mathbf{M}}_\varepsilon^{-1/2} \cdot (\underline{\mathbf{M}}_\mu^{-1/2} \cdot \underline{\mathbf{C}} \cdot \underline{\mathbf{M}}_\varepsilon^{-1/2})^T \cdot (\underline{\mathbf{M}}_\mu^{-1/2} \cdot \underline{\mathbf{C}} \cdot \underline{\mathbf{M}}_\varepsilon^{-1/2}) \cdot \underline{\mathbf{M}}_\varepsilon^{1/2}$. Therefore all the eigenvalues λ_i are real-valued and nonnegative numbers and all the eigensolutions correspond either to static solutions or to nondissipative and nongrowing oscillations with a real angular frequency $\omega_i = \sqrt{\lambda_i}$. This is the proof for the space stability of the MGEs.

From the double-curl equation it follows that, $\underline{\tilde{\mathbf{C}}} \cdot \underline{\mathbf{M}}_\mu^{-1} \cdot \underline{\mathbf{C}} \cdot \mathbf{V} = \omega^2\underline{\mathbf{M}}_\varepsilon \cdot \mathbf{V}$. If this is multiplied from the left by $\underline{\tilde{\mathbf{S}}}$ and then the topological property $\underline{\tilde{\mathbf{S}}} \cdot \underline{\tilde{\mathbf{C}}} = \underline{\mathbf{0}}$ is applied, the relation $\omega^2\underline{\tilde{\mathbf{S}}} \cdot \underline{\mathbf{M}}_\varepsilon \cdot \mathbf{V} = \omega^2\underline{\tilde{\mathbf{S}}} \cdot \mathbf{\Psi} = \mathbf{0}$ is obtained. So we have only two distinct cases for the solutions: static solutions with $\underline{\tilde{\mathbf{S}}} \cdot \mathbf{\Psi} \neq \mathbf{0}$ or time-harmonic solutions with $\underline{\tilde{\mathbf{S}}} \cdot \mathbf{\Psi} = \mathbf{0}$. The particular topological properties of the two cell complexes enforce the solenoidal condition for time-harmonic fields, thus excluding any irrotational spurious mode.

The discrete formulation of Maxwell's equations can be used both for solving problems in the frequency domain and, after a discretization of the time derivatives, for simulations in the time domain. This is useful in many practical problems where a pure time- or frequency-domain algorithm is not sufficient. Although several time-marching schemes are available, the leapfrog scheme is usually applied, since it is fully explicit in most of the cases. The values of the electric voltage V and of the magnetic flux Φ are sampled at times separated by half a time step (as in space, a primary and dual temporal grids are built up, and a bijective mapping exists between time intervals and dual time instants [88]), and in a lossless medium the MGEs reduce to

$$\begin{aligned} \mathbf{V}^{n+1} &= \mathbf{V}^n + \Delta t \underline{\mathbf{M}}_\varepsilon^{-1} \cdot \left(\underline{\mathbf{C}}^T \cdot \underline{\mathbf{M}}_\mu^{-1} \cdot \mathbf{\Phi}^{n+1/2} - \mathbf{I}^{n+1/2} \right) \\ \mathbf{\Phi}^{n+1/2} &= \mathbf{\Phi}^{n-1/2} - \Delta t \underline{\mathbf{C}} \cdot \mathbf{V}^n. \end{aligned} \tag{5.51}$$

Expression (5.51) can be recast in the compact recursive matrix form as

$$\mathbf{f}^{i+1} = \underline{\mathbf{A}} \cdot \mathbf{f}^{i} + \mathbf{s}^{i} \tag{5.52}$$

with

$$\mathbf{f}^{n} = \begin{pmatrix} \mathbf{\Phi}^{n+1/2} \\ \mathbf{V}^{n} \end{pmatrix}, \quad \underline{\mathbf{A}} = \begin{pmatrix} \underline{\mathbf{I}} & -\Delta t \underline{\mathbf{C}} \\ \Delta t \underline{\mathbf{M}}_{\varepsilon}^{-1} \cdot \underline{\mathbf{C}}^{T} \cdot \underline{\mathbf{M}}_{\mu}^{-1} & \underline{\mathbf{I}} \end{pmatrix}, \quad \mathbf{s} = \begin{pmatrix} \mathbf{0} \\ -\Delta t \underline{\mathbf{M}}_{\varepsilon}^{-1} \cdot \mathbf{I}^{n+1/2} \end{pmatrix}. \tag{5.53}$$

It can be easily shown that in the standard case of Cartesian grids, the FIT scheme is computationally equivalent to the FDTD method. The recursion is stable if all the eigenvalues of the matrix $\underline{\mathbf{A}}$ lie within the unit circle in the complex plane. This leads to the necessary and sufficient condition that $\Delta t \leq 2/\sqrt{\lambda_{\max}}$, where $\lambda_{\max}$ is the maximum eigenvalue of the matrix $c^2 \underline{\mathbf{M}}_{\mu}^{-1} \cdot \underline{\mathbf{C}} \cdot \underline{\mathbf{M}}_{\varepsilon}^{-1} \cdot \underline{\mathbf{C}}^{T}$, where c is the speed of light in vacuum. This is the generalized Courant criterion for the leapfrog algorithm, which exactly describes the stability limit for unstructured meshes and inhomogeneous material distributions. It reduces to the CFL condition for the simplest case of an homogeneous medium and constant mesh step. The FIT scheme (5.52) shows a second-order convergence, as the classical FDTD method.

According to equations (5.51) the material matrices that directly appear in the time-stepping scheme are $\underline{\mathbf{M}}_{\varepsilon}^{-1}$ and $\underline{\mathbf{M}}_{\mu}^{-1}$. As mentioned above, the necessary features of the constitutive matrices, and consequently of their inverse, are symmetry, positive definiteness, and consistency. However, for an algorithm to be fully explicit, the inverse material matrices $\underline{\mathbf{M}}_{\varepsilon}^{-1}$ and $\underline{\mathbf{M}}_{\mu}^{-1}$ should be known a priori and should be sparse for efficiency; otherwise, the scheme would be implicit and a linear system should be solved at each time step. For this purpose it is necessary to make a crucial distinction for the numerical computation. In general, it is possible to build up a sparse material matrix $\underline{\mathbf{M}}_{\alpha}$ from local considerations and derive, as a consequence, the inverse matrix $\underline{\mathbf{M}}_{\alpha}^{-1}$, which is generally a full matrix. Inversely, from similar local considerations, it is often desirable to build up directly the inverse matrix denoted as $\underline{\mathbf{M}}_{\alpha^{-1}}$, which in general is not equal to $\underline{\mathbf{M}}_{\alpha}^{-1}$. Thus the bottleneck for the general application of the FIT to non-orthogonal grids is development of a general way to directly build inverse constitutive matrices $\underline{\mathbf{M}}_{\varepsilon^{-1}}$ and $\underline{\mathbf{M}}_{\mu^{-1}}$ that are symmetric, positive definite, consistent, and sparse. This is a subject that has sparked much research interest, and several schemes have been proposed, such as the microcell interpolation scheme [90] together with appropriate symmetrization methods [92]. It has been recently shown that if it is possible to find a sparse matrix $\underline{\mathbf{M}}_{\mu^{-1}}$ satisfying all the criterions above on a tetrahedral grid, generally it is not possible to find a sparse matrix $\underline{\mathbf{M}}_{\varepsilon^{-1}}$ satisfying all the criterions on the dual grid [91]. This implies that in the state-of-art, scheme (5.51) is implicit and comparable to that of FEM in the time domain.

The FIT has been widely applied both in time and frequency domains and on orthogonal meshes, where it maintains strong similarities to the FDTD method, where the latter is seen as an interlinked array of Faraday's and Ampere's law contours in the 3D space. Over the years the FIT has been provided with the PML to simulate open-boundary problems, with a perfectly boundary approximation (PBA) technique to simulate rounded boundaries on orthogonal meshes with second-order accuracy, with stable subgridding techniques to achieve accurate models of small structure details, and with methods to include dispersive and anisotropic media. Dispersion characteristics and numerical accuracy (second-order) have been widely studied and are well assessed.

5.5 TRANSMISSION-LINE MATRIX METHOD

The transmission-line matrix (TLM) method [93] belongs to the wide class of time-domain differential methods, although applications in the frequency domain also exist. Its distinctive feature lies in the fact that it is a numerical technique for solving field problems with reference to circuit analogies. It is based on the well-established equivalence between Maxwell's equations for the electric and magnetic fields of a 1D wave and the telegraphers' equations for voltages and currents along a continuous two-wire transmission line. Although the TLM method requires a discretization process of the computational space, unlike others methods, this discretization approach is physical and not mathematical: each constitutive block in space is in fact replaced by a physical network of transmission lines suitably connected to form a constitutive node.

The method was originally developed to study the propagation of transverse-magnetic (TM) and transverse-electric (TE) waves in a 2D homogeneous medium [94,95]. The computational space is divided into a Cartesian mesh of open two-wire transverse-electromagnetic (TEM) transmission lines of equal length Δl that are parallel to the coordinate axes. Each junction between a pair of transmission lines forms a *shunt node* in the mesh, where an impedance discontinuity is lumped. Since Δl is usually smaller than one-tenth of the wavelength at the highest frequency of operation, the elementary length of the transmission line between two nodes is represented by lumped inductors and capacitors, given the per unit length (p.u.l.) inductance L and capacitance C of the line. A direct comparison between the Kirchhoff voltage and current laws and the Maxwell equations for a 2D TM wave allows equivalences to be established between electric field and voltage, magnetic field and current, and to correlate the dielectric permittivity and the magnetic permeability to the p.u.l. capacitance and inductance, respectively. When dealing with a TE wave, the concept of duality may be applied to reverse the roles of electric and magnetic fields or a dual mesh made with the connection of *series nodes* may be used.

To carry out the solution of the wave propagation in the time domain, the Huygens principle is applied, according to which a wavefront consists of a number of secondary sources that radiate spherical waves, whose envelope forms a new wavefront that, in turn, gives rise to new secondary sources, and so on. By assuming a delta impulse incident at one port of a node, with unit-magnitude energy, the energy is isotropically scattered in all the the four directions, giving rise at the four ports to three radiated and one reflected pulses, each one carrying one-forth of the incident energy. By applying the superposition principle, the more general case of four incident delta voltage impulses can be studied [94]. By subdividing the time axis in discrete time steps Δt, it is possible to construct, for each node, a suitable 4×4 local scattering matrix $\underline{\mathbf{S}}_L$, relating the four reflected voltages V_{r} at the time step $n+1$ to the four incident voltages V_{inc} at the previous time step n. By rearranging all the local matrices into a single global matrix $\underline{\mathbf{S}}_G$ and ordering all the unknown voltages into vectors, it is possible to formally write

$$\mathbf{V}_{\mathrm{r}}^{n+1} = \underline{\mathbf{S}}_G \cdot \mathbf{V}_{\mathrm{inc}}^{n}. \tag{5.54}$$

The computed reflected pulses become the new pulses incident into the adjacent nodes at the new time step, that is,

$$\mathbf{V}_{\mathrm{inc}}^{n+1} = \underline{\mathbf{C}} \cdot \mathbf{V}_{\mathrm{r}}^{n+1}, \tag{5.55}$$

where $\underline{\mathbf{C}}$ is the connection matrix which describes pulse transmission between nodes. The TLM algorithm is thus the repetition of these two steps as long as the transient wave propagation is not extinguished. At each time step, the electric- and magnetic-field components may be directly obtained from the voltages and currents at each node.

The method was extended to 3D problems in [96,97]. In order to support all the components of electric and magnetic fields, a new TLM cell is developed, consisting of three shunt and three series nodes. The shunt and series nodes alternate along the coordinate directions, spaced from each other by $\Delta l/2$. The voltages at the shunt nodes represent the electric field, while the currents at the series nodes correspond to the magnetic field. Since the nodes where different field components are calculated are $\Delta l/2$ spatially separated (thus resulting in a half time step delay), the resulting network is called an *expanded-node* (ExpN) network. To overcome the outstanding disadvantages of this network approach (e.g., complicated graph, difficulties in modeling of boundaries, and, consequently, liability to errors), the *symmetrical condensed-node* (SCN) network was proposed in [98], where the scattering process takes place at the same point in space and time for all the nodes. Along each coordinate direction the two TM and TE polarizations are carried on two pairs of transmission lines that do not intersect with each other. Consequently the new node is formed by the intersection of six transmission lines, presenting two ports in each coordinate direction. The scattering of a delta impulse voltage incident into a port is studied by a direct comparison between Kirchhoff's current and voltage laws written for the condensed shunt and series nodes, respectively, and Maxwell's equations. The analogy is repeated to all the 12 ports so that a local 12×12 scattering matrix $\underline{\mathbf{S}}_L$ for the 3D case can be written and the previously described TLM algorithm can be applied.

When dealing with inhomogeneous materials and/or graded meshes, extra capacitances and inductances are required to model the EM properties of the block

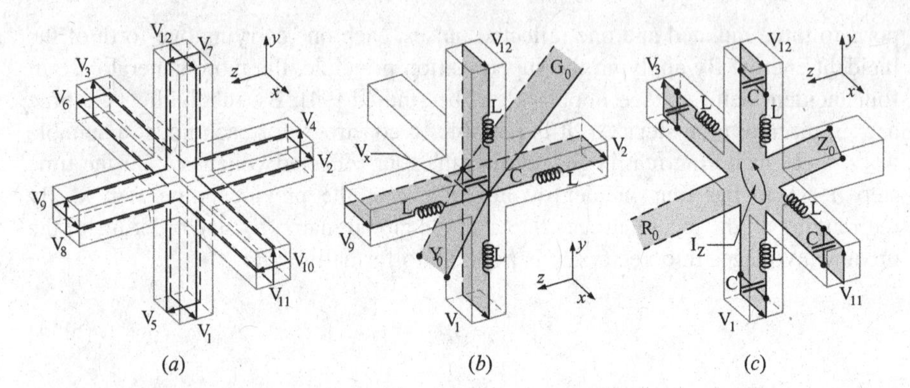

FIGURE 5.9 Stubbed SCN (*a*); shunt node (*b*); and series node (*c*).

of space represented by the node and to ensure that the same propagation delay occurs in all directions [99,100]. The *stubbed SCN*, illustrated in Figure 5.9, maintains the line impedance of the 12 link lines equal to the free-space impedance and introduces the required additional parameters in the form of stubs. Any deficiency in capacitance (inductance), due to dielectric (magnetic) media or to grading, is compensated by loading shunt nodes (series nodes) with open-circuited (short-circuited) reactive stubs of suitable normalized characteristic admittance Y_0 (reactance X_0) and length equal to half the mesh spacing Δl. Similarly losses in a dielectric (magnetic) medium are inserted by loading shunt (series) nodes with a shunt (series) line of infinite length with suitable normalized characteristic admittance G_0 (impedance R_0), which extracts energy at each time iteration. The general local scattering matrix $\underline{\mathbf{S}}_L$ for the stubbed SCN has dimension 24×18, since no reflection occurs from the infinite matched transmission lines.

The numerical characteristics of the TLM method has been extensively investigated. The mesh generally presents an upper cutoff frequency (at which the propagation velocity drops to zero), slow-wave dispersion characteristics, and a maximum allowable time step (which depends on the material properties and the space discretization).

In the last decade further developments have stemmed from the first introduction of the stubbed SCN, with the aim of always improving the numerical features (e.g., improving efficiency, increasing maximum permissible time step, and decreasing numerical dispersion). By relaxing the condition that all the link lines should have the same impedance, two hybrid symmetrical condensed nodes (HSCN) were proposed, where the characteristic impedances of the lines are varied to account for various mesh grading and to model some of the medium parameters. As a result the number of required stubs is reduced if compared with the stubbed SCN, which directly implies improved dispersive properties and generally a larger maximum time step. A first hybrid method, called type-1 HSCN, was introduced in [101], where the characteristic impedances of the 12 link lines are varied to model the correct inductance required by the medium; the capacitance of the lines are then chosen to maintain synchronism in the three coordinate directions and three parallel open-circuited stubs with suitable characteristic admittances are used to account for deficiency in capacitance. A complement of the method, called type-II HSCN was proposed in [102], where the characteristic admittances of the lines are varied to model the correct capacitance required by the medium; the inductances are chosen to achieve synchronism and inductance deficiencies are compensated through the use of series short-circuited stubs. Later a completely stubless SCN, called *symmetrical supercondensed node* (SSCN) was presented in [103], where all the effects due to materials and grading are incorporated into the link lines. The SSCN has shown substantial improvement in storage, efficiency, and maximum allowable time step. The magnitude of the propagation error is larger than in the stubbed SCN but smaller than in the graded HSCN meshes. Nevertheless, the method does not suffer of bilateral dispersion, presenting a unique and unilateral dispersion characteristic. Finally, further investigations have led to a *general symmetrical condensed node* (GSCN) [104] where it is possible to select, with the maximum flexibility, a suitable proportion of material and geometrical properties modeled separately by stubs and by link lines, achieving nodes with optimum dispersion characteristics. These schemes have been optimized through appropriate weighting functions leading to the *adaptable symmetrical condensed node* (ASCN) [105].

Boundaries, such as electric and magnetic walls, are generally placed half way between two adjacent nodes and modeled by enforcing appropriate reflection and transmission coefficients to compute the impulses at the successive time step. In modeling problems that involve wave propagation in an open space (e.g., radiation and scattering problems) artificial boundaries are enforced to truncate the solution domain while creating the numerical illusion of an infinite space. The ABCs developed through the last two decades for the FDTD method have been successfully applied in the TLM method as well, such as the PML technique [106,107].

For extra flexibility and efficiency, multi-grid TLM [108] schemes have been developed to overcome the usual limitations encountered when applying a uniform mesh. As for the FDTD method, sub-cell models have been developed to account for small geometric features such as thin wires, narrow slots, and thin panels [109].

5.6 PARTIAL ELEMENT EQUIVALENT CIRCUIT METHOD

The partial element equivalent circuit (PEEC) method [110] is a 3D full-wave modeling method that has been proved to be very effective for combined EM and circuit analysis for both the frequency and the time domain. The same as the MoM, the PEEC method starts from an integral equation, but its distinguishing feature is that the integral equation is reduced to an equivalent circuit for the basic PEEC cell, which results in a complete equivalent circuit for 3D geometries. Thus the PEEC method can be considered a circuit-based formulation of the integral equation that provides a circuit interpretation of the integral equation governing the EM model, thus enabling an intuitive understanding of the problem. The PEEC method is a full-wave and full-spectrum method; the full-wave aspect refers to the fact that up to a high-frequency limit, all modes of propagation are calculated. The full-spectrum label means that unlike the MoM, the method does not have a low-frequency limit; the solution is valid from dc to the maximum frequency given by the meshing.

Once the EM problem is reduced to an equivalent circuit, additional circuit elements can be easily included. The solution techniques available in most Spice-type circuit solvers for both the time and frequency domain can be used. For these reasons PEEC is a very flexible approach for solving mixed EM and circuit problems.

The PEEC formulation below is initially presented in the time domain only for conductors. Then further extensions are introduced.

The goal of the PEEC approach is to build up an equivalent circuit for the 3D geometry. The integral equation is interpreted as a Kirchhoff voltage law applied to the basic PEEC cell, while the continuity equation is enforced by a Kirchhoff current law at the level of the circuit solution. The starting point of the PEEC model is the sum of all the sources of the electric field at any point on a conductor:

$$\mathbf{E}^{\mathrm{inc}}(\mathbf{r},t) = \frac{\mathbf{J}(\mathbf{r},t)}{\sigma} + \frac{\partial}{\partial t}\mathbf{A}(\mathbf{r},t) + \nabla\Phi(\mathbf{r},t), \tag{5.56}$$

where $\mathbf{E}^{\mathrm{inc}}$ is the incident electric field, $\mathbf{J}$ is the current density in a conductor, $\mathbf{A}$ and Φ are the magnetic vector and electric scalar potentials, respectively. The magnetic vector potential $\mathbf{A}$ and the electric scalar potential Φ at a point $\mathbf{r}$ are given by

$$\begin{aligned} \mathbf{A}(\mathbf{r},t) &= \mu_0 \iiint_{V_\beta} G(\mathbf{r},\mathbf{r}')\mathbf{J}(\mathbf{r}',t_{\mathrm{d}})\mathrm{d}V', \\ \Phi(\mathbf{r},t) &= \frac{1}{\varepsilon_0} \iiint_{V_\beta} G(\mathbf{r},\mathbf{r}')\rho_{\mathrm{e}}(\mathbf{r}',t_{\mathrm{d}})\mathrm{d}V', \end{aligned} \tag{5.57}$$

where V_β is the volume of the material in which the source current $\mathbf{J}$ and the charge density ρ_{e} are localized and the delay time $t_{\mathrm{d}} = t - |\mathbf{r} - \mathbf{r}'|/c$ is the free space travel time between the source point $\mathbf{r}'$ and the observation point $\mathbf{r}$. Both the delay and the Green function are free-space quantities, with $G(\mathbf{r},\mathbf{r}') = 1/(4\pi|\mathbf{r} - \mathbf{r}'|)$. Finally, by inserting the equations above into (5.56), an integral equation for the electric field at a point $\mathbf{r}$ located on the conductor can be formulated as

$$\mathbf{E}^{\mathrm{inc}}(\mathbf{r},t) = \frac{\mathbf{J}(\mathbf{r},t)}{\sigma} + \mu_0 \iiint_{V_\beta} G(\mathbf{r},\mathbf{r}')\frac{\partial \mathbf{J}(\mathbf{r}',t_{\mathrm{d}})}{\partial t}\mathrm{d}V' + \frac{\nabla}{\varepsilon_0}\left[\iiint_{V_\beta} G(\mathbf{r},\mathbf{r}')\rho_{\mathrm{e}}(\mathbf{r}',t_{\mathrm{d}})\mathrm{d}V'\right]. \tag{5.58}$$

To numerically solve the integral equation, appropriate approximations for the current and charge densities are chosen. Unlike the usual MoM solution, the continuity equation $\nabla \cdot \mathbf{J} + \partial \rho_e / \partial t = 0$ is not used to replace all the charge variables with the current ones. Instead, the PEEC solution works with both current and charge as unknowns to cover both the time and frequency domain. The continuity equation is implemented in the form of a Kirchhoff current law in the circuit solution at each node. When dealing with finite-thickness conductors, no high-frequency skin-effect approximation is used, but a more accurate approximation of the internal current flow is introduced. The finite-thickness conductor is subdivided in basic building blocks (referred to as slabs) that are discretized into inductive cells, while neglecting any capacitive effect since displacement currents are negligible in good conductors.

The vector quantities are discretized into orthogonal components in the coordinate system as $\mathbf{J} = J_x \mathbf{u}_x + J_y \mathbf{u}_y + J_z \mathbf{u}_z$. Since the free charge is restricted to the outside surfaces of all the conductors, a surface layer charge density ρ_S is considered rather than a volume quantity. Substituting these relationships into (5.58) results in a set of three coupled equations that are identical in form with the exception of the space directions $\gamma = x, y, z$:

$$E_\gamma^{\text{inc}}(\mathbf{r}, t) = \frac{J_\gamma(\mathbf{r}, t)}{\sigma} + \mu_0 \iiint_{V_\beta} G(\mathbf{r}, \mathbf{r}') \frac{\partial J_\gamma(\mathbf{r}', t_d)}{\partial t} dV' + \frac{1}{\varepsilon_0} \frac{\partial}{\partial \gamma} \left[\oint_{S_\beta} G(\mathbf{r}, \mathbf{r}') \rho_s(\mathbf{r}', t_d) dS' \right]. \tag{5.59}$$

The structure under analysis is discretized into inductive-resistive cells and capacitive cells as illustrated in Figure 5.10. This is accomplished by specifying the number of nodes within the conductor that will be the network nodes of the equivalent circuit, and by defining nonoverlapping rectangular bricks for the three

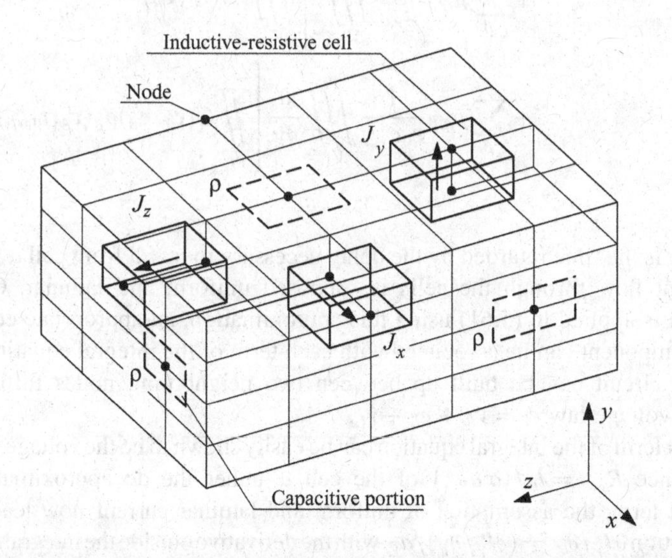

FIGURE 5.10 Volume cells for currents and surface cells for charges.

components of the current and the surface charge. The unknowns $\mathbf{J}$ and ρ_S are expanded into a series of pulse basis functions over the cells with unknown amplitudes, as $J_\gamma = \sum_{k=1}^{N_\gamma} p_{i\gamma,k} J_{\gamma,k}$ (where N_γ is the number of cells for the γ component of the current) and $\rho_S = \sum_{k=1}^{N_q} p_{q,k} \rho_{S,k}$.

Because of the pulse basis-function expansion, the geometry must be discretized into small enough cells so that accurate results can be obtained. Pulse functions are also selected as testing functions for a Galerkin formulation of the solution. The inner product is formed with respect to a cell and is defined as

$$\langle f(\mathbf{r}), g(\mathbf{r}) \rangle = \frac{1}{a_{\gamma,\alpha}} \iiint_{V_\alpha} f(\mathbf{r}) g(\mathbf{r})\, \mathrm{d}V, \tag{5.60}$$

where V_α and $a_{\gamma,\alpha}$ are the volume and the cross section perpendicular to the current flow of the testing cell, respectively, $f(\mathbf{r})$ is the integrand, and $g(\mathbf{r})$ is the pulse testing function.

As was already mentioned, the goal of the PEEC formulation is to convert the field equations into an equivalent circuit, where I and Q (or the corresponding potentials Φ) are the unknowns. Inserting the expansions for the unknowns in (5.59) and performing the inner product with respect to an inductive cell α leads to

$$\begin{aligned}
\frac{1}{a_{\gamma,\alpha}} \iiint_{V_\alpha} E_\gamma^{\mathrm{inc}}(\mathbf{r}_\alpha, t) \mathrm{d}V &= \frac{1}{a_{\gamma,\alpha}} \iiint_{V_\alpha} \frac{J_{\gamma,\alpha}(\mathbf{r}_\alpha, t)}{\sigma} \mathrm{d}V \\
&+ \sum_\beta p_{i\gamma,\beta} \frac{\mu_0}{a_{\gamma,\alpha}} \iiint_{V_\alpha} \left[\iiint_{V_\beta} G(\mathbf{r}_\alpha, \mathbf{r}'_\beta) \frac{\partial J_{\gamma,\beta}(\mathbf{r}'_\beta, t_{\mathrm{d},\alpha\beta})}{\partial t} \mathrm{d}V' \right] \mathrm{d}V \\
&+ \sum_\beta p_{q,\beta} \frac{1}{\varepsilon_0 a_{\gamma,\alpha}} \iiint_{V_\alpha} \frac{\partial}{\partial \gamma} \left[\oiint_{S_\beta} G(\mathbf{r}_\alpha, \mathbf{r}'_\beta) \rho_\beta(\mathbf{r}'_\beta, t_{\mathrm{d},\alpha\beta}) \mathrm{d}S' \right] \mathrm{d}V,
\end{aligned} \tag{5.61}$$

where $t_{\mathrm{d},\alpha\beta}$ is the time retarded by the delay necessary to travel from cell α to cell β. The current flow through the cells is supposed uniform and laminar. Once the integration is applied to (5.61) using this approximation, an appropriate equivalent lumped component can be associated with each term of the integral equation. So an equivalent circuit can be built up between two neighboring nodes fulfilling the Kirchhoff voltage law $e_i = v_R + v_L + v_C$.

The first term of the integral equation can be easily shown to be the voltage v_R across the resistance $R_{\gamma,\alpha} = l_\alpha / (\sigma\, a_{\gamma,\alpha})$ of the cell α under the dc approximation. For the second term, the assumption of uniform and laminar current flow leads to the approximation $\partial J_\gamma / \partial \gamma = (\partial I_\gamma / \partial \gamma) / a_\gamma$, with the derivative outside the integral. Thus the second term represents the partial self-inductance of the conductor cell α plus the partial mutual inductances between parallel cells α and β. The partial inductance is given by

$$L_{p\gamma,\alpha\beta} = \frac{\mu}{a_{\gamma,\alpha}a_{\gamma,\beta}} \iiint\limits_{V_\alpha} \left[\iiint\limits_{V_\beta} \frac{1}{4\pi|\mathbf{r}_\alpha - \mathbf{r}_\beta|} \mathrm{d}V_\beta \right] \mathrm{d}V_\alpha. \tag{5.62}$$

For an orthogonal coordinate system all the nonparallel cells are perpendicular so that the partial mutual inductances are zero. Hence the inductive term is simply interpreted as $v_L = L_{p\gamma,\alpha\alpha}\mathrm{d}I_\alpha(t)/\mathrm{d}t + \sum_\beta L_{p\gamma,\alpha\beta}\,\mathrm{d}I_\beta(t_{\mathrm{d},\alpha\beta})/\mathrm{d}t$ where the first term is the partial self-inductance of the cell α and the other terms represent the inductive coupling to cell α from the current in cell β. This term is computed at the retarded time $t_{\mathrm{d},\alpha\beta} = t - R_{\alpha\beta}/c$, which accounts for the delay between the two cells (given by the time required by the field to travel along the center-to-center distance $R_{\alpha\beta}$ between cell α and β). An important fact to note is that the effect of the retardation must be taken outside the integral in order to rewrite the contributions of the vector potential in terms of the definition for partial inductance. The retarded time corresponds to the time delay between the centers of the two cells, which is a valid approximation when the dimensions of the cells are small compared to the minimum wavelength. The partial inductances with retardation can be handled by introducing a partial self-inductance $L_{p\gamma,\alpha\alpha}$ in series with a voltage-controlled voltage source $V_{p\gamma,\alpha}(t) = \sum_{\beta\neq\alpha}\left(L_{p\gamma,\alpha\beta}/L_{p\gamma,\beta\beta}\right)v_{L\gamma,\beta}(t_{\mathrm{d},\alpha\beta})$, where $v_{L\gamma,\beta}(t_{\mathrm{d},\alpha\beta})$ is the voltage at the partial self-inductance of the cell β computed at the retarded time [111].

As concerns the third term of the integral equation, it represents a pseudo-capacitive coupling. By defining the inside integral as

$$F(\mathbf{r}) = \oiint\limits_{S_\beta} G(\mathbf{r}_\alpha, \mathbf{r}'_\beta)\rho_\beta(\mathbf{r}'_\beta, t_{\mathrm{d},\alpha\beta})\mathrm{d}S', \tag{5.63}$$

we can then approximate

$$\frac{1}{a_{\gamma,\alpha}} \iiint\limits_{V_\alpha} \frac{\partial}{\partial\gamma} F(\mathbf{r})\mathrm{d}V \simeq F(\mathbf{r} + \Delta\mathbf{l}_{\gamma,\alpha}/2) - F(\mathbf{r} + \Delta\mathbf{l}_{\gamma,\alpha}/2), \tag{5.64}$$

where $\Delta\mathbf{l}_{\gamma,\alpha}$ is the length of cell α in the γ direction. Applying the preceding results leads to

$$v_C = \sum_\beta p_{q,\beta} Q_\beta\left(\mathbf{r}'_\beta, t_{\mathrm{d},\alpha\beta}\right)\left(\frac{1}{\varepsilon_0 S_\beta} \oiint\limits_{S_\beta} G(\mathbf{r}^+_\alpha, \mathbf{r}'_\beta)\mathrm{d}S' - \frac{1}{\varepsilon_0 S_\beta} \oiint\limits_{S_\beta} G(\mathbf{r}^-_\alpha, \mathbf{r}'_\beta)\,\mathrm{d}S'\right), \tag{5.65}$$

where $Q_\beta\left(\mathbf{r}'_\beta, t_{\mathrm{d},\alpha\beta}\right)$ is the total charge on a capacitive cell, S_β is the surface of the capacitive cell β over which the uniform charge density is located, $\mathbf{r}^+_\alpha$ and $\mathbf{r}^-_\alpha$ are the positive and negative ends, respectively, of the testing inductive cell α in the γ direction. The ends of the cell are nodes of the mesh and will be nodes of the equivalent network. Since the capacitive cells are shifted from the inductive cells by half the size of a cell, these two points are at the centers of the neighboring capacitive cells along the γ direction. Thus the potential coefficient between the source node β and the testing node α can be computed as an average over the capacitive testing cell α, resulting in the final expression for the retarded potential coefficient as

$$pp_{\alpha\beta} = \frac{1}{\varepsilon_0 S_\alpha S_\beta} \oiint_{S_\alpha} \left[\oiint_{S_\beta} \frac{1}{4\pi |\mathbf{r}_\alpha - \mathbf{r}_\beta|} \mathrm{d}S_\beta \right] \mathrm{d}S_\alpha. \tag{5.66}$$

The capacitive term is interpreted as $v_C = pp_{\alpha\alpha}\, Q_\alpha(t) + \sum_\beta pp_{\alpha\beta}\, Q_\beta(t_{\mathrm{d},\alpha\beta})$, where $pp_{\alpha\beta}$ is the coefficient of the potential between cells α and β. Since the charges reside on the conductor surfaces, the potentials are only due to nodes external to the conductors.

A key issue in the PEEC modeling approach is to know how these capacitances are included in the equivalent circuit model. The concept of capacitance in fact is invalid for the case where retardation is not negligible, so the matrix of retarded coefficients of potential $\underline{\mathbf{P}}$ cannot be inverted into the short-circuit capacitance matrix. In this case the retarded coefficients of potential need to be modeled as a controlled source. The controlled sources are derived from the equation $\mathbf{\Phi} = \underline{\mathbf{P}} \cdot \mathbf{Q}$ and from the constitutive relationship between current and charge, namely $i = \mathrm{d}q/\mathrm{d}t$. The formulation presented in [112] builds up the circuit in terms of pseudocapacitances $C'_\alpha = 1/p_{\alpha\alpha}$ in series to voltage-controlled voltage sources $U_\alpha(t) = \sum_{\beta\neq\alpha} (p_{\alpha\beta}/p_{\beta\beta}) v_{C,\beta}(t_{\mathrm{d},\alpha\beta})$, where $v_{C,\beta}(t_{\mathrm{d},\alpha\beta})$ is the potential of pseudocapacitance of cell β computed at the retarded time. Alternatively, the formulation presented in [113] builds up the circuit in terms of pseudocapacitances in parallel to current-controlled current sources $I_\alpha(t) = \sum_{\beta\neq\alpha} (p_{\alpha\beta}/p_{\alpha\alpha}) i_{C,\beta}(t_{\mathrm{d},\alpha\beta})$, where $i_{C,\beta}(t_{\mathrm{d},\alpha\beta})$ is the total current flowing from node β toward earth at the retarded time.

The RHS of the integral equation is due to the incident electric field. An equivalent induced voltage source is obtained by performing the inner product [114]. The induced voltage source acts on each cell α with amplitude

$$V_{p\gamma,\alpha}(t_\alpha) = \frac{1}{a_\alpha} \iiint_{V_\alpha} E_\gamma^{\mathrm{inc}}(\mathbf{r}_\alpha, t_\alpha) \mathrm{d}V, \tag{5.67}$$

where t_α is the time instant in which the impinging field arrives at cell α. The resulting equivalent circuits are represented in Fig. 5.11. By the described procedure, a complex network can be developed in the 3D space.

As concerns the inclusion of dielectrics into the PEEC formulation, any displacement current due to the bound charges for the dielectrics with $\varepsilon_\mathrm{r} > 1$ is treated separately from the conducting currents due to the free charges. Hence the dielectrics are treated separately as an equivalent current in a free-space environment. The computation is accomplished by adding and subtracting a term $\varepsilon_0 \partial\mathbf{E}/\partial t$ in Maxwell's equation for $\mathbf{H}$, by which there can be defined an equivalent polarization current $\varepsilon_0(\varepsilon_\mathrm{r} - 1)\partial\mathbf{E}/\partial t$ due to the dielectrics. Consequently the total current is defined as the sum of the conducting current plus this equivalent

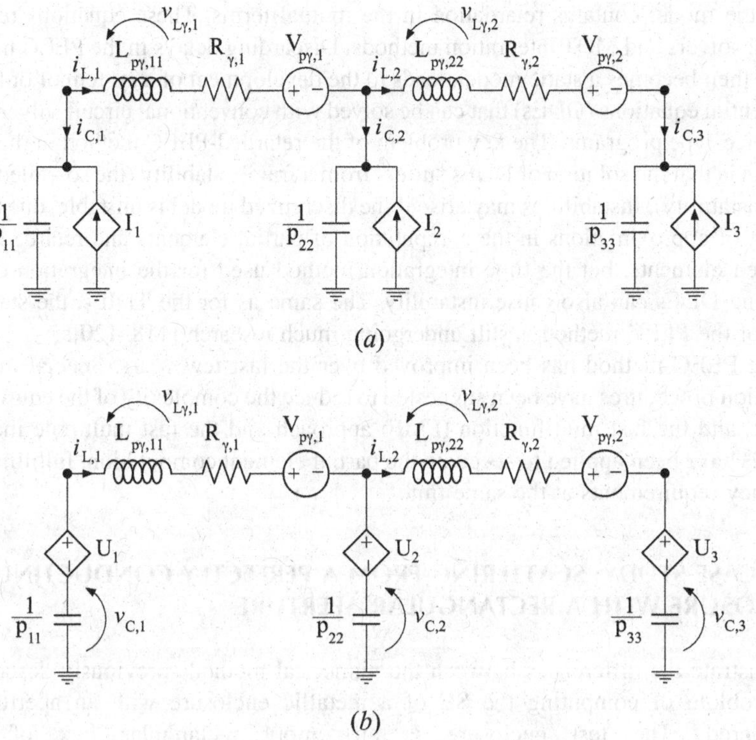

(a)

(b)

FIGURE 5.11 Equivalent PEEC circuits.

polarization current. The dielectric regions are discretized into cells as the conductors and are represented with additional circuit elements [111]. The excess capacitance of the dielectric cell is defined as $C_{\gamma,\alpha} = \varepsilon_0(\varepsilon_r - 1)\, a_\alpha / l_\alpha$, where l_α is the length of the cell, so the equivalent circuit is given by a partial inductance $L_{p,\alpha\alpha}$ in series to the capacitance $C_{\gamma,\alpha}$. The coupling to other cells is taken into account through the partial mutual inductances $L_{p,\alpha\beta}$, as done for the conductor cell.

Most of the early work in PEEC modeling was restricted to rectangular Manhattan geometries, since the method made use of rectangular bricks. Recently a nonorthogonal formulation has been developed, using quadrilateral and hexahedral cells [115] to model more complex and generally shaped geometries. The method can be applied in both the time and frequency domain. For performing a simulation in the frequency domain, the equivalent circuit of the PEEC cell remains the same as in the time domain. The retardation in time in the controlled voltage sources is accounted for by a phase term $\exp(-jkR_{\alpha\beta})$, where k is the free-space wavenumber.

Once the equivalent circuit model has been formed, the circuit equations can be set up by a standard systematic network solution like the modified nodal analysis (MNA) [116] or the modified loop analysis (MLA) [117]; this results in fewer unknowns but is only valid for the frequency domain. The distinguishing feature of the resulting equations of the PEEC method is that they are delay differential equations (DDEs), since the model contains retardation in the mutual terms. These equations require special solvers and MOT integration methods. Discarding delays in the PEEC model, which then becomes a static model, leads to the development of a system of ordinary differential equations (ODEs) that can be solved with conventional circuit solvers like the Spice-type programs. The key problem of the retarded-PEEC method in the time domain is that the solution of DDEs suffers from a fragile stability (the so-called late-time instability). Instabilities may arise if the discretized model is unstable, due to the impact of approximations in the computation of partial elements and retarded time between elements, but the time integration method used for the integration of the resulting DDEs can also cause instability. The same as for the TDIEs, the stability issue of the PEEC method is still undergoing much research [118–120].

The PEEC method has been improved over the last few years. Special model-reduction procedures have been suggested to reduce the complexity of the equivalent circuit, and the fast multifunction (FMF) approach and the fast multipole method (FMM) have been applied to speed up the partial element computation, fulfilling the accuracy requirements at the same time.

5.7 CASE STUDY: SCATTERING FROM A PERFECTLY CONDUCTING ENCLOSURE WITH A RECTANGULAR APERTURE

To illustrate the differences between the numerical methods previously described, the problem of computing the SE of a metallic enclosure with an aperture is considered. The test enclosure is an empty rectangular box of size $A \times B \times C = 30 \times 12 \times 30$ cm with a thin rectangular aperture of dimensions $a \times b = 10 \times 0.5$ cm located at the center of one of its side walls, as shown in Figure 5.12. A uniform plane wave is assumed to impinge orthogonally to the front side containing the aperture, with the electric-field vector of amplitude 1 V/m linearly polarized along the shortest side of the aperture. Being a worst case, this configuration may be most suitable for an investigation of the shielding properties of

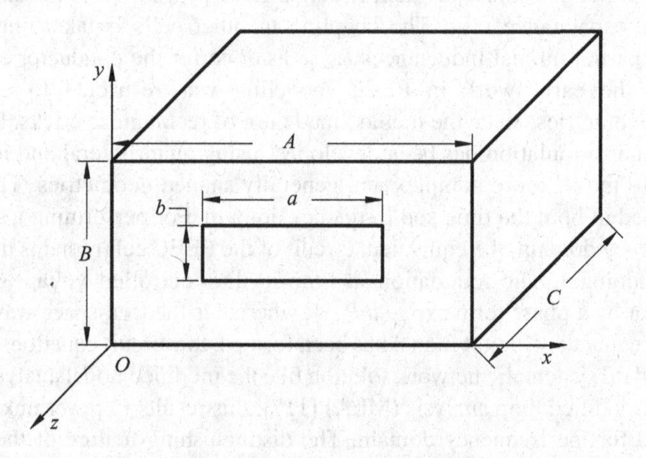

FIGURE 5.12 Geometry of the test enclosure.

the enclosure. The classical electric SE is computed in the frequency window from 100 MHz to 3 GHz, with the observation point at the center of the enclosure.

The FEM, MoM, FIT, and the TLM methods have been applied to predict the electric SE of the test enclosure. The walls of the enclosure are modeled as PEC. Because of the problem symmetry, only a quarter of the enclosure needs to be simulated, with the electric and magnetic symmetry on the xz and yz plane, respectively. In all the differential methods, the enclosure is surrounded by 30 cm of vacuum and the computational domain is terminated with a perfectly matched medium, to absorb the outgoing waves.

The FEM model uses first-order linear vector curl-conforming elements (edge elements) defined on an unstructured tetrahedral mesh (60,841 tetrahedra) locally refined inside the enclosure and near the aperture (refining edge factor 2). The maximum allowable element length is set to $\lambda_{\min}/10$, where $\lambda_{\min}$ is the minimum wavelength of interest in free space (10 cm at 3 GHz). The solution is worked out in the frequency domain, by a linear discrete sweep from 100 MHz to 3 GHz with 291 points (frequency step equal to 10 MHz).

The FIT model is based on a nonuniformly graded orthogonal structured mesh. It is thus equivalent to the classical FDTD method, except for its implementation based on a finite technique. The used mesh is composed of 163,840 cells presenting at least 10 lines per wavelength $\lambda_{\min}$ and a limit cell deformation factor of 10:1. The mesh is locally refined near the aperture with an edge refinement factor of 2. The excitation signal is a Gaussian modulated impulse with frequency content from 100 MHz to 3 GHz. The time step is $\Delta t = 2.45$ ps. To extinguish the transient inside the resonant cavity, 982,832 time steps have been necessary. The results in the time domain are then Fourier transformed in the frequency window of interest, using 1001 frequency points.

The TLM model uses the stubbed symmetrical condensed node (SCN). The mesh is structured with a nonuniform grading, with characteristics similar to that of the FIT model. The number of cells is 212,350, the time step is $\Delta t = 2.30$ ps, and the number of time steps required to reach the steady state is 1,054,350. The time-domain results are Fourier transformed using again 1001 frequency points.

The MoM model uses first-order linear vector div-conforming elements (RWG basis functions) defined on an unstructured surface triangular mesh (2980 triangles). The maximum edge length for triangular patches is set to $\lambda_{\min}/10$ and the mesh is refined in the neighborhood of the aperture with an edge-based refinement process, in order to obtain triangles with maximum edge length of 5 mm on the border of the aperture. The MoM code is based on the EFIE and makes full use of the electric and magnetic symmetries of the problem. The solution is worked out in the frequency domain, using a linear discrete sweep equal to that used in the FEM model.

The results are shown in Figure 5.13*a*. In Figure 5.13*b* the resonant frequencies and modes of the box without the aperture are also reported (see Chapter 7). From the results it can be seen that the electric SE values obtained with the different methods show a good agreement in the frequency range of interest, generally within 3 dB of difference.

The main differences between the methods are in the required CPU time and computational resources. The FIT takes about 13 hours to reach the steady state on

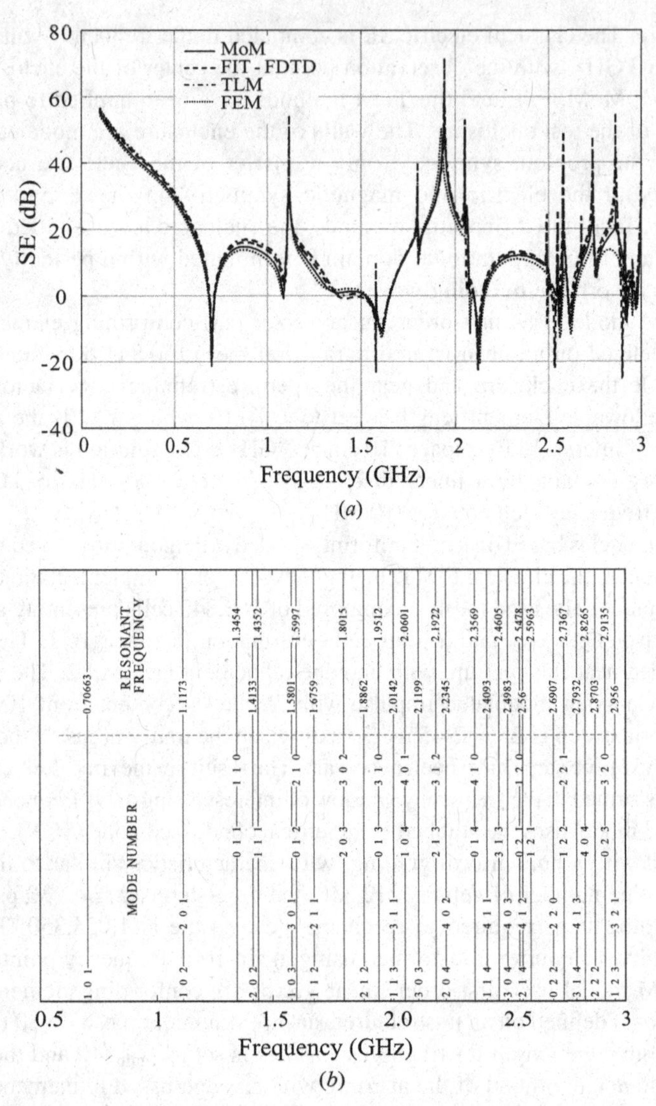

FIGURE 5.13 Comparison of electric SE results from different codes (*a*); resonant frequencies and modes of the box (*b*).

a dual Xeon Workstation with 4 GB RAM (with parallelization module), while the TLM method is a bit faster, saving two hours of computation. These long simulations are mainly due to the high-resonant behavior of the enclosure whose walls are considered lossless and to the fact that the aperture is very thin. The memory requirement of the two codes is about 300 MB. The introduction of a subgrid on the aperture in the FIT model with a volume refinement factor of 3 does not substantially reduce the computational time but only the memory requirement, since the cell number in the coarse mesh is reduced. Anyway, this is not a problem that can be classified as critical for the memory requirement. As concerns the FEM, it takes about 14.5 hours to compute the span over 291 frequency points, which is about 3 minutes for each frequency. This is the time necessary to build up the banded matrix at each frequency and to solve it by means of the iterative conjugate gradient method. In addition the memory required by the method is huge, around 1.25 GB, due to the banded matrix to be stored. The method is attractive for simulating a structure at a single frequency, but it cannot be considered efficient when a frequency sweep is needed, especially in a wideband. In this case special techniques for a fast interpolating sweep in conjunction with a model-order reduction should be used to speed up the simulation. As regards to the MoM, it takes about 3.5 hours to compute a sweep on 291 frequency points (about 40 seconds for single frequency) and requires about 530 MB. The method cannot be considered very efficient from the storage point of view: it needs to store a full-populated matrix, but it is fast. In fact the calculation of the matrix elements only requires the evaluation of the free-space Green function, and the final matrix can be efficiently solved with the direct Gauss elimination method. It is necessary to point out that the MoM is fast but computes the SE only on 291 frequencies, while the time-domain methods allow practically any number of samples to be computed from the time-domain results through the Fourier transform. The only limit is given by the Nyquist criterion, but the sampling period of the time-domain methods is usually much smaller than that required by the Nyquist rate, due to stability and accuracy reasons.

REFERENCES

[1] E. K. Miller. "A selective survey of computational electromagnetics." *IEEE Trans. Antennas Propagat.*, vol. 36, no. 9, pp. 1281–1305, Sep. 1988.

[2] P. Hammond. "Some thoughts on the numerical modeling of electromagnetic processes." *Int. J. Numer. Mod.*, vol. 1, pp. 3–6, 1988.

[3] O. C. Zienkiewicz and Y. K. Cheung. "Finite elements in the solution of field problems." *Engineer*, vol. 220, pp. 507–510, 1965.

[4] P. P. Silvester. "Finite-element solution of homogeneous waveguide problems." *Alta Frequenza*, vol. 38, pp. 313–317, 1969.

[5] A. M. Winslow. "Magnetic field calculations in an irregular triangular mesh." Lawrence Radiation Laboratory, Livermore, CA, UCRL-7784-T, Rev. 1, 1965.

[6] G. Strang and G. J. Fix. *An Analysis of the Finite Element Method*. Englewood Cliffs, NJ: Prentice-Hall, 1973.

[7] J. T. Oden and J. N. Reddy. *An Introduction to the Mathematical Theory of Finite Elements*. New York: Wiley, 1976.

[8] P. P. Silvester and R. L. Ferrari. *Finite Elements For Electrical Engineers*, 3rd ed. Cambridge: Cambridge University Press, 1996.

[9] J. M. Jin. *The Finite Element Method in Electromagnetics*. New York: Wiley, 1993.

[10] O. C. Zienkiewicz and R. L. Taylor. *The Finite Element Method*. London: McGraw-Hill, 1988.

[11] P. P. Silvester and G. Pelosi. *Finite Elements for Wave-Problems*. New York: IEEE Press, 1994.

[12] S. G. Mikhlin. *Variational Methods in Mathematical Physics*. New York: Macmillan, 1964.

[13] A. Konrad. "Triangular finite elements for vector fields in electromagntics." PhD Thesis, *Department of Electrical Engineering*, McGill University, 1974.

[14] A. Konrad. "Vector variational formulation of electromagnetic fields in anisotropic media." *IEEE Trans. Microwave Theory Tech.*, vol. 24, no. 9, pp. 553–559, Sep. 1976.

[15] B. M. A. Rahman and J. B. Davies. "Penalty function improvement of waveguide solution by finite elements." *IEEE Trans. Microwave Theory Tech.*, vol. 32, no. 8, pp. 922–928, Aug. 1984.

[16] M. Hano. "Finite-element analysis of dielectric-loaded waveguides." *IEEE Trans. Microwave Theory Tech.*, vol. 32, no. 10, pp. 1275–1279, Oct. 1984.

[17] S. H. Wong and Z. J. Cendes. "Combined finite element-modal solution of three-dimensional eddy current problems." *IEEE Trans. Magn.*, vol. 24, no. 6, pp. 2685–2687, Nov. 1988.

[18] S. H Wong and Z. J. Cendes. "Numerically stable finite-element methods for the Galerkin solution of eddy current problems." *IEEE Trans. Magn.*, vol. 25, no. 4, pp. 3019–3021, Jul. 1989.

[19] J. C. Nedelec. "Mixed finite elements in R3." *Numer. Mathematik*, vol. 35, no. 3, pp. 315–341, 1980.

[20] M. L. Barton and Z. J. Cendes. "New vector finite elements for three-dimensional magnetic-field computation." *J. Appl. Phys.*, vol. 61, pp. 3919–3921, Apr. 1987.

[21] A. Bossavit and I. Mayergoyz. "Edge-elements for scattering problems." *IEEE Trans. Magn.*, vol. 25, no. 4, pp. 2816–2821, Jul. 1989.

[22] Z. J. Cendes. "Vector finite elements for electromagnetic field computation." *IEEE Trans. Magn.*, vol. 27, no. 5, pp. 3958–3966, Sep. 1991.

[23] A. Bossavit. "Whitney forms: A class of finite elements for three-dimensional computations in electromagnetism." *IEE Proc., Pt. A: Sci., Meas. Tech.*, vol. 135, no. 8, pp. 493–500, Nov. 1988.

[24] H. Whitney. *Geometric Integration Theory*. Princeton: Princeton University Press, 1957.

[25] C. R. I. Emson. "Methods for the solution of open-boundary electromagnetic-field problems." *IEE Proc., Pt. A: Sci., Meas. Tech.*, vol. 135, no. 3, pp. 151–158, Mar. 1988.

[26] R. Mittra, O. Ramahi, A. Khebir, R. Gordon, and A. Kouki. "A review of absorbing boundary conditions for two and three-dimensional electromagnetic scattering problems." *IEEE Trans. Magn.*, vol. 25, no. 4, pp. 3034–3039, Jul. 1989.

[27] Q. Chen and A. Konrad. "A review of finite element open boundary techniques for static and quasi-static electromagnetic field problems." *IEEE Trans. Magn.*, vol. 33, no. 1, pp. 663–676, Jan. 1997.

[28] P. P. Silvester, D. A. Lowther, C. J. Carpenter, and E. A. Wyatt. "Exterior finite elements for 2-dimensional field problems with open boundaries." *IEE Proc.*, vol. 124, pp. 1267–1270, Dec. 1977.

[29] P. Bettess and J. A. Bettess. "Infinite elements for static problems." *Eng. Comput.*, vol. 1, no. 1, pp. 4–16, Mar. 1984.

[30] X. Yuan, D. R. Linch, and J. W. Strohbehn. "Coupling of finite element and moment methods for electromagnetic scattering from inhomogeneous objects." *IEEE Trans. Antennas Propagat.*, vol. 38, no. 3, pp. 386–393, Mar. 1990.

[31] C. A. Brebbia. *The Boundary Element Method for Engineers*. London: Pentech Press, 1978.

[32] A. Bayliss, M. Gunzburger, and E. Turkel. "Boundary conditions for the numerical solution of elliptic equations in exterior regions." *SIAM J. Appl. Math.*, vol. 42, no. 2, pp. 430–451, Apr. 1982.

[33] J. P. Berenger. "A perfectly matched layer for the absorption of electromagnetic waves." *J. Comput. Phys.*, vol. 114, pp. 185–200, Oct. 1994.

[34] J. P. Berenger. "Three-dimensional perfectly matched layer for the absorption of electromagnetic waves." *J. Comput. Phys.*, vol. 127, pp. 363–379, Sep. 1996.

[35] G. Mur. "The finite element modeling of three-dimensional time-domain electromagnetic fields in strongly inhomogeneous media." *IEEE Trans. Magn.*, vol. 28, no. 2, pp. 1130–1133, Mar. 1992.

[36] S. D. Gedney and U. Navsariwala. "An unconditionally stable finite-element time-domain solution of the vector wave equation." *IEEE Microwave Guided Wave Lett.*, vol. 5, no. 10, pp. 332–334, May 1995.

[37] R. F. Harrington. *Field Computation by Moment Methods*. New York: Macmillan, 1968.

[38] J. Moore and R. Pizer. *Moment Methods in Electromagnetics: Techniques and Applications*. New York: Wiley, 1984.

[39] N. Morita, N. Kumagai, and J. R. Mautz. *Integral Equation Methods for Electromagnetics*. Norwood, MA: Artech House, 1990.

[40] E. K. Miller, L. Medgyesi-Mitschang, and E. H. Newman. *Computational Electromagnetics: Frequency-Domain Method of Moments*. New York: IEEE Press, 1992.

[41] A. F. Peterson, S. L. Ray, and R. Mittra. *Computational Methods for Electromagnetics*. New York: IEEE Press, 1998.

[42] R. F. Harrington. "Matrix methods for field problems." *IEEE Proc.*, vol. 55, no. 2, pp. 136–149, Feb. 1967.

[43] A. W. Maue. "Toward formulation of a general diffraction problem via an integral equation." *Zeitschrift für Physik*, vol. 126, pp. 601–618, 1949.

[44] J. Van Bladel. *Electromagnetic Fields*. New York: Hemisphere Publishing, 1985.

[45] D. S. Jones. "A critique of the variational method in scattering problems." *IRE Trans. Antennas Propagat.*, vol. AP-4, pp. 297–301, Jul. 1965.

[46] V. H. Rumsey. "Reaction concept in electromagnetic theory." *Phys. Rev.*, vol. 94, no. 6, pp. 1483–1491, Jun. 1954.

[47] S. M. Rao, D. R. Wilton, and A. W. Glisson. "Electromagnetic scattering by surfaces of arbitrary shape." *IEEE Trans. Antennas Propagat.*, vol. 30, no. 3, pp. 409–418, May 1982.

[48] R. D. Graglia, D. R. Wilton, and A. F. Peterson. "Higher order interpolatory vector bases for computational electromagnetics." *IEEE Trans. Antennas Propagat.*, vol. 45, no. 3, pp. 329–342, Mar. 1997.

[49] Z. Wang, J. Volakis, K. Saitou, and K. Kurabayashi. "Comparison of semi-analytical formulations and gaussian-quadrature rules for quasi-static double-surface potential integrals." *IEEE Antennas Propagat. Mag.*, vol. 45, no. 6, pp. 96–102, Dec. 2003.

[50] J. R. Mautz and R. F. Harrington. "H-field, E-field and Combined-field solutions for conducting bodies of revolution." *AEÜ*, vol. 32, no. 4, pp. 157–163, Apr. 1978.

[51] J. R. Mautz and R. F. Harrington. "A combined-source solution for radiation and scattering from a perfectly conductor body." *IEEE Trans. Antennas Propagat.*, vol. 27, no. 4, pp. 445–454, Jul. 1982.

[52] W. L. Wu, A. Glisson, and D. Kajfez. "A study of two numerical solution procedures for the electric field integral equation at low frequency." *ACES J.*, vol. 10, no. 3, pp. 69–80, Nov. 1995.

[53] G. Vecchi. "Loop-star decomposition of basis functions in the discretization of the EFIE." *IEEE Trans. Antennas Propagat.*, vol. 47, no. 2, pp. 339–346, Feb. 1999.

[54] W. C. Chew, J. M. Jin, E. Michielssen, and J. M. Song. *Fast and Efficient Algorithms in Computational Electromagnetics*. Norwood, MA: Artech House, 2001.

[55] V. Rokhlin. "Rapid solution of integral equations of scattering theory in two dimensions." *J. Comput. Phys.*, vol. 86, no. 2, pp. 414–439, 1990.

[56] V. Rokhlin. "Diagonal form of translation operators for the Helmholtz equation in three dimensions." *Appl. Computat. Harmon. Analysis*, vol. 1, no. 1, pp. 82–93, Dec. 1993.

[57] R. Coifman, V. Rokhlin, and S. Wandzura. "The fast multipole method for the wave equation: A pedestrian prescription." *IEEE Antennas Propagat. Mag.*, vol. 15, no. 3, pp. 7–12, Jun. 1993.

[58] S. M. Rao. *Time Domain Electromagnetics*. New York: Academic Press, 1999.

[59] A. Taflove and S. Hagness. *Computational Electrodynamics: The Finite-Difference Time-Domain Method.* Boston: Artech House, 2005.

[60] K. S. Kunz and R. J. Luebbers. *The Finite Difference Time Domain Method for Electromagnetics*. Boca Raton, FL: CRC Press, 1993.

[61] D. Sullivan. *Electromagnetic Simulation Using the FDTD Method.* New York: IEEE Press, 2000.

[62] W. Yu, R. Mittra, T. Su, Y. Liu, and X. Yang. *Parallel Finite-Difference Time-Domain Method.* Boston: Artech House, 2006.

[63] K. S. Yee. "Numerical solution of initial boundary value problems involving Maxwell's equations in isotropic media." *IEEE Trans. Antennas Propagat.*, vol. 14, no. 3, pp. 302–307, Mar. 1966.

[64] K. R. Umashankar and A. Taflove. "A novel method to analyze electromagnetic scattering of complex objects." *IEEE Trans. Electromagn. Compat.*, vol. 24, no. 4, pp. 397–405, Nov. 1982.

[65] J. L. Young and R. O. Nelson. "A summary and systematic analysis of FDTD algorithms for linear dispersive media." *IEEE Antennas Propagat. Mag.*, vol. 43, no. 1, pp. 61–77, Feb. 2001.

[66] J. A. Pereda, L. A. Vielva, M. A. Solano, A. Vegas, and A. Prieto. "FDTD analysis of magnetized ferrites: application to the calculation of dispersion characteristics of ferrite-loaded waveguides." *IEEE Trans. Microwave Theory Tech.*, vol. 43, no. 2, pp. 350–357, Feb. 1995.

[67] R. M. Josep and A. Taflove. "FDTD Maxwell's equations models for nonlinear electrodynamics and optics." *IEEE Trans. Antennas Propagat.*, vol. 43, no. 3, pp. 364–374, Mar. 1997.

[68] G. D. Smith. *Numerical Solution of Partial Differential Equations*. Oxford: Oxford University Press, 1978.

[69] F. Edelvik, R. Schuhmann, and T. Weiland. "A general stability analysis of FIT/FDTD applied to lossy dielectrics and lumped elements." *Int. J. Numer. Model.*, vol. 17, pp. 407–419, Jul./Aug. 2004.

[70] P. G. Petropoulos. "Stability and phase error analysis of FD-TD in dispersive dielectrics." *IEEE Trans. Antennas Propagat.*, vol. 42, no. 1, pp. 62–69, Jan. 1994.

[71] K. L. Shlager and J. B. Schneider. "Comparison of the dispersion properties of several low-dispersion finite-difference time-domain algorithms." *IEEE Trans. Antennas Propagat.*, vol. 51, no. 3, pp. 642–653, Mar. 2003.

[72] P. Monk. "Sub-gridding FDTD schemes." *ACES J.*, vol. 11, no. 1, pp. 37–46, Jan. 1996.

[73] M. Okoniewski, E. Okoniewska, and M. A. Stuchly. "Three-dimensional subgridding algorithm for FDTD." *IEEE Trans. Antennas Propagat.*, vol. 45, no. 3, pp. 422–429, Mar. 1997.

[74] M. W. Chevalier, R. J. Luebbers, and V. P. Cable. "FDTD local grid with material transverse." *IEEE Trans. Antennas Propagat.*, vol. 45, no. 3, pp. 411–421, Mar. 1997.

[75] P. Thoma and T. Weiland. "A consistent subgridding scheme for the finite difference time domain method." *Int. J. Numer. Model.*, vol. 9, no. 5, pp. 359–374, May 1996.

[76] D. W. Peaceman and H. H. Rachford Jr. "The numerical solution of parabolic and elliptic differential equations." *SIAM J.*., vol. 3, no. 1, pp. 28–41, Jan. 1955.

[77] J. Douglas. "On the numerical integration of $U_{xx} + U_{yy} = U_t$ by implicit methods." *SIAM J.*, vol. 3, no. 1, pp. 42–65, Jan. 1955.

[78] R. Holland. "Implicit three-dimensional finite differencing of Maxwell's equations." *IEEE Trans. Nucl. Sci.*, vol. 31, pp. 1323–1326, Dec. 1984.

[79] T. Namiki. "A new FDTD algorithm based on alternating-direction implicit method." *IEEE Trans. Microwave Theory Tech.*, vol. 47, no. 10, pp. 2003–2007, Oct. 1999.

[80] T. Namiki. "3-D ADI-FDTD method—Unconditionally stable time-domain alghorithm for solving full vector Maxwell's equations." *IEEE Trans. Microwave Theory Tech.*, vol. 48, no. 10, pp. 1743–1748, Oct. 2000.

[81] F. Zheng, Z. Chen, and J. Zhang. "Toward the development of a three-dimensional unconditionally stable finite-difference time-domain method." *IEEE Trans. Microwave Theory Tech.*, vol. 48, no. 9, pp. 1550–1558, Oct. 1999.

[82] I. Ahmed and Z. Chen. "Error reduced ADI-FDTD methods." *IEEE Antennas Wireless Propagat. Lett.*, vol. 4, pp. 323–325, 2005.

[83] T. Weiland. "A discretization method for the solution of Maxwell's equations for six-component fields." *AEÜ*, vol. 31, no. 3, pp. 116–120, Mar. 1977.

[84] E. Tonti. "On the geometrical structure of electromagnetism." In *Gravitation, Electromagnetism and Geometrical Structures for the 80th birthday of A. Lichnerowicz*, G. Gerrarese, ed. Bologna: Pitagora Editrice, 1995, pp. 281–308.

[85] E. Tonti. "Finite formulation of the electromagnetic field." *Progress in Electromagn. Res.*, vol. 32, pp. 1–44, 2001.

[86] M. Clemens and T. Weiland. "Discrete electromagnetism with the finite integration technique." *Progress in Electromagn. Res.*, vol. 32, pp. 65–87, 2001.

[87] R. Schuhmann and T. Weiland. "Conservation of discrete energy and related laws in the finite integration technique." *Progress in Electromagn. Res.*, vol. 32, pp. 301–316, 2001.

[88] A. Bossavit. "Computational electromagnetism and geometry: Building a finite-dimensional "Maxwell's house." (1) Network equations, (2) Network constitutive laws, (3) Convergence, (4) From degrees of freedom to fields, (5) The Galerkin hodge, (6) Some questions and answers," *Journal of Japan Society of Applied Electromagnetics and Mechanics*, (1) vol. 7, pp. 150–159, 1999, (2) vol. 7, pp. 294–301, 1999, (3) vol. 7, pp. 401–408, 1999, (4) vol. 8, pp. 102–109, 2000, (5) vol. 8, pp. 203–209, 2000, (6) vol. 8, pp. 372–377, 2000.

[89] E. Tonti. "A direct discrete formulation of field laws: The cell method." *CMES*, vol. 2, no. 2, pp. 237–258, 2001.

[90] M. Marrone. "Computational aspects of cell method in electrodynamics." *Progress in Electromagn. Res.*, vol. 32, pp. 317–356, 2001.

[91] M. Cinalli, F. Edelvik, R. Schuhmann, and T. Weiland. "Consistent material operators for tetrahedral grids based on geometrical principles." *Int. J. Numeric. Model.*, vol. 17, pp. 487–507, 2004.

[92] M. Marrone. "A new consistent way to build symmetric constitutive matrices on general 2-D grids." *IEEE Trans. Magn.*, vol. 40, no. 2, pp. 1420–1423, Mar. 2004.

[93] C. Christopolous. *The Trasmission-Line Modeling (TLM) Method.* Piscataway, NJ: IEEE Press, 1995.

[94] P. B. Johns and R. L. Beurle. "Numerical solution of 2-dimensional scattering problem using a transmission-line matrix." *Proc. IEE*, vol. 118, no. 9, pp. 1203–1208, Sep. 1971.

[95] S. Akhtarzad and P. B. Johns. "Generalized elements for TLM method of numerical analysis." *Proc. IEE*, vol. 122, no. 12, pp. 1349–1352, Dec. 1975.

[96] S. Akhtarzad and P. B. Johns. "Solution of Maxwell's equations in three space dimensions and time by the TLM method of numerical analysis." *Proc. IEE*, vol. 122, no. 12, pp. 1344–1348, Dec. 1975.

[97] S. Akhtarzad and P. B. Johns. "Three-dimensional transmission-line matrix computer analysis of microstrip resonators." *IEEE Trans. Microwave Theory Tech.*, vol. 23, no. 12, pp. 990–997, Dec. 1975.

[98] P. B. Johns. "A symmetrical condensed node for the TLM method." *IEEE Trans. Microwave Theory Tech.*, vol. 35, no. 4, pp. 370–377, Apr. 1987.

[99] P. B. Johns. "The solution of inhomogeneous waveguide problems using a transmission-line matrix." *IEEE Trans. Microwave Theory Tech.*, vol. 22, no. 3, pp. 209–215, Mar. 1974.

[100] R. Allen, A. Mallik, and P. B. Johns. "Numerical results for the symmetrical condensed TLM node." *IEEE Trans. Microwave Theory Tech.*, vol. 35, no. 4, pp. 378–382, Mar. 1987.

[101] R. A. Scaramuzza and A. J. Lowery. "Hybrid symmetrical condensed node for TLM method." *Electron. Lett.*, vol. 26, no. 23, pp. 1947–1949, Nov. 1990.

[102] P. Berrini and K. Wu. "A pair of hybrid symmetrical condensed TLM nodes." *IEEE Microwave Guided Wave Lett.*, vol. 4, no. 7, pp. 244–246, Jul. 1994.

[103] V. Trenkic, C. Christopoulos, and T. M. Benson. "Theory of the symmetrical super-condensed node for the TLM method." *IEEE Trans. Microwave Theory Tech.*, vol. 43, no. 6, pp. 1342–1348, Jun. 1995.

[104] V. Trenkic, C. Christopoulos, and T. M. Benson. "Optimization of TLM schemes based on the general symmetrical condensed node." *IEEE Trans. Antennas Propagat.*, vol. 45, no. 3, pp. 457–465, Mar. 1997.

[105] V. Trenkic, C. Christopoulos, and T. M. Benson. "Advanced node formulation in TLM—The adaptable symmetric condensed node." *IEEE Trans. Microwave Theory Tech.*, vol. 44, no. 12, pp. 2473–2478, Dec. 1996.

[106] J. Paul, C. Christopoulos, and D. W. P. Thomas. "Perfectly matched layer for transmission line modeling (TLM) method." *Electron. Lett.*, vol. 33, no. 9, pp. 729–730, Apr. 1997.

[107] J.-L. Dubard and D. Pompei. "Optimization of the PML efficiency in 3-D TLM method." *IEEE Trans. Microwave Theory Tech.*, vol. 48, no. 7, pp. 1081–1088, Jul. 2000.

[108] J. L. Herring and C. Christopoulos. "Solving electromagnetic field problems using a multiple grid transmission-line modeling method." *IEEE Trans. Antennas Propagat.*, vol. 42, no. 12, pp. 1654–1658, Dec. 1992.

[109] P. Argus, P. Fischer, A. Konrad, and A. J. Schwab. "Efficient modeling of apertures in thin conducting screens by the TLM method." *Proc. 2000 IEEE Int. Symp. Electromagnetic Compatibility*, Washington, DC, 21-25 Aug. 2000, pp. 101–106.

[110] A. E. Ruehli. "Equivalent circuit models for three-dimensional multiconductor systems." *IEEE Trans. Microwave Theory Tech.*, vol. 22, no. 3, pp. 216–221, Mar. 1974.

[111] A. E. Ruehli and H. Heeb. "Circuit models for three-dimensional geometries including dielectrics." *IEEE Trans. Microwave Theory Tech.*, vol. 40, no. 7, pp. 1507–1516, Jul. 1992.

[112] H. Heeb and A. E. Ruehli. "Approximate time-domain models of three-dimensional interconnects." *Proc. 1990 IEEE Int. Conf. Computer Design*. Cambridge, MA, 17–19 Sep. 1990, pp. 201–205.

[113] A. E. Ruehli. "Partial element equivalent circuit (PEEC) method and its application in the frequency and time domain." *Proc. 1996 IEEE Int. Symp. Electromagnetic Compatibility*. Santa Clara, CA, 19–23 Aug. 1996, pp. 128–133.

[114] A. E. Ruehli. "Circuit models for 3D structures with incident fields." *Proc. 1993 IEEE Int. Symp. Electromagnetic Compatibility*. Dallas, TX, 9–13 Aug. 1993, pp. 28–32.

[115] A. E. Ruehli, G. Antonini, J. Ekman, and A. Orlandi. "Nonorthogonal PEEC formulation for time- and frequency-domain EM and circuit modeling." *IEEE Trans. Electromagn. Compat.*, vol. 45, no. 2, pp. 167–176, May 2003.

[116] C. W. Ho, A. E. Ruehli, and P. A. Brennan. "The modified nodal approach to network analysis." *IEEE Trans. Circuits Syst.*, vol. 22, no. 6, pp. 504–509, Jun. 1975.

[117] J. Garrett, A. Ruehli, and C. Paul. "Efficient frequency domain solutions for PEEC EFIE for modeling 3D geometries." *Proc. 1995 Int. Zurich Symp. Electromagnetic Compatibility*. Zurich, Switzerland, Mar. 1995, pp. 179–184.

[118] J. Ekman, G. Antonini, A. Orlandi, and A. E. Ruehli. "Impact of partial element accuracy on PEEC model stability." *IEEE Trans. Electromagn. Compat.*, vol. 48, no. 1, pp. 19–32, Feb. 2006.

[119] A. E. Ruehli, U. Miekkala, and H. Heeb. "Stability of discretized partial element equivalent EFIE circuit models." *IEEE Trans. Antennas Propagat.*, vol. 43, no. 6, pp. 553–559, Jun. 1995.

[120] J. E. Garret, A. E. Ruehli, and C. R. Paul. "Accuracy and stability improvements of integral equation models using the partial element equivalent circuit (PEEC) approach." *IEEE Trans. Antennas Propagat.*, vol. 46, no. 12, pp. 1824–1832, Dec. 1998.

CHAPTER SIX

Apertures in Planar Metal Screens

The transmission of the EM field through an aperture in a planar conducting screen is a canonical problem that has attracted a great attention in the EM community (see [1,2], and references therein). This problem is in fact a first step in the study of many practical issues arising in different areas of electromagnetics, and in particular, in EMC. In the framework of EMC, such a classical problem is directly related to the design of an efficient shielding metallic enclosure. Shielding enclosures are usually employed to protect against radiation from external EM fields and leakage effects from interior components. However, the efficiency of these enclosures is often compromised by apertures and slots located on the walls of the enclosure; such apertures may be intentional (e.g., necessary for ventilation purposes or access to interior equipments) or not (e.g., cracks around plates covering access ports). In any case, such openings allow coupling between external and internal fields, thus affecting and possibly deteriorating the desired performance of the enclosure. Of course, realistic problems in EMC involve apertures in finite, possibly nonplanar, screens and loaded by a cavity or other 3D objects. However, very often, the consideration of an infinite flat screen in place of a finite curved one may not considerably affect the accuracy of the solution. The problem of loading is a very serious one and will be treated in the next chapter. In this chapter we aim at studying the effects of apertures on the SE of an infinite planar conducting screen subject to a time-harmonic EM field.

As it can be expected, even such a simple problem depends on a number of factors, such as, the shape of the aperture, the size of the aperture, the frequency of operation, and the characteristics of the incident EM field. After many years such a problem still remains, in the most general case, a quite challenging one [3–5] whose analytical solution is available only for an incident plane wave impinging on an infinitesimally

---

Electromagnetic Shielding by Salvatore Celozzi, Rodolfo Araneo and Giampiero Lovat

thin circular aperture (in this case the solution is expressed as an expansion in terms of oblate spheroidal vector wavefunctions [6]). An excellent discussion of this fundamental problem and of its classical formulations is given in [2].

Usually apertures of interest in EMC are electrically small (i.e., the maximum aperture dimension is small compared to the operating wavelength). The "small-aperture problem" is quite well understood and, subject to a "low-frequency" approximation, some general conclusions can be drawn. However, as can be guessed, the field penetration increases with increasing frequency, and it is maximum at the resonant frequency of the aperture (i.e., when the aperture acts as a resonant slot antenna).

6.1 HISTORICAL BACKGROUND

The problem of EM transmission through an aperture in an infinite planar conducting screen is essentially a direct EM scattering (diffraction) problem. From a mathematical point of view, such a problem is a well-posed boundary-value problem; that is, it always admits a unique solution, and the involved operator is continuous. As is well known, there is not a general procedure for solving a boundary-value problem, so several numerical or approximate analytical methods have been developed and used through the years. These methods can be usefully classified by dividing the frequency domain into three ranges:

1. The *high-frequency region*, where the wavelength is small compared with the aperture dimensions.
2. The *intermediate- (resonance-) frequency region*, where the wavelength is of the same order as the aperture dimensions.
3. The *low-frequency region*, where the aperture dimensions are small with respect to the wavelength.

In the high-frequency region (i.e., very large apertures), the first systematic attempt to treat the aperture diffraction problem was made by Kirchhoff [7], based on Huygens's principle. Such an approach, although mathematically inconsistent in its initial formulation, works pretty well in the frequency domain of interest; the mathematical inconsistencies were then soon removed by Sommerfeld [8] and Rayleigh [9]. However, since the theory is essentially scalar, Kirchhoff's approach cannot account for the polarization of the EM field. A vector equivalent of the scalar Kirchhoff's formulation was introduced by Stratton and Chu [10] and Schelkunoff [11]; Smythe was then the first to obtain a solution for the diffracted electric field in terms of the tangential electric field in the aperture of a perfectly conducting screen [12]. Other techniques based on Kottler's theory [13] introduce additional contour integrals along the rim of the aperture (representing the effects of fictitious line charges) to take into account the vectorial nature of the problem. Usually, in the high-frequency region, asymptotic techniques are widely used (e.g., physical optics, geometric theory of diffraction [14], uniform theory of diffraction [15], and physical theory of diffraction [16]).

In the low-frequency region (small apertures), Lord Rayleigh was the first to propose a solution of the problem [17]: the solution procedure was based on a series expansion in ascending powers of the wavenumber of certain quantities (the so-called Rayleigh's series) and it has been shown that it leads to a sequence of simple integral equations with a kernel of the electrostatic type [2]. In a famous paper Bethe studied the low-frequency EM scattering by a small circular hole cut in an infinite perfectly conducting plane [18], and by using a scalar potential approach, he derived the leading terms of the Rayleigh series. Bouwkamp studied the same problem in a more rigorous way using a complicated system of integro-differential equations, and he found some errors and incorrect results in Bethe's solution [1]. An alternative use of Rayleigh's series expansion has been discussed by Stevenson [19], Kleinman (who corrected and simplified Stevenson's work) [20], and Eggimann [21]. An interestingly elegant variational formulation of EM diffraction problems for planar apertures, which allowed for approximate but accurate numerical evaluations of the scattered fields in a wide range of frequencies, was provided by Levine and Schwinger [22].

In the intermediate- (resonance-) frequency region, where a rigorous full-wave analysis is needed, the integral-equation method is the most common approach since both the radiation and the boundary conditions are implicitly taken into account in the formulation. Usually most of the relevant numerical techniques are based on the application of the method of moments [2].

6.2 STATEMENT OF THE PROBLEM

The basic EM problem that we refer to consists in finding the transmitted EM field through a planar infinitesimally thin and perfectly conducting screen S of infinite extent perforated with a finite aperture A, as shown in Figure 6.1, and due to

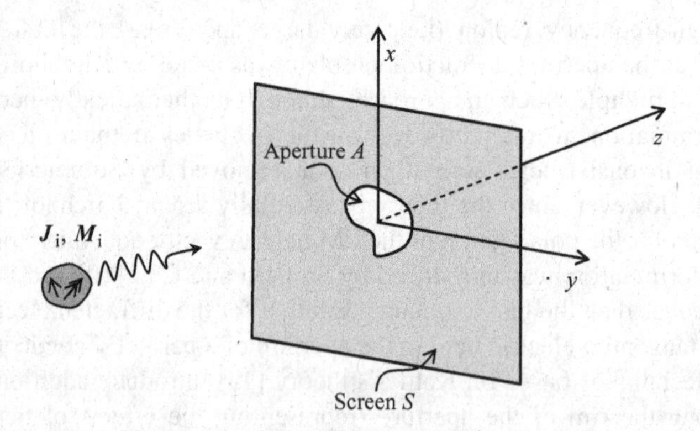

FIGURE 6.1 Transmission of the EM field through a finite aperture A cut in a planar perfectly conducting screen S of infinite extent and zero thickness.

time-harmonic electric and magnetic impressed sources ($\mathbf{J}_i$ and $\mathbf{M}_i$, respectively): in particular, the screen *S* coincides with the *xy* plane of a Cartesian coordinate system and the EM sources are assumed located in the half-space $z < 0$. It should be noted that consideration of a perfectly conducting screen is not a limiting factor in the analysis of the shielding properties of a planar metal screen; the field transmitted through a small aperture is in fact usually much larger than the field penetrating the screen due to a finite conductivity of the screen itself.

Electromagnetic theory requires that the solution of the scattering problem in the half-space $z > 0$ (i.e., the transmitted EM field) satisfy Maxwell's equations, boundary conditions on the screen and on the aperture, and the radiation (Silver-Müller) condition:

$$\lim_{r\to\infty} r(\mathbf{E} \times \mathbf{u}_r + \eta_0 \mathbf{H}) = \mathbf{0} \qquad \forall \mathbf{u}_r. \tag{6.1}$$

6.3 LOW-FREQUENCY ANALYSIS: TRANSMISSION THROUGH SMALL APERTURES

As mentioned above, the problem of EM transmission through a small circular aperture in a planar screen was solved by Bethe by means of electrostatic and magnetostatic approximations of the fields at the aperture [18]. In particular, in Bethe's theory for small apertures, the field scattered by the aperture (the first-order diffracted field) can be represented as the superposition of an electric-dipole field and a magnetic-dipole field due to electric and magnetic dipole sources, respectively, both located at the center of the aperture; the electric dipole $\mathbf{p}_e$ is proportional to the component normal to the screen of the electric field, while the magnetic dipole $\mathbf{p}_m$ is proportional to the component parallel to the screen of the magnetic field through the so-called *electric* and *magnetic aperture polarizabilities*, respectively.

In Bethe's approximation, the dipole moments can be determined from the knowledge of the incident field, and therefore the diffracted field due to the presence of the small aperture can easily be determined. In particular, for the problem of Figure 6.1, where *A* is a small aperture, the electric polarizability α_e is defined by

$$\mathbf{p}_e = \mathbf{u}_z p_e = \mathbf{u}_z \varepsilon_0 \alpha_e E_z^{\text{inc}}(\mathbf{0}), \tag{6.2}$$

and the magnetic polarizability $\underline{\boldsymbol{\alpha}}_m$ by

$$\mathbf{p}_m = -\mu_0 \underline{\boldsymbol{\alpha}}_m \cdot \mathbf{H}^{\text{inc}}(\mathbf{0}). \tag{6.3}$$

Note that while α_e is a scalar quantity, $\underline{\boldsymbol{\alpha}}_m$ is, in general, a dyadic. If the aperture has lines of symmetry, the magnetic-polarizability dyadic is diagonal. Polarizabilities have been numerically studied for small apertures of different shapes [23,24]. In

(6.2) and (6.3), the fields $\mathbf{E}^{\text{inc}}$ and $\mathbf{H}^{\text{inc}}$ are the *incident* electric and magnetic field, namely, those fields due to $\mathbf{J}_{\text{i}}$ and $\mathbf{M}_{\text{i}}$ that would exist in the absence of the screen. It should be pointed out that expressions (6.2) and (6.3) give moments of the equivalent dipoles *radiating in free space*.

It should also be noted that Bethe's theory does not provide a solution for the radiation conductance of the aperture. When one tries to determine an equivalent network to represent the small aperture, a physically meaningful representation cannot be derived because power is not conserved; to overcome such difficulties, a solution has been proposed by Collin [25] and by Harrington and Mautz [26].

As mentioned in Section 6.2, an exact solution for the circular-aperture case was found by Flammer [6]. Bouwkamp [1] also showed that the approximate Bethe's solution gives correct results for the distant field. However, he also showed that Bethe's solution does not reproduce the correct behavior of the EM field at the edges and does not provide a correct approximate field near the aperture.

According to Bethe's theory, for observation points far from A (i.e., for distances large compared to the aperture dimensions), the electric and magnetic fields in the half-space $z > 0$ are approximately

$$\mathbf{E} = \mathbf{E}_{\text{e}} + \mathbf{E}_{\text{m}}, \quad \mathbf{H} = \mathbf{H}_{\text{e}} + \mathbf{H}_{\text{m}}, \tag{6.4}$$

where $\mathbf{E}_{\text{e}}$ and $\mathbf{E}_{\text{m}}$ and $\mathbf{H}_{\text{e}}$ and $\mathbf{H}_{\text{m}}$ are the electric and magnetic fields due to the equivalent electric and magnetic dipole moments, respectively, given by

$$\mathbf{E}_{\text{e}}(\mathbf{r}) = \frac{p_{\text{e}}}{4\pi\varepsilon_0}\left(k_0^2\mathbf{u}_z + \nabla\frac{\partial}{\partial z}\right)\frac{e^{-jk_0r}}{r} \tag{6.5}$$

and

$$\mathbf{E}_{\text{m}}(\mathbf{r}) = -\frac{j\omega}{4\pi}\left(\nabla\frac{e^{-jk_0r}}{r}\right) \times \mathbf{p}_{\text{m}}. \tag{6.6}$$

6.4 THE SMALL CIRCULAR-APERTURE CASE

To gain a physical insight into the penetration of the EM field through a small aperture in a perfectly conducting planar screen, let us consider the particular case where the small aperture has a circular shape. As mentioned above, such a problem has been solved in an exact way by means of an expansion in spheroidal vector wavefunctions [6]. Basically the complementary disk problem is first solved (the disk is viewed as a limiting case of an ellipsoid), and next, by an application of the Babinet principle [27], the solution for the circular aperture is obtained. The circular aperture is the one with the most symmetric shape; in such a special case, the magnetic-polarizability dyadic reduces to a simple scalar quantity. Moreover, expressions for the electric and magnetic polarizabilities are known in a simple closed form. In particular, for a circular aperture of radius a the results are

$$\alpha_{\mathrm{e}} = \frac{8}{3}a^3, \quad \alpha_{\mathrm{m}_{xx}} = \alpha_{\mathrm{m}_{yy}} = \frac{16}{3}a^3, \quad \alpha_{\mathrm{m}_{xy}} = \alpha_{\mathrm{m}_{yx}} = 0. \tag{6.7}$$

We will consider a uniform plane wave impinging on a perfectly conducting planar screen with a small circular aperture of radius a. As is well known, any uniform plane wave can be decomposed into the sum of one TE and one TM uniform plane wave (in this case the attributes TE and TM are referred to the normal to the screen, i.e., the z axis). The problem is depicted in Figure 6.2*a*, with the relevant physical and geometrical parameters.

According to Bethe's theory, TE polarization gives rise to the presence of an equivalent magnetic dipole only, whereas TM polarization produces both types of equivalent dipole moments. From (6.2), (6.3), and (6.7) we have

$$\begin{aligned} p_{\mathrm{e}}^{\mathrm{TE}} &= 0, \qquad \mathbf{p}_{\mathrm{m}}^{\mathrm{TE}} = -\mu_0 \frac{16}{3} a^3 H_{\mathrm{inc}}^{\mathrm{TE}} \cos\theta_{\mathrm{inc}}^{\mathrm{TE}} \mathbf{u}_{\mathrm{t}}, \\ p_{\mathrm{e}}^{\mathrm{TM}} &= -\varepsilon_0 \frac{8}{3} a^3 E_{\mathrm{inc}}^{\mathrm{TM}} \sin\theta_{\mathrm{inc}}^{\mathrm{TM}}, \qquad \mathbf{p}_{\mathrm{m}}^{\mathrm{TM}} = -\mu_0 \frac{16}{3} a^3 \mathbf{H}_{\mathrm{inc}}^{\mathrm{TM}}, \end{aligned} \tag{6.8}$$

where $\mathbf{u}_{\mathrm{t}}$ is the unit vector along the direction of the magnetic-field component tangential to the aperture plane.

From (6.5) and (6.6) (and the dual ones for the magnetic field), the EM field transmitted through the small circular aperture in the half-space $z > 0$ can easily be calculated, both in the near and in the far field. In particular, following Jaggard [28], the ratio of the near-field transmitted energy density W^{tr} to the

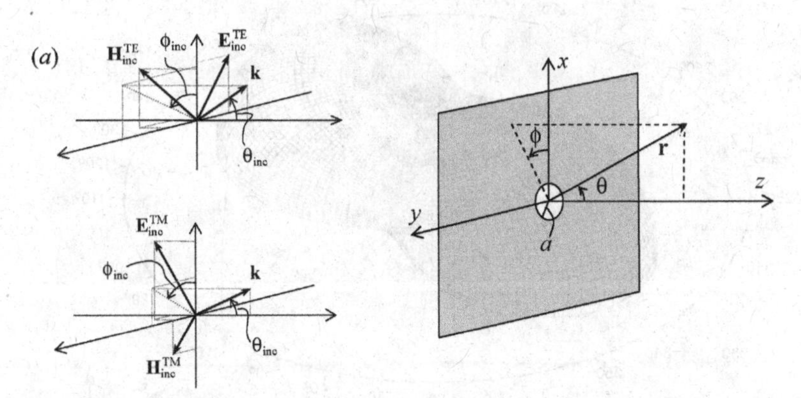

FIGURE 6.2 TE and TM uniform plane-wave incidence on a perfectly conducting infinitesimally thin planar screen of infinite extent with a small circular aperture. (*a*) Near-field energy distribution for TE (*b*) and TM (*c*) plane-wave incidence. SE for TE (*d*) and TM (*e*) plane-wave incidence.

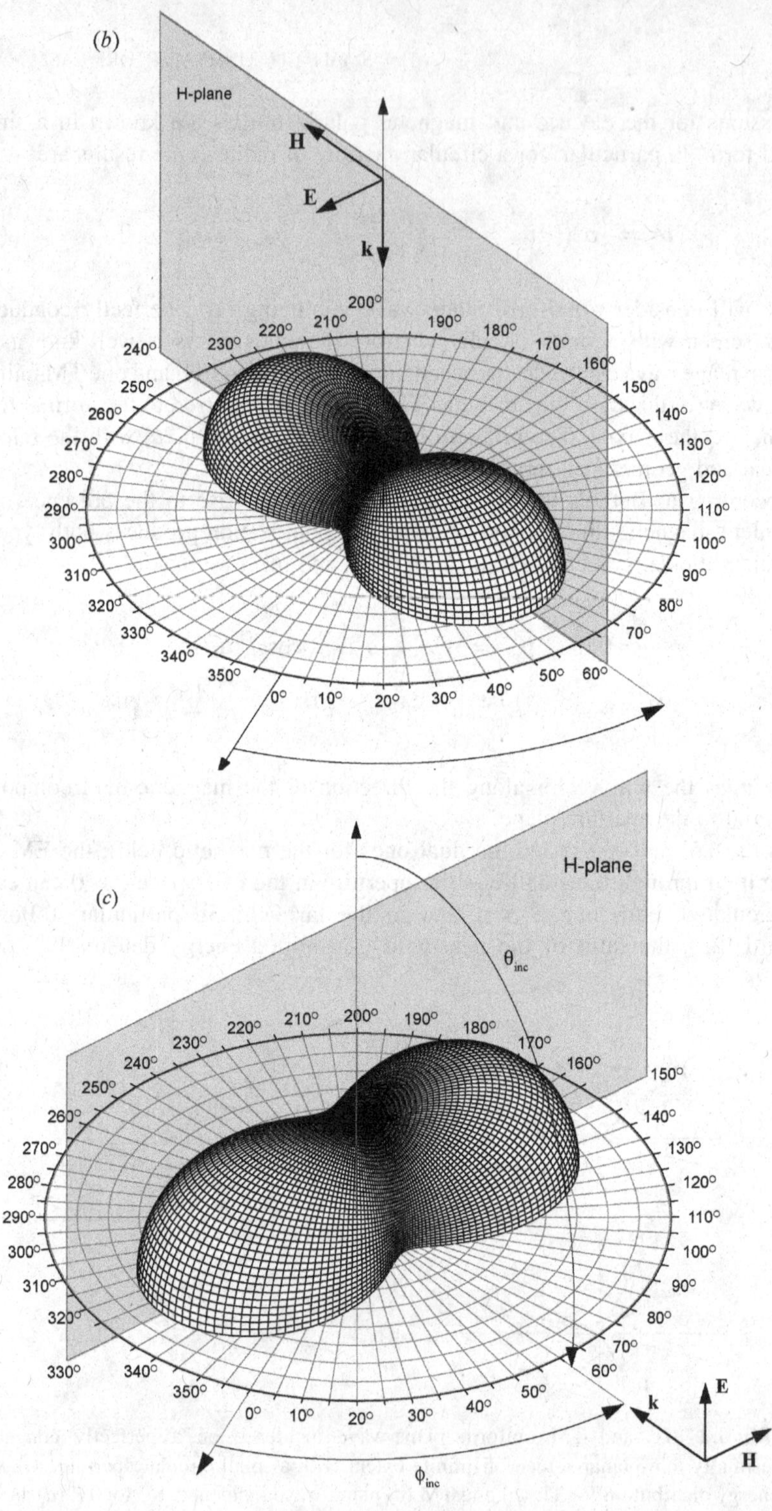

FIGURE 6.2 (*Continued*)

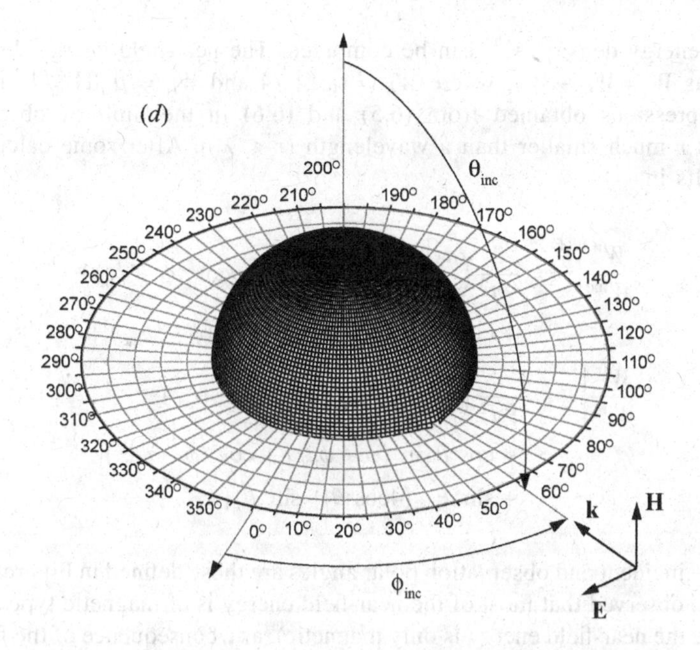

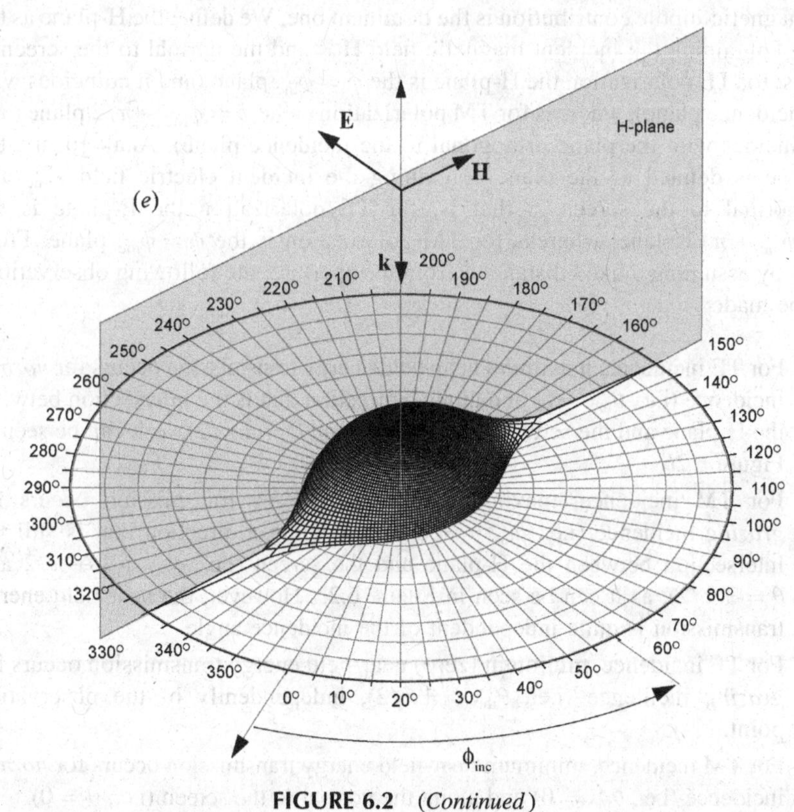

FIGURE 6.2 (*Continued*)

incident energy density W^{inc} can be computed. The near-field energy density is defined as $W = W_E + W_H$, where $W_E = \varepsilon_0|\mathbf{E}|^2/4$ and $W_H = \mu_0|\mathbf{H}|^2/4$ and with fields expressions obtained from (6.5) and (6.6) in the limit of observation distances r much smaller than a wavelength ($r \ll \lambda_0$). After some calculations this results in

$$\begin{aligned}
\left.\frac{W^{\text{tr}}}{W^{\text{inc}}}\right|^{\text{TE}} &= \frac{2}{9\pi^2}\left(\frac{a}{r}\right)^6 \{4\cos^2\theta_{\text{inc}}[4\sin^2\theta\cos^2(\phi-\phi_{\text{inc}}) \\
&\quad + \cos^2\theta\cos^2(\phi-\phi_{\text{inc}}) + \sin^2(\phi-\phi_{\text{inc}})]\}, \\
\left.\frac{W^{\text{tr}}}{W^{\text{inc}}}\right|^{\text{TM}} &= \frac{2}{9\pi^2}\left(\frac{a}{r}\right)^6 \{4[4\sin^2\theta\sin^2(\phi-\phi_{\text{inc}}) \\
&\quad + \cos^2\theta\sin^2(\phi-\phi_{\text{inc}}) + \cos^2(\phi-\phi_{\text{inc}})] \\
&\quad + \sin^2\theta_{\text{inc}}(4\cos^2\theta + \sin^2\theta)\},
\end{aligned} \tag{6.9}$$

where the incident and observation polar angles are those defined in Figure 6.2*a*. It should be observed that most of the near-field energy is of magnetic type (for TE incidence the near-field energy is only magnetic), as a consequence of the fact that the magnetic-dipole contribution is the dominant one. We define the H-plane as the plane containing the incident magnetic field $\mathbf{H}_{\text{inc}}$ and the normal to the screen z: that is, for TE polarization the H-plane is the $\phi = \phi_{\text{inc}}$ plane (and it coincides with the incidence plane), whereas for TM polarization is the $\phi = \phi_{\text{inc}} + \pi/2$ plane (and it coincides with the plane orthogonal to the incidence plane). Analogously, the E-plane is defined as the plane containing the incident electric field $\mathbf{E}_{\text{inc}}$ and the normal to the screen z: that is, for TE polarization the E-plane is the $\phi = \phi_{\text{inc}} + \pi/2$ plane, whereas for TM polarization is the $\phi = \phi_{\text{inc}}$ plane. From (6.9), by assuming a fixed distance r from the aperture, the following observations can be made:

1. For TE incidence, maximum near-field energy transmission occurs for *normal* incidence (i.e., $\theta_{\text{inc}} = 0$) and along a direction that is the intersection between the H-plane and the screen (i.e., $\phi = \phi_{\text{inc}}$ and $\theta = \pm\pi/2$), as it can be seen in Figure 6.2*b*.
2. For TM incidence, maximum near-field energy transmission occurs for *grazing* incidence (i.e., $\theta_{\text{inc}} = \pm\pi/2$) and along a direction that is still the intersection between the H-plane and the screen (i.e., $\phi = \phi_{\text{inc}} + \pi/2$ and $\theta = \pm\pi/2$), as it can be seen in Figure 6.2*c*. However, the near-field energy transmission is quite independent of the incidence angle.
3. For TE incidence, minimum (zero) near-field energy transmission occurs for *grazing* incidence (i.e., $\theta_{\text{inc}} = \pm\pi/2$), independently of the observation point.
4. For TM incidence, minimum near-field energy transmission occurs for *normal* incidence (i.e., $\theta_{\text{inc}} = 0$) and along the normal to the screen (i.e., $\theta = 0$).

As was already pointed out, Bethe's theory is not rigorously correct for points too close to the aperture. However, Jaggard has shown that for observation distances larger than the aperture diameter and in the normal-incidence case, the approximate results above are in good agreement with the exact ones obtained by Bouwkamp (in particular, the maximum error is less than 2% for $r > 2a$ on the E- and H-planes). Moreover the validity of the approximate expressions holds also for arbitrary polarizations and angles of incidence [28].

The SE can be considered as a figure of merit to quantify the penetration of the EM field through the aperture. In particular, it should be recalled that the SE at a certain point of the space is defined as the ratio (in dB) of the incident electric (or magnetic) field (i.e., the field in the absence of the screen) divided by the electric (or magnetic) field in the presence of the perforated screen. Since Bethe's solution is certainly correct for large distances from the aperture (i.e., $r \gg \lambda_0$), accurate approximate expressions for the SE of the screen can be derived. In particular, from (6.5) and (6.6), in the limit $r \gg \lambda_0$, after lengthy calculations there results

$$\mathrm{SE}^{\mathrm{TE}} = -20\log\left[\frac{4k_0^2a^3}{3\pi r}\cos\theta_{\mathrm{inc}}\sqrt{1-\sin^2\theta\cos^2(\phi-\phi_{\mathrm{inc}})}\right],$$

$$\mathrm{SE}^{\mathrm{TM}} = -20\log\left\{\frac{4k_0^2a^3}{3\pi r}\right.$$

$$\left.\cdot\sqrt{\cos^2\theta+\sin^2\theta\left[\cos^2(\phi-\phi_{\mathrm{inc}})+\frac{1}{4}\sin^2\theta_{\mathrm{inc}}\right]-\sin\theta\cos^2(\phi-\phi_{\mathrm{inc}})\sin\theta_{\mathrm{inc}}}\right\}. \tag{6.10}$$

Since a nonzero electric field beyond the screen (i.e., in the half-space $z > 0$) is due only to the presence of the aperture, (6.10) can be taken as the expressions of the SE of a small circular aperture. It is immediate to see that the opposites of (6.10) represent the far-field energy distribution under TE and TM incidence, respectively. Based on (6.10), the following observations can be made:

1. For TE incidence, the minimum SE (maximum energy transmission) occurs for *normal* incidence (i.e., $\theta_{\mathrm{inc}} = 0$) and along two main directions: normal to the screen (i.e., $\theta = 0$) or along the intersection between the E-plane and the screen (i.e., $\phi = \phi_{\mathrm{inc}} + \pi/2$ and $\theta = \pm\pi/2$).
2. For TM incidence, the minimum SE (maximum energy transmission) occurs for grazing incidence (i.e., $\theta_{\mathrm{inc}} = \pi/2$), on the E-plane (i.e., $\phi = \phi_{\mathrm{inc}}$), and in a direction opposite to the direction of incidence (i.e., $\theta = -\theta_{\mathrm{inc}} = -\pi/2$).
3. For TE incidence, the maximum SE (minimum energy transmission) occurs for *grazing* incidence (i.e., $\theta_{\mathrm{inc}} = \pi/2$) independently of the observation

point, as it can be seen in Figure 6.2*d*. In particular, for this case the energy transmission is zero.

4. For TM incidence, the maximum SE (minimum energy transmission) occurs for *normal* incidence (i.e., $\theta_{\text{inc}} = 0$) and along a direction that is the intersection between the plane orthogonal to the incidence plane and the screen (i.e., $\phi = \phi_{\text{inc}} + \pi/2$ and $\theta = \pi/2$), as it can be seen in Figure 6.2*e*. Also for this case the energy transmission is zero.
5. For both polarizations, the far-field energy distribution is dramatically different from the near-field energy distribution.

6.5 SMALL NONCIRCULAR APERTURES

Expressions for the aperture polarizabilities of elliptical apertures have been derived in closed form, although they involve elliptical integrals of first and second kind. The relevant expressions are reported in Figure 6.3 together with the relevant geometrical parameters.

Much simpler expressions can be obtained under the assumption of narrow ellipses [2]. However, for other aperture shapes, simple expressions for the magnetic and electric polarizabilities are not available and therefore numerical evaluations are usually required [23,24]. In addition the magnetic polarizability is in general a nondiagonal dyadic. In any case, it can be shown that minimum shielding always occurs when the incident magnetic field is polarized along the major axis of the aperture (i.e., for TM polarization) for *grazing* incidence and in a direction *opposite* to that of the incident wave (i.e., $\theta = -\theta_{\text{inc}} = -\pi/2$). For *unpolarized* fields (i.e., EM waves in which the two perpendicular components of the electric field have equal average magnitudes and a random relative phase difference), lower and upper bounds for the polarizabilities can be derived [28] that can be effectively used to study the influence of the aperture shape on the SE of the perforated screen. In particular, it has been shown that for given area

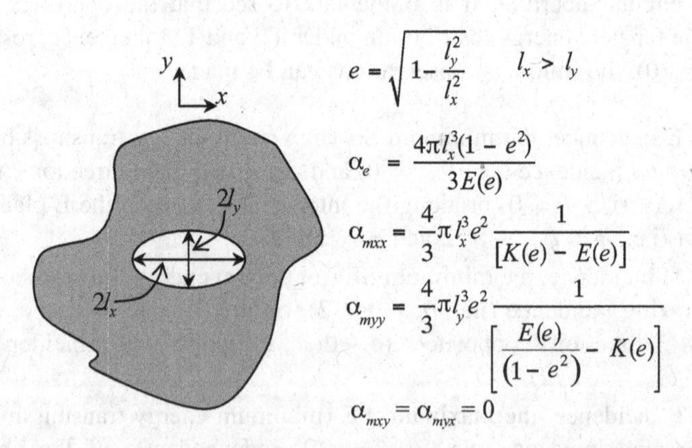

FIGURE 6.3 Electric and magnetic polarizabilities of an elliptical aperture.

and eccentricity (which has to be suitably defined for non-elliptical shapes), the aperture with the largest perimeter gives rise to the minimum SE [28]; the circular shape is thus the one which provides maximum SE.

6.6 FINITE NUMBER OF SMALL APERTURES

Let us now consider the presence of more apertures on a planar metal screen. Obviously the simplest treatment of a number of apertures is to consider them as being independent. However, it is clear that apertures in close proximity interact through a coupling mechanism. In particular, each aperture is actually excited not only by the incident field but also by the field diffracted by all the other apertures.

As a first example, let us a consider a screen perforated with only two equal and symmetrical closely spaced apertures (i.e., both characterized by the same scalar electric polarizability and by the same diagonal magnetic-polarizability dyadic, and placed at a distance d such that $k_0 d \ll 1$). Under suitable assumptions several interesting conclusions can be drawn. These simplifying assumptions are the following: First, the near fields of the aperture are assumed to be dipole near fields (for circular apertures of radius a, this assumption is adequate for $d \geq 3a$, while for noncircular apertures the approximation holds for $d \geq 1.5L$, where L is the typical dimension of the aperture). Second, in calculating the interaction fields, intermediate and far fields are neglected. Third, the interaction is assumed to be small. Finally, the interaction field is assumed to be constant over the aperture. It thus follows [28] that the problem can be described by characterizing each aperture with noninteracting dipole moments associated with polarizabilities slightly modified with respect to the case in which only one aperture is present; in particular, the EM coupling produces crossed magnetic polarizabilities (so that the magnetic-polarizability dyadic is not diagonal anymore). Depending on the polarization of the incident field, the coupling can either increase or decrease the components of the magnetic polarizability dyadic, while it always increases the scalar electric polarizability. The SE of each aperture can thus be either increased or decreased with respect to its value when each aperture acts individually. In particular, when the incident magnetic field is parallel to the line connecting the apertures, the SE of each aperture decreases, whereas when the incident magnetic field is orthogonal to the line connecting the apertures, the SE of each aperture increases.

The previous conclusions can be generalized to include the case of transmission through many small apertures. In particular, a row of N identical equidistant apertures can be considered, as shown in Figure 6.4*a*. When considering this problem, two main differences arise with respect to the two-aperture case. One is obvious, that the interaction among *all* the apertures has to be taken into account. This gives rise to an additional modification of the aperture polarizabilities. However, it should be noted that most of the coupling is due to adjacent apertures. Therefore, with respect to the two-apertures case, the coupling is approximately doubled. The further effect of all the other

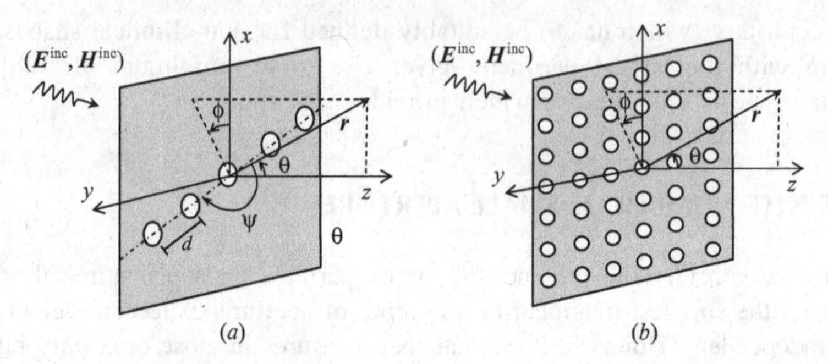

FIGURE 6.4 Row of N equidistant circular small apertures in a planar perfectly conducting screen of infinite extent (*a*) and double periodic array of small circular apertures (*b*).

apertures can then be taken into account through an effective coupling coefficient [28]. The second and most important difference lies in the fact that the radiated far field is essentially due to an *array* of apertures, and this can lead to dramatically different results with respect to the two-aperture case. In fact, while in the presence of only two closely spaced apertures (i.e., $k_0d \ll 1$), the far-field energy distribution is similar to that of a single aperture, for a large number N of apertures it can result $k_0Nd \geq 1$, so a crucial role can be played by phase effects. In such a case the transmission pattern of the single aperture (which already takes into account the presence of other apertures through modified polarizabilities) needs to be multiplied by a suitable array factor AF. For a large number N of apertures a simple array factor $\mathrm{AF} \simeq \sin\,(N\psi/2)/\sin(\psi/2)$ can be introduced [28] to adapt the results of the single aperture to the case of N apertures lying along a line that makes an angle ψ with the observation vector $\mathbf{r}$. A general observation can be made for arrays of apertures:

1. For apertures with small eccentricity, the SE is minimum for *grazing* incidence with the incident magnetic field polarized along the line of the array.
2. For apertures with large eccentricity, the SE is minimum for *grazing* incidence with the incident magnetic field polarized along the major axis of the apertures.

A type of metal screen perforated with more than one aperture is the 2D periodically perforated screen (i.e., a doubly periodic array of apertures) shown in Figure 6.4*b*. For this problem a basic step is to use Bethe's theory in the low-frequency region to replace each aperture with the corresponding electric and magnetic dipoles. As in the two-aperture case different approximations can be used to estimate the coupling effect. As a very first approximation, interactions between neighboring elements are simply neglected; for a more accurate analysis, in the case of small aperture spacing, the interaction field can be approximated by a static field [29]. A more refined model, useful when the aperture dimensions are an appreciable fraction of the wavelength, requires to evaluate the dynamic interaction field, and this needs the expression of higher order approximations for the aperture polarizabilities [30]. Numerical methods are needed to investigate the properties of periodically perforated screen in the intermediate- and high-frequency regions: this is an important topic closely related to frequency selective surfaces, subject of Chapter 10.

6.7 RIGOROUS ANALYSIS FOR APERTURES OF ARBITRARY SHAPE: INTEGRAL EQUATION FORMULATION

This section illustrates a general procedure used to derive an integral equation whose solution allows for the calculation of the EM field (due to time-harmonic sources) penetrating an aperture of arbitrary shape in a planar perfectly conducting infinitesimally thin screen of infinite extent in any frequency range [2]. The transverse aperture electric field $\mathbf{E}_t^{ap}$ (i.e., the component parallel to the screen of the *total* electric field at the aperture location) is chosen as the unknown of such an integral equation. Based on the equivalence principle, equivalent magnetic current densities $\mathbf{M}_S$ (proportional to $\mathbf{E}_t^{ap}$) are introduced, and the problem is split into two half-space problems (for $z < 0$ and $z > 0$, respectively). The magnetic field on both sides of the screen can thus be expressed as a function of $\mathbf{M}_S$ (and thus of $\mathbf{E}_t^{ap}$) through a superposition integral involving the half-space Green function of the electric vector potential. Such a representation is particularly convenient because it guarantees that Maxwell's equations, radiation condition, and boundary conditions on the screen are satisfied. Moreover the continuity of the tangential component of the electric field through the aperture is also automatically fulfilled (the tangential electric field at the aperture location is in fact $\mathbf{E}_t^{ap}$ for both the involved half-space problems). Therefore the only condition that must be imposed is the continuity of the magnetic field through the aperture: the sought-for integral equation follows directly from such a constraint.

The original problem is sketched in Figure 6.5*a*; based on equivalence principle, in Figure 6.5*b*, the aperture A is *short-circuited* (i.e., completely replaced by a perfectly conducting plate), and it constitutes the domain for the equivalent surface magnetic current density $\mathbf{M}_S = \mathbf{u}_z \times \mathbf{E}_t^{ap}$, which accounts for a nonzero value of the electric field at $z = 0^+$. By means of image theory, the problem of Figure 6.5*b* is then transformed into the one of Figure 6.5*c*. Since the problem of Figure 6.5*c* represents currents that radiate in free space, the magnetic field $\mathbf{H}^+$ in the half-space $z > 0$ can be written as

$$\mathbf{H}^+(\mathbf{r}) = \iint_A \underline{\mathbf{G}}^{HM}(\mathbf{r}, \mathbf{r}') \cdot 2\mathbf{M}_S(\mathbf{r}')\mathrm{d}S', \qquad z > 0, \qquad (6.11)$$

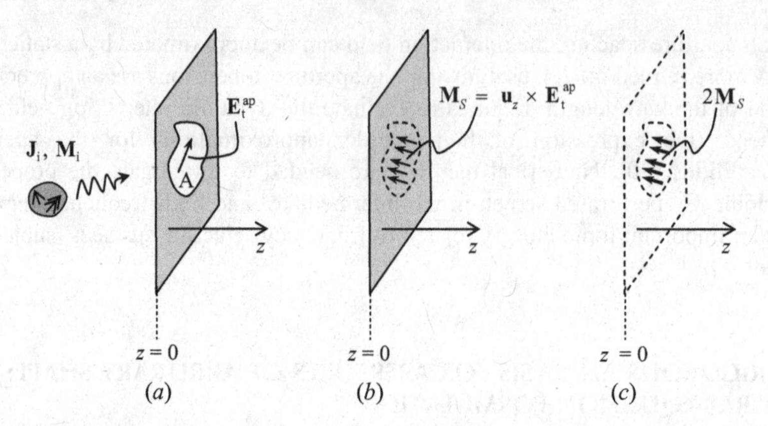

FIGURE 6.5 Original problem (*a*); application of equivalence principle (*b*); an application of image theory (*c*).

where $\underline{\mathbf{G}}^{HM}$ is the free-space dyadic Green function of the HM type (magnetic field **H** due to magnetic current **M**). By means of a similar argument, the magnetic field $\mathbf{H}^-$ in the half-space $z < 0$ can be written as

$$\mathbf{H}^-(\mathbf{r}) = \mathbf{H}^{sc}(\mathbf{r}) + \iint_A \underline{\mathbf{G}}^{HM}(\mathbf{r}, \mathbf{r}') \cdot [-2\mathbf{M}_S(\mathbf{r}')]\, dS', \qquad z < 0, \tag{6.12}$$

where $\mathbf{H}^{sc}$ is the *short-circuited* magnetic field, which is the field (due to the sources $\mathbf{J}_i$ and $\mathbf{M}_i$) that would be present in the half-space $z < 0$ with the aperture short-circuited; the current density $-2\mathbf{M}_S$ is instead the equivalent magnetic current for the half-space problem $z < 0$ (the normal to the aperture that defines the magnetic current density is reversed with respect to the half-space problem $z > 0$).

As mentioned above, the key condition that has to be imposed is the continuity of the tangential magnetic field through the aperture. This is accomplished by enforcing

$$\lim_{z \to 0^-} [\mathbf{H}^-(\mathbf{r}) \times \mathbf{u}_z] = \lim_{z \to 0^+} [\mathbf{H}^+(\mathbf{r}) \times \mathbf{u}_z] \qquad \forall \mathbf{r} \in A, \tag{6.13}$$

which, from (6.11) and (6.12), can be expressed as

$$2\left[\iint_A \underline{\mathbf{G}}^{HM}(\mathbf{r}, \mathbf{r}') \cdot \mathbf{M}_S(\mathbf{r}')\, dS'\right] \times \mathbf{u}_z = \mathbf{H}^{inc}(\mathbf{r}) \times \mathbf{u}_z \qquad \forall \mathbf{r} \in A. \tag{6.14}$$

Note that the magnetic field in (6.14) is the *incident* magnetic field, since, by virtue of image theory, $\mathbf{H}^{sc} \times \mathbf{u}_z = 2\mathbf{H}^{inc} \times \mathbf{u}_z$. Equation (6.14) represents the sought-for integral equation whose solution allows for evaluating the EM field everywhere in space.

An equivalent but more tractable integral equation can be derived by using the convenient formalism of the mixed potentials, thus obtaining the so-called mixed potential integral equation (MPIE) [31]. In this representation the magnetic field is expressed in terms of an electric vector potential $\mathbf{F}$ and a scalar magnetic potential W, associated with transverse magnetic current density $\mathbf{M}_S$ and magnetic charge density $\rho_{mS} = -\nabla \cdot \mathbf{M}_S/(j\omega)$, respectively. Usually the MPIE formulation of the problem is preferred because of the lower order singularity in the integral kernel. In particular, there results

$$\begin{aligned}\mathbf{H}(\mathbf{r}) &= \iint_A \underline{\mathbf{G}}^{HM}(\mathbf{r}, \mathbf{r}') \cdot \mathbf{M}_S(\mathbf{r}')\, dS' = -j\omega\varepsilon_0 \mathbf{F}(\mathbf{r}) - \nabla W(\mathbf{r}) \\ &= -j\omega\varepsilon_0 \iint_A \underline{\mathbf{G}}^F(\mathbf{r}, \mathbf{r}') \cdot \mathbf{M}_S(\mathbf{r}')\, dS' - \nabla \iint_A G^W(\mathbf{r}, \mathbf{r}') \frac{\rho_{mS}(\mathbf{r}')}{\mu_0}\, dS'.\end{aligned} \tag{6.15}$$

As is well known, the mixed-potential Green functions (see (1.40)) in free space are [31]

$$\underline{\mathbf{G}}^F(\mathbf{r}, \mathbf{r}') = \underline{\mathbf{I}} G(\mathbf{r}, \mathbf{r}'), \qquad G^W(\mathbf{r}, \mathbf{r}') = G(\mathbf{r}, \mathbf{r}'), \tag{6.16}$$

where $\underline{\mathbf{I}}$ is the identity dyadic and $G(\mathbf{r}, \mathbf{r}')$ is the scalar free-space Green function (see (1.40)) given by

$$G(\mathbf{r}, \mathbf{r}') = \frac{e^{-jk_0|\mathbf{r}-\mathbf{r}'|}}{4\pi|\mathbf{r}-\mathbf{r}'|}. \tag{6.17}$$

Equation (6.14) can thus be rewritten as

$$\begin{aligned}&\left[-j\omega\varepsilon_0 \iint_A \underline{\mathbf{G}}^F(\mathbf{r}, \mathbf{r}') \cdot \mathbf{M}_S(\mathbf{r}')\, dS' + \frac{1}{j\omega\mu_0} \nabla \iint_A G^W(\mathbf{r}, \mathbf{r}') \nabla' \cdot \mathbf{M}_S(\mathbf{r}') dS'\right] \times \mathbf{u}_z \\ &= \mathbf{H}^{\mathrm{inc}}(\mathbf{r}) \times \mathbf{u}_z, \quad \forall \mathbf{r} \in A\end{aligned} \tag{6.18}$$

which can be efficiently solved with standard MoM techniques. Recently a modified version of this integral-equation technique proved to be efficient in dealing with screens of finite thickness [4].

An important point to note is that edge conditions for the EM field require that the component of $\mathbf{M}_S$ *normal* to the aperture-screen edge approach zero as the square root of the distance between the observation point and the edge. However, the *tangential* component of $\mathbf{M}_S$ diverges as the inverse of this square-root distance, so suitable basis functions should be adopted to efficiently expand the unknown $\mathbf{M}_S$ [32].

Finally, once $\mathbf{M}_S$ has been determined from the solution of (6.14) or (6.18), the transmitted electric field can be calculated as

$$\mathbf{E}^{+}(\mathbf{r}) = -\nabla \times \mathbf{F}(\mathbf{r}) = \nabla \times \iint_A \underline{\mathbf{G}}^F(\mathbf{r}, \mathbf{r}') \cdot \mathbf{M}_S(\mathbf{r}')\, dS', \qquad z > 0. \tag{6.19}$$

6.8 RULES OF THUMB

This section illustrates a qualitative way of seeing the aperture effects on the shielding properties of a planar metal screen and also a way to obtain a rough estimate of the SE of a perforated metal screen. As is well known, when an incident field impinges on a perfectly conducting metal screen, electric currents are induced on the screen that can be seen to be origin of the reflected field (the introduction of a reflected field is necessary in order to satisfy the boundary conditions on the perfect conductor). The more the induced currents are perturbed by the presence of an aperture, the more the shielding performance of the screen will, of course, be affected. In particular, if a slot is cut on the screen in a direction *orthogonal* to the current flow, the current will be abruptly interrupted, causing radiation (as shown in Figure 6.6*a*) and thus dramatically decreasing the SE of the screen. However, if the slot is placed along a direction *parallel* to the current flow, its effect on the current distribution (and therefore on the shielding) is greatly reduced (see Figure 6.6*b*). Since, in general, it is not possible to determine a priori the direction of the induced current (and therefore to choose the best slot orientation), a large number of small holes are usually employed, as reported in Figure 6.6*c*. The use of small holes in fact gives rise to a much less perturbing effect on the current distribution with respect to a slot of equal area. As a consequence a large number of small holes can be expected to cause less radiation than a large aperture of the same total area.

By Babinet's principle [27], a narrow slot behaves like a thin dipole, so maximum radiation (i.e., minimum shielding) occurs when its length is equal to a

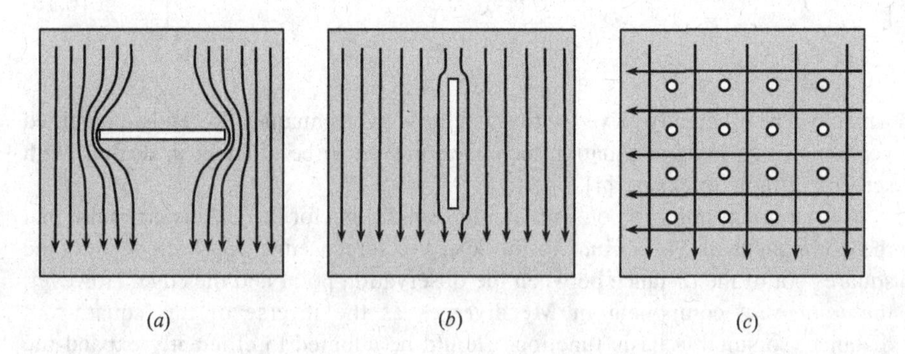

FIGURE 6.6 Qualitative effects of apertures on the surface electric current distributions.

half-wavelength. For slots with a length d equal or less than half-wavelength, the SE is approximately equal to

$$SE = 20\log\left(\frac{\lambda_0}{2d}\right). \tag{6.20}$$

In general, more than one aperture decreases the SE. The amount of reduction depends on a variety of factors, such as the number of apertures, the apertures' shape, the apertures' dimensions, the spacing between apertures, and frequency. However, when N identical apertures are placed in sufficiently close proximity, the increasing in the transmitted power (compared with the single-aperture case) can be regarded as roughly proportional to the number of apertures. In other words, the reduction of the SE is approximately proportional to the square root of the number of apertures N. Therefore, from (6.20), the SE due to N slots of length d is roughly

$$SE = 20\log\left(\frac{\lambda_0}{2d}\right) - 20\log\sqrt{N}. \tag{6.21}$$

Finally, when a *finite thickness* t of the screen needs to be considered, its effect on the SE of a single slot can be accounted for by the "waveguide below cutoff" principle, according to which a thick aperture behaves as a waveguide below cutoff. Therefore the following approximation holds for the SE of the single slot in a thick screen:

$$SE = 20\log\left(\frac{\lambda_0}{2d}\right) + 20\log e^{\pi d/t}. \tag{6.22}$$

REFERENCES

[1] C. J. Bouwkamp. "Diffraction theory." *Rep. Prog. Phys.*, vol. 17, pp. 35–100, 1954.

[2] C. M. Butler, Y. Rahmat-Samii, and R. Mittra. "Electromagnetic penetration through apertures in conducting surfaces." *IEEE Trans. Antennas Propagat.*, vol. AP-26, no. 1, pp. 82–93, Jan. 1978.

[3] K. Hongo and H. Serizawa. "Diffraction of electromagnetic plane wave by a rectangular plate and a rectangular hole in the conducting plate." *IEEE Trans. Antennas Propagat.*, vol. 47, no. 6, pp. 1029–1041, Jun. 1999.

[4] J. R. Mosig. "Scattering by arbitrarily-shaped slots in thick conducting screens: An approximate solution." *IEEE Trans. Antennas Propagat.*, vol. 52, no. 8, pp. 2109–2117, Aug. 2004.

[5] A. Drezet, J. C. Woehl, and S. Huant. "Diffraction of light by a planar aperture in a metallic screen" *J. Math. Phys.*, vol. 47, 072901-1–072901-10, 2006.

[6] C. Flammer. "The vector wave function solution of the diffraction of electromagnetic waves by circular disks and apertures. II. The diffraction problems," *J. Appl. Phys.*, vol. 24, no. 9, pp. 1224–1231, Sep. 1953.

[7] G. R. Kirchhoff. "Zur Theorie der Lichtstrahlen." *Ann. Phys. (Leipzig)*, vol. 18, pp. 663–695, 1883.

[8] A. Sommerfeld. "Zur mathematischen Theorie der Beugungserscheinungen," *Nachr. Kgl. Akad. Wiss. Göttingen*, vol. 4, pp. 338–342, 1894.

[9] J. W. Strutt Lord Rayleigh. "On the passage of waves through apertures in plane screens, and allied problems." *Philos. Mag.*, vol. 43, pp. 259–272, 1897.

[10] J. A. Stratton and L. J. Chu. "Diffraction theory of electromagnetic waves." *Phys. Rev.*, vol. 56, pp. 99–107, Jul. 1939.

[11] S. A. Schelkunoff. "On diffraction and radiation of electromagnetic waves." *Phys. Rev.*, vol. 56, pp. 308–316, Aug. 1939.

[12] W. R. Smythe. "The double current sheet in diffraction." *Phys. Rev.*, vol. 72, no. 11, pp. 1066–1070, Dec. 1947.

[13] F. Kottler. "Diffraction at a black screen, Part 2: Electromagnetic theory." *Prog. Optics*, vol. 6, pp. 335–377, 1967.

[14] J. B. Keller, "Geometrical theory of diffraction," *J. Opt. Soc. Amer.*, vol. 52, no. 2, pp. 116–130, Feb. 1962.

[15] R. G. Kouyoumjian and P. H. Pathak. "A uniform geometrical theory of diffraction for and edge in a perfectly conducting surface." *Proc. IEEE*, vol. 62, no. 11, pp. 1448–1461, Nov. 1974.

[16] P. Ya. Ufimtsev. "Elementary edge waves and the physical theory of diffraction." *Electromagn.*, vol. 11, no. 2, pp. 125–160, Apr./Jun. 1991.

[17] J. W. Strutt Lord Rayleigh. "On the incidence of aerial and electric waves upon small obstacles in the form of ellipsoids or elliptic cylinders, and on the passage of electric waves through a circular aperture in a conducting screen." *Phil. Mag.*, vol. 44, pp. 28–52, Jul. 1897.

[18] H. A. Bethe. "Theory of diffraction by small holes." *Phys. Rev.*, vol. 66, no. 7–8, pp. 163–182, Oct. 1944.

[19] A. F. Stevenson. "Solution of electromagnetic scattering problems as power series in the ratio (dimensions scatterer)/wavelength." *J. Appl. Phys.*, vol. 24, no. 9, pp. 1134–1142, Sep. 1953.

[20] A. F. Kleinman. "Low frequency solutions of electromagnetic scattering problems." In *Electromagnetic Wave Theory*, J. Brown, ed. London: Pergamon Press, 1967, pp. 891–905.

[21] W. H. Eggimann. "Higher-order evaluation of electromagnetic diffraction by circular disks." *IRE Trans. Microwave Theory Tech.*, vol. MTT-9, no. 9, pp. 408–418, Sep. 1961.

[22] H. Levine and J. Schwinger. "On the theory of electromagnetic wave diffraction by an aperture in an infinite plane conducting screen." *Comm. Pure Applied Math.*, vol. 3, pp. 355–391, 1950.

[23] F. De Meulenaere, and J. Van Bladel. "Polarizability of some small apertures." *IEEE Trans. Antennas Propagat.*, vol. AP-25, no. 3, pp. 198–205, Mar. 1977.

[24] E. E. Okon and R. F. Harrington. "The polarizabilities of electrically small apertures of arbitrary shape." *IEEE Trans. Electromagn. Compat.*, vol. EMC-23, pp. 359–366, Nov. 1981.

[25] R. E. Collin. "Rayleigh scattering and power conservation." *IEEE Trans. Antennas Propagat.*, vol. AP-29, no. 5, pp. 795–798, Sep. 1981.

[26] R. F. Harrington. "Resonant behavior of a small aperture backed by a conducting body." *IEEE Trans. Antennas Propagat.*, vol. AP-30, no. 2, pp. 205–212, Mar. 1982.

[27] R. E. Collin. *Field Theory of Guided Waves*, 2nd ed. Piscataway, NJ: Wiley-IEEE Press, 1991.

[28] D. L. Jaggard. "Transmission through one or more small apertures of arbitrary shape." *AFWL Interaction Note 323*, Sep. 1977.

[29] J. Brown and W. Jackson. "The relative permittivity of tetragonal arrays of perfectly conducting thin disks." *Proc. IEE*, vol. 102, Pt. B, pp. 37–42, Jan. 1955.

[30] R. E. Collin and W. H. Eggimann. "Dynamic interaction fields in a two-dimensional lattice." *IRE Trans. Microwave Theory Tech.*, vol. MTT-9, no. 2, pp. 110–115, Mar. 1961.

[31] J. R. Mosig. "Integral equation technique." In *Numerical Techniques for Microwave and Millimeter Wave Passive Structures*, T. Itoh, ed. New York: Wiley, 1989, ch. 3.

[32] T. Anderson. "Moment-method calculations on apertures using basis singular functions." *IEEE Trans. Antennas Propagat.*, vol. 41, no. 12, pp. 1709–1716, Dec. 1993.

CHAPTER SEVEN

Enclosures

An analysis of the interaction between an EM field and an enclosure is particularly relevant in shielding problems because metallic housings are necessary for the reduction of the EM coupling between the internal apparatus and systems and the external environment. The first important assessment concerns whether the solid walls of the enclosure under consideration are penetrable by the EM field. If they are not, the penetration of the EM field into the enclosure occurs through other coupling paths, always present in actual enclosures.

Before attempting any systematic study of the coupling process between the interior of the enclosure and the external environment, it is extremely important to first understand the characteristics of the field inside uncoupled enclosures (i.e., perfectly conducting EM cavities) and then introduce coupling effects (e.g., lossy walls and/or apertures) and small loadings. A metallic enclosure consists of an empty cavity with metallic walls. Even in the absence of EM sources, oscillations can take place at certain frequencies (resonant frequencies) and with certain spatial distributions (resonant modes) that depend on the geometry of the enclosure. The number of resonant frequencies and resonant modes is a countable infinity.

Similarly to what happens in any resonant system, when an enclosure is excited by a time-harmonic source, the closer the operating frequency is to the resonant frequency, the larger is the amplitude of the oscillations. Moreover the spatial distribution of the EM field excited inside the enclosure resembles that of the corresponding resonant mode. The resonant modes are part of a complete set of vector functions for the expansion of an arbitrary EM field excited by an arbitrary source inside the enclosure and thus provide an excellent tool to investigate the properties of the considered enclosure.

7.1 MODAL EXPANSION OF ELECTROMAGNETIC FIELDS INSIDE A METALLIC ENCLOSURE

As is well known, for the expansion of functions in a region V enclosed by a surface S, a complete set of orthogonal (or orthonormal) functions in V has to be determined. In particular, the scalar eigenfunctions ψ_m of the Laplace operator $\nabla^2[\cdot]$ with *Dirichlet* boundary conditions, defined as

$$\begin{aligned} \nabla^2\psi_m(\mathbf{r}) + k^2_{\psi m}\psi_m(\mathbf{r}) &= 0 \quad \text{in } V, \\ \psi_m(\mathbf{r}) &= 0 \quad \text{on } S, \end{aligned} \tag{7.1}$$

form a complete set of orthonormal functions in the Hilbert space $L^2(V)$, and $k_{\psi m}$ are the associated eigenvalues. Another complete set of orthonormal functions in $L^2(V)$ consists of the scalar eigenfunctions ϕ_m of the Laplace operator $\nabla^2[\cdot]$ with *Neumann* boundary conditions and eigenvalues $k_{\phi m}$, defined as

$$\begin{aligned} \nabla^2\phi_m(\mathbf{r}) + k^2_{\phi m}\phi_m(\mathbf{r}) &= 0 \quad \text{in } V, \\ \frac{\partial\phi_m(\mathbf{r})}{\partial n} &= 0 \quad \text{on } S, \end{aligned} \tag{7.2}$$

where n indicates the direction normal to the boundary S. Therefore an arbitrary scalar function belonging to $L^2(V)$ can be expanded using either the set $\{\psi_1, \psi_2, \ldots\}$ or the set $\{\phi_1, \phi_2, \ldots\}$ [1].

In an analogous way, two complete sets of orthonormal vector functions are available for the expansion of an arbitrary *vector* function in $L^2(V)$. One set consists of the solutions of the following boundary-value problem:

$$\begin{aligned} \nabla^2\mathbf{\Psi}_m(\mathbf{r}) + K^2_{\Psi m}\mathbf{\Psi}_m(\mathbf{r}) &= \mathbf{0} \quad \text{in } V, \\ \mathbf{u}_n \times \mathbf{\Psi}_m(\mathbf{r}) &= \mathbf{0} \quad \text{on } S, \\ \nabla \cdot \mathbf{\Psi}_m(\mathbf{r}) &= 0 \quad \text{on } S, \end{aligned} \tag{7.3}$$

where $\mathbf{u}_n$ indicates the outward normal at the boundary S. The other set is formed by the solutions of

$$\begin{aligned} \nabla^2\mathbf{\Phi}_m(\mathbf{r}) + K^2_{\Phi m}\mathbf{\Phi}_m(\mathbf{r}) &= \mathbf{0} \quad \text{in } V, \\ \mathbf{u}_n \times \nabla \times \mathbf{\Phi}_m(\mathbf{r}) &= \mathbf{0} \quad \text{on } S, \\ \mathbf{u}_n \cdot \mathbf{\Phi}_m(\mathbf{r}) &= 0 \quad \text{on } S. \end{aligned} \tag{7.4}$$

It should be noted that *two* boundary conditions are necessary to define the vector eigenfunctions. Some of the eigenfunctions $\mathbf{\Psi}_m$ and $\mathbf{\Phi}_m$ have the same eigenvalues $K_{\Phi m} = K_{\Psi m} = k_m \neq 0$ and are related to each other. They are also indicated as $\mathbf{E}_m$

and $\mathbf{H}_m$, respectively, for which

$$\begin{aligned} k_m \mathbf{E}_m(\mathbf{r}) &= \nabla \times \mathbf{H}_m(\mathbf{r}), \\ k_m \mathbf{H}_m(\mathbf{r}) &= \nabla \times \mathbf{E}_m(\mathbf{r}). \end{aligned} \tag{7.5}$$

From (7.5) it follows that $\nabla \cdot \mathbf{E}_m = \nabla \cdot \mathbf{H}_m = 0$ in V, meaning they are solenoidal [1]. Such eigenfunctions are also called *short-circuit modes*, and in particular, the $\mathbf{E}_m$ eigenfunctions are the *electric-field modes* (since they satisfy the boundary condition of perfectly conducting wall $\mathbf{u}_n \times \mathbf{E}_m = \mathbf{0}$ on S), while the $\mathbf{H}_m$ eigenfunctions are the *magnetic-field modes* (since they satisfy the boundary condition $\mathbf{u}_n \cdot \mathbf{H}_m = 0$ on S) [2]. All the other $\boldsymbol{\Psi}_m$ and $\boldsymbol{\Phi}_m$ eigenfunctions that are not solenoidal are indicated as $\mathbf{F}_m$ and $\mathbf{G}_m$, respectively, and it can be shown that they are irrotational (or lamellar, i.e., $\nabla \times \mathbf{F}_m = \nabla \times \mathbf{G}_m = \mathbf{0}$). Moreover $\mathbf{F}_m$ and $\mathbf{G}_m$ with nonzero eigenvalues are given by

$$\mathbf{F}_m = \frac{\nabla \psi_m}{k_{\psi m}}, \quad \mathbf{G}_m = \frac{\nabla \phi_m}{k_{\phi m}}, \tag{7.6}$$

and their eigenvalues coincide with $k_{\psi m}$ and $k_{\phi m}$, respectively (the same eigenvalues of the scalar eigenfunctions ψ_m and ϕ_m). All the eigenfunctions $\mathbf{E}_m$, $\mathbf{H}_m$, $\mathbf{F}_m$, and $\mathbf{G}_m$ can be chosen to be *real* vector functions. It should be noted that $\mathbf{F}_m$ eigenfunctions with zero eigenvalue can exist if the closed region V has more than one boundary (e.g., V is the cavity between two concentric spheres; see Figure 7.1*a*), and in this case they can be set equal to the gradient of a scalar function, meaning $\mathbf{F}_0 = \nabla \psi$. In particular, if the enclosure is bounded by $P + 1$ separate boundaries, there exist a number P of different $\mathbf{F}_0$ eigenfunctions (indicated as $\mathbf{F}_0^{(1)}, \mathbf{F}_0^{(2)}, \ldots, \mathbf{F}_0^{(P)}$). On the other hand, $\mathbf{G}_m$ eigenfunctions with zero eigenvalue can exist if V is multiply connected (e.g., V is the cavity between two closed coaxial cylinders; see Figure 7.1*b*), and in this case they can be set equal to the curl of a vector function. Alternatively, also $\mathbf{G}_0$ can be set equal to the gradient of a scalar function, but such a function must be multivalued. In particular, if the volume V is a $Q + 1$ connected region, there exists a number Q of different $\mathbf{G}_0$ eigenfunctions (indicated as $\mathbf{G}_0^{(1)}, \mathbf{G}_0^{(2)}, \ldots, \mathbf{G}_0^{(Q)}$) [3].

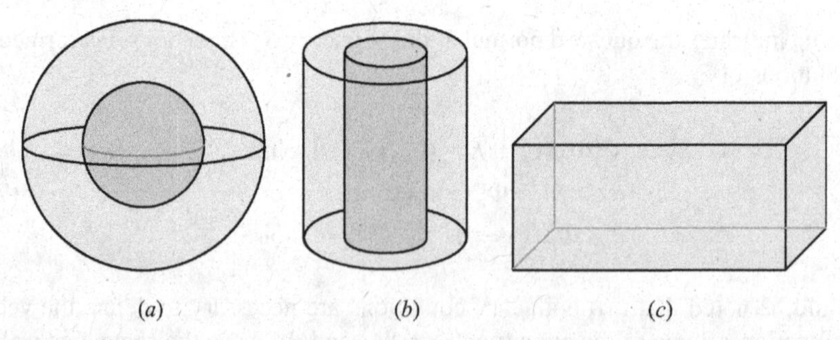

FIGURE 7.1 Examples of enclosures: Closed region with more than one boundary (*a*); multiply connected enclosure (*b*); simply connected enclosure with only one boundary (*c*).

The eigenfunctions $\mathbf{F}_0^p$ and $\mathbf{G}_0^q$ have both zero curl and zero divergence and are called *harmonic* eigenfunctions. In a simply connected region with only one boundary (e.g., a simple rectangular cavity; see Figure 7.1*c*), such eigenfunctions do not exist [2].

A general EM field in the presence of sources has an electric and a magnetic field (**E** and **H**) with nonzero curl and nonzero divergence. Therefore both solenoidal and irrotational vector functions have to be used in the eigenfunction expansion [3]. For the expansion of the electric field **E**, the solenoidal eigenfunctions $\mathbf{E}_m$ and the irrotational eigenfunctions $\mathbf{F}_m$ are used, since they have boundary conditions similar to those of the actual electric field **E** in a cavity (where the $\mathbf{F}_0^p$ modes exist, they represent the static electric field that may exist between conducting surfaces at different potentials). Of course, nondegenerate $\mathbf{E}_m$ and $\mathbf{F}_m$ modes (i.e., modes with different eigenvalues) are orthogonal among themselves and with each other. For degenerate modes, a Gram-Schmidt orthonormalization procedure can be used to obtain a set of orthonormal modes. Analogously, for the expansion of the magnetic field **H**, the solenoidal eigenfunctions $\mathbf{H}_m$ and the irrotational eigenfunctions $\mathbf{G}_m$ are used (where the $\mathbf{G}_0^q$ modes exist, they represent the static magnetic field due to dc currents flowing along the boundaries). Again, nondegenerate $\mathbf{H}_m$ and $\mathbf{G}_m$ modes are orthogonal among themselves and with each other [4].

An arbitrary EM field inside the enclosure can thus be represented as

$$\begin{aligned} \mathbf{E}(\mathbf{r}) &= \sum_{m=1}^{\infty} e_m \mathbf{E}_m(\mathbf{r}) + \sum_{m=1}^{\infty} f_m \mathbf{F}_m(\mathbf{r}) + \sum_{p=1}^{P} f_0^p \mathbf{F}_0^p(\mathbf{r}), \\ \mathbf{H}(\mathbf{r}) &= \sum_{m=1}^{\infty} h_m \mathbf{H}_m(\mathbf{r}) + \sum_{m=1}^{\infty} g_m \mathbf{G}_m(\mathbf{r}) + \sum_{q=1}^{Q} g_0^q \mathbf{G}_0^q(\mathbf{r}), \end{aligned} \tag{7.7}$$

where the zero-frequency modes $\mathbf{F}_0^p$ and $\mathbf{G}_0^q$ have to be taken into account only for regions having more than one boundary and for multiply connected regions, respectively. The coefficients of the expansion are given by

$$\begin{aligned} e_m &= \iiint_V \mathbf{E} \cdot \mathbf{E}_m dV', \quad f_m = \iiint_V \mathbf{E} \cdot \mathbf{F}_m dV', \quad f_0^p = \iiint_V \mathbf{E} \cdot \mathbf{F}_0^p dV', \\ h_m &= \iiint_V \mathbf{H} \cdot \mathbf{H}_m dV', \quad g_m = \iiint_V \mathbf{H} \cdot \mathbf{G}_m dV', \quad g_0^q = \iiint_V \mathbf{H} \cdot \mathbf{G} dV'. \end{aligned} \tag{7.8}$$

In this chapter only simply connected enclosures with a unique boundary will be considered for simplicity. The contribution of the eigenfunctions $\mathbf{F}_0^p$ and $\mathbf{G}_0^q$ will be thus disregarded so that the EM-field expansion takes the form

$$\begin{aligned} \mathbf{E}(\mathbf{r}) &= \sum_{m=1}^{\infty} e_m \mathbf{E}_m(\mathbf{r}) + \sum_{m=1}^{\infty} f_m \mathbf{F}_m(\mathbf{r}), \\ \mathbf{H}(\mathbf{r}) &= \sum_{m=1}^{\infty} h_m \mathbf{H}_m(\mathbf{r}) + \sum_{m=1}^{\infty} g_m \mathbf{G}_m(\mathbf{r}), \end{aligned} \tag{7.9}$$

with the e_m, f_m, h_m, and g_m coefficients still given by (7.8).

The expansion (7.9) is especially important, and it will be used in a next section to study the EM field excited inside an enclosure by a finite inner source or by a small aperture that allows for the penetration of an external field.

7.2 OSCILLATIONS INSIDE AN IDEAL SOURCE-FREE ENCLOSURE

An ideal enclosure consists of a cavity with perfectly conducting walls and filled with a linear, isotropic, homogeneous, stationary, and lossless medium (characterized by its dielectric permittivity $\varepsilon = \varepsilon_0 \varepsilon_r$ and magnetic permeability $\mu = \mu_0 \mu_r$). The volume of the enclosure is indicated with V, while its delimiting surface is indicated with S. In the absence of sources, EM fields can exist inside the enclosure if, at certain frequencies $f_m = \omega_m/(2\pi)$, the source-free Maxwell equations and boundary conditions are satisfied:

$$\begin{aligned} \nabla \times \mathbf{E}(\mathbf{r}, \omega_m) &= -j\omega_m \mu \mathbf{H}(\mathbf{r}, \omega_m) && \text{in } V, \\ \nabla \times \mathbf{H}(\mathbf{r}, \omega_m) &= j\omega_m \varepsilon \mathbf{E}(\mathbf{r}, \omega_m) && \text{in } V, \\ \mathbf{u}_n \times \mathbf{E}(\mathbf{r}, \omega_m) &= \mathbf{0} && \text{on } S, \end{aligned} \tag{7.10}$$

where $\mathbf{u}_n$ indicates the outward normal at the boundary S. It should be clear from (7.10) that the sought-for EM field has both electric and magnetic fields that are solenoidal (i.e., $\nabla \cdot \mathbf{E} = 0$ and $\nabla \cdot \mathbf{H} = 0$). It is useful to introduce the vector $\hat{\mathbf{H}} = -j\eta\mathbf{H}$ (being $\eta = \sqrt{\mu/\varepsilon}$ the intrinsic impedance of the medium filling the cavity) so that the Maxwell equations in (7.10) can be modified to assume a symmetric form:

$$\begin{aligned} \nabla \times \mathbf{E}(\mathbf{r}, \omega_m) &= k_m \hat{\mathbf{H}}(\mathbf{r}, \omega_m) && \text{in } V, \\ \nabla \times \hat{\mathbf{H}}(\mathbf{r}, \omega_m) &= k_m \mathbf{E}(\mathbf{r}, \omega_m) && \text{in } V, \end{aligned} \tag{7.11}$$

where $k_m = \omega_m \sqrt{\mu\varepsilon}$ is a positive unknown parameter [3]. The existence of oscillating fields is associated with the existence of positive values of k_m for which the system (7.11) (together with the appropriate boundary conditions) admits nontrivial solutions. Such nontrivial solutions are the *solenoidal* eigenfunctions $(\mathbf{E}_m, \mathbf{H}_m)$ of the Laplace operator $\nabla^2[\cdot]$ studied in the previous section. In particular, there exists a sequence of real and positive eigenvalues $\{k_1, k_2, \ldots\}$ with two associated sequences of real orthonormal vector eigenfunctions $\{\mathbf{E}_1, \mathbf{E}_2, \ldots\}$ and $\{\mathbf{H}_1, \mathbf{H}_2, \ldots\}$ [3]. Therefore, at the frequency $f = f_m$, the solution of the EM problem inside the source-free enclosure is given by

$$\mathbf{E}(\mathbf{r}) = A_m \mathbf{E}_m(\mathbf{r}), \quad \mathbf{H}(\mathbf{r}) = j\frac{A_m}{\eta} \mathbf{H}_m(\mathbf{r}), \tag{7.12}$$

which represents a *normal* (*resonant*) *mode* of the cavity (A_m is an arbitrary complex number). So there exists a countable infinity of resonant modes, each associated with a pair of solenoidal eigenfunctions $\{\mathbf{E}_m, \mathbf{H}_m\}$. Since the eigenfunctions are real functions, in the time domain the EM field is given by

$$\begin{aligned} \mathbf{e}(\mathbf{r}, t) &= |A_m|\mathbf{E}_m(\mathbf{r})\cos(\omega_m t + \mathrm{Arg}[A_m]), \\ \mathbf{h}(\mathbf{r}, t) &= -\frac{|A_m|}{\eta}\mathbf{H}_m(\mathbf{r})\sin(\omega_m t + \mathrm{Arg}[A_m]). \end{aligned} \tag{7.13}$$

From (7.13) it can be seen that the spatial distribution of both the electric and the magnetic field of a resonant mode depends only on the solenoidal eigenfunctions $\{\mathbf{E}_m, \mathbf{H}_m\}$ and does not vary with time. The electric and magnetic fields oscillate in quadrature.

7.3 THE ENCLOSURE DYADIC GREEN FUNCTION

As is well known, the Green function is the response of a linear system to a point source of unit strength. In electromagnetics, the linear system is usually described either by the vector wave equation for the electric **E** and magnetic **H** fields or by the vector (scalar) Helmholtz equation for the vector (**A, F**) and scalar (*V, W*) potentials. The impressed source can be an electric current density $\mathbf{J}_\mathrm{i}$, an electric charge density ρ_ei, a magnetic current density $\mathbf{M}_\mathrm{i}$, or a magnetic charge density ρ_mi. Obviously, the appropriate boundary conditions of the considered EM problem have also to be taken into account. From a different point of view, the Green function is the kernel of an integral operator that transforms the source and the boundary conditions into the response of the system. Depending on the quantity of interest and on the type of source, different types of Green functions can be defined. In what follows we will refer to the electric and magnetic fields due to electric and magnetic current densities. In these cases, since each component of the current gives rise to a vector field, the general relation between the current and the field is a *dyadic* relation: the involved Green function is thus a dyadic quantity [5].

Let us thus consider an enclosure filled with a medium with constitutive parameters $\mu = \mu_0\mu_\mathrm{r}$ and $\varepsilon = \varepsilon_0\varepsilon_\mathrm{r}$ and excited by a time-harmonic source. The electric field $\mathbf{E}(\mathbf{r})$ inside the enclosure due to an electric current density $\mathbf{J}_\mathrm{i}(\mathbf{r})$ is the solution of

$$\nabla \times \nabla \times \mathbf{E}(\mathbf{r}) - k^2\mathbf{E}(\mathbf{r}) = -j\omega\mu\mathbf{J}_\mathrm{i}(\mathbf{r}), \tag{7.14}$$

where $k^2 = \omega^2\mu\varepsilon$. This equation can be solved by using the expansion (7.9) for the electric field. Substituting (7.9) into (7.14) one obtains

$$\sum_{m=1}^{\infty}[(k_m^2 - k^2)e_m\mathbf{E}_m(\mathbf{r}) - k^2 f_m\mathbf{F}_m(\mathbf{r})] = -j\omega\mu\mathbf{J}_\mathrm{i}(\mathbf{r}). \tag{7.15}$$

The coefficients e_m and f_m can be derived by scalar multiplying (7.15) by $\mathbf{E}_m$ and $\mathbf{F}_m$, respectively, and by integrating over the volume V of the enclosure. Because of the orthonormality of the eigenfunctions one in fact obtains

$$\begin{aligned}(k_m^2 - k^2)e_m &= -j\omega\mu \iiint_V \mathbf{E}_m(\mathbf{r}') \cdot \mathbf{J}_\mathrm{i}(\mathbf{r}')\,\mathrm{d}V', \\ -k^2 f_m &= -j\omega\mu \iiint_V \mathbf{F}_m(\mathbf{r}') \cdot \mathbf{J}_\mathrm{i}(\mathbf{r}')\,\mathrm{d}V'.\end{aligned} \tag{7.16}$$

So, from (7.9) and (7.16), the electric field $\mathbf{E}$ can be finally expressed as

$$\mathbf{E}(\mathbf{r}) = -j\omega\mu \sum_{m=1}^{\infty} \iiint_V \left[\frac{\mathbf{E}_m(\mathbf{r})\mathbf{E}_m(\mathbf{r}')}{k_m^2 - k^2} - \frac{\mathbf{F}_m(\mathbf{r})\mathbf{F}_m(\mathbf{r}')}{k^2}\right] \cdot \mathbf{J}_\mathrm{i}(\mathbf{r}')\,\mathrm{d}V'. \tag{7.17}$$

From (7.17) it follows immediately that the *EJ*-type dyadic Green function $\underline{\mathbf{G}}^{EJ}$ (electric field $\mathbf{E}$ due to electric current $\mathbf{J}$) is given by

$$\underline{\mathbf{G}}^{EJ}(\mathbf{r}, \mathbf{r}') = \sum_{m=1}^{\infty} \left[\frac{\mathbf{E}_m(\mathbf{r})\mathbf{E}_m(\mathbf{r}')}{k_m^2 - k^2} - \frac{\mathbf{F}_m(\mathbf{r})\mathbf{F}_m(\mathbf{r}')}{k^2}\right]. \tag{7.18}$$

The vector eigenfunctions $\mathbf{E}_m$ and $\mathbf{F}_m$ can be determined analytically only for enclosures of very simple shape (e.g., rectangular, cylindrical, and spherical cavities), and in these cases the solenoidal electric-field modes $\mathbf{E}_m$ can be split into the usual $\mathbf{M}_m$- and $\mathbf{N}_m$-type modes types used in waveguide problems, as it will be shown later. The irrotational $\mathbf{F}_m$ eigenfunctions correspond to the $\mathbf{L}_m$-type modes [4]. The explicit analytic determination of the vector eigenfunctions is based on the method of separation of variables, and therefore the index m in the series above actually represents a *triple set* of integers, by which the series becomes a triple series, as it will be shown in the particular case of a rectangular enclosure treated in Section 7.8. Finally, since for a cavity the unit dyadic source $\underline{\mathbf{I}}\delta(\mathbf{r} - \mathbf{r}')$ can be represented as [4]

$$\underline{\mathbf{I}}\delta(\mathbf{r} - \mathbf{r}') = \sum_{m=1}^{\infty} [\mathbf{E}_m(\mathbf{r})\mathbf{E}_m(\mathbf{r}') + \mathbf{F}_m(\mathbf{r})\mathbf{F}_m(\mathbf{r}')], \tag{7.19}$$

the $\underline{\mathbf{G}}^{EJ}$ Green function can also be represented in terms of the electric-field modes only:

$$\underline{\mathbf{G}}^{EJ}(\mathbf{r}, \mathbf{r}') = \sum_{m=1}^{\infty} \frac{k_m^2}{k^2(k_m^2 - k^2)} \mathbf{E}_m(\mathbf{r})\mathbf{E}_m(\mathbf{r}') - \frac{1}{k^2}\underline{\mathbf{I}}\delta(\mathbf{r} - \mathbf{r}'), \tag{7.20}$$

where $\underline{\mathbf{I}}$ is the (3×3) identity dyadic. In (7.20) a singularity appears explicitly in the Green function when the source $(\mathbf{r}')$ and observation $(\mathbf{r})$ points coincide whereas in (7.18) the singularity is embedded in the irrotational term. Therefore, when observation and source points coincide, the Green function series does not converge. It should be pointed out that in general, several different representations of the Green function exist, and depending on the problem, they may or may not be computationally efficient. In addition to the eigenfunction-expansion method presented here, the method of images (a special case of the method of multiple-reflected waves) is widely used [4]. Other methods combine the eigenfunction representation (or spectral series representation) and the method of images (leading to the so-called spatial series representation) to obtain hybrid representations having faster convergence properties. A review of such methods for enclosure problems can be found in [6].

The magnetic field $\mathbf{H}(\mathbf{r})$ inside the enclosure due to a magnetic current density $\mathbf{M}_{\mathrm{i}}(\mathbf{r})$ can be determined in a similar way. $\mathbf{H}(\mathbf{r})$ is in fact the solution of

$$\nabla \times \nabla \times \mathbf{H}(\mathbf{r}) - k^2\mathbf{H}(\mathbf{r}) = -j\omega\varepsilon\mathbf{M}_{\mathrm{i}}(\mathbf{r}), \tag{7.21}$$

which can be solved by using the expansion (7.9) for the magnetic field. In particular, the coefficients h_m and g_m can be determined by the same steps used above to determine the e_m and f_m coefficients. Dually to (7.17), it turns out that

$$\mathbf{H}(\mathbf{r}) = -j\omega\varepsilon \sum_{m=1}^{\infty} \iiint_V \left[\frac{\mathbf{H}_m(\mathbf{r})\mathbf{H}_m(\mathbf{r}')}{k_m^2 - k^2} - \frac{\mathbf{G}_m(\mathbf{r})\mathbf{G}_m(\mathbf{r}')}{k^2}\right] \cdot \mathbf{M}_{\mathrm{i}}(\mathbf{r}')\, dV'. \tag{7.22}$$

So the HM-type dyadic Green function $\underline{\mathbf{G}}^{HM}$ can be expressed as

$$\underline{\mathbf{G}}^{HM}(\mathbf{r},\mathbf{r}') = \sum_{m=1}^{\infty} \left[\frac{\mathbf{H}_m(\mathbf{r})\mathbf{H}_m(\mathbf{r}')}{k_m^2 - k^2} - \frac{\mathbf{G}_m(\mathbf{r})\mathbf{G}_m(\mathbf{r}')}{k^2}\right]. \tag{7.23}$$

The magnetic field $\mathbf{H}(\mathbf{r})$ due to an electric current density $\mathbf{J}_{\mathrm{i}}(\mathbf{r})$ can be derived from (7.17) and the Maxwell equation $\mathbf{H} = -\nabla \times \mathbf{E}/(j\omega\mu)$. Then from (7.5) and the fact that the $\mathbf{F}_m$ eigenfunctions are irrotational, it follows that

$$\mathbf{H}(\mathbf{r}) = \sum_{m=1}^{\infty} \iiint_V \left[\frac{k_m\mathbf{H}_m(\mathbf{r})\mathbf{E}_m(\mathbf{r}')}{k_m^2 - k^2}\right] .\mathbf{J}_{\mathrm{i}}(\mathbf{r}')\, dV' \tag{7.24}$$

and the HJ-type dyadic Green function $\underline{\mathbf{G}}^{HJ}(\mathbf{r},\mathbf{r}')$ is given by

$$\underline{\mathbf{G}}^{HJ}(\mathbf{r},\mathbf{r}') = \sum_{m=1}^{\infty} \frac{k_m}{k_m^2 - k^2}\mathbf{H}_m(\mathbf{r})\mathbf{E}_m(\mathbf{r}'). \tag{7.25}$$

Similarly the electric field $\mathbf{E}(\mathbf{r})$ due to a magnetic current density $\mathbf{M}_{\mathrm{i}}(\mathbf{r})$ can be derived from (7.22) and thus expressed as

$$\mathbf{E}(\mathbf{r}) = -\sum_{m=1}^{\infty} \iiint\limits_V \left[\frac{k_m \mathbf{E}_m(\mathbf{r}) \mathbf{H}_m(\mathbf{r}')}{k_m^2 - k^2} \right] \cdot \mathbf{M}_{\mathrm{i}}(\mathbf{r}')\, \mathrm{d}V'. \tag{7.26}$$

The EM-type dyadic Green function $\underline{\mathbf{G}}^{EM}(\mathbf{r}, \mathbf{r}')$ is given by

$$\underline{\mathbf{G}}^{EM}(\mathbf{r}, \mathbf{r}') = -\sum_{m=1}^{\infty} \frac{k_m}{k_m^2 - k^2} \mathbf{E}_m(\mathbf{r}) \mathbf{H}_m(\mathbf{r}'). \tag{7.27}$$

7.4 EXCITATION OF A METALLIC ENCLOSURE

Based on equations (7.17), (7.22), (7.24), and (7.26), a general expression can be obtained for the EM field excited inside an ideal metallic enclosure by impressed electric and magnetic current densities. In particular, by a comparison of the above-mentioned equations with (7.9), the coefficients of the expansion (7.9) are found to be as

$$e_m = -\frac{k_m}{k_m^2 - k^2} \left[\iiint\limits_V \mathbf{H}_m(\mathbf{r}') \cdot \mathbf{M}_{\mathrm{i}}(\mathbf{r}')\, \mathrm{d}V' \right] - \frac{j\omega\mu}{k_m^2 - k^2} \left[\iiint\limits_V \mathbf{E}_m(\mathbf{r}') \cdot \mathbf{J}_{\mathrm{i}}(\mathbf{r}')\, \mathrm{d}V' \right],$$

$$f_m = -\frac{1}{j\omega\varepsilon} \iiint\limits_V \mathbf{F}_m(\mathbf{r}') \cdot \mathbf{J}_{\mathrm{i}}(\mathbf{r}')\, \mathrm{d}V', \tag{7.28}$$

and

$$h_m = \frac{k_m}{k_m^2 - k^2} \left[\iiint\limits_V \mathbf{E}_m(\mathbf{r}') \cdot \mathbf{J}_{\mathrm{i}}(\mathbf{r}')\, \mathrm{d}V' \right] - \frac{j\omega\varepsilon}{k_m^2 - k^2} \left[\iiint\limits_V \mathbf{H}_m(\mathbf{r}') \cdot \mathbf{M}_{\mathrm{i}}(\mathbf{r}')\, \mathrm{d}V' \right],$$

$$g_m = -\frac{1}{j\omega\mu} \iiint\limits_V \mathbf{G}_m(\mathbf{r}') \cdot \mathbf{M}_{\mathrm{i}}(\mathbf{r}')\, \mathrm{d}V'. \tag{7.29}$$

From equations (7.28) and (7.29) two important qualitative observations can now be made:

1. The coefficients of the electric (magnetic) field modes $\mathbf{E}_m$ ($\mathbf{H}_m$) are proportional to $1/(k_m^2 - k^2)$; that is, as the operating frequency f approaches the resonant frequency $f_m = ck_m/(2\pi)$, the field amplitude becomes infinite. This

behavior is clearly due to the assumption of ideal enclosure. In *real* enclosures, such amplitudes are bounded by the losses due to the finite conductivity of the walls or of the medium filling the cavity, as it can be seen in (7.28) and (7.29) if complex values of the wavenumber k are considered (corresponding, e.g., to a lossy filling material). However, for low-loss enclosures, in the neighborhood of the resonant frequency f_m, the contribution of the resonant mode $(\mathbf{E}_m, \mathbf{H}_m)$ is clearly dominant (if excited), and it provides an accurate representation of the actual excited EM field $(\mathbf{E}, \mathbf{H})$, independently of the source.

2. The amplitude of each resonant mode $(\mathbf{E}_m, \mathbf{H}_m)$ depends on the value of the reaction integral between the source system $(\mathbf{J}_\mathrm{i}, \mathbf{M}_\mathrm{i})$ and the resonant mode itself. In particular, it can be seen that a resonant mode is not excited if the electric-field mode $\mathbf{E}_m$ is orthogonal to the electric source $\mathbf{J}_\mathrm{i}$ and the magnetic-field mode $\mathbf{H}_m$ is orthogonal to the magnetic source $\mathbf{M}_\mathrm{i}$.

7.5 DAMPED OSCILLATIONS INSIDE ENCLOSURES WITH LOSSY WALLS AND QUALITY FACTOR

As is well known, when the walls of the enclosure are characterized by a finite conductivity σ, the approximate impedance boundary condition

$$\mathbf{u}_n \times \mathbf{E} = Z_S \mathbf{H}_\mathrm{t} \quad \text{on } S \tag{7.30}$$

can be used, where $Z_S = R_S + jX_S = (1+j)/(\sigma\delta)$ is the surface impedance of the walls (and $\delta = 1/\sqrt{\pi\mu_0 f \sigma}$ is the skin depth), $\mathbf{H}_\mathrm{t}$ is the component of the magnetic field tangential to the surface S, and $\mathbf{u}_n$ is the corresponding normal unit vector pointing outward.

As a consequence of the losses present in the walls, some differences arise with respect to the ideal-enclosure results [7]. First of all, because of losses, the energy associated with each resonant mode decreases by increasing time, and the mode oscillations are damped. Second, the electric current induced in the walls is now a volume current, which penetrates into the walls up to a depth of the order of the skin depth δ. Some of the energy is dissipated in the walls. Last, as noted in the previous section, the field amplitude is not infinite at the resonances, since the finite conductivity of the walls makes the eigenvalues k_m have a small imaginary part. In any case, for not too low values of conductivity σ, the field amplitude is still very large and the resonant mode associated with the considered resonance still accurately represents the total field. In what follows, these conclusions are quantified by way of the energy-balance principle [4].

Let us consider a real enclosure. For highly conducting walls and low material losses, we can assume that, near the resonant frequency f_m, the field is essentially that of the nondegenerate mth mode. Since the conductivity of the walls usually has large values, the time-domain expression of the mth resonant mode can still be expressed by (7.13), where $|A_m|$ and $\mathrm{Arg}[A_m]$ are now slowly varying functions of time t. So in the time interval $T_m = 2\pi/\omega_m$, the resonant-mode time variation can be still considered approximately sinusoidal. Because of the orthonormality of the eigenfunctions, the time-averaged energy W stored in the cavity is

$$W = \frac{1}{4}\iiint_V [\varepsilon|\mathbf{E}(\mathbf{r})|^2 + \mu|\mathbf{H}(\mathbf{r})|^2]\,\mathrm{d}V \simeq \frac{1}{4}\iiint_V \left[\varepsilon|A_m\mathbf{E}_m(\mathbf{r})|^2 + \mu\left|\frac{jA_m}{\eta}\mathbf{H}_m(\mathbf{r})\right|^2\right]\mathrm{d}V$$
$$= \frac{\varepsilon}{2}|A_m|^2 = W_m, \tag{7.31}$$

where W_m indicates the time-averaged energy associated with the mth resonant mode. On the other hand, the power loss P_L due to the finite conductivity of the walls can be calculated as

$$P_\mathrm{L} = \frac{1}{2}\mathrm{Re}\left\{\oiint_S \mathbf{u}_n \cdot [\mathbf{E}(\mathbf{r}) \times \mathbf{H}^*(\mathbf{r})]\,\mathrm{d}S\right\} = \frac{1}{2}\mathrm{Re}\left\{Z_S \oiint_S |\mathbf{H}_\mathrm{t}(\mathbf{r})|^2\,\mathrm{d}S\right\}$$
$$\simeq \frac{R_S|A_m|^2}{2\eta^2}\oiint_S |\mathbf{H}_{mt}(\mathbf{r})|^2\,\mathrm{d}S = P_{\mathrm{L}m}, \tag{7.32}$$

where $P_{\mathrm{L}m}$ indicates the power loss of the mth resonant mode.

From (7.31), it follows that $|A_m|^2 = 2W_m/\varepsilon$. Therefore $P_{\mathrm{L}m} = 2\alpha_m W_m$, where

$$\alpha_m = \frac{R_S}{2\mu}\oiint_S |\mathbf{H}_{mt}(\mathbf{r})|^2\,\mathrm{d}S. \tag{7.33}$$

By virtue of the energy-balance principle, the rate of decrease of the average stored energy must be equal to the power loss:

$$-\frac{\mathrm{d}W_m}{\mathrm{d}t} = P_{\mathrm{L}m} = 2\alpha_m W_m, \tag{7.34}$$

whose solution is $W(t) = W(0)\,e^{-2\alpha_m t}$. Consequently a free oscillation in the enclosure has the time dependence $E_m \propto e^{j\omega_m t - \alpha_m t} = e^{j\omega_m^\mathrm{L} t}$, where $\omega_m^\mathrm{L} = \omega_m + j\alpha_m$ is the perturbed resonant angular frequency.

The quality factor Q_m of the enclosure for the mth resonant mode is defined as

$$Q_m = \omega_m \frac{\text{average stored energy}}{\text{energy loss per second}} = \frac{\omega_m W_m}{P_m} = \frac{\omega_m}{2\alpha_m} = \left(\frac{R_S}{\omega_m \mu}\oiint_S |\mathbf{H}_{mt}(\mathbf{r})|^2\,\mathrm{d}S\right)^{-1}. \tag{7.35}$$

It can thus be seen that the larger Q_m, the smaller is the power loss and the slower is the field decay. In effect the quality factor gives a measure of how much a real enclosure resembles an ideal enclosure. The perturbed eigenvalue k_m^{L} of the mth resonant mode can be expressed in terms of the quality factor as

$$k_m^{\mathrm{L}} = k_m\left(1 + j\frac{1}{2Q_m}\right). \tag{7.36}$$

A more refined analysis, based on a variational formulation [4], provides also a small correction to the real part of the eigenvalues:

$$k_m^{\mathrm{L}} \simeq k_m\left(1 + j\frac{1}{2Q_m} - \frac{1}{2Q_m}\right). \tag{7.37}$$

It can then be seen that at the resonant frequency f_m of the lossless enclosure, the amplitude of the excited mth resonant mode inside the lossy enclosure is no longer infinity. In general, it is proportional to the $[(k_m^{\mathrm{L}})^2 - k^2]^{-1}$ factor, which has a finite maximum at the perturbed resonant frequency $f_m^{\mathrm{L}} \simeq f_m - 1/(2Q_m)$. In particular, at $f = f_m^{\mathrm{L}}$, the amplitude of the mth resonant mode is proportional to jQ_m/k_m^2, and in this sense, the quality factor is also a measure of the excitation of the corresponding mode near the resonant frequency. Moreover it can be shown that for frequencies close to the resonant frequency, the larger the quality factor, the sharper is the resonance.

Additionally from (7.37) it can be seen that in the presence of losses, the resonant frequency of a given mode slightly decreases with respect to that of the corresponding ideal enclosure. This can be intuitively explained by taking into account the virtual enlargement of the dimensions of the enclosure due to the penetration of the EM field inside the metallic walls (alternatively, it can be viewed as a consequence of the additional magnetic energy stored in the inductive reactance of the surface impedance [4]).

7.6 APERTURES IN PERFECTLY CONDUCTING ENCLOSURES

Shielding enclosures are usually employed to protect against radiation from external EM fields and leakage effects from interior components. However, as mentioned in the previous chapter, the efficiency of these enclosures is often compromised by apertures and slots located on the walls of the enclosure. Therefore it is extremely important to include an evaluation of the aperture effects in determining the shielding properties of an enclosure. In principle, both analytical and numerical methods can be used to evaluate the SE of such "partially open" enclosures. However, closed-form expressions are available only for apertures of simple shapes and under certain limiting assumptions.

The first analytical approach, proposed by Mendez, is based essentially on Bethe's theory of small holes described in the previous chapter. Its validity is therefore restricted to very small apertures in a suitable low-frequency range [8].

A very simple, but effective and accurate method based on a TL analogy was developed in [9], where the aperture (assumed to be rectangular) is represented as a length of coplanar stripline shorted at each end. However, such an approach is also subject to many severe limitations: only rectangular boxes with a centrally placed narrow rectangular aperture can be studied, under normal plane-wave incidence, and with the incident electric field parallel to the shortest aperture side. Only points in front of the aperture can be considered, and moreover all the analysis is valid only below the first resonance of the enclosure. This method has been further extended to take into account general angles of incidence and polarizations and higher order modes of the rectangular cavity [10,11].

The above-mentioned analytical investigations have been supplemented by a good many numerical approaches. It is extremely important to have efficient, accurate, and reliable codes, because a rigorous full-wave solution usually requires large computation time. However, different numerical approaches have led, in critical regions, to different results. Among the many methods proposed in the last few years are finite-difference time-domain methods (e.g., [12,13]), finite-element methods ([14,15]), method of moments (e.g., [16–19]), and hybrid methods (e.g., [20]). This is currently a hot topic of research.

7.6.1 Small-Aperture Approximation

As mentioned above, the first and simplest analytical approach for solving the problem of a metallic enclosure with an aperture cut in its walls consists in applying Bethe's theory to represent the effects of such an aperture. Recall from the previous chapter that the main assumption behind Bethe's theory is that the aperture has to be small (i.e., its maximum dimension has to be much smaller than the wavelength), so that its validity is restricted to the low-frequency region. The small-aperture approximation approach is very similar to that described in the previous chapter in connection with an infinite perfectly conducting screen having a small hole.

Let us thus consider an ideal enclosure, having a small aperture *A* cut in one of its walls that allows for coupling with the external environment (e.g., free space), and let us assume that an EM source (e.g., an electric dipole) is placed inside the enclosure, as shown in Figure 7.2*a*. In the absence of the aperture, the EM field produced by the source would be confined inside the enclosure and (in the absence of other excitations outside the cavity) be identically zero outside. However, the presence of the small hole allows the source to produce a field that "escapes" through the hole, so it has nonzero values in the external environment. This external field is the quantity of interest that needs to be evaluated.

By Bethe's theory, the effects of the aperture can be represented by an electric and a magnetic dipole moment $\mathbf{p}_e = \varepsilon_0 \underline{\boldsymbol{\alpha}}_e \cdot \mathbf{E}^{inc}$ and $\mathbf{p}_m = -\mu_0 \underline{\boldsymbol{\alpha}}_m \cdot \mathbf{H}^{inc}$, respectively, where $\underline{\boldsymbol{\alpha}}_e$ and $\underline{\boldsymbol{\alpha}}_m$ are the electric and magnetic polarizability (2×2) tensors representing the small aperture, while $\mathbf{E}^{inc}$ and $\mathbf{H}^{inc}$ are the electric and magnetic fields incident upon the aperture (i.e., those fields that exist at the aperture location when the aperture is short-circuited). It should be pointed out that another fundamental approximation consists in assuming the wall where the aperture is cut

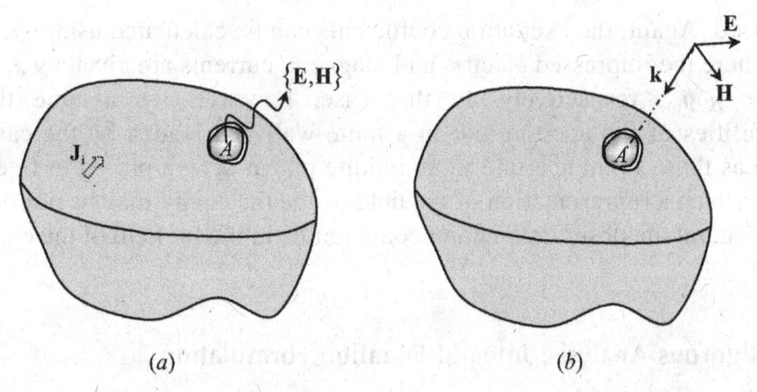

FIGURE 7.2 Enclosure with an aperture excited by an internal source (*a*) and by an external source (*b*).

to be of infinite extent (this is required in order to apply Bethe's theory). The incident fields can be obtained from (7.9), where the amplitude coefficients of each eigenfunction are given by (7.28) and (7.29). For example, if the source is an ideal infinitesimal electric dipole placed in $\mathbf{r}' = \mathbf{r}_0$ and directed along an arbitrary unit vector $\mathbf{u}_d$ (i.e., $\mathbf{J}_\mathrm{i}(\mathbf{r}') = \delta(\mathbf{r}' - \mathbf{r}_0)\mathbf{u}_d$), from (7.28) and (7.29) it simply results in

$$\begin{aligned} e_m &= \frac{j\omega\mu}{k_m^2 - k^2}\mathbf{E}_m(\mathbf{r}_0)\cdot\mathbf{u}_d, \\ f_m &= -\frac{1}{j\omega\varepsilon}\mathbf{F}_m(\mathbf{r}_0)\cdot\mathbf{u}_d, \\ h_m &= \frac{k_m}{k_m^2 - k^2}\mathbf{E}_m(\mathbf{r}_0)\cdot\mathbf{u}_d, \\ g_m &= 0. \end{aligned} \tag{7.38}$$

An example of closed-form expressions for the enclosure eigenfunctions will be given in Section 7.8, where the rectangular enclosure will be studied in detail.

Finally, once the incident field and the aperture polarizabilities are known, the corresponding electric and magnetic dipole moments can be used to evaluate the EM field outside the enclosure by assuming that they radiate in free space (the obtained results are thus valid only for the half-space that does not include the enclosure and bounded by an infinite plane containing the aperture).

Another basic problem consists in evaluating the field inside the enclosure when a small aperture in one of its walls is illuminated by an external incident plane wave, as shown in Figure 7.2*b*. Also in this case the aperture effects can be represented by the equivalent electric and magnetic dipole moments $\mathbf{p}_\mathrm{e} = \varepsilon_0\underline{\boldsymbol{\alpha}}_\mathrm{e}\cdot\mathbf{E}^\mathrm{inc}$ and $\mathbf{p}_\mathrm{m} = -\mu_0\underline{\boldsymbol{\alpha}}_\mathrm{m}\cdot\mathbf{H}^\mathrm{inc}$. Differently from the previous situation (source inside the enclosure) the incident field is now a plane-wave field. To evaluate the field radiated by these dipoles inside the enclosure, the representation

(7.9) is used. Again, the excitation coefficients can be calculated using (7.28) and (7.29), where the impressed electric and magnetic currents are given by $\mathbf{J}_\mathrm{i} = j\omega\mathbf{p}_\mathrm{e}$ and $\mathbf{M}_\mathrm{i} = j\omega\mathbf{p}_\mathrm{m}$, respectively. In this case, however, we assume that the polarizabilities of the aperture cut in a finite wall and loaded by the cavity are the same as those of an aperture in an infinite planar screen placed in free space. Moreover, such a representation of the field inside the cavity may be not rigorous, since in general the observation point could not be in the far field of the equivalent dipoles.

7.6.2 Rigorous Analysis: Integral-Equation Formulation

A general procedure can be derived that leads to the formulation of an integral equation whose solution furnishes the exact field radiated by a finite source placed inside an enclosure having an arbitrary aperture A cut in one of its walls (or the exact field penetrating inside the enclosure and due to an external source). Such a procedure is similar to that described in Chapter 6. In the general case, a system of sources is placed inside the enclosure and another system of sources is placed in the external environment. Then the transverse aperture electric field $\mathbf{E}_\mathrm{t}^\mathrm{ap}$ (i.e., the component parallel to the aperture of the *total* electric field at the aperture location) is chosen as the unknown of the integral equation. Based on the equivalence principle, equivalent magnetic current densities $\mathbf{M}_S$ (proportional to $\mathbf{E}_\mathrm{t}^\mathrm{ap}$) are introduced, and the problem is split into two problems (inside V and outside V, respectively). The magnetic field inside and outside the enclosure can be expressed as a function of $\mathbf{M}_S$ (and thus of $\mathbf{E}_\mathrm{t}^\mathrm{ap}$) through a superposition integral involving the Green function of the enclosure and of the external environment, respectively. The integral equation is obtained by enforcing the continuity of the tangential magnetic field through the aperture.

The original problem is sketched in Figure 7.3*a*, where $\{\mathbf{J}_\mathrm{i}^\mathrm{int}, \mathbf{M}_\mathrm{i}^\mathrm{int}\}$ is the system of sources placed inside the enclosure, whereas $\{\mathbf{J}_\mathrm{i}^\mathrm{ext}, \mathbf{M}_\mathrm{i}^\mathrm{ext}\}$ is the system of sources placed outside. Based on the equivalence principle, the aperture is *short-circuited* and the equivalent surface magnetic current density $\mathbf{M}_S = -\mathbf{u}_n \times \mathbf{E}_\mathrm{t}^\mathrm{ap}$ is introduced ($\mathbf{u}_n$ is the unit vector normal to the aperture pointing outwards, i.e., from the enclosure to the external environment). This way the problem is transformed into that of Figure 7.3*b* and Figure 7.3*c*, where two subproblems are considered: the closed cavity excited by the sources $\{\mathbf{J}_\mathrm{i}^\mathrm{int}, \mathbf{M}_\mathrm{i}^\mathrm{int}\}$ and $-\mathbf{M}_S$ and the external environment (i.e., the free space with the enclosure as an obstacle) excited by the sources $\{\mathbf{J}_\mathrm{i}^\mathrm{ext}, \mathbf{M}_\mathrm{i}^\mathrm{ext}\}$ and $\mathbf{M}_S$ (the fact that the equivalent magnetic current has opposite sign in the two regions ensures the continuity of the tangential component of the electric field across the aperture).

By following the same steps as in Chapter 6 for the screen problem, the magnetic field outside the enclosure $\mathbf{H}^\mathrm{ext}$ can be written as

$$\mathbf{H}^\mathrm{ext}(\mathbf{r}) = \mathbf{H}_\mathrm{i}^\mathrm{ext}(\mathbf{r}) + \iint_A \underline{\mathbf{G}}_\mathrm{ext}^{HM}(\mathbf{r}, \mathbf{r}') \cdot \mathbf{M}_S(\mathbf{r}')\mathrm{d}S', \qquad \mathbf{r} \notin V, \tag{7.39}$$

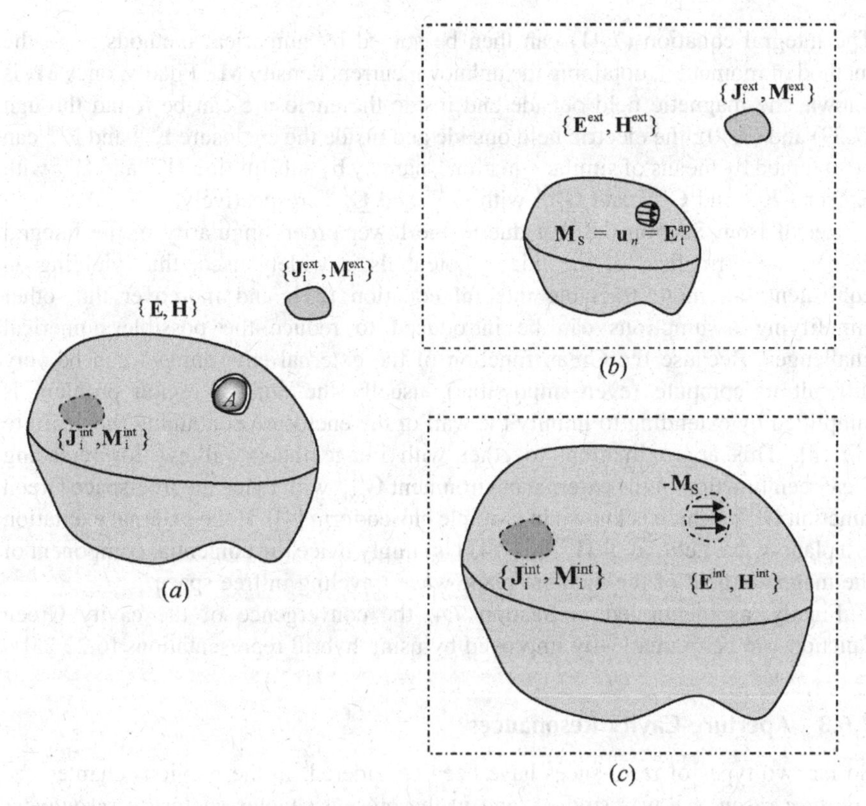

FIGURE 7.3 Original problem (*a*) and application of the equivalence principle (*b*, *c*).

where $\mathbf{H}_i^{ext}$ is the magnetic field due to the external sources $\mathbf{J}_i^{ext}$ and $\mathbf{M}_i^{ext}$ that would be present outside the enclosure with the aperture short-circuited, while $\underline{\mathbf{G}}_{ext}^{HM}$ is the *HM*-type dyadic Green function of the external environment (when a plane wave is considered as an external excitation, $\mathbf{H}_i^{ext}$ is the magnetic field due to the plane-wave excitation in the presence of the closed cavity). In the same way the magnetic field inside the enclosure $\mathbf{H}^{int}$ can be expressed as

$$\mathbf{H}^{int}(\mathbf{r}) = \mathbf{H}_i^{int}(\mathbf{r}) + \iint_A \underline{\mathbf{G}}_C^{HM}(\mathbf{r}, \mathbf{r}') \cdot [-\mathbf{M}_S(\mathbf{r}')] \mathrm{d}S', \qquad \mathbf{r} \in V, \tag{7.40}$$

where $\mathbf{H}_i^{int}$ is the magnetic field due to the internal sources $\mathbf{J}_i^{int}$ and $\mathbf{M}_i^{int}$ that would be present inside the enclosure with the aperture short-circuited, while $\underline{\mathbf{G}}_C^{HM}$ is the *HM*-type dyadic Green function of the enclosure (see (7.23)).

From (7.39) and (7.40), by imposing the continuity of the tangential magnetic field through the aperture, we obtain

$$\mathbf{u}_n \times \iint_A [\underline{\mathbf{G}}_C^{HM}(\mathbf{r}, \mathbf{r}') + \underline{\mathbf{G}}_{ext}^{HM}(\mathbf{r}, \mathbf{r}')] \cdot \mathbf{M}_S(\mathbf{r}') \, \mathrm{d}S' = \mathbf{u}_n \times [\mathbf{H}_i^{ext}(\mathbf{r}) - \mathbf{H}_i^{int}(\mathbf{r})], \quad \forall \mathbf{r} \in A. \tag{7.41}$$

The integral equation (7.41) can then be solved by numerical methods (e.g., the method of moments), obtaining the unknown current density $\mathbf{M}_S$. Finally, once $\mathbf{M}_S$ is known, the magnetic field outside and inside the enclosure can be found through (7.39) and (7.40); the electric field outside and inside the enclosure $\mathbf{E}^{\mathrm{ext}}$ and $\mathbf{E}^{\mathrm{int}}$ can be obtained by means of similar equations, namely by substituting $\mathbf{H}_{\mathrm{i}}^{\mathrm{ext}}$ and $\mathbf{H}_{\mathrm{i}}^{\mathrm{int}}$ with $\mathbf{E}_{\mathrm{i}}^{\mathrm{ext}}$ and $\mathbf{E}_{\mathrm{i}}^{\mathrm{int}}$ and $\underline{\mathbf{G}}_{\mathrm{ext}}^{HM}$ and $\underline{\mathbf{G}}_{\mathrm{C}}^{HM}$ with $\underline{\mathbf{G}}_{\mathrm{ext}}^{EM}$ and $\underline{\mathbf{G}}_{\mathrm{C}}^{EM}$, respectively.

Recall from Section 6.8 that due to the lower order singularity in the integral kernel, the formalism of the mixed potentials is widely used, thus yielding an equivalent but more tractable integral equation [21], and moreover that other simplifying assumptions can be introduced to reduce the possible numerical challenges. Because the Green function of the external environment can be very difficult to compute (even impossible), usually the outside region problem is simplified by extending to infinity the wall of the enclosure containing the aperture [17,18]. This approximation, together with image theory, allows for replacing the Green function of the external environment $\underline{\mathbf{G}}_{\mathrm{ext}}^{HM}$ with twice the free-space Green function $\underline{\mathbf{G}}_{\mathrm{free}}^{HM}$ (which is known in a simple closed form [4]). If the external excitation is a plane-wave field, $\mathbf{u}_n \times \mathbf{H}_{\mathrm{i}}^{\mathrm{ext}}$ in (7.41) is simply twice the tangential component of the magnetic field of the incident plane wave traveling in free space.

Finally, as mentioned in Section 7.4, the convergence of the cavity Green function can be dramatically improved by using hybrid representations [6,22,23].

7.6.3 Aperture-Cavity Resonances

So far two types of resonances have been considered. In the previous chapter, the *aperture resonance* was studied, and in the present chapter *enclosure resonances* have been analyzed in some detail. The characteristics of the involved phenomena are, however, similar. Under resonant conditions it has been shown that in fact either the amplitude of the equivalent magnetic current on the aperture cut in a perfectly conducting screen or the amplitude of a resonant mode inside an enclosure becomes very large. Of course, when an enclosure is coupled with the external environment through an aperture cut in one of its walls, both phenomena (aperture and enclosure resonances) are present, although they cannot be studied separately because the two problems are coupled. The structure can be regarded either as an aperture loaded by a cavity or as an enclosure perturbed by the presence of an aperture. In any case, we have to consider (and to define) the resonances of the system as a whole. Such resonances will be referred to as *aperture-cavity resonances*. In connection with the formalism developed in the previous subsection, the aperture-cavity resonance phenomenon occurs when the operating frequency is such that the amplitude of the equivalent magnetic current representing the aperture becomes very large, so that the amplitude of the field it radiates either inside or outside the enclosure is also very large (if compared with that radiated at other frequencies). In these cases it is obvious that the EM penetration is maximum and the SE can dramatically decrease. In order to rigorously define such aperture-cavity resonances, the concept of aperture admittance is introduced and the network formulation of Harrington and Mautz [24] is briefly resumed.

Let us consider the problem of Figure 7.3*a*, where an enclosure V is coupled with the external environment through an aperture A cut in its boundary S. Electromagnetic sources are present both inside ($\{\mathbf{J}_i^{\text{int}}, \mathbf{M}_i^{\text{int}}\}$) and outside ($\{\mathbf{J}_i^{\text{ext}}, \mathbf{M}_i^{\text{ext}}\}$) the enclosure. By applying the equivalence principle and by introducing the equivalent magnetic current density $\mathbf{M}_S$, the problem can be split into the two subproblems of Figure 7.3*b*. As already expressed in detail in the previous subsection, the integral equation (7.41) can be derived, which is rewritten here as

$$\mathbf{H}_t^{\text{int}}(\mathbf{M}_S) + \mathbf{H}_t^{\text{ext}}(\mathbf{M}_S) = \mathbf{H}_{i_t}^{\text{ext}} - \mathbf{H}_{i_t}^{\text{int}}, \qquad \text{on } A, \tag{7.42}$$

where the subscript t indicates the component of the magnetic fields tangential to the aperture. The method of moments can be used to solve equation (7.42); in particular, a set of expansion functions $\{\mathbf{M}_{S1}, \mathbf{M}_{S2}, \ldots, \mathbf{M}_{SN}\}$ can be introduced so that

$$\mathbf{M}_S = \sum_{n=1}^{N} V_n \mathbf{M}_{Sn}, \tag{7.43}$$

where V_n are unknown coefficients to be determined. By substituting (7.43) in (7.42) and by performing a Galerkin's testing procedure, the following set of equations is obtained:

$$\sum_{n=1}^{N} V_n \left[\langle \mathbf{M}_{Sm}, \mathbf{H}_t^{\text{int}}(\mathbf{M}_{Sn}) \rangle + \langle \mathbf{M}_{Sm}, \mathbf{H}_t^{\text{ext}}(\mathbf{M}_{Sn}) \rangle \right] = \left\langle \mathbf{M}_{Sm}, \mathbf{H}_{i_t}^{\text{ext}} \right\rangle - \left\langle \mathbf{M}_{Sm}, \mathbf{H}_{i_t}^{\text{int}} \right\rangle, \tag{7.44}$$

where $m = 1, 2, \ldots, N$, and the inner product $\langle \mathbf{u}, \mathbf{v} \rangle$ is defined as

$$\langle \mathbf{u}, \mathbf{v} \rangle = \iint_A \mathbf{u} \cdot \mathbf{v}^* \, dS'. \tag{7.45}$$

The system (7.44) can be solved to determine the coefficients V_n; these coefficients in turn are used to calculate the equivalent magnetic current $\mathbf{M}_S$, and then the fields. The system (7.44) can also be concisely written as

$$(\underline{\mathbf{Y}}^{\text{int}} + \underline{\mathbf{Y}}^{\text{ext}}) \cdot \mathbf{V} = \mathbf{I}_i, \tag{7.46}$$

where an admittance matrix $\underline{\mathbf{Y}}^{\text{int}}$ is defined for the enclosure

$$Y_{mn}^{\text{int}} = \langle \mathbf{M}_{Sm}, \mathbf{H}_t^{\text{int}}(\mathbf{M}_{Sn}) \rangle, \tag{7.47}$$

an admittance matrix $\underline{\mathbf{Y}}^{\text{ext}}$ is defined for the external environment

$$Y_{mn}^{\text{ext}} = \langle \mathbf{M}_{Sm}, \mathbf{H}_{\text{t}}^{\text{ext}}(\mathbf{M}_{Sn}) \rangle, \tag{7.48}$$

and a source vector $\mathbf{I}_{\text{i}}$ and a coefficient vector $\mathbf{V}$ are introduced as

$$\mathbf{I}_{\text{i}} = \left[\left\langle \mathbf{M}_{Sm}, \mathbf{H}_{\text{i}_\text{t}}^{\text{ext}} \right\rangle - \left\langle \mathbf{M}_{Sm}, \mathbf{H}_{\text{i}_\text{t}}^{\text{int}} \right\rangle \right]_{N\times 1}, \qquad \mathbf{V} = [V_m]_{N\times 1}. \tag{7.49}$$

The matrix equation (7.46) can thus be interpreted in terms of generalized networks as two networks (described by the admittance matrices $\underline{\mathbf{Y}}^{\text{int}}$ and $\underline{\mathbf{Y}}^{\text{ext}}$, respectively) in parallel with the current source $\mathbf{I}_{\text{i}}$.

To clarify the physics, let us assume to use only one basis function for the equivalent magnetic current density, meaning $\mathbf{M}_S = V_1 \mathbf{M}_{S1}$. The set of testing functions also consists of the only testing function $\mathbf{M}_{S1}$ and the matrix equation (7.46) reduces to the scalar equation

$$\left(Y^{\text{int}} + Y^{\text{ext}} \right) V_1 = I_{\text{i}}, \tag{7.50}$$

where

$$I_{\text{i}} = \iint_A [\mathbf{M}_{S1}(\mathbf{r}') \cdot \mathbf{H}_{\text{i}_\text{t}}^{\text{ext}}(\mathbf{r}') - \mathbf{M}_{S1}(\mathbf{r}') \cdot \mathbf{H}_{\text{i}_\text{t}}^{\text{int}}(\mathbf{r}')] \, \mathrm{d}S', \tag{7.51}$$

while the admittances are

$$Y^{\text{int}} = \iint_A [\mathbf{M}_{S1} \cdot \mathbf{H}_{\text{t}}^{\text{int}}(\mathbf{M}_{S1})] \, \mathrm{d}S', \quad Y^{\text{ext}} = \iint_A [\mathbf{M}_{S1} \cdot \mathbf{H}_{\text{t}}^{\text{ext}}(\mathbf{M}_{S1})] \, \mathrm{d}S'. \tag{7.52}$$

The total admittance seen by the current source is

$$Y_1 = Y^{\text{int}} + Y^{\text{ext}} = \left(G^{\text{int}} + jB^{\text{int}} \right) + \left(G^{\text{ext}} + jB^{\text{ext}} \right) = G_1 + jB_1, \tag{7.53}$$

where $G_1 = G^{\text{int}} + G^{\text{ext}}$ and $B_1 = B^{\text{int}} + B^{\text{ext}}$. It is immediate to see that the magnitude of the magnetic current density $\mathbf{M}_S = V_1 \mathbf{M}_{S1}$ is at a maximum when $|V_1|$ is at a maximum. The conductance G_1 can be assumed to be mildly dependent on frequency, so that the maxima occur when the condition $\text{Im}\{Y_1\} = B_1 = 0$ is satisfied; such a condition allows the *aperture-cavity resonances* to be defined. According to the perturbation theory, if the aperture is not too large, such resonances appear as small perturbations of the enclosure resonances. Finally, it should be noted that the aperture resonance considered in the previous chapter is defined by the condition $\text{Im}\{Y^{\text{ext}}\} = 0$. In fact there results $Y^{\text{int}} = Y^{\text{ext}}$ when only an aperture in a perfectly conducting screen is considered, so that from (7.53) $\text{Im}\{Y_1\} = 0$ implies

$\text{Im}\{Y^{\text{ext}}\} = 0$. Contrary to what happens for cavity resonances, the aperture resonance can be greatly perturbed by the loading cavity [25]. A detailed description of the aperture-cavity resonances, also in the presence of losses in the cavity walls, can be found in [26] and [27].

7.7 SMALL LOADING EFFECTS

Another issue that certainly deserves some considerations consists in the effects that the presence of some small object placed inside the enclosure can produce on the resonant frequencies of the cavity, with respect to the completely empty enclosure.

Let us consider an enclosure operating close to its mth resonant frequency that contains a small object (dielectric, magnetic, or conducting) inside. The main assumption of the following theory consists in considering the object small, so that it can be characterized by its dielectric and magnetic dipole moments $\mathbf{p}_{\text{e}} = \varepsilon \underline{\boldsymbol{\alpha}}_{\text{e}} \cdot \mathbf{E}^{\text{inc}}$ and $\mathbf{p}_{\text{m}} = -\mu_0 \underline{\boldsymbol{\alpha}}_{\text{m}} \cdot \mathbf{H}^{\text{inc}}$, respectively, similarly to what has been shown in the previous section for a small aperture. However, for a small 3D object, $\underline{\boldsymbol{\alpha}}_{\text{e}}$ and $\underline{\boldsymbol{\alpha}}_{\text{m}}$ are the electric and magnetic polarizability (3×3) tensors, respectively. For instance, the polarizabilities of a dielectric sphere of radius r_0 and permittivity $\varepsilon_{\text{S}} = \varepsilon_0 \varepsilon_{\text{Sr}}$ are

$$\underline{\boldsymbol{\alpha}}_{\text{e}} = 4\pi r_0^3 \frac{\varepsilon_{\text{Sr}} - 1}{\varepsilon_{\text{Sr}} + 2} \underline{\mathbf{I}}, \qquad \underline{\boldsymbol{\alpha}}_{\text{m}} = \underline{\mathbf{0}}, \tag{7.54}$$

while if the sphere is perfectly conducting there results

$$\underline{\boldsymbol{\alpha}}_{\text{e}} = 4\pi r_0^3 \underline{\mathbf{I}}, \qquad \underline{\boldsymbol{\alpha}}_{\text{m}} = 2\pi r_0^3 \underline{\mathbf{I}}, \tag{7.55}$$

where $\underline{\mathbf{I}}$ is the (3×3) identity dyadic [4]. Expressions for the polarizabilities of other objects can be found in [4].

Once the incident field $\{\mathbf{E}^{\text{inc}}, \mathbf{H}^{\text{inc}}\}$ (i.e., the field at the obstacle location in the absence of the obstacle) is known, the polarizability tensors can be used to obtain the electric and magnetic dipole moments of the obstacle ($\mathbf{p}_{\text{e}}$ and $\mathbf{p}_{\text{m}}$), which in turn are equivalent to electric and magnetic currents $\mathbf{J}_{\text{ob}} = j\omega \mathbf{p}_{\text{e}}$ and $\mathbf{M}_{\text{ob}} = j\omega \mathbf{p}_{\text{m}}$. Such currents can then be used in (7.28) and (7.29) in place of $\mathbf{J}_{\text{i}}$ and $\mathbf{M}_{\text{i}}$, respectively, in order to obtain the field scattered by the object inside the cavity (the incident field is assumed to be the mth resonant mode of the unloaded cavity). In particular, it can be shown that a pair of homogeneous equations for the coefficients e_m and h_m can be derived [4]. By equating the determinant of such a homogeneous system to zero, the perturbed resonant frequency of the mode is obtained as a function of the perturbation terms $\Delta_{\text{e}} = \mathbf{E}_m \cdot \underline{\boldsymbol{\alpha}}_{\text{e}} \cdot \mathbf{E}_m$ and

$\Delta_{\text{m}} = \mathbf{E}_m \cdot \underline{\boldsymbol{\alpha}}_{\text{m}} \cdot \mathbf{E}_m$ [4]. In the general case of an enclosure with lossy walls, there results

$$k_m^{\text{pert}} \simeq k_m \left(1 - \frac{1-j}{2Q_m} - \frac{\Delta_{\text{e}} + \Delta_{\text{m}}}{2}\right), \tag{7.56}$$

Since the perturbation terms are of order V_{ob}/V (where V_{ob} is the volume occupied by the small obstacle), the change in the resonant frequency also is of order V_{ob}/V. Interestingly, however, if the small object is placed in a null of the unperturbed mth resonant mode, the resonant frequency of the enclosure does not change, at least at the first order of approximation.

7.8 THE RECTANGULAR ENCLOSURE

As mentioned above, the calculation of the resonant modes of a given cavity can be performed analytically only in very few cases and for very simple geometries. In general, numerical approaches are needed. In this section we aim at specializing the cavity theory, deriving the expressions of the electric and magnetic fields of the resonant modes for the most common enclosure, which is the rectangular enclosure.

The rectangular enclosure is a particular type of cylindrical enclosure whose cross section is a rectangle, as depicted in Figure 7.4. Without loss of generality it is assumed that the edges of the cavity are directed along the axes x, y, and z, having lengths l_x, l_y, and l_z, respectively. In particular, a metallic uniform cylindrical enclosure can be seen as a metallic uniform cylindrical waveguide

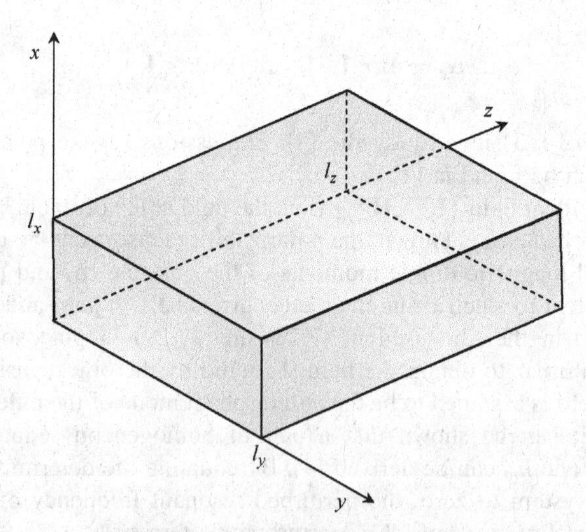

FIGURE 7.4 Rectangular enclosure.

whose terminations along the longitudinal direction are closed off by two metallic plates; in this case it is well known that the modes supported by the structure can be split in TE and TM modes with respect to the longitudinal direction. For the rectangular enclosure each edge can actually be considered as the longitudinal direction of the original rectangular waveguide, so the modes of the rectangular enclosure can be split in $\{TE^{(x)}, TM^{(x)}\}$ modes or in $\{TE^{(y)}, TM^{(y)}\}$ modes, or even in $\{TE^{(z)}, TM^{(z)}\}$ modes.

For cylindrical enclosures the solenoidal vector eigenfunctions $\mathbf{E}_m$ (i.e., the resonant modes) can be divided into two classes of modes: the $\mathbf{M}_m$ modes and the $\mathbf{N}_m$ modes [4]. The $\mathbf{M}_m$ modes (which give rise to TE fields) are given by $\mathbf{M}_m = \nabla \times \mathbf{u}_i \phi_m$ (where $i = x, y, z$ is the direction with respect to which the field is TE), while the $\mathbf{N}_m$ modes (which give rise to TM fields) are given by $\mathbf{N}_m = \nabla \times \nabla \times \mathbf{u}_i \psi_m$. The functions ψ_m and ϕ_m are the eigenfunctions of the boundary-value problems (7.1) and (7.2), respectively. Because of the simple geometry of the rectangular enclosure, such boundary-value problems can be solved in a simple closed-form by means of the method of separation of variables [4]. Without loss of generality we can choose $\mathbf{u}_i = \mathbf{u}_z$, which results in

$$\phi_m(\mathbf{r}) = \phi_{m_x,m_y,m_z}(x,y,z) = \Phi_{m_x,m_y,m_z} \cos\left(\frac{m_x\pi}{l_x}x\right) \cos\left(\frac{m_y\pi}{l_y}y\right) \sin\left(\frac{m_z\pi}{l_z}z\right) \quad (7.57)$$

and

$$\psi_m(\mathbf{r}) = \psi_{m_x,m_y,m_z}(x,y,z) = \Psi_{m_x,m_y,m_z} \sin\left(\frac{m_x\pi}{l_x}x\right) \sin\left(\frac{m_y\pi}{l_y}y\right) \cos\left(\frac{m_z\pi}{l_z}z\right), \quad (7.58)$$

where $m = \{m_x, m_y, m_z\}$ is a triple set of nonnegative integers. The coefficients Φ_{m_x,m_y,m_z} and Ψ_{m_x,m_y,m_z} that ensure the correct normalization are

$$\Phi_{m_x,m_y,m_z} = \sqrt{\frac{\varepsilon_{0m_x}\varepsilon_{0m_y}\varepsilon_{0m_z}}{l_x l_y l_z}} \sqrt{\frac{1}{(m_x\pi/l_x)^2 + (m_y\pi/l_y)^2}}, \quad \Psi_{m_x,m_y,m_z} = \frac{\Phi_{m_x,m_y,m_z}}{k_{m_x,m_y,m_z}}, \quad (7.59)$$

where ε_{0j} is the Neumann symbol (i.e., $\varepsilon_{0j} = 1$ for $j = 0$ and $\varepsilon_{0j} = 2$ for $j \neq 0$) and

$$k_{m_x,m_y,m_z} = k_m = k_{\psi m} = k_{\phi m} = \sqrt{\left(\frac{m_x\pi}{l_x}\right)^2 + \left(\frac{m_y\pi}{l_y}\right)^2 + \left(\frac{m_z\pi}{l_z}\right)^2} \quad (7.60)$$

are the corresponding eigenvalues. More precisely, to avoid trivial eigenfunctions, for $k_{\phi m}$, there has to be $m_z > 0$ and either m_x or m_y being zero but not both, whereas for $k_{\phi m}$, there has to be $m_x > 0$ and $m_y > 0$. The electric and magnetic fields of the $\mathrm{TE}^{(z)}_{m_x,m_y,m_z}$ resonant mode are thus

$$\begin{aligned}\mathbf{E}^{\mathrm{TE}}_{m_x,m_y,m_z}(\mathbf{r}) = \Phi_{m_x,m_y,m_z}\Bigg[&\left(\frac{m_y\pi}{l_y}\right)\cos\left(\frac{m_x\pi}{l_x}x\right)\sin\left(\frac{m_y\pi}{l_y}y\right)\sin\left(\frac{m_z\pi}{l_z}z\right)\mathbf{u}_x \\ &-\left(\frac{m_x\pi}{l_x}\right)\sin\left(\frac{m_x\pi}{l_x}x\right)\cos\left(\frac{m_y\pi}{l_y}y\right)\sin\left(\frac{m_z\pi}{l_z}z\right)\mathbf{u}_y\Bigg]\end{aligned} \tag{7.61}$$

and

$$\begin{aligned}\mathbf{H}^{\mathrm{TE}}_{m_x,m_y,m_z}(\mathbf{r}) = j\frac{\Psi_{m_x,m_y,m_z}}{\eta}\Bigg\{&-\left(\frac{m_x\pi}{l_x}\right)\left(\frac{m_z\pi}{l_z}\right)\sin\left(\frac{m_x\pi}{l_x}x\right)\cos\left(\frac{m_y\pi}{l_y}y\right)\cos\left(\frac{m_z\pi}{l_z}z\right)\mathbf{u}_x \\ &-\left(\frac{m_y\pi}{l_y}\right)\left(\frac{m_z\pi}{l_z}\right)\cos\left(\frac{m_x\pi}{l_x}x\right)\sin\left(\frac{m_y\pi}{l_y}y\right)\cos\left(\frac{m_z\pi}{l_z}z\right)\mathbf{u}_y \\ &+\left[\left(\frac{m_x\pi}{l_x}\right)^2+\left(\frac{m_y\pi}{l_y}\right)^2\right]\cos\left(\frac{m_x\pi}{l_x}x\right)\cos\left(\frac{m_y\pi}{l_y}y\right)\sin\left(\frac{m_z\pi}{l_z}z\right)\mathbf{u}_z\Bigg\}.\end{aligned} \tag{7.62}$$

The electric and magnetic fields of the $\mathrm{TM}^{(z)}_{m_x,m_y,m_z}$ resonant mode are instead

$$\begin{aligned}\mathbf{E}^{\mathrm{TM}}_{m_x,m_y,m_z}(\mathbf{r}) = \Psi_{m_x,m_y,m_z}\Bigg\{&-\left(\frac{m_x\pi}{l_x}\right)\left(\frac{m_z\pi}{l_z}\right)\cos\left(\frac{m_x\pi}{l_x}x\right)\sin\left(\frac{m_y\pi}{l_y}y\right)\sin\left(\frac{m_z\pi}{l_z}z\right)\mathbf{u}_x \\ &-\left(\frac{m_y\pi}{l_y}\right)\left(\frac{m_z\pi}{l_z}\right)\sin\left(\frac{m_x\pi}{l_x}x\right)\cos\left(\frac{m_y\pi}{l_y}y\right)\sin\left(\frac{m_z\pi}{l_z}z\right)\mathbf{u}_y \\ &+\left[\left(\frac{m_x\pi}{l_x}\right)^2+\left(\frac{m_y\pi}{l_y}\right)^2\right]\sin\left(\frac{m_x\pi}{l_x}x\right)\sin\left(\frac{m_y\pi}{l_y}y\right)\cos\left(\frac{m_z\pi}{l_z}z\right)\mathbf{u}_z\Bigg\}\end{aligned} \tag{7.63}$$

and

$$\begin{aligned}\mathbf{H}^{\mathrm{TM}}_{m_x,m_y,m_z}(\mathbf{r}) = j\frac{\Phi_{m_x,m_y,m_z}}{\eta}\Bigg[&\left(\frac{m_y\pi}{l_y}\right)\sin\left(\frac{m_x\pi}{l_x}x\right)\cos\left(\frac{m_y\pi}{l_y}y\right)\cos\left(\frac{m_z\pi}{l_z}z\right)\mathbf{u}_x \\ &-\left(\frac{m_x\pi}{l_x}\right)\cos\left(\frac{m_x\pi}{l_x}x\right)\sin\left(\frac{m_y\pi}{l_y}y\right)\cos\left(\frac{m_z\pi}{l_z}z\right)\mathbf{u}_y\Bigg],\end{aligned} \tag{7.64}$$

where m_x, m_y, and m_z are nonnegative integers. In particular, from (7.61) and (7.62) it can be seen that in order to have a nontrivial solution, $\mathrm{TE}^{(z)}$ resonant modes must be characterized by a triple set of nonnegative integers $\{m_x, m_y, m_z\}$ such that that $m_z > 0$ and either m_x or m_y are zero, but not both. On the other hand, from (7.63) and (7.64),

$TM^{(z)}$ nonzero resonant modes are characterized by a triple set of nonnegative integers $\{m_x, m_y, m_z\}$ such that m_z may be zero, but m_x and m_y must be greater than zero. According to the formulas above, the integer m_i $(i = x, y, z)$ indicates the number of half-sinusoid variations of the standing-wave pattern of the field along the i direction.

From (7.60) it follows that the resonant frequencies of the rectangular enclosures are

$$f_{m_x,m_y,m_z} = \frac{c}{2\pi} k_{m_x,m_y,m_z} = \frac{c}{2} \sqrt{\left(\frac{m_x}{l_x}\right)^2 + \left(\frac{m_y}{l_y}\right)^2 + \left(\frac{m_z}{l_z}\right)^2} \tag{7.65}$$

for both TE and TM modes. Interestingly, except for those modes that have a zero m_i $(i = x, y, z)$ index, TE and TM modes corresponding to the same indexes $\{m_x, m_y, m_z\}$ are always degenerate. Moreover, if the edge lengths are in the ratio of integers, many other different TE and TM modes can share the same resonant frequency (degenerate modes), as it will be shown in detail in the next section.

Finally, the irrotational eigenfunctions $\mathbf{F}_m$ (corresponding to the longitudinal $\mathbf{L}_m$ modes [4]) are given by $\mathbf{F}_m = \nabla \chi_m$, where

$$\chi_m(\mathbf{r}) = \chi_{m_x,m_y,m_z}(x,y,z) = \mathrm{X}_{m_x,m_y,m_z} \sin\left(\frac{m_x\pi}{l_x}x\right) \sin\left(\frac{m_y\pi}{l_y}y\right) \sin\left(\frac{m_z\pi}{l_z}z\right), \tag{7.66}$$

and the coefficient X_{m_x,m_y,m_z} is

$$\mathrm{X}_{m_x,m_y,m_z} = \frac{1}{k_{m_x,m_y,m_z}} \sqrt{\frac{\varepsilon_{0m_x}\varepsilon_{0m_y}\varepsilon_{0m_z}}{l_x l_y l_z}}. \tag{7.67}$$

7.8.1 Symmetry Considerations

To better understand the numerical results that will be presented in the next section, it is useful to give some consideration to the symmetry of the modes of a rectangular enclosure. Let us consider again the rectangular enclosure of Figure 7.4. The planes $i = l_i/2$ will be indicated here as π_i planes $(i = x, y, z)$. According to expressions (7.61) to (7.64), both components of the electric field tangential to the π_i plane vanish on the π_i plane if the index m_i is an even integer, while both components of the magnetic field tangential to the π_i plane vanish on the π_i plane if the index m_i is an odd integer. This means that the π_i planes are symmetry planes. In particular, they are PEC planes if the mode has an even m_i index, and PMC planes if the mode has an odd m_i index. Let us consider, for example, the TE_{101} mode. Based on the considerations above, it can immediately be inferred that both the π_x and the π_z planes are PMC planes (i.e., $H_y = H_z = 0$ on π_x and $H_x = H_y = 0$ on π_z), while the π_y plane is a PEC plane (i.e., $E_x = E_z = 0$ on π_y). These considerations can be useful when we consider the problem of the mode excitation. So let us consider an infinitesimal electric dipole placed in the middle of the enclosure (i.e., at $\mathbf{r}_0 = (l_x/2, l_y/2, l_z/2)$) and directed along one of the principal axes, for example, the x axis. Based on (7.28), such a dipole can excite only modes having a nonzero component E_x of the electric field at $\mathbf{r}_0$, so all the modes characterized by even indexes m_y and m_z will not be excited (TE_{102}, TM_{120}, TE_{120}, TE_{022}, etc.). Analogously, let us consider an infinitesimal magnetic dipole (e.g., a very small loop) placed at $\mathbf{r}_0$ and directed along the z axis. Based on (7.29), such a source can excite only modes having a nonzero component H_z of the magnetic field at $\mathbf{r}_0$, so all the modes characterized by odd indexes m_x and m_y will be not excited (TE/TM_{110}, TE/TM_{111}, TE/TM_{130}, etc.).

All these considerations still remain valid when an aperture is introduced in a wall of the cavity or an external source is considered, provided that some geometrical symmetry of the structure is maintained. For instance, suppose that a circular aperture is cut in the center of the face lying on the $z = 0$ plane and a uniform plane wave impinges normally to the aperture with the electric field polarized along the x axis. It can be immediately deduced that the π_x and π_z planes are PEC and PMC planes, respectively. This means that the resonant modes having different types of symmetry planes (i.e., the modes having either even m_x or odd m_y indexes) will not be excited. Other examples will be shown in the next section.

7.9 SHIELDING EFFECTIVENESS OF A RECTANGULAR ENCLOSURE WITH A CIRCULAR HOLE

In this section we provide numerical results for two typical situations in EM shielding problems. In particular, we consider a perfectly conducting rectangular enclosure with edges along the x, y, and z axes, having lengths l_x, l_y, and l_z, respectively. Such an enclosure has a circular aperture with radius R cut in the wall $z = l_z$ centered on this face; the π_x and π_z planes are thus symmetry planes. Both external sources (impinging uniform plane waves) and internal sources (elemental electric and magnetic dipoles) will be considered. For external sources, the electric SE is evaluated inside the enclosure, whereas for internal sources, it is evaluated outside the enclosure. The dimensions of the enclosure under test are $l_x = 20$ cm, $l_y = 40$ cm, and $l_z = 60$ cm, and the radius of the circular aperture is $R = 1.5$ cm. In the frequency range from dc to $f = 1$ GHz, the closed cavity presents seven different resonant frequencies associated with the different modes:

$$
\begin{aligned}
f_{011}^{TE} &= 450.38\,\text{MHz},\\
f_{012}^{TE} &= 624.57\,\text{MHz},\\
f_{101}^{TE} &= f_{021}^{TE} = 790.02\,\text{MHz},\\
f_{013}^{TE} &= f_{110}^{TM} = 837.94\,\text{MHz},\\
f_{111}^{TE} &= f_{111}^{TM} = 874.39\,\text{MHz},\\
f_{102}^{TE} &= f_{022}^{TE} = 900.76\,\text{MHz},\\
f_{112}^{TE} &= f_{112}^{TM} = 975.56\,\text{MHz}.
\end{aligned}
\tag{7.68}
$$

In the next subsections the effects of such resonances on the shielding performance of the perforated enclosure will be described.

7.9.1 External Sources: Plane-Wave Excitation

A uniform plane wave is assumed to impinge on the considered rectangular enclosure with incident electric field

$$\mathbf{E}^{\mathrm{inc}}(\mathbf{r}) = \left(\mathbf{u}_\theta \cos\alpha + \mathbf{u}_\phi \sin\alpha\right) e^{j\mathbf{k}^{\mathrm{inc}}\cdot\mathbf{r}}, \tag{7.69}$$

where α is the real polarization angle, $\mathbf{r}$ is the observation point, and $\mathbf{k}^{\mathrm{inc}}$ is the incident wavevector. The latter has rectangular components $k_x^{\mathrm{inc}} = k_0 \sin\theta^{\mathrm{inc}} \cos\phi^{\mathrm{inc}}$, $k_y^{\mathrm{inc}} = k_0 \sin\theta^{\mathrm{inc}} \sin\phi^{\mathrm{inc}}$, and $k_z^{\mathrm{inc}} = -k_0 \cos\theta^{\mathrm{inc}}$, while the incidence angles θ^{inc} and ϕ^{inc} also define the relevant unit vectors $\mathbf{u}_\theta$ and $\mathbf{u}_\phi$. The relevant EM problem is sketched in Figure 7.5. The electric SE calculated at the center of the enclosure (i.e., at $\mathbf{r} = (l_x/2, l_y/2, l_z/2))$ is reported in Figure 7.6 for different polarizations and incident angles.

The plane of incidence is the $\phi = \pi/2$ plane; the results for TE_z polarizations (i.e., $\alpha = \pi/2$) are shown in Figure 7.6*a*, while those for TM_z polarizations (i.e., $\alpha = 0$) are reported in Figure 7.6*b*. From Figure 7.6*a* it can be seen in the shown frequency range that the electric SE of the enclosure for TE_z polarization improves by increasing the incidence angle from $\theta^{\mathrm{inc}} = 0$, whereas an almost opposite trend can be observed from Figure 7.6*b* for TM_z polarization. Interestingly, for rectangular

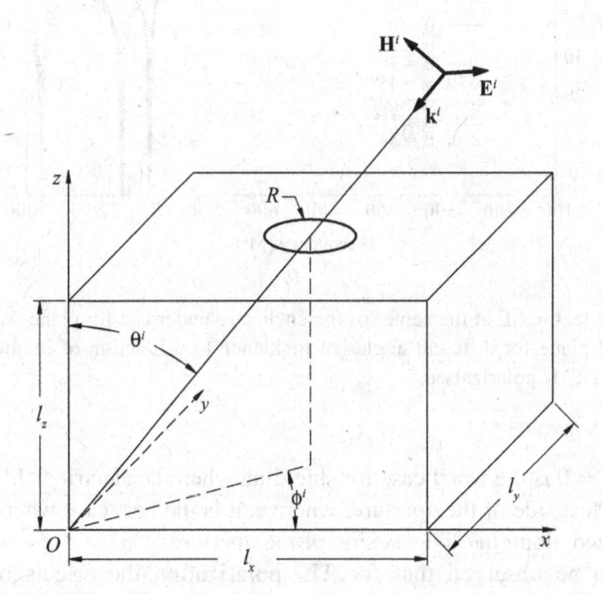

FIGURE 7.5 Uniform plane wave impinging on a rectangular enclosure with a circular aperture cut in one of its walls.

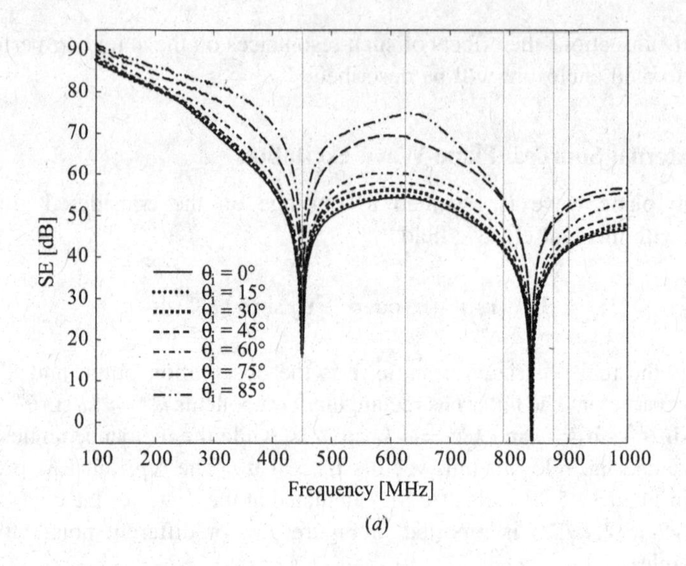

(*a*)

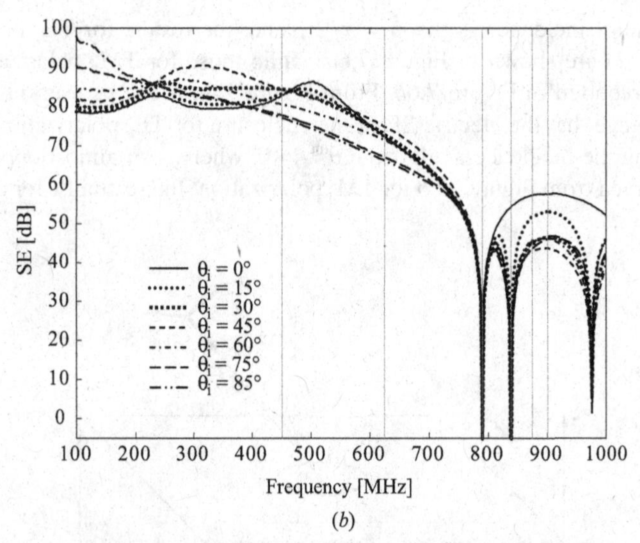

(*b*)

FIGURE 7.6 Electric SE at the center of the enclosure under test for plane-wave incidence on the $\phi = \pi/2$ plane for different angles of incidence as a function of frequency. (*a*) TE_z polarization; (*b*) TM_z polarization.

apertures, $\theta^{\text{inc}} = 0$ is the worst case for shielding when the electric field is polarized along the shortest side of the aperture, whereas it is the best case when the electric field is polarized along the longest side of the aperture.

It can also be observed that for TE_z polarization the effects of only two resonances are visible. In particular, such resonances are those corresponding to the TE_{011} and TE_{013} modes, which are the only excited resonant modes with a nonzero electric field at the observation point $\mathbf{r} = (l_x/2, l_y/2, l_z/2)$. Because the incident electric field is directed along $\mathbf{u}_x$, for symmetry considerations the π_x plane is in fact a PEC plane, so only modes with zero y and z components of the electric field at $x = l_x/2$ can be excited (i.e., the TE_{011}, TE_{012}, TE_{021}, TE_{013}, and TE_{022} modes; see (7.61)–(7.64)). However, at the observation point, the TE_{012}, TE_{021}, and TE_{022} modes have a zero electric field, and their effect on the electric SE at that point is zero (their effect would instead be visible in the magnetic SE, since their magnetic field at $\mathbf{r} = (l_x/2, l_y/2, l_z/2)$ is different from zero). Moreover it is important that for observation points different from $\mathbf{r} = (l_x/2, l_y/2, l_z/2)$ the resonance effects of the other modes can be pronounced and the electric SE dramatically reduced. It is thus evident that the evaluation of the SE at only one point of the enclosure (especially if it lies on some symmetry plane) can give rise to unreliable predictions.

It can be observed that for TM_z polarization the effects of three resonances are visible. In particular, such resonances are those corresponding to the TE_{101}, TM_{110}, and TM_{112} modes, which are the only excited resonant modes with a nonzero electric field at the observation point $\mathbf{r} = (l_x/2, l_y/2, l_z/2)$. Since the incident magnetic field is directed along $\mathbf{u}_x$, for symmetry considerations the π_x plane is now a PMC plane, so only modes with zero y and z components of the magnetic field at $x = l_x/2$ can be excited, meaning the TE_{101}, TM_{110}, TE_{112}, and TM_{112} (see (7.61)–(7.64)). However, at the observation point, the TE_{112} mode has a zero electric field, and its effect on the electric SE at that point is zero (its effect would instead be significant in the magnetic SE, since its magnetic field at $\mathbf{r} = (l_x/2, l_y/2, l_z/2)$ is different from zero). Also in this case it is important to note that for observation points different from $\mathbf{r} = (l_x/2, l_y/2, l_z/2)$ the resonance effects of the other modes could be clearly visible and the electric SE dramatically reduced. The effects of mode resonances on the spatial distribution of the EM field inside the enclosure can be clearly appreciated by comparing Figure 7.7 and Figure 7.8.

In Figure 7.7 the spatial distribution of the absolute value of the electric field excited inside the enclosure by an incident TE_z plane wave at the resonant frequency $f_{011}^{TE} = 450.38$ MHz is reported in the planes π_x, π_y, and π_z (Figure 7.7*a*, *b*, and *c*, respectively). It can be seen that the field distribution is that of the TE_{011} mode, thus revealing the dominant character of such a mode at its resonant frequency.

In Figure 7.8 the same spatial distribution as in Figure 7.7 is reported but at $f = 200$ MHz, which is a frequency far below the first resonant frequency. Two main differences can be noted: first, at $f = 200$ MHz the electric-field distribution does not resemble any of the resonant modes of the cavity and, second, at $f = 200$ MHz the peak-field values are much smaller than those found at $f_{011}^{TE} = 450.38$ MHz (by approximately three orders of magnitude). The former difference is due to the fact that all the modes contribute to the field distribution, none of them being dominant, and their superposition does not have a modal configuration; the latter is a consequence of the nonresonant behavior of the field whose amplitude is instead extremely large at the resonant frequency (ideally infinite, for a closed cavity).

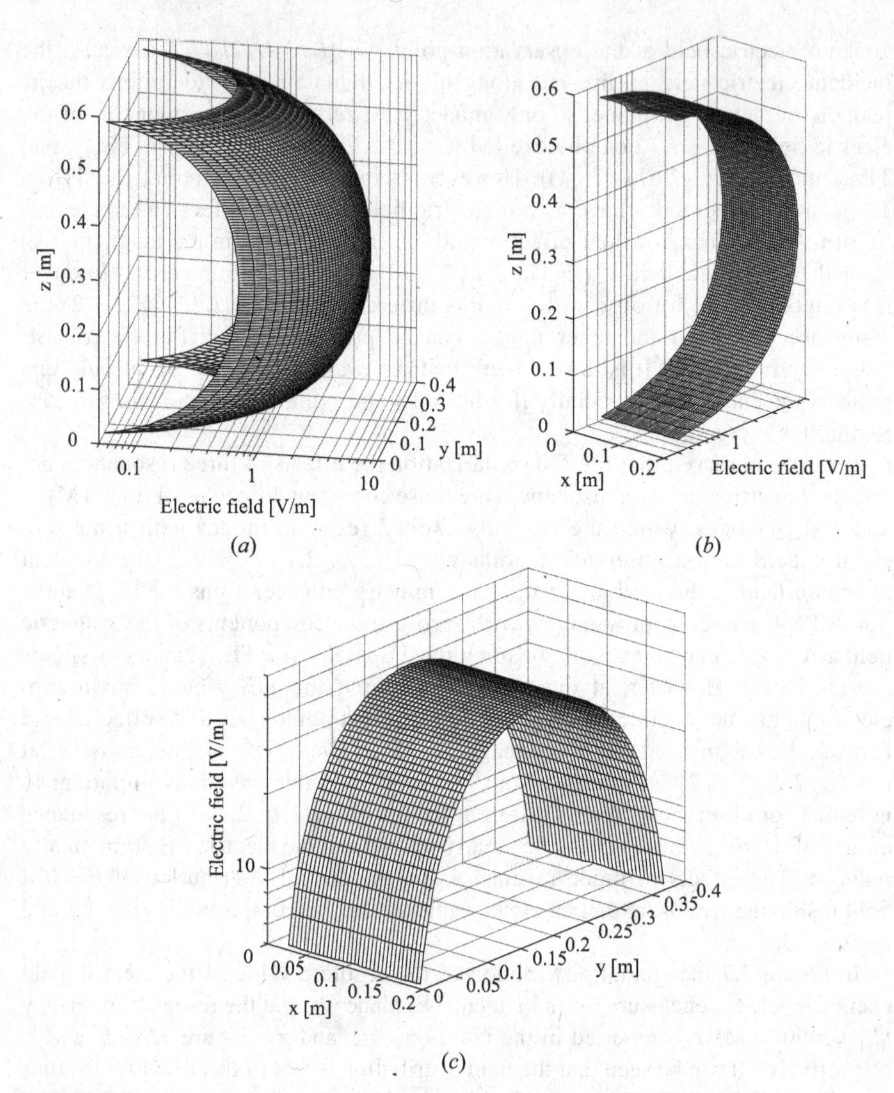

FIGURE 7.7 Absolute value of the electric field inside the enclosure under TE_z plane-wave incidence at the frequency $f_{011}^{\text{TE}} = 450.38$ MHz. (*a*) π_x plane; (*b*) π_y plane; (*c*) π_z plane.

7.9.2 Internal Sources: Electric and Magnetic Dipole Excitations

Elemental electric or magnetic dipoles can be assumed to be internal sources, each having the mathematical representation

$$\begin{aligned} \mathbf{J}_{\mathrm{i}}(\mathbf{r}) &= \mathbf{u}_i \delta(x - x')\delta(y - y')\delta(z - z'), \\ \mathbf{M}_{\mathrm{i}}(\mathbf{r}) &= \mathbf{u}_i \delta(x - x')\delta(y - y')\delta(z - z'), \end{aligned} \tag{7.70}$$

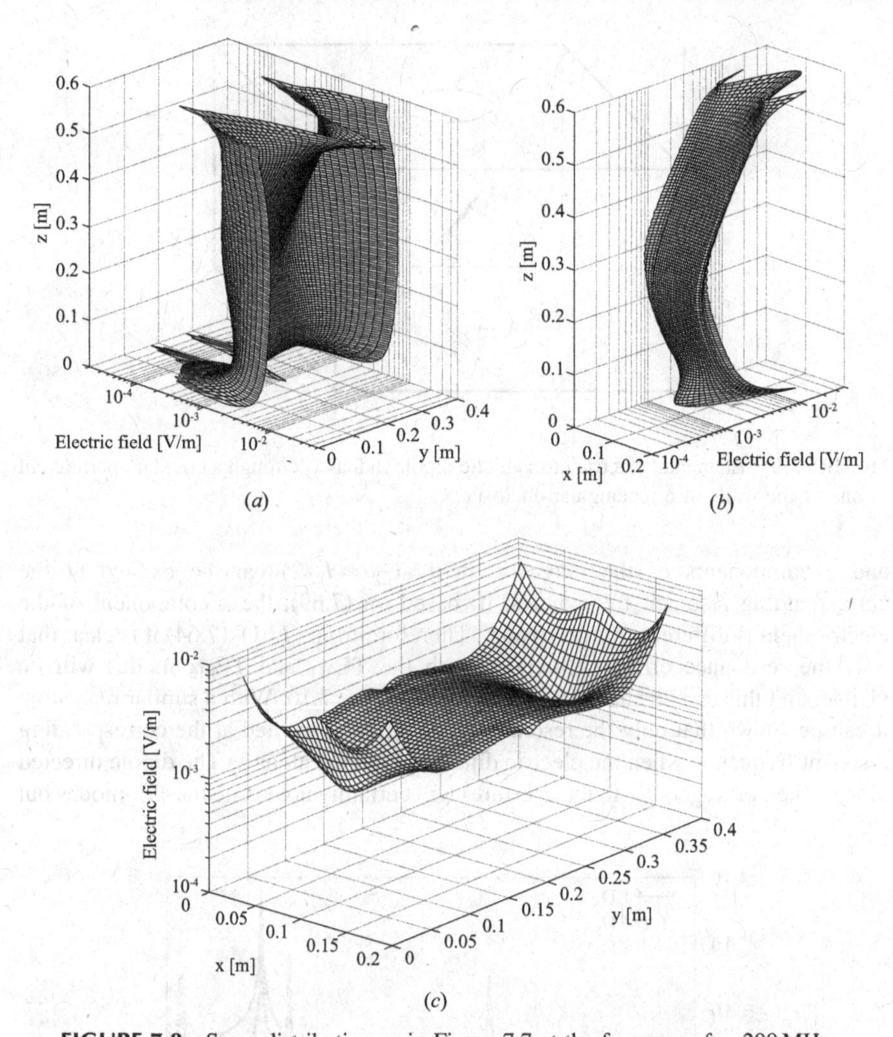

FIGURE 7.8 Same distribution as in Figure 7.7 at the frequency $f = 200$ MHz.

where $\mathbf{u}_i$ $(i = x, y, z)$ indicates the direction of the source. The relevant EM problem is sketched in Figure 7.9. It is interesting to observe how the position and the orientation of the dipoles affect the behavior of the electric field outside the enclosure.

In Figure 7.10 the magnitude of the electric field radiated at the point $\mathbf{r} = (l_x/2, l_y/2, l_z + \Delta z)$ by an electric dipole placed at the center of the enclosure (i.e., $\mathbf{r}' = (l_x/2, l_y/2, l_z/2)$) is reported as a function of frequency. In particular, three possible orientations along the principal axes are considered besides $\Delta z = 3$ m. Let us consider first the electric dipole directed along $\mathbf{u}_x$. The symmetry planes π_x and π_y are PEC and PMC planes, respectively. So only the modes satisfying such boundary conditions (i.e., zero y and z components of the electric field at $x = l_x/2$ and zero x

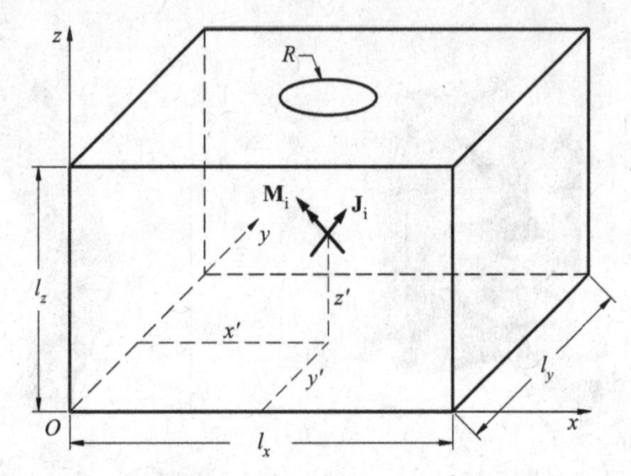

FIGURE 7.9 Elemental electric or magnetic dipole radiating through a circular aperture cut in one of the walls of a rectangular enclosure.

and z components of the magnetic field at $y = l_y/2$) can be excited at the corresponding resonant frequencies, if, based on (7.69), the x component of the electric field is different from zero at $\mathbf{r}'$. Therefore from (7.61)–(7.64) it is clear that only the resonance effects associated with the TE_{011} and TE_{013} modes will be visible, and this can be checked by looking at Figure 7.10. With a similar reasoning it can be shown that only the resonant mode TE_{101} is excited at the corresponding resonant frequency when the electric dipole is directed along y. The dipole directed along z (i.e., orthogonally to the aperture) will certainly not excite the TE_z modes but

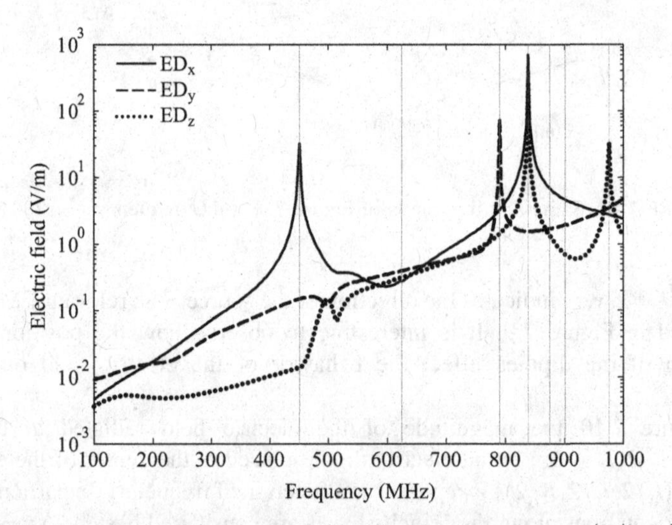

FIGURE 7.10 Magnitude of the electric field radiated at the point $\mathbf{r} = (l_x/2, l_y/2, l_z + \Delta z)$ ($\Delta z = 3$ m) by an elemental electric dipole located at the center of the metallic rectangular enclosure under test $\mathbf{r}' = (l_x/2, l_y/2, l_z/2)$. Three possible orientations of the dipole are considered.

rather, by the usual considerations, only the resonant modes TM_{110} and TM_{112} (in this case both the π_x and π_y planes are PMC planes). Finally, from Figure 7.10, it can be seen that according to Bethe's theory, the best case for shielding at low frequencies (before any resonance effect) is the electric-dipole orientation orthogonal to the aperture (this configuration in fact gives rise to a zero incident tangential magnetic field). Importantly, if the dipole source is moved off the center of the enclosure, some symmetry properties will be lost, and this effect implies that other modes can give rise to resonant effects, as shown in Figure 7.11. From (7.28) and (7.29) it is evident that the presence of such resonances can be established by observing whether the corresponding modes (expressed by (7.61) through (7.64)) have a nonzero component of the electric field along the dipole direction at the source point.

In Figure 7.12 an elemental magnetic dipole placed at the center of the enclosure is considered as a source and the magnitude of the electric field radiated at the point $\mathbf{r} = (l_x/2, l_y/2, l_z + \Delta z)$ (still with $\Delta z = 3$ m) is reported as a function of frequency. When the magnetic dipole is directed along x, the symmetry planes π_x and π_y are PMC and PEC planes, respectively, and among the modes that satisfy such boundary conditions, only those having a nonzero x component of the magnetic field at $\mathbf{r}' = (l_x/2, l_y/2, l_z/2)$ will be excited at the resonant frequencies. From (7.61)–(7.64) it is clear that only the TE_{102} mode satisfies such conditions and its effect is visible in the peak of the field at $f_{102}^{TE} = 900.76$ MHz. On the other hand, when the centered magnetic dipole is directed along y, the symmetry planes π_x and π_y are PEC and PMC planes, respectively, and therefore, by (7.61)–(7.64), the only excited mode that gives a resonant contribution in the frequency range between dc and $f = 1$ GHz is the TE_{012} mode (whose effect is visible in Figure 7.11 when the peak of the field is at $f_{012}^{TE} = 624.57$ MHz). Finally, if the centered magnetic dipole is orthogonal to the aperture (i.e., along the z

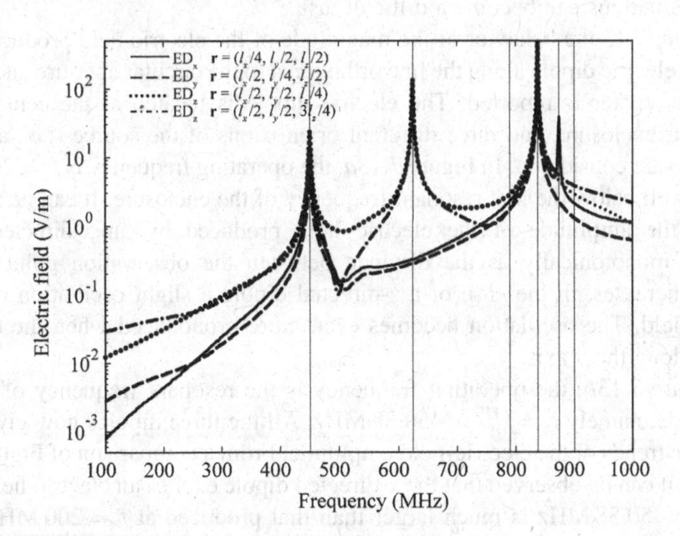

FIGURE 7.11 Same distribution as in Figure 7.10 but with the electric dipole placed off-center in the enclosure.

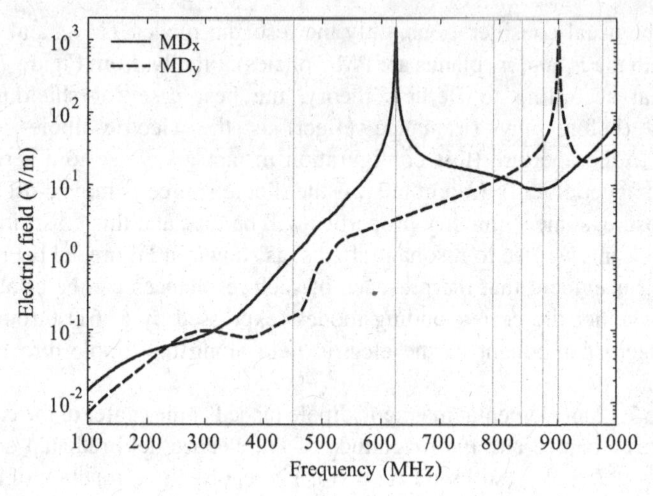

FIGURE 7.12 Same distribution as in Figure 7.10 but with an elemental magnetic dipole source.

direction), both the π_x and π_y planes are PEC planes, so at the observation point (which lies along the intersection between such planes) no electric field can be radiated.

Another important issue concerns the behavior of the electric field as a function of the distance from the aperture outside the enclosure. Although in the far field only the radiating contribution is present (so a monotonically decreasing field amplitude can be expected by increasing the distance of the observation point from the aperture), close to the aperture capacitive and inductive contributions can in fact give rise to oscillations in the field amplitude. Then the determination of the worst point for SE evaluations can become a difficult task.

In Figure 7.13 the behavior of the magnitude of the electric field produced by an elemental electric dipole along the line orthogonal to the circular aperture and passing through its center is reported. The electric dipole is located at the center of the rectangular enclosure, and three different orientations of the source (i.e., along the main axes) are considered. In Figure 7.13*a*, the operating frequency is $f = 200$ MHz, which is well below the first resonant frequency of the enclosure. It can be seen that although the amplitude of the electric field produced by an *x*-directed dipole decreases monotonically as the distance between the observation point and the aperture increases, in the case of a *z*-directed dipole a slight oscillation occurs in the near field. The oscillation becomes even more pronounced when the dipole is directed along the *y* axis.

In Figure 7.13*b*, the operating frequency is the resonant frequency of the first TE_{011} mode, namely $f = f_{011}^{TE} = 450.38$ MHz. All the three dipoles now give rise to monotonic trends of the electric-field amplitude. From a comparison of Figure 7.13*a* and 7.13*b* it can be observed that the *x*-directed dipole excites an electric field that at $f = f_{011}^{TE} = 450.38$ MHz is much larger than that produced at $f = 200$ MHz, while the fields produced by *y*- and *z*-directed dipoles in the two cases have values of the same order of magnitude. This effect is consistent with the fact that only the centered *x*-directed dipole can excite a resonant TE_{011} mode.

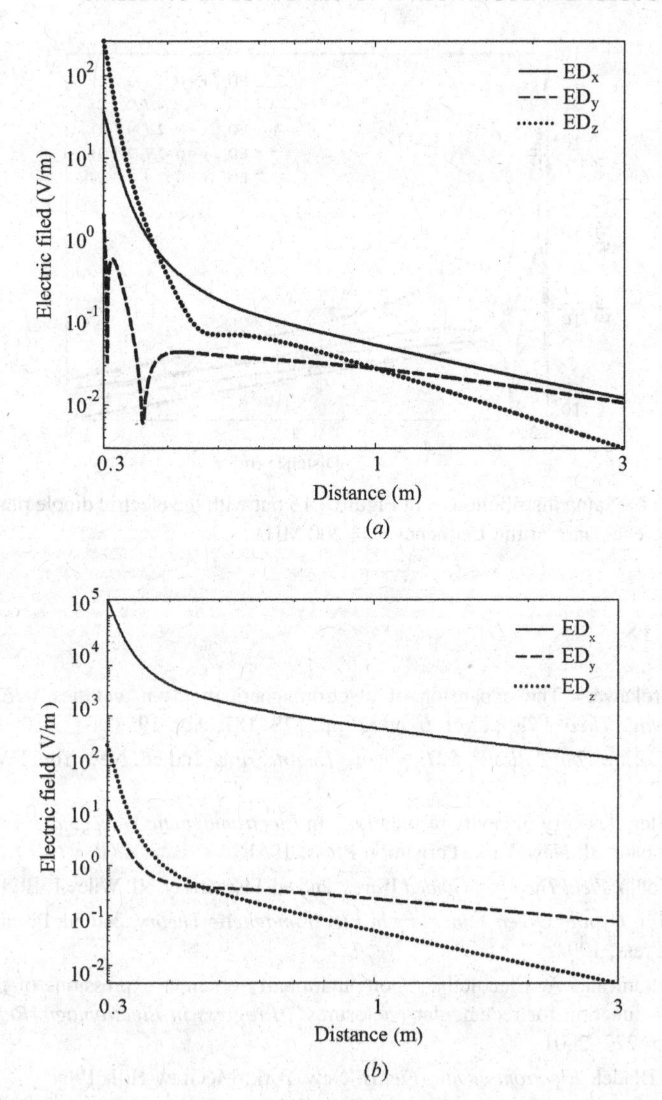

FIGURE 7.13 Magnitude of the electric field as a function of z $(x = l_x/2, y = l_y/2)$ produced by an elemental electric dipole located at the center of the metallic rectangular enclosure under the test $\mathbf{r}' = (l_x/2, l_y/2, l_z/2)$. Three possible orientations of the dipole are considered. (*a*) $f = 200$ MHz; (*b*) $f = f_{011}^{TE} = 450.38$ MHz.

However, the presence of oscillations in the field amplitude close to the aperture cannot be easily predicted. Figure 7.14 shows several curves that represent the magnitude of the electric field excited by a y-directed dipole as a function of z along the $(x = l_x/2, y = l_y/2)$ line at the operating frequency $f = 200$ MHz, when the dipole source is moved off the center of the enclosure. Note how the field behavior depends on the source location, especially in the near-field region.

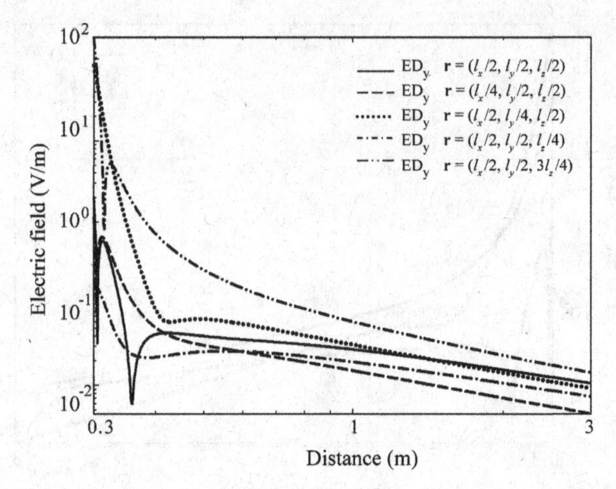

FIGURE 7.14 Same distribution as in Figure 7.13 but with the electric dipole placed off the center of the enclosure at the frequency $f = 200$ MHz.

REFERENCES

[1] K. Kurokawa. "The expansion of electromagnetic fields in cavities." *IEEE Trans. Microwave Theory Tech.*, vol. 6, no. 2, pp. 178–187, Apr. 1958.

[2] R. E. Collin. *Foundations of Microwave Engineering*, 2nd ed. New York: Wiley-IEEE Press, 2001.

[3] R. Müller. "Theory of cavity resonators." In *Electromagnetic Waveguides and Cavities*, G. Goubau, ed. New York: Pergamon Press, 1961.

[4] R. E. Collin. *Field Theory of Guided Waves*, 2nd ed. Piscataway, NJ: Wiley-IEEE Press, 1991.

[5] C.-T. Tai. *Dyadic Green Functions in Electromagnetic Theory*, 2nd ed. Piscataway, NJ: IEEE Press, 1994.

[6] F. Marliani and A. Ciccolella. "Computationally efficient expressions of the Dyadic Green's function for rectangular enclosures." *Progress in Electromagn. Res.*, vol. 31, pp. 195–223, 2001.

[7] J. Van Bladel. *Electromagnetic Fields*. New York: McGraw-Hill, 1964.

[8] H. A. Mendez. "On the theory of low-frequency excitation of cavity resonances." *IEEE Trans. Microwave Theory Tech.*, vol. MTT–18, no. 8, pp. 444–448, Aug. 1970.

[9] M. P. Robinson, T. M. Benson, C. Christopoulos, J. F. Dawson, M. D. Ganley, A. C. Marvin, S. J. Porter, and D. W. P. Thomas. "Analytical formulation for the shielding effectiveness of enclosures with apertures." *IEEE Trans. Electromagn. Compat.*, vol. 40, no. 3, pp. 240–248, Aug. 1998.

[10] R. Azaro, S. Caorsi, M. Donelli, and G. L. Gragnani. "A circuital approach to evaluating the electromagnetic field on rectangular apertures backed by rectangular cavities." *IEEE Trans. Microwave Theory Tech.*, vol. 50, no. 10, pp. 2259–2266, Oct. 2002.

[11] T. Konefal, J. F. Dawson, A. C. Marvin, M. P. Robinson, and S. J. Porter. "A fast multiple mode intermediate level circuit model for the prediction of shielding effectiveness of a rectangular box containing a rectangular aperture." *IEEE Trans. Electromagn. Compat.*, vol. 47, no. 4, pp. 678–691, Nov. 2005.

[12] M. Li, J. Nuebel, J. L. Drewniak, T. H. Hubing, R. E. DuBroff, and T. P. Van Doren. "EMI from cavity modes of shielding enclosures—FDTD modeling and measurements." *IEEE Trans. Electromagn. Compat.*, vol. 42, no. 1, pp. 29–38, Feb. 2000.

[13] S. V. Georgakapoulos, C. R. Birtcher, and C. A. Balanis. "HIRF penetration through apertures: FDTD versus measurements." *IEEE Trans. Electromagn. Compat.*, vol. 43, no. 3, pp. 282–294, Aug. 2001.

[14] W. P. Carpes Jr., L. Pinchon, and A. Razek. "Analysis of the coupling of an incident wave with a wire inside a cavity using an FEM in frequency and time domains." *IEEE Trans. Electromagn. Compat.*, vol. 44, no. 3, pp. 470–475, Aug. 2002.

[15] S. Benhassine, L. Pinchon, and W. Tabbara. "An efficient finite-element time-domain method for the analysis of the coupling between wave and shielded enclosure." *IEEE Trans. Magn.*, vol. 38, no. 2, pp. 709–712, Mar. 2002.

[16] G. Cerri, R. De Leo, and V. M. Primiani. "Theoretical and experimental evaluation of the electromagnetic radiation from apertures in shielded enclosures." *IEEE Trans. Electromagn. Compat.*, vol. 34, no. 4, pp. 423–432, Nov. 1992.

[17] F. Olyslager, E. Laermans, D. De Zutter, S. Criel, R. De Smedt, N. Lietaert, and A. De Clercq. "Numerical and experimental study of the shielding effectiveness of a metallic enclosure." *IEEE Trans. Electromagn. Compat.*, vol. 41, no. 3, pp. 202–213, Aug. 1999.

[18] W. Wallyn, D. De Zutter, and H. Rogier. "Prediction of the shielding and resonant behavior of multisection enclosures based on magnetic current modeling." *IEEE Trans. Electromagn. Compat.*, vol. 44, no. 1, pp. 130–138, Feb. 2002.

[19] Z. A. Khan, C. F. Bunting, and M. D. Deshpande. "Shielding effectiveness of metallic enclosures at oblique and arbitrary polarizations." *IEEE Trans. Electromagn. Compat.*, vol. 47, no. 1, pp. 112–122, Feb. 2005.

[20] C. Feng and Z. Shen. "A hybrid FD–MoM technique for predicting shielding effectiveness of metallic enclosures with apertures." *IEEE Trans. Electromagn. Compat.*, vol. 47, no. 3, pp. 456–462, Aug. 2005.

[21] J. R. Mosig. "Integral equation technique." In *Numerical Techniques for Microwave and Millimeter Wave Passive Structures*, T. Itoh, ed. New York: Wiley, 1989, ch. 3.

[22] M.-J. Park, J. Park, and S. Nam. "Efficient calculation of the Green's function for the rectangular cavity." *IEEE Microwave Guided Wave Lett.*, vol. 8, no. 3, pp. 124–126, Mar. 1998.

[23] M. G. Silveirinha and C. A. Fernandes. "A new acceleration technique with exponential convergence rate to evaluate periodic Green functions." *IEEE Trans. Antennas Propagat.*, vol. 53, no. 1, pp. 347–355, Jan. 2005.

[24] R. F. Harrington and J. R. Mautz. "A generalized network formulation for aperture problems." *IEEE Trans. Antennas Propagat.*, vol. AP-24, no. 6, pp. 870–872, Nov. 1976.

[25] R. F. Harrington. "Resonant behavior of a small aperture backed by a conducting body." *IEEE Trans. Antennas Propagat.*, vol. AP-30, no. 2, pp. 205–212, Mar. 1982.

[26] C.-H. Liang and D. K. Cheng. "Electromagnetic fields coupled into a cavity with a slot-aperture under resonant conditions." *IEEE Trans. Antennas Propagat.*, vol. AP–30, no. 4, pp. 664–672, Jul. 1982.

[27] B. Ma and D. K. Cheng. "Resonant electromagnetic field coupled into a lossy cavity through a slot aperture." *IEEE Trans. Antennas Propagat.*, vol. AP–35, no. 9, pp. 1074–1077, Sep. 1987.

CHAPTER EIGHT

Cable Shielding

Shielded cables are so largely diffused components that they are entitled to specific analysis. Moreover, their aspect ratio is very often potentially responsible of an antenna-like behavior, and thus of the resulting emission and susceptibility problems in almost all installations. Unfortunately, since shielded cables are components of larger, more complex systems their EM performance is strongly dependent not only on their own characteristics but also on those of the rest of the system.

The roots of any work on cable shielding are in the pioneering researches conducted by Schelkunoff [1], where the basic configuration of a driven coaxial cable of infinite length is analyzed. The cable shield was considered solid, while the effects on the coupling due to holes, defects, or braids have been the subject of several *Interaction Notes* of the Air Force Research Laboratory [2,3]. From these starting points stem a number of issues and configurations that have been deeply analyzed, and several books devoted to the analysis of this very special component have been published [4,5]. The interested reader should be aware that in the following, for obvious reasons of conciseness, only some of the most important aspects are pointed out. On this topic, the cited literature is representative but not exhaustive of the large amount of research work and, unfortunately, neither updated, because of the continuous progress in this field and consequently of the endless updating of knowledge on cable-related phenomena.

The preliminary problems to be assessed in cable shielding are basically two: whether the concern is emission or immunity and which type of shielded cable needs analysis.

Emission or immunity shielding is directly related to the source characteristics. When immunity is the primary goal of cable shielding, the external EM field may be considered as a uniform plane wave, whereas when emission from the cable is of

Electromagnetic Shielding by Salvatore Celozzi, Rodolfo Araneo and Giampiero Lovat

concern, the EM problem is that of a near-field source. As a consequence the same shielded cable with the same installation characteristics may behave very differently in immunity and emission tests.

The type of shielded cable has direct consequences on the mechanisms of penetration of the EM field through the shield: apertures (e.g., typical of braided shields) and number of shields strongly affect the performance. Other relevant issues, not analyzed in the following, concern the number of wire conductors in the shielded regions [6], the grounding conditions of the terminal ends of the cable shield(s) [7], and the connectors and the junctions between them and the wires at the wire ends [8,9]. The latter point is often the most serious and important coupling mechanism between the inside and the outside of the shield. Obviously in immunity problems the orientation and the path of the shielded cable with respect to the external EM field is fundamental in a quantitative assessment of coupling, but the worst case is usually considered, since in general the incident field is not predictable with precision.

This chapter will be limited to single coaxial shields solid or braided; the configurations with multiple shields [10] or multiple wires are left to the cited references. Linear materials are considered because nonlinear shields are generally of interest in low-frequency applications [11]. Moreover anisotropic materials will be not considered, but an example of analysis in such conditions can be found in [12].

8.1 TRANSFER IMPEDANCE IN TUBULAR SHIELDED CABLES AND APERTURE EFFECTS

The most used shielding parameter in shielded-cable evaluation is represented by the *transfer impedance* Z_t, introduced by Schelkunoff. It is defined as the ratio between the voltage, per unit length, arising on the internal surface of the shield when a current flows on its external side, or by reciprocity, as the ratio between the voltage appearing on the external side of the shield as a consequence of the current flowing on the internal surface.

With reference to Figure 8.1, the equations describing the voltage on the two surfaces of the cable shield may be written in the form:

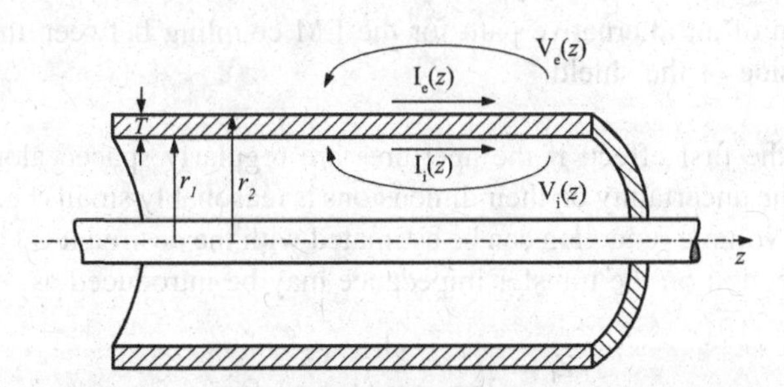

FIGURE 8.1 Coupling through the shield of a coaxial line.

$$V_i = Z_{int} I_i + Z_t I_e, \qquad V_e = Z_t I_i + Z_{ext} I_e, \tag{8.1}$$

where V_i and V_e are the longitudinal (i.e., parallel to the cable axis) per-unit-length voltages along the internal and the external surface of the shield, respectively. The per-unit-length impedances are

$$\begin{aligned} Z_{int} &= \sqrt{\frac{j\omega\mu_s}{\sigma_s}} \frac{1}{2\pi r_1 D} [I_0(\gamma_s r_1) K_1(\gamma_s r_2) + I_1(\gamma_s r_2) K_0(\gamma_s r_1)], \\ Z_{ext} &= \sqrt{\frac{j\omega\mu_s}{\sigma_s}} \frac{1}{2\pi r_2 D} [I_0(\gamma_s r_2) K_1(\gamma_s r_1) + I_1(\gamma_s r_1) K_0(\gamma_s r_2)], \\ Z_t &= \frac{1}{2\pi\sigma_s \, r_1 r_2 \, D}, \end{aligned} \tag{8.2}$$

where $D = [I_1(\gamma_s r_2) K_1(\gamma_s r_1) - I_1(\gamma_s r_1) K_1(\gamma_s r_2)]$ and $\gamma_s = \sqrt{j\omega\mu_s(\sigma_s + j\omega\varepsilon_s)}$; the subscript "s" stands for shield, while r_1 and r_2 are the inner and outer radius of the shield, respectively. The functions I_n and K_n are the nth-order modified Bessel functions of the first and second kind, respectively.

Equations (8.1) relate the voltages and the currents on the two shield surfaces. The transfer impedance Z_t thus represents a measure of the separation between the two domains provided by the shield cable; the lower its value, the greater is the efficiency. Of course, in such a TL approach, it is implicitly assumed that all the hypotheses at the basis of the analysis of distributed-parameter circuits are satisfied.

In Figure 8.2 the frequency-dependence of the transfer impedance of various shields is reported. As can be seen, for actual cables with tubular solid shields, the transfer impedance tends to zero above a certain frequency belonging to the Megahertz range. Various approximations (yielding errors between less than 1% [1] up to 10% [13] and above) have been introduced in the past to avoid the use of Bessel functions. Nowadays the exact expressions are easily handled, and there is no need of approximate expressions in terms of hyperbolic trigonometric functions.

Apertures in the shield of cables cause three main effects that increase coupling and thus worsen performance:

1. Reduction in the volume of the conductive material.
2. Perturbation of the linear path of the current-density distribution.
3. Creation of an alternative path for the EM coupling between the inside and the outside of the shield.

As concerns the first effect, if the apertures are regularly spaced along the cable shield and if the uncertainty on their dimensions is reasonably small (i.e., the amount of conductive volume reduction can be estimated with the desired level of accuracy), a simple correction on the transfer impedance may be introduced as

$$Z_t = \frac{1}{2\pi\sigma_s \, r_1 r_2 (1 - \tau)} \frac{1}{D}, \tag{8.3}$$

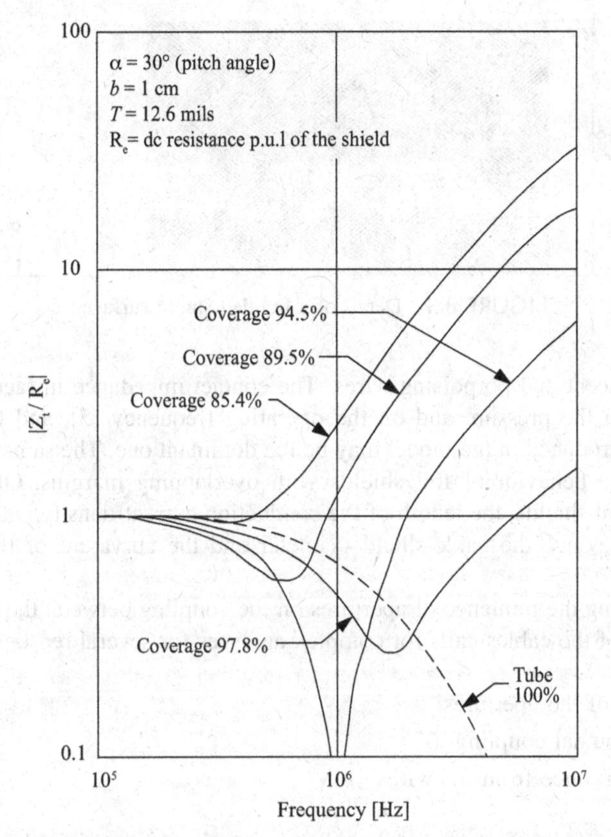

FIGURE 8.2 Normalized transfer impedance of tubular and braided shielded cables.

where τ, *transparency* of shield according to Vance [3] terminology, is a function of the number n of apertures per unit length of the cable shield and of their cross section S_a:

$$\tau = \frac{nS_a}{r_1 + r_2}. \tag{8.4}$$

Unfortunately, this correction works only if the apertures are realized in a solid tubular shield, which is not the shape most adopted for such cables. Mainly because of flexibility and weight considerations, the shield is generally realized by means of a mesh or by means of multiple wires helicoidally woven in alternate clockwise and counterclockwise way to form the so-called braided shield, a portion of which is shown in Figure 8.3.

In mesh shields the increase in length of the current-density path is not easily predicted because it depends on the shape of the apertures. In braided shields, although the length of the paths can be estimated by assuming the woven wires to be insulated from each other, an additional issue (uncertainty) arises about contacts

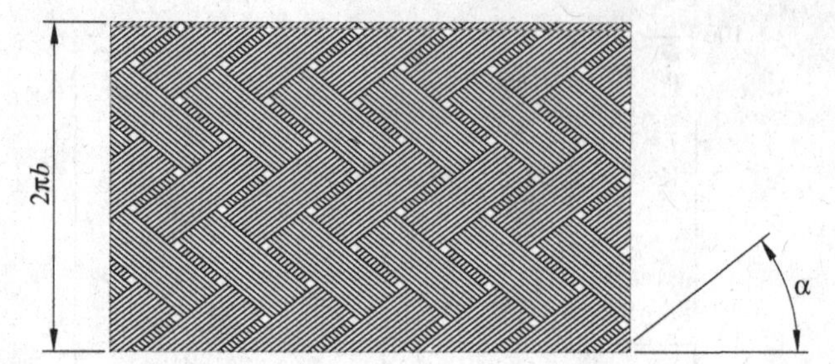

FIGURE 8.3 Detail of a braided shield surface.

between adjacent and porpoising wires. The contact impedance in fact is strongly dependent on the pressure and on the operating frequency [5], and this random parameter introduced in the model may be the dominant one. The same uncertainty influences the behavior of foil shields with overlapping margins. Of course, in compact spiral shields, the length of the conduction current-density path can easily be computed, once the cable-shield diameter and the curvature of the helix are known.

Determining the influence of apertures on the coupling between the interior and the exterior of the cables calls for complex analyses for several reasons:

- Shapes of the apertures
- Their mutual coupling
- Their distance from the wires

Usually the overall effect of distributed apertures is accounted for by means of a correction term in the transfer impedance represented by a mutual inductance and by means of a mutual capacitance [3]. The mutual capacitance is due to the direct coupling between the inner wire conductor and the external return path. Such a mutual capacitance can be introduced into the equivalent circuit of an elemental length of cable as shown in the pictorial sketch of Figure 8.4.

A number of expressions have been proposed for these two correction terms; their validities, however, have not always been confirmed by experimental results. In particular, for mesh shields, the mutual inductance and capacitance are given by [3,4]:

$$M_{12} = n\,\mu_0 \frac{d_{\mathrm{eq}}^3}{6\pi^2 D^2}, \tag{8.5a}$$

$$C_{12} = \frac{n}{\varepsilon_0 \varepsilon_{\mathrm{r}}} \frac{C_1 C_2 d_{\mathrm{eq}}^3}{6\pi^2 D^2}, \tag{8.5b}$$

where d_{eq} is the equivalent diameter of the apertures, D is the cable shield diameter, ε_{r} is the relative permittivity of the inner insulating material, and C_1 and C_2 are the per-unit-length capacitance between the inner wire conductor and the shield and that

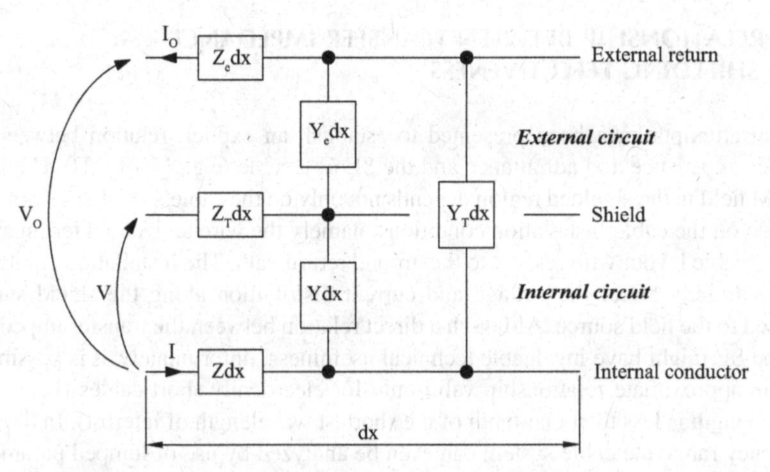

FIGURE 8.4 Equivalent circuit of an elemental length of coaxial cable.

between the shield and the outer return path for the current, respectively. It should be noted that the mutual capacitance is not an intrinsic parameter of the cable; it depends on C_2, which is function of the cable installation. For this reason two alternative intrinsic parameters are used in lieu of the transfer admittance $Y_t = j\omega C_{12}$; they are called *through elastance* K_T [14] and *radial electric coupling coefficient* ξ_R [6], and defined as

$$K_T = \frac{C_{12}}{C_1 C_2}, \tag{8.6a}$$

$$\xi_R = \frac{C_{12}}{C_1}. \tag{8.6b}$$

For a braided shield the EM model is a bit more complex and the approximations somewhat crude with respect to the phenomena actually occurring in this shield arrangement [3]. Usually it is assumed that the wires are isolated from each other (i.e., the current flowing in each wire is so well confined that no proximity effect is taken into account) and that the apertures are regularly spaced and rhomboidal. Another type of approximation involves replacement of the rhomboidal apertures with elliptical ones having the same values for the major and the minor axes. Expressions can be found in cited references, where the influence of various geometrical and physical parameters on the transfer impedance and admittance is also investigated. They are not reported here because of their unpredictable accuracy in the absence of experimental confirmation, which is a consequence of the several strict assumptions necessary for their derivation. Comparisons between predicted and measured data have been presented in the literature for a number of cable shield types (e.g., see [5,15–18]).

8.2 RELATIONSHIP BETWEEN TRANSFER IMPEDANCE AND SHIELDING EFFECTIVENESS

Several attempts have been presented to establish an explicit relation between the transfer impedance and admittance and the SE of a cable (e.g., [5,19–21]). However, the EM field in the shielded region depends not only on the cable's shield performance but also on the cable installation conditions, namely the wire and shield terminations and the cable layout with respect to the ground return path. The installation conditions in fact directly affect the voltage and current distribution along the shield surface exposed to the field source. Although a direct relation between the transfer impedance and the SE might have invaluable technical usefulness, unfortunately, it is possible to give an approximate relationship valid only for electrically short cables (i.e., cables whose length is less than one-tenth of the shortest wavelength of interest). In the low-frequency range the cable system can even be analyzed by use of lumped parameters instead of distributed ones or, in more complex situations, by general field equations and the consequent boundary-value problem. It should be obvious that a lumped-parameter approximation calls for an upper frequency limit on the order of a few (tens of) MHz in actual configurations.

If in immunity problems the induced voltage or current in the shielded wire are taken as a measure of the penetrated EM field, then the lumped parameter approximation provides very simple expressions for the shielding effectiveness SE_{LP} provided that the incident electric field is parallel to the cable axis and the magnetic field is transverse. If we assume that for a wire parallel to the ground the induced voltage in the absence of the shield V_i^0 has the same value of that induced in the shield when the latter is present, the lumped-parameter approximation leads to the following expression for the SE_{LP}:

$$SE_{LP} = 20 \log \left| \frac{V_i^0}{V_i^S} \right|, \tag{8.7}$$

where the induced voltage in the inner wire in the presence of the shield V_i^S is a function of the current I_S flowing on the external shield surface:

$$V_i^S = Z_t \ell I_S \tag{8.8}$$

and

$$I_S = \frac{V_i^0}{Z_{ext\_loop}}, \tag{8.9}$$

where Z_{ext_loop} is the total impedance of the current loop formed up by the shield and the ground return path while ℓ is the length of the cable. From (8.8) and (8.9), it follows that

$$SE_{LP} = 20 \log \left| \frac{Z_{ext\_loop}}{Z_t \ell} \right|. \tag{8.10}$$

A definition of SE_{LP} different from (8.8) is also possible, such as in terms of the currents induced in the shield and in the internal wire, respectively [19]. Improvements in the accuracy of the relationship between the SE and the transfer impedance have been shown to explicitly account for terminal conditions [20]. However, in all cases where the lumped-parameter approximation is still adopted, the increase in the complexity of the expressions (which now take into account both the reflection coefficients at the ends of the cable line and the diagonal terms of the characteristic-impedance matrix of the shielded cable) makes them of little use. Unfortunately, perfect matching at both ends is generally not achieved because of both the frequency dependence of the characteristic-impedance matrix and the usual choice for the cable shield ends, left open or short-circuited to the ground reference [22]. Losses may contribute to damping possible resonances [23], but these circumstances result in standing wave patterns of voltages and currents that impair any effort toward a simple link between intrinsic cable parameters and emission or susceptibility performance.

Finally, in emission analyses the boundary-value problem may be solved exactly by resorting to the lengthy but straightforward procedure described in [24] both for a shield of infinite length and for a shield of finite length.

8.3 ACTUAL CABLES AND HARNESSES

For the reasons above the measurement of the transfer impedance by line-injection methods can become extremely difficult at frequencies above several hundreds of MHz. So the measurement of the cable SE by means of a mode-stirred chamber is generally preferred at frequencies above the cited practical threshold [25]. A simple relationship between the transfer impedance and the SE has been also derived [26], indicating that the SE can decrease in frequency faster than 20 dB/decade, while the cautionary increase in the transfer impedance of coaxial cables is 20 dB/decade [14].

A very important fact that has been reported is that unfortunately, cables suffer from aging, exhibiting an increase in the transfer impedance values over a lifetime that cannot easily be predicted and may reach one order of magnitude [20,27], in part due to the porpoising effect in braided cables [28]. Furthermore a certain sensitivity (a few dB) of measured performance has been observed with respect to handling manipulations [29], like stretching before installation.

Multiconductor shielded cables have several peculiarities compared with coaxial cables: not only, as mentioned earlier, do they present different coupling modes [6] (thus requiring the introduction of specific parameters), they additionally exhibit an appreciable increase (up to one order of magnitude) of the values of the transfer impedance depending on the eccentricity of the considered conductors [30]. Apart from the complexities associated with the multiconductor-cable configuration, other issues may arise, such as in the evaluation of the response because of the mixing of propagation modes. The mixing of modes calls for the use of sophisticated algorithms for the assessment of the cable shielding performance [31].

It should be noted that grounding and earthing the shielded cable is sometimes mandatory for safety considerations: the effects are strongly dependent on several factors, such as frequency and the p.u.l. cable parameters. Interestingly, on one hand, grounding is responsible for ground loop emission; on the other hand, if left open, the shield may demonstrate, at certain frequencies, an antenna-like behavior and thus increase the EM radiation, which can be reduced by use of ferrites along the cable's length [5,7,20]. Periodic or multiple grounding and bondings can further present advantages and disadvantages with respect to open or shorted-at-ends solutions, depending on the circumstances and the frequency spectrum of interest.

REFERENCES

[1] S. A. Schelkunoff. "The electromagnetic theory of coaxial transmission lines and cylindrical shields." *Bell Syst. Tech. J.*, vol. 13, pp. 532–579, Oct. 1934.

[2] R. W. Latham. "Small holes in cable shields." *AFWL Interaction Note 118*, Sep. 1972.

[3] E. F. Vance. "Shielding effectiveness of braided wire shields." *AFWL Interaction Note 172*, Apr. 1974.

[4] E. F. Vance. *Coupling to Shielded Cables*. New York: Wiley, 1978.

[5] A. Tsaliovich. *Cable Shielding for Electromagnetic Compatibility*. New York: Van Nostrand Reinhold, 1995.

[6] F. Broydé and E. Clavelier. "Comparison of coupling mechanisms on multiconductor cables." *IEEE Trans. Electromagn. Compat.*, vol. 35, no. 4, pp. 409–416, Nov. 1993.

[7] R. R. Morrison and W. H. Lewis. *Grounding and Shielding in Facilities*. New York: Wiley, 1990.

[8] B. T. Szentkuti. "Shielding quality of cables and connectors: Some basics for better understanding of test methods." *Proc. 1992 IEEE Int. Symp. Electromagnetic Compatibility.* Anaheim, CA, 17–21 Aug. 1992, pp. 294–301.

[9] C. R. Paul. "Effect of pigtails on crosstalk to braided-shield cables." *IEEE Trans. Electromagn. Compat.*, vol. 22, no. 3, pp. 161–172, Aug. 1980.

[10] B. Demoulin, P. Degauque, M. Cauterman, and R. Gabillard. "Shielding performance of triply shielded coaxial cables." *IEEE Trans. Electromagn. Compat.*, vol. 22, no. 3, pp. 173–180, Aug. 1980.

[11] D. E. Merewether. "Analysis of the shielding characteristics of saturable ferromagnetic cable shields." *IEEE Trans. Electromagn. Compat.*, vol. 12, no. 3, pp.134–137, Aug. 1970.

[12] R. W. Latham. "An approach to certain cable shielding calculations." *AFWL Interaction Note 90*, Jan. 1972.

[13] K. F. Casey and E. F. Vance. "EMP coupling through cable shields." *IEEE Trans. Antennas Propagat.*, vol. 26, no. 1, pp. 100–106, Jan. 1978.

[14] E. P. Fowler and L. K. Halme. "State of art in cable screening measurements." *Proc. 1991 Int. Zurich Symp. Electromagnetic Compatibility.* Zurich, Switzerland, Mar. 1991, pp. 151–158.

[15] T. Kley. "Measuring the coupling parameters of shielded cables." *IEEE Trans. Electromagn. Compat.*, vol. 35, no. 1, pp. 10–20, Feb. 1993.

[16] S. Sali. "Cable shielding measurements at microwave frequencies." *IEE Proc. Sci. Meas. Technol.*, vol. 151, no. 4, pp. 235–243, Jul. 2004.

[17] M. Taghivand. "Correlation between shielding effectiveness and transfer impedance of shielded cable." *Proc. 2004 IEEE Int. Symp. Electromagnetic Compatibility.* Santa Clara, CA, 9–13 Aug. 2004, pp. 942–945.

[18] R. De Leo, G. Cerri, V. Mariani Primiani, and R. Botticelli. "A simple but effective way for cable shielding measurement." *IEEE Trans. Electromagn. Compat.*, vol. 41, no. 3, pp. 175–179, Aug. 1999.

[19] E. D. Knowles and L. W. Olson. "Cable shielding effectiveness testing." *IEEE Trans. Electromagn. Compat.*, vol. 16, no. 1, pp. 16–23, Feb. 1974.

[20] D. R. J. White and M. Mardiguian. *Electromagnetic Shielding: A Handbook Series on EMI and Compatibility.* vol. 3. Gainsville, VA: Interference Control Technology, Inc., 1988.

[21] R. J. Peel. "Simple relations between shielding effectiveness and transfer impedance/admittance for cables." *Proc. 1988 IEEE Int. Symp. Electromagnetic Compatibility.* Seattle, WA, 2–4 Aug. 1988, pp. 134–139.

[22] R. M. Whitmer. "Cable shielding performance and CW response." *IEEE Trans. Electromagn. Compat.*, vol. 15, no. 4, pp. 180–187, Nov. 1973.

[23] F. A. Benson, P. A. Cudd, and J. M. Tealby. "Leakage from coaxial cables." *IEE Proc. Sci. Meas. Technol.*, vol. 139, no. 6, pp. 285–303, Nov. 1992.

[24] J. A. Tegopoulos and E. E. Kriezis. *Eddy Currents in Linear Conducting Media.* Amsterdam: Elsevier, 1985.

[25] B. Eicher and L. Boillot. "Very low frequency to 40 GHz screening measurements on cables and connectors; Line injection method and mode stirred chamber." *Proc. 1992 IEEE Int. Symp. Electromagnetic Compatibility.* Anaheim, CA, 17–21 Aug. 1992, pp. 302–307.

[26] L. O. Hoeft. "A simplified relationship between surface transfer impedance and mode stirred chamber shielding effectiveness of cables and connectors." *Proc. Int. Eur. Electromagn. Compat. Symp.*, Sorrento, Italy, Sep. 2002, pp. 441–446.

[27] IEEE Std. 1143-1994. *IEEE Guide on Shielding Practice for Low Voltage Cables.* 1994.

[28] P. J. Madle. "Contact resistance and porpoising effects in braid shielded cables." *Proc. 1980 IEEE Int. Symp. Electromagnetic Compatibility.* Baltimore, MD, 7–9 October 1980, pp. 206–210.

[29] A. P. C. Fourie, O. Givati, and A. R. Clark. "Simple technique for the measurement of the transfer impedance of variable length coaxial interconnecting leads." *IEEE Trans. Electromagn. Compat.*, vol. 40, no. 2, pp. 163–166, May 1998.

[30] M. J. A. M. van Helvoort, A. P. J. van Deursen, and P. C. T. van der Laan. "The transfer impedance of cables with a nearby return conductor and a noncentral inner conductor." *IEEE Trans. Electromagn. Compat.*, vol. 37, no. 2, pp. 301–306, May 1995.

[31] J. Beilfuss, A. Bell, B. Gray, and R. Hamrick. "Multiconductor cable response dependency on propagation modes." *Proc. 1988 IEEE Int. Symp. Electromagnetic Compatibility.* Seattle, WA, 2–4 Aug. 1988, pp. 118–123.

[32] F. C. Yang. "Cable shields with periodic bondings." *AFWL Interaction Note 401*, Jan. 1981.

CHAPTER NINE

Components and Installation Guidelines

Components' adequacy and installation conditions are key to any successful shielding installation, and these factors are in strict correlation. Components affect shielding performance, and they can create a discontinuity in the shield's integrity; installation conditions are important for both the main shielding structures and the components. Apart from functional apertures, such as those due to viewing needs, air ventilation, or pass-through cables, the most important and tricky aspect is represented by the joints between adjacent parts of the main shielding structure and between the structure and the installed components (switches, lamps, etc.). Joints are classified as permanent, semipermanent, or frequently operated joints [1]. The last two types are usually provided with gaskets as described in the following section; the permanent joints are often fabricated by means of soldering, screws, or rivets. Continuous soldering should be preferred over (in order of performance) spot welding, screws with cap nuts or in blind holes, and rivets. Any protrusion of metallic parts that radiates toward the shielded region must be absolutely avoided. If use of such parts is unavoidable, adequate grounding is necessary [2].

For all the components briefly described in the following, the reader is invited to analyze the huge amount of data available from manufacturers.

9.1 GASKETS

Electromagnetic-interference gaskets are an important class of components installed in the joints between panels to reduce the EM-field penetration. They are manufactured

Electromagnetic Shielding by Salvatore Celozzi, Rodolfo Araneo and Giampiero Lovat

in a very wide variety of materials and shapes to meet their assigned levels of performance. Gasket performance is usually specified in terms of their EM sealing capability and their mechanical strength and resistance to chemical agents, depending on the function and application. Several considerations need to be accounted for in the choice of the most suitable version for any specific application, as follows (not in order of importance):

- Severity of environmental conditions (temperature, presence of corrosive agents, pressure conditions, vibrations, etc.)
- Amount of shielding required
- Uneveness of the mating surfaces to be joint
- Class of use (i.e., number and frequency of operation)
- Cost

Among the main types of constructive materials are conductive polymers or rubbers, wire-mesh flexible fabrics, and finger-shaped gaskets. Beryllium-copper, steel, and tin-coated phosphor bronze are among the most used metallic materials. Figure 9.1 shows several types of gaskets.

The EM performance of a gasket relates to its capability of restoring the electrical continuity between the two adjacent mating surfaces. Therefore it is very useful to have knowledge of the correct equivalent circuit and to understand the frequency behavior of any gasket configuration. Additionally it is useful to be aware of uncertainties in the values of components in an equivalent circuit (shown in Figure 9.2) that are due to installation conditions, together with their degradation over time. So a reliable approach depends mainly on graphs and measurement that describe the EM performance of any gasket type. The IEEE Standard 1302 gives guidelines for the EM characterization of conductive gaskets in the frequency range of dc to 18 GHz [3].

FIGURE 9.1 Different types of shielded gaskets (courtesy of Chomerics).

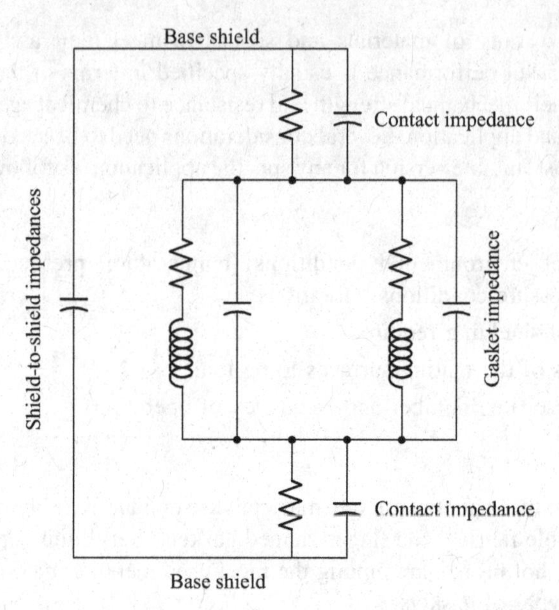

FIGURE 9.2 Equivalent circuit for a portion of a shielded gasket.

A very approximate method for a rough estimate of the gasket's SE is based on the TL approximation [1] and leads to the following approximate expression:

$$\mathrm{SE} \simeq R = 20\log\frac{\eta_0}{4\eta_\mathrm{g}}, \tag{9.1}$$

where a uniform plane wave normally incident onto the gasket is assumed as the incident field, η_0 is the free-space impedance, and η_g is the intrinsic impedance of the gasket, assumed to be equal to the total impedance (resistance, in the cited reference) offered by the gasket to a current flowing through it from one surface to the other:

$$\eta_\mathrm{g} \simeq R_\mathrm{c1} + R_\mathrm{g} + R_\mathrm{c2}. \tag{9.2}$$

In (9.2), R_c1 and R_c2 are the contact resistance (impedance) existing between the gasket and the two mating surfaces, while R_g is its internal resistance (impedance). The modeling of inductive behavior is not a simple task, and in an oversimplified method it is generally neglected. Two other flaws are present in such an approximation: first, (9.1) has been deduced for a planar shield of infinite extension and second, the gasket impedance given by (9.2) is generally different from the intrinsic impedance of the gasket material to be used in (9.1), which does not depend on gasket shape and dimensions. For all these reasons, the use of expression (9.1) is not recommended.

Another approximate approach is based on the introduction of a transfer impedance that relates the per-unit-length current density J_S on the gasket side exposed to the incident uniform plane wave to the voltage drop V_0 across the gasketed seam on the opposite side:

$$Z_t = \frac{V_0}{J_S}. \tag{9.3}$$

The units of Z_t are $[\Omega \cdot m]$. Because of the assumption of a normally incident plane wave, the amplitude H^{inc} of the incident magnetic field is related to the amplitude of the induced current density as $J_S \simeq 2H^{inc}$ and to the amplitude E^{inc} of the incident electric field as $E^{inc} = \eta_0 H^{inc}$. On the other hand, the voltage drop on the nonexposed side is almost equal to the product of the gasket thickness t by the transmitted electric field E^{tr}. For gasket lengths much smaller than the operating wavelength, it can be assumed that the following approximation holds for the SE of the gasket:

$$\mathrm{SE} = 20 \log \left| \frac{E^{inc}}{E^{tr}} \right| \simeq 20 \log \left| \frac{\eta_0 t}{2 Z_t} \right|. \tag{9.4}$$

If the incident EM field is not a plane wave and its wave impedance is known, expression (9.4) can still be used by adopting the appropriate values $Z_{w1,2}$ instead of η_0 (e.g., see Section 4.6). In most cases, however, the wave impedance is not unique and/or is not known. Given this drawback, together with the limitation to the gasket length in comparison with the wavelength, a more general method is called for to characterize gasket behavior, namely a radiated testing technique performed by means of different antenna types, like magnetic-field loop antennas, electric-field monopole antennas, and plane-wave radiators. These antenna techniques usually adopt a metal cavity for the EM power to penetrate through a gasketed seam; the measurement performed in the absence of the gasket offers a reference value for the evaluation of performance. It should be noted that such measurement methods based on standardized apertures are strongly influenced by several factors such as the mutual positions of the antennas-seam system and the surrounding environments on the two sides of the gasketed aperture. The values of the aperture dimensions are obviously also crucial.

Nested reverberation chambers can also be used to estimate the power beyond the gasketed components ([4]; see also Appendix C) and that in the source region. More sophisticated models of perforated gaskets (knitted-mesh type, finger-shaped, etc.) are based on the evaluation (analytical or numerical) of the EM-field penetration through the holes of the gasket [5]. This is an open field of research. The reader is reminded to carefully check the conditions under which the gaskets have been tested by manufacturers, and additionally not to neglect two especially important considerations: the pressure agent on the gasket and its potentially evolutive performance because of corrosion, galvanic action, and so forth.

9.2 SHIELDED WINDOWS

Shielded windows are very common components in shielding applications. The visibility of displays or status lamps and leds is needed but is often accompanied by the introduction of a discontinuity in the shield, with the consequent detriment of performance. Shielded windows are a way to reduce such performance degradation and are usually applied through a gasket on their contour, although other techniques are also available, such as those based on permanent soldering or welding.

Various nonelectromagnetic constraints are critical to the choice of shielded windows: transparency in the frequency spectrum corresponding to wavelengths between 400 and 700 nm, anti-glare and anti-scratch tractability, mechanical and thermal properties, resistance to chemical agents, and physical dimensions of the window. Other elements that influence the choice are the refractive index of the transparent material and the conductive busbar and gasketing available or necessary for the termination of the shielded window.

Unfortunately, window dimensions affect performance measurements. Therefore all the manufacturers' data can be compared only when the sample dimensions are equal. In general, the smaller the dimensions of the sample under test, the higher is the performance that it exhibits.

Materials available for shielded windows range from conductive glasses or plastics (conductivity is achieved through embedded metallic particles or thin films on the surfaces; see also Chapter 2) to composites, frequently in the form of wire grids, woven or knitted, embedded in an optically transparent host.

Performance is strongly dependent on frequency, so reference should be made to manufacturers' data. Additionally some general expressions suitable for a rough estimation of the expected capabilities exist, usually based on the assumption of infinite extension of the structure. For instance, a shielded window realized by means of a woven wire grid and excited by a plane-wave field offers a SE given by the expression [1]

$$\mathrm{SE} \simeq 164 - 20\log(df), \tag{9.5}$$

where d is the distance between adjacent wires and f is the operating frequency. However, this expression is useful only for an idea about the most important quantities affecting performance; it is not adequate for an accurate prediction, which involves dimensions of the shielded window, material characteristics, wires diameter, and so forth. Typically 80 to 200 openings per inch (OPI) are considered for woven grid meshes, consisting of wires whose diameter is in the range of 20 to 100 μm. The SE that can be achieved (as per manufacturers' declarations) is generally between 50 and 80 dB under plane-wave excitation at the frequency of 1 GHz. Sample dimensions are usually between 50 × 50 mm and 300 × 300 mm, and the optical transparency is often between 60% and 90%.

Knitted wire meshes generally offer worse SE performance than woven grids, with a typical SE ranging between 20 and 40 dB under plane-wave excitation at the frequency of 1 GHz. The number of OPI is generally between 10 and 30 for wire diameters being the same as those adopted in woven wire grids.

For a conductive glass or plastic, the expressions given in Chapter 4 may be considered. Typical values (i.e., those declared by manufacturers) of SE achievable by means of conductive glasses or plastics are in the range of 40 and 60 dB under plane-wave excitation at the frequency of 1 GHz. The thickness of the compact thin films is often in the range of 50 to 200 nm, with an optical transparency in the order of 90%.

9.3 ELECTROMAGNETIC ABSORBERS

Electromagnetic absorbers are used for a number of applications. Most common are the anechoic chambers used in compliance testing of apparatus and systems in lieu of open area test sites. Other applications range from the reduction of radar cross sections of objects to the improvement of the radiation pattern of antennas, to resonances damping in shielding applications. Of course, in the framework of EM shielding, the last application is of evident usefulness in the improvement of the enclosure's performance. The high conductivity of shield walls works to trap the EM energy penetrated through the shield discontinuities (which may be viewed as traveling back and forth within the shielded volume). Materials capable of effectively dissipating such energy in the shielded region can considerably help in the improvement of shielding performance.

The shape and material of EM absorbers vary according to EM constraints, environmental peculiarities, and manufacturer design. Generally, they are used to cover a metallic surface in such a way that:

1. the reflected field is as low as possible;
2. the reflected field phase is opposite, as much as possible, to that of the incident field;
3. the transmitted field is attenuated as much as possible while traveling through the absorber material and before being reflected by the conductive surface over which the EM absorber has been installed.

The first and third conditions are those that offer the best opportunities in broadband unknown polarization situations. The second is more suitable for frequency and polarization selective performance because of its resonant nature.

The most common shapes for EM absorbers are pyramidal cones, possibly truncated or twisted, whose dimensions may range between few centimeters and few meters, and tiles of homogeneous or multilayer material whose thickness is in the order of few millimeters. Material compositions are often not fully publicized by manufacturers, but graphites, iron oxides, ferrites, urethanes, and conductive foams are among the most used.

The prediction of absorber performance is rather complex [6–7], depending on the geometrical shape and on the physical parameters of the materials adopted.

Approximations based on equivalent TLs or homogenization approaches have been considered to circumvent the need of a numerical analysis for accurate predictions. Information on the frequency dependence of permittivity and permeability of materials is generally not available from manufacturers, and for this reason prediction of EM absorber behavior under conditions different from those used to test it could be tricky.

Under plane-wave illumination the comparison of EM absorber performance is generally based on the reflection coefficient Γ (which is a complex quantity), defined as

$$\Gamma = \frac{E^{\mathrm{r}}}{E^{\mathrm{inc}}}, \tag{9.6}$$

where E^{r} is the (complex) amplitude of the (linearly polarized) reflected electric field. Alternatively, the reflectivity R_{a} (sometimes termed in other ways) can be used:

$$R_{\mathrm{a}} = 20\log(|\Gamma|). \tag{9.7}$$

Typical values of reflectivity of commercially available absorbers range between -10 and -50 (or better) dB for normally incident plane waves, at frequencies between about 30 MHz and well above the GHz. For oblique incidence, performance is generally worse.

9.4 SHIELDED CONNECTORS

Connectors are generally installed to allow the entrance of cables and optic fibers into the shielded region. They are especially important in determining the shielding performance. Several types of shielded connectors are commercially available, and the reader will find plentiful information from manufacturers. The best connectors for EMC purposes are those with a shielded backshell. The use of a shielded cable whose shield is adequately terminated at the connector input is often advisable as well as the installation of a filter to limit the antenna behavior of the portion of cable present in the shielded region. Ways of properly grounding the cable shield are found in [2].

9.5 AIR-VENTILATION SYSTEMS

Almost all the electric and electronic systems present some thermal requirements, generally satisfied by means of air-ventilation ducts with or without fans. Such apertures represent serious shield discontinuities if not adequately shielded. They are usually of two types: those obtained by covering the apertures by means of meshed panels and those obtained by the use of honeycomb apertures exploiting the waveguide below cut-off attenuation. The first solution has the advantage of keeping out dust at the cost of an extra resistance to the air flow, while the second has better performance in terms of SE and drop of air pressure, but it presents the drawback of requiring an adequate dust filter. Typical performance (at the frequency of 1 GHz) range between 50 and 75 dB against a uniform plane-wave field for the first type, and between 75 and over 100 dB for the second one. Various shapes of honeycomb openings have been designed for the improvement of performance, and their description is beyond the scope of this chapter. The interested reader will find all the relevant information in manufacturers' data sheets. Various models exist for the estimation of the SE provided by a metallic honeycomb, either approximate [1,9], such as

$$\mathrm{SE} = 27\frac{\ell}{a} - 20\log N, \tag{9.8}$$

where ℓ is the length of the waveguide (i.e., the thickness of the honeycomb cell) and a is the radius of the N apertures. Or, more accurate [10], is

$$\mathrm{SE} = 27.3\frac{\ell}{a} - 20\log\frac{2ka}{\pi}\cos\phi, \tag{9.9}$$

which accounts also for the angle of incidence ϕ of the external TM-polarized plane-wave field.

9.6 FUSES, SWITCHES, AND OTHER SIMILAR COMPONENTS

Control and protection devices directly accessible from the exterior of the shielded housing represent another class of coupling paths that may deteriorate shielding performance. Actual shielded components for control, protection, and operation are generally reliable as concerns their performance. However, a gasket between the main body of the component and the internal side of the panel and an adequate nut are usually recommended to limit the discontinuity in the shield integrity. Moreover nonconductive shafts for switches, potentiometers, and the like are useful to avoid a guided path between the source region and the shielded one.

REFERENCES

[1] D. R. J. White and M. Mardiguian. *Electromagnetic Shielding: A Handbook Series on EMI and Compatibility*, vol. 3. Gainsville, VA: Interference Control Technology, 1988.

[2] R. R. Morrison and W. H. Lewis. *Grounding and Shielding in Facilities*. New York: Wiley, 1990.

[3] IEEE Std. 1302–1998. *IEEE Guide for the Electromagnetic Characterization of Conductive Gaskets in the Frequency Range of DC to 18 GHz, IEEE, Inc.*, 345 E 47th Street, New York, NY 10017, 1998.

[4] C. L. Holloway, D. A. Hill, J. Ladbury, G. Koepke, and R. Garzia. "Shielding effectiveness measurements of materials using nested reverberation chambers." *IEEE Trans. Electromagn. Compat.*, vol. 45, no. 2, pp. 350–356, May 2003.

[5] D. Pouhé and G. Monich. "Assessment of shielding effectiveness of gaskets by means of the modified Bethe's coupling theory." Accepted for publication on *IEEE Transactions on Electromagnetic Compatibility.*

[6] E. Kuester and C. Holloway. "A low-frequency model for wedge pyramid absorber arrays—I: Theory." *IEEE Trans. Electromagn. Compat.*, vol. 36, no. 4, pp. 300–306, Nov. 1994.

[7] C. Holloway and E. Kuester. "A low-frequency model for wedge or pyramid absorber arrays—II: Computed and measured results." *IEEE Trans. Electromagn. Compat.*, vol. 36, no. 4, pp. 307–313, Nov. 1994.

[8] M. Li, J. L. Drewniak, T. Hubing, R. E. DuBroff, and T. P. Van Doren."Slot and aperture coupling for airflow aperture arrays in shielding enclosure designs." *Proc. 1999 IEEE Int. Symp. Electromagnetic Compatibility*, Seattle, WA, 2–6 Aug. 1999, pp. 35–39.

[9] D. J. Angelakos. "Radio frequency shielding properties of metal honeycomb materials and of wire-mesh." *IEEE Int. Symp. Electromagnetic Compatibility*, pp. 265–280, 1960.

[10] W. A. Bereuter and D. C. Chang. "Shielding effectiveness of metallic honeycombs." *IEEE Trans. Electromagn. Compat.*, vol. 24, no. 1, pp. 58–61, Feb. 1982.

CHAPTER TEN

Frequency Selective Surfaces

As was mentioned in Chapter 4, a planar screen can be designed to present certain selective properties. Ideally the screen should be completely transparent to the EM radiation in a part of the frequency range and completely opaque in the remaining part of the frequency spectrum. Alternatively, the screen might be designed to reflect the EM field in a given frequency interval only. The screens that are able to perform a frequency discrimination through their reflecting/transmitting properties are known as frequency selective surfaces (FSSs), and basically they work as EM filters. As a consequence they are usually designed to provide four standard spectral responses: high pass, low pass, band stop, and band pass [1–3].

Historically Rittenhouse is believed to have been the first, at the end of 1700, to observe that a noncontinuous surface can present different reflecting/transmitting properties for different operating frequencies. The first patent, however, was not issued until 1919, to Marconi and Franklin. They built a parabolic reflector consisting of a noncontinuous surface made of horizontal wires. Later, in the 1960s, FSSs started to become the subject of intensive research for military applications. Today applications of FSSs can be found in the design of radomes, dichroic subreflectors, and reflect-array lenses, and more recently, in RFID tags, collision avoidance systems, RCS augmentation, EMI protection, selective EM shielding, and EM absorbers.

The required frequency-selective behavior of a screen is usually obtained by means of periodic structures in either one or two dimensions that, thanks to their particular geometry, provide a filter operation. These array structures mainly consist of thin conducting elements periodically arranged on a given lattice forming a rectangular array or, more in general, a triangular array, as depicted in Figure 10.1*a*. The conducting elements may be printed on a dielectric support either for practical reasons or for a desired performance. Alternatively to conducting elements, such arrays can

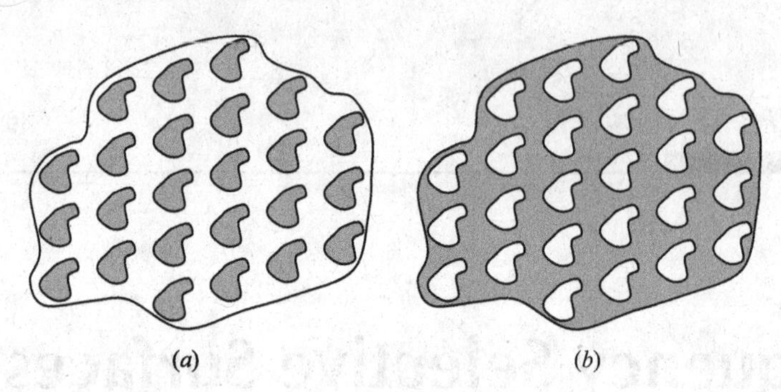

FIGURE 10.1 Example of 2D array structures of metallic elements (*a*) and of apertures in a conducting plane (*b*).

instead be designed to consist of periodic apertures (cut in a conducting plane, as shown in Figure 10.1*b*), to obtain reflecting/transmitting properties that are in some way complementary to the conducting-elements array, as it will be discussed later.

In general, more than one element (conducting or aperture) can be used to form the *unit cell*, which is defined as the smallest portion of the structure that is periodically repeated in space in order to form the entire periodic structure. More than one layer of arrays can be used to further improve the filtering properties of an FSS, possibly obtaining a more broadband behavior.

10.1 ANALYSIS OF PERIODIC STRUCTURES

The periodic nature of FSS structures allows for an enormous simplification of the EM problem analysis. This is because the computational domain can be restricted to only one period of the structure geometry (*unit cell*) by enforcing the so-called periodic boundary conditions [4]. In addition to FSSs, periodic structures are important in electromagnetics for a variety of applications that range from antenna arrays to artificial materials. In what follows, a brief summary of the fundamentals of the periodic-structure theory is presented.

10.1.1 Floquet's Theorem and Spatial Harmonics

Let us consider, for simplicity, a one-dimensional periodic structure, whose geometry is characterized by a spatial periodicity along one dimension, say x, with spatial period p_x. The Floquet theorem states that in such a periodic structure, time-harmonic *modal* EM fields $\mathbf{U}(x, y, z)$ (which therefore are source-free fields) have the property

$$\mathbf{U}(x + p_x, y, z) = \mathbf{U}(x, y, z)e^{-jk_{x0}p_x}, \qquad (10.1)$$

where $k_{x0} = \beta_{x0} - j\alpha_x$ is a complex wavenumber (fundamental propagation constant) that describes the phase shift and the attenuation of the field between different cells of the periodic structure. The Floquet theorem can also be expressed by stating that such modal fields $\mathbf{U}(x, y, z)$ have the property

$$\mathbf{U}(x,y,z) = \mathbf{P}(x,y,z)e^{-jk_{x0}x}, \tag{10.2}$$

where $\mathbf{P}$ is a periodic vector function such that

$$\mathbf{P}(x \pm mp_x, y, z) = \mathbf{P}(x,y,z) \tag{10.3}$$

and m is an integer. A function $\mathbf{U}(x, y, z)$ having the property (10.1) (or (10.2) and (10.3)) is also called *pseudoperiodic* or *Floquet periodic*. According to (10.1), the field distribution in the yz plane remains unchanged under an axial translation of the observation point along x through a period p_x; the only change is in the (complex) amplitude of the field, which is multiplied by a factor $e^{-jk_{x0}p_x}$. On the other hand, (10.2) and (10.3) express the fact that based on the knowledge of the field inside a unit cell and of the fundamental propagation constant, the field distribution in the whole space is uniquely determined. Moreover it can be shown that in the presence of a Floquet-periodic source (e.g., a plane wave) the excited EM field is Floquet periodic and has therefore a representation as in (10.1) [4]. The complex wavenumber k_{x0} can then be an unknown of the problem (e.g., in the absence of sources, when the modes of the periodic structures need to be determined) or can be impressed by the external source (e.g., by an impinging plane wave, as it will be discussed later).

Since the vector function $\mathbf{P}$ is periodic, it can be expanded in a Fourier series,

$$\mathbf{P}(x,y,z) = \sum_{m=-\infty}^{+\infty} \mathbf{a}_m(y,z)e^{-j(2\pi m/p_x)x}, \tag{10.4}$$

where the coefficients $\mathbf{a}_m$ are given by

$$\mathbf{a}_m(y,z) = \frac{1}{p_x}\int_{-p_x/2}^{+p_x/2} \mathbf{P}(x,y,z)e^{j(2\pi m/p_x)x}\mathrm{d}x. \tag{10.5}$$

By inserting (10.4) into (10.2), we obtain

$$\mathbf{U}(x,y,z) = \sum_{m=-\infty}^{+\infty} \mathbf{a}_m(y,z)e^{-jk_{xm}x}, \tag{10.6}$$

where

$$k_{xm} = k_{x0} + \frac{2\pi m}{p_x} = \left(\beta_{x0} + \frac{2\pi m}{p_x}\right) - j\alpha_x = \beta_{xm} - j\alpha_x. \tag{10.7}$$

Therefore the considered EM field **U** can be expressed as a sum of an infinite number of traveling waves of the form $\mathbf{a}_m(y,z)e^{-jk_{xm}x}$, called *spatial (Floquet) harmonics.* It should be mentioned that because of the convergence properties of the Fourier series, the amplitude of the coefficients $|\mathbf{a}_m|$ tends to zero as $|m|$ tends to infinity. Moreover a single spatial harmonic does not satisfy the boundary conditions of the structure and therefore cannot constitute, by itself, a mode of the periodic structure; rather, an infinite superposition of spatial harmonics can represents a mode (called *Floquet mode* or *Bloch wave*).

To calculate the coefficients $\mathbf{a}_m$, we need to solve the problem in the unit cell by treating the unit cell as a parallel-plate waveguide of width p_x with phase-shift walls (i.e., walls defined by the boundary conditions that the Floquet expansion dictates; see (10.1)) and choosing a direction of propagation transverse to x.

10.1.2 Plane-Wave Incidence on a Planar 1D Periodic Structure

The simplest periodic structure is perhaps the strip grating. Its 2D geometry consists of a planar periodic arrangement of infinitely long and infinitesimally thin conducting strips standing in free space, as shown in Figure 10.2*a* with the relevant coordinate system. Let us assume that the structure is excited by a uniform plane wave of the form

$$\mathbf{E}^{\mathrm{inc}}(x,y,z) = \mathbf{u}_y E_y^{\mathrm{inc}}(x,z) = \mathbf{u}_y E_0 e^{jk_0(x\sin\theta^{\mathrm{inc}} + z\cos\theta^{\mathrm{inc}})}. \qquad (10.8)$$

As can be seen, the incident field satisfies (10.1). This means that it can be considered as a particular case of the Floquet field having the only nonzero fundamental harmonic characterized by the real propagation constant $k_{x0} = -k_0 \sin\theta^{\mathrm{inc}}$. Since the structure is periodic in the x direction and uniform in the y direction and the excitation is Floquet periodic and independent of y, the scattered field $\mathbf{E}^{\mathrm{s}} = \mathbf{u}_y E_y^{\mathrm{s}}$ can be expressed as a superposition of Floquet harmonics independent of y:

$$E_y^{\mathrm{s}}(x,z) = \sum_{m=-\infty}^{+\infty} e_m(z) e^{-jk_{xm}x}. \qquad (10.9)$$

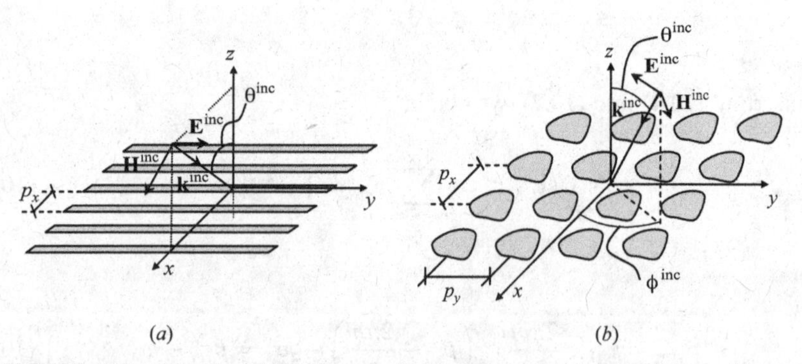

FIGURE 10.2 1D (*a*) and 2D (*b*) periodic structures under plane-wave incidence.

Because each Floquet harmonic is a solution of the Helmholtz equation, there results

$$e_m(z) = E_m e^{\pm jk_{zm}z}, \tag{10.10}$$

where

$$k_{zm} = \sqrt{k_0^2 - k_{xm}^2}. \tag{10.11}$$

It can thus be seen that for $k_{xm}^2 < k_0^2$, the mth Floquet harmonic propagates away from the strip grating, whereas for $k_{xm}^2 > k_0^2$, it decays exponentially in the z direction (the branch of the square root is chosen accordingly, i.e., $\mathrm{Im}[k_{zm}] < 0$, while the plus or minus sign in (10.10) is used for $z < 0$ or $z > 0$, respectively). Therefore in the Floquet expansion (10.6) only few harmonics are propagating plane waves in the xz plane; all the others are evanescent plane waves. As it can be seen from (10.9)–(10.11) and (10.6), the number of how many harmonics propagate depends on frequency and on incidence angle. In particular, as long as $k_{x1}^2 > k_0^2$, only the fundamental harmonic can propagate; therefore, the frequency at which the first-order harmonic starts to propagate is such that

$$\left| k_{x0} + \frac{2\pi}{p_x} \right| = \frac{2\pi}{\lambda_0} \quad \Rightarrow \quad p_x = \frac{\lambda_0}{1 + \sin\theta^{\mathrm{inc}}}. \tag{10.12}$$

The onset of the first-order harmonic thus occurs for grazing incidence when the spatial period is half-wavelength and for normal incidence when the spatial period is a wavelength long.

10.1.3 Plane-Wave Incidence on a Planar 2D Periodic Structure

The previous discussion can easily be extended to planar structures that are periodic in two dimensions, say along x and y, with spatial periods p_x and p_y, respectively, and excited by arbitrary uniform plane waves, as shown in Figure 10.2*b*. As is well known, an arbitrary uniform plane wave can be decomposed into the sum of a TE_z and a TM_z plane wave, and the scattering problem solved for each of these waves. In general, the scattered field can be expressed in terms of Floquet harmonics as

$$\mathbf{E}^{s}(x, y, z) = \sum_{n=-\infty}^{+\infty} \sum_{m=-\infty}^{+\infty} \mathbf{e}_{mn}(z) e^{-jk_{xm}x} e^{-jk_{yn}y}. \tag{10.13}$$

By setting $k_{x0} = -k_0 \cos\phi \sin\theta$ and $k_{y0} = -k_0 \sin\phi \sin\theta$ for the wavenumber of the incident plane wave along x and y, respectively, we can write the phasing along x (k_{xm}) and y (k_{yn}) of each Floquet harmonic as

$$k_{xm} = k_{x0} + \frac{2\pi m}{p_x}, \quad k_{yn} = k_{y0} + \frac{2\pi n}{p_y}. \tag{10.14}$$

The scattered field can then be expressed as

$$\mathbf{E}^{s}(x, y, z) = \sum_{n=-\infty}^{+\infty} \sum_{m=-\infty}^{+\infty} \mathbf{E}_{mn} e^{-jk_{xm}x} e^{-jk_{yn}y} e^{\pm jk_{zmn}z}, \tag{10.15}$$

where

$$k_{zmn} = \sqrt{k_0^2 - k_{xm}^2 - k_{yn}^2}. \tag{10.16}$$

The branch of the square root is chosen so that $\text{Im}[k_{zmn}] < 0$ (the plus or minus sign in (10.15) is used for $z < 0$ or $z > 0$, respectively). Therefore, for two-dimensional periodic screens, the field can be written as a superposition of Floquet harmonics, which are TE_z or TM_z plane waves either propagating or evanescent in the direction normal to the screen.

This way it is possible to associate an equivalent transmission line with each of these plane waves, the same as shown in Chapter 4, with the FSS represented by a multiport network (see Figure 10.3*a*). However, for sufficiently low frequencies (e.g., for $\min\{p_x, p_y\} < \lambda_0/2$) only the fundamental harmonic ($m = n = 0$) can propagate. The combined effect of all higher order harmonics can be lumped together in a 2×2 equivalent network (represented by a dyadic admittance $\underline{\mathbf{Y}}_{\text{FSS}}$) that couples the transmission lines associated with the TE_z and TM_z components of the fundamental harmonic (see Figure 10.3*b*). If we assume that no TE_z–TM_z coupling occurs and that the FSS is infinitesimally thin, this process will lead to the network representation of the FSS in terms of the *homogenized admittance* Y_{FSS} (see Figure 10.3*c*), which is generally dependent on frequency, polarization, and the angle of incidence [5].

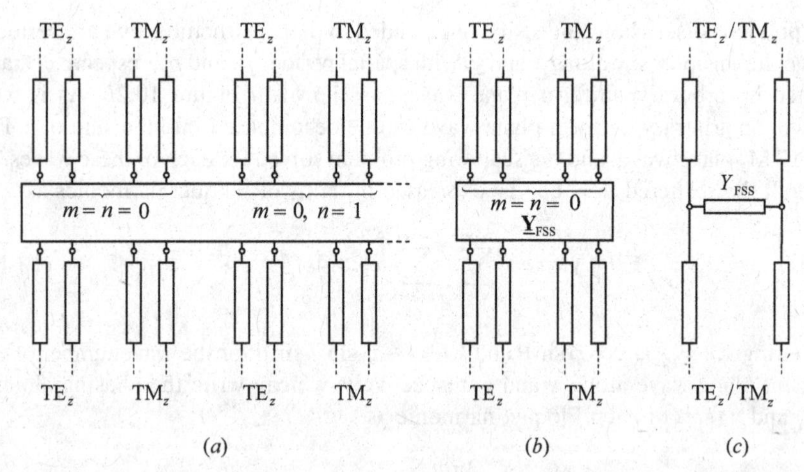

FIGURE 10.3 Multiport network representing a 2D periodic structure under plane-wave incidence (*a*), equivalent circuits at low frequencies (*b*) and (*c*).

By simple geometric considerations, the Floquet expansion theory can be generalized to handle 2D periodic structures whose axes of periodicity are not necessarily orthogonal [2].

10.2 HIGH- AND LOW-PASS FSSs

An example of high-pass FSS consists of a periodic arrangement of infinitely long conducting elements (e.g., infinitesimally thin and perfectly conducting narrow strips), as shown in Figure 10.4*a*. This screen in fact behaves as a shunt inductance for the incident plane waves with the electric field polarized along the conductive elements (the y direction in Figure 10.4*a*). This can be established in a simple manner. As described in the previous section, a uniform plane wave impinging on a periodic structure excites a countable infinity of plane waves (Floquet harmonics). However, for sufficiently low frequencies only one of such harmonics (the fundamental harmonic) can propagate, the others being evanescent. It is thus possible to associate an equivalent transmission line with this propagating plane wave. In this framework the periodic structure is represented by means of a shunt reactance in which all higher order harmonics effects are lumped. The shunt nature of the discontinuity (i.e., the screen) is due to the assumption of infinitesimally thin conducting elements which leads to a continuous transverse electric field (if the elements were assumed to have a finite thickness the proper network representation of the discontinuity would be a T or Π network); the absence of a resistive part is due to the assumption of lossless elements. Independently of the reactance value (which can be obtained, e.g., by a variational method or by a full-wave analysis), the capacitive or inductive nature of the discontinuity can be simply predicted. For instance, the uniform plane wave with the electric field polarized along y of Figure 10.4*a* is a TE_z uniform plane wave that induces an electric current on the strip that is y directed and independent of y; the field scattered by the strip grating is thus a

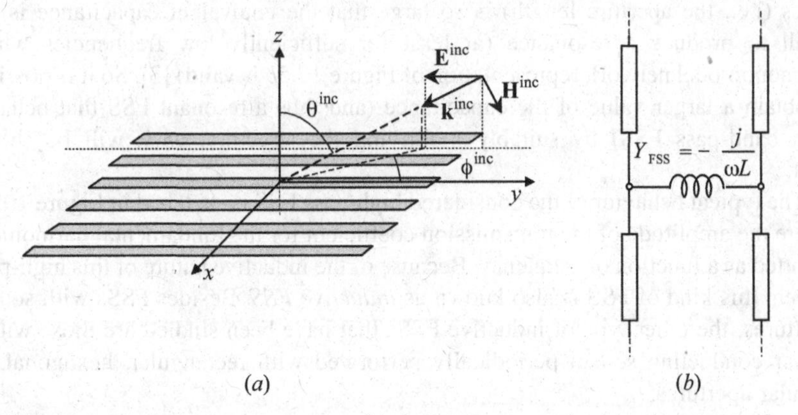

FIGURE 10.4 Periodical arrangement of infinitely long conducting elements under TE_z plane-wave incidence (*a*) and its equivalent circuit (*b*).

TE_z field, as well (i.e., all the harmonics are TE_z plane waves). It is well known that the magnitude of the reactance is inversely proportional to the difference $W_H - W_E$ between the stored magnetic and electric energies in the neighborhood of the discontinuity. As can be seen from the formulas derived in Chapter 4, the impedance associated with an evanescent TE_z plane wave has an inductive nature, and the stored energy will be predominantly magnetic. This means that the susceptance representing the periodic screen has a negative sign, and the screen consequently acts as a shunt inductance for the incident uniform plane wave with the electric field polarized along the conductive elements. The equivalent network is represented in Figure 10.4*b*.

At sufficiently low frequencies an inductor in parallel to the load of a transmission line acts like a short circuit, and at high frequencies like an open circuit. More precisely, at sufficiently low frequencies, the strip grating reflects almost totally the y-polarized uniform plane wave, whereas at high frequencies almost no reflection occurs (at high frequencies other harmonics may, however, propagate, and the monomodal network representation will then lose its validity).

By similar reasoning it can be shown that at sufficiently low frequencies, the same strip grating is transparent for incident plane waves with the electric field polarized along the x direction. Therefore, in order to achieve a high-pass behavior also for x-polarized uniform plane waves, a grid of strips must be directed also along the x direction. This is done by way of the *wire mesh* shown in Figure 10.5*a*. The wire mesh appears as a conducting screen periodically perforated with square apertures: a more refined network representation of such a screen would require as well the presence of a capacitance element in parallel with the inductor, whereby any adjacent pair of wires acts as a shunt capacitance for waves with the electric field polarized in the direction orthogonal to the wires on the screen plane. A more correct network representation for the 2D array of square apertures is then that of Figure 10.5*b*. With this model it is apparent that the structure no longer behaves as a high-pass filter because the network representation is an LC parallel that behaves as a band-pass filter when it resonates. Nevertheless, the distance between the wires (i.e., the aperture length) is so large that the equivalent capacitance is too small to produce a resonance (at least for sufficiently low frequencies where the monomodal network representation of Figure 10.5*b* is valid) [3]. So it is possible to obtain a larger value of the capacitance (and thus a resonant FSS that behaves as a band-pass FSS) by suitably modifying the structure, as it will be shown later.

The typical behavior of the considered high-pass FSS is sketched in Figure 10.5*c*, where the amplitude of the transmission coefficient for the fundamental harmonic is reported as a function of frequency. Because of the inductive nature of this high-pass screen, this kind of FSS is also known as *inductive FSS*. Besides FSSs with square apertures, the other type of inductive FSSs that have been studied are those with a planar conducting screen periodically perforated with rectangular, hexagonal, or circular apertures.

In the literature a reference to *complementary* FSSs is often encountered. Basically complementary arrays are patch or slot arrays (consisting of elements

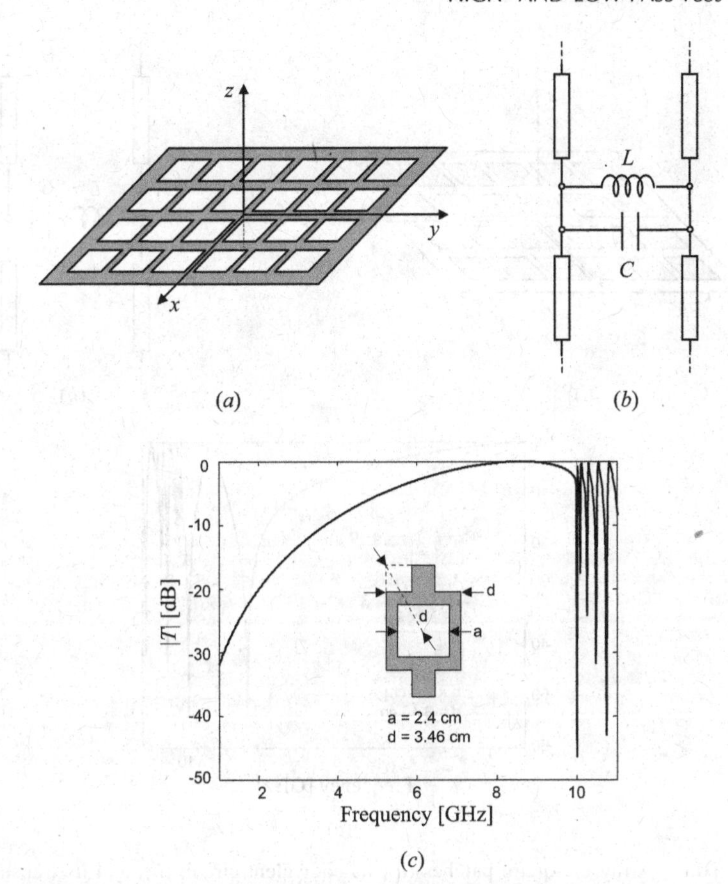

FIGURE 10.5 Wire-mesh (*a*), its equivalent circuit (*b*), and its transmission coefficient (*c*).

having an arbitrary shape) such that if the complementary FSSs are superimposed, an infinitesimally thin perfectly conducting plane is obtained, as shown in Figure 10.1 for complementary arrays of slots and patches. An important consequence of complementarity is that, based on the Babinet principle, the transmission coefficient for one FSS is equal to the reflection coefficient of the complementary FSS, when dual polarizations of the incident plane waves are considered. Therefore, based on the concept of complementary FSSs, a low-pass FSS can be easily imagined as being simply the complementary FSS of a high-pass FSS. An example is illustrated in Figure 10.6, where a bidimensional array of perfectly conducting square patches is shown (Figure 10.6*a*), together with its network representation (Figure 10.6*b*), and with the typical behavior of its transmission coefficient (Figure 10.6*c*). Because of the capacitive nature of this low-pass screen, this kind of FSS is also known as *capacitive FSS*. It should be noted that if the screen is not perfectly conducting, or it has a finite thickness, or it is loaded with a thin layer of dielectric, the transmission and reflection coefficients of complementary FSSs are no longer equal (the definition

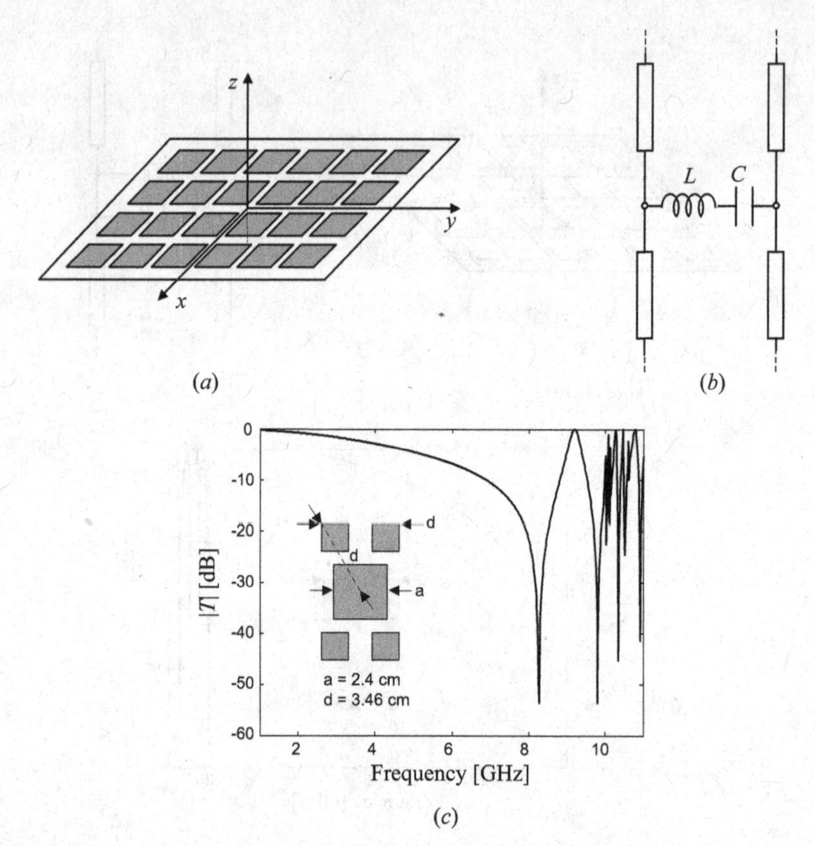

FIGURE 10.6 Array of square patches (*a*), its equivalent circuit (*b*), and its transmission coefficient (*c*).

"complementary FSSs" is no longer correct either); they show different resonant frequencies and different bandwidths.

10.3 BAND-PASS AND BAND-STOP FSSs

For the FSS to perform as a band-pass or a band-stop filter, it needs to possess resonant properties, meaning both an inductor and a capacitor must be present in its equivalent network representation and the inductance and capacitance values must be such that the resonant frequency be smaller than the frequency at which the first higher order Floquet harmonic starts to propagate. Such values can be obtained by suitably choosing the elements of the considered FSS. In general, when presenting resonance (at frequency f_R), the *conducting-element arrays present band-stop behavior.* Band-stop behavior occurs when the induced electric currents values are high enough to make a structure act as a conducting plate. However, when presenting resonance (at frequency f_R), the *aperture-element arrays present band-pass behavior* rising from the high values of the equivalent magnetic currents in the apertures that make the structure transparent to the incident field. The bandwidth performances are usually defined by a -10 dB level for the band-stop FSSs and by a -0.5 dB level for the band-pass FSSs [2]. In general, an FSS presents an infinite number of alternating stop and pass bands, but only the first bands can be used because of the onset of higher order harmonics that dramatically deteriorate the spectral performance.

One parameter that is critical to determining the bandwidth of an FSS is the interelement spacing. In general, larger spacing obtains a narrower bandwidth, earlier onset of higher order harmonics, and greater sensitivity of the resonant frequency on the angle of incidence and on polarization [3].

Of course, the geometry (shape and dimensions) of the elements characterizing the unit cell of the periodic screen can also dramatically influence the spectral properties of an FSS. For simplicity, the elements of the FSS are supposed to consist of straight conducting sections (plates) and straight conducting segments (dipoles having widths much smaller than their lengths) with or without round corners. Their complementary aperture elements should also be considered. A possible classification of elements based on their geometry, as suggested and analyzed in [3], is described below.

10.3.1 Center-Connected Elements or N-Pole Elements

In this category are elements consisting in a connected union of dipoles. Examples of such elements are the simple dipole (usually arranged in the so-called Gangbuster surface [3]), the tripole, the anchor, the Jerusalem cross, and the square spiral [3].

The dipole (shown in Figure 10.7*a*) is perhaps the simplest element. All other elements (also those belonging to categories other than the center connected) can be regarded as dipole combinations (with the exception of curved elements). As is well known, dipoles are resonant elements with very small bandwidths. However, when dipoles are arranged in a periodic fashion, the bandwidth of the corresponding FSS is much larger than that of the single element and can be made very large by packing the elements very close together [6]. A possible drawback is that the reflecting/transmitting properties greatly depend on the polarization of the incident plane-wave field. In particular, in the case of conducting dipoles, the electric field has to be polarized along the dipole length, whereas for the corresponding aperture element,

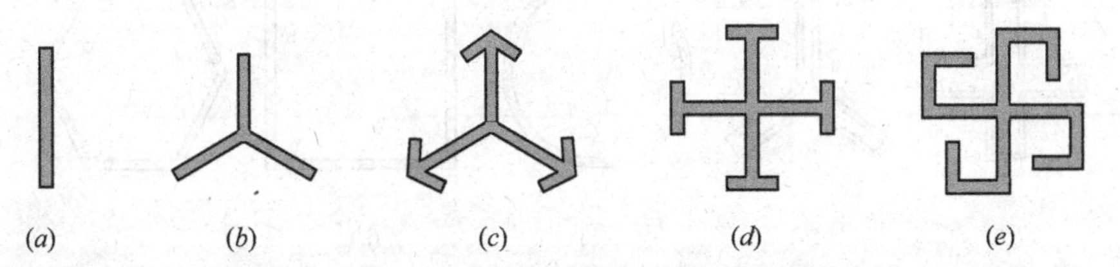

FIGURE 10.7 Examples of center-connected elements.

the polarization is along the slot width. Therefore dipole arrays can handle only a particular type of linear polarization.

The tripole consists of three dipoles connected through one of their endings and rotated one to the other by 120°, as shown in Figure 10.7*b*. In general, the tripole FSS has lower cross-polarization levels and a larger bandwidth compared to the dipole FSS [7,8].

The anchor element can be obtained from the tripole by adding a capacitance at the end of each arm, as it is shown in Figure 10.7*c*. This provides a larger bandwidth and a delayed onset of the first higher order harmonic with respect to other center-connected designs [3].

The Jerusalem cross (Figure 10.7*d*) is one of the oldest considered elements to form an FSS. It consists of two crossed dipoles with end loadings [9,10]. The Jerusalem cross is mainly used for narrowband applications.

Finally, the square-spiral (Figure 10.7*e*) can have a very large bandwidth. The square-spiral is claimed to be the best element for many applications, likely because of the very small interelement spacing [3].

10.3.2 Loop-Type Elements

A number of elements consists of closed loops. Examples are the four-legged loaded element [3], the three-legged loaded element [3], the square loop (single and concentric) [11,12], the ring (single and concentric) [13–18], and the hexagonal element [3], shown in Figure 10.8*a–e*. This kind of elements seems to be the most used, and it can provide a wide range of bandwidths, depending on the element (e.g., narrow bandwidths with three- and four-legged loaded elements and large bandwidths with hexagonal elements). Because they resonate when their total length is approximately equal to one wavelength, their size is often smaller than one-third the wavelength, which allows for the small interelement spacings.

10.3.3 Solid-Interior-Type Elements

Elements that consist of plates of simple shapes (squares, rectangles, hexagons, circles, etc.) have been considered in previous sections where low-pass and high-pass FSSs were described. In particular, it was shown how difficult it is to make the structure resonate because of the usually intrinsically low values of shunt

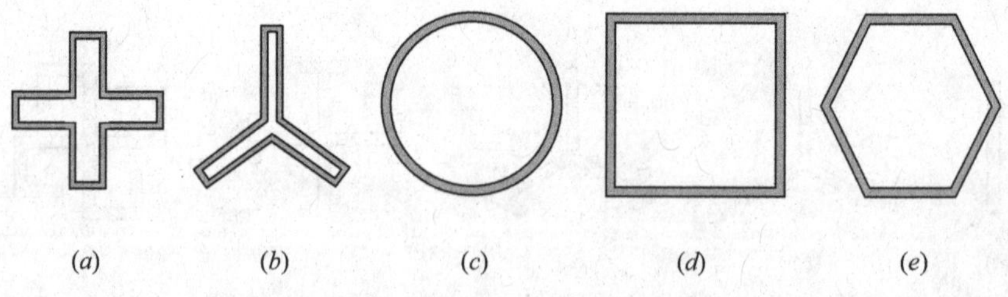

FIGURE 10.8 Examples of loop-type elements.

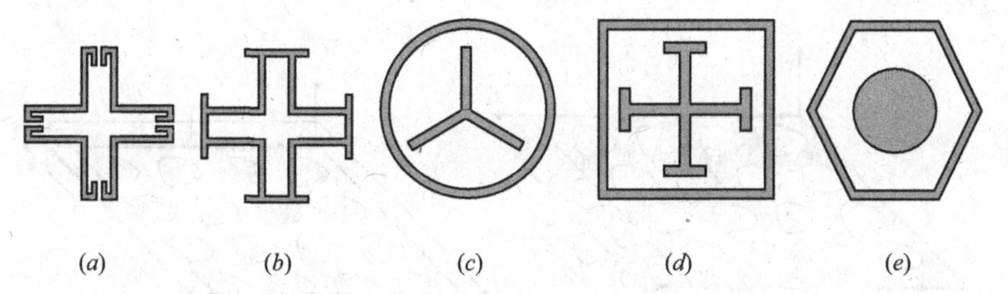

FIGURE 10.9 Examples of various combined elements.

capacitance or inductance in the equivalent network model. However, such structures can be modified to increase such values. For instance equivalent capacitance in the square aperture FSS can be increased by adding a square patch inside the aperture (this way the larger value of the capacitance is due to the smaller distance between the conductors). Eventually this element resembles the square loop aperture element, in that the final element belongs to the loop-type category. Despite their drawbacks as resonant FSSs, this kind of periodic structure was the first to be analyzed in detail [19–21] and still remains the classic example of both high-pass and low-pass surfaces.

10.3.4 Combinations and Fractal Elements

As can be expected, all the above-mentioned elements have been suitably combined to form a new element in order to improve the performance of the purely center-connected element, loop-type element, and solid-interior-type element FSSs. Examples of such combinations are shown in Figure 10.9*a–e*.

Attempts have been made to reduce the dimension of the unit cell maintaining fixed the resonant frequency (and thus the length of the element) in order to implement combined elements on curved structures and reduce the possible effects of a finite curvature. To this aim, convoluted [22,23] and fractal [24–26] elements have been recently proposed that show additionally the possibility of multi-band frequency operation.

10.4 DEGREES OF FREEDOM IN DESIGNING FSSs

As described in the previous section, the performance of an FSS can be evaluated in terms of bandwidth, sensitivity to angle of incidence and polarization, and onset of higher order harmonics. Other quality factors include the cross-polarization level, the band spacing ratio, and, when shields of finite extent are considered, edge diffraction effects.

The so-called angular stability can be improved by a suitable choice of the element of the array, of the lattice type, and of the lattice spacings (i.e., the spatial periods) [3]. Moreover the resonant curve of a transmission or reflection coefficient is usually desired to present a top part as flat as possible and steep descents. Starting

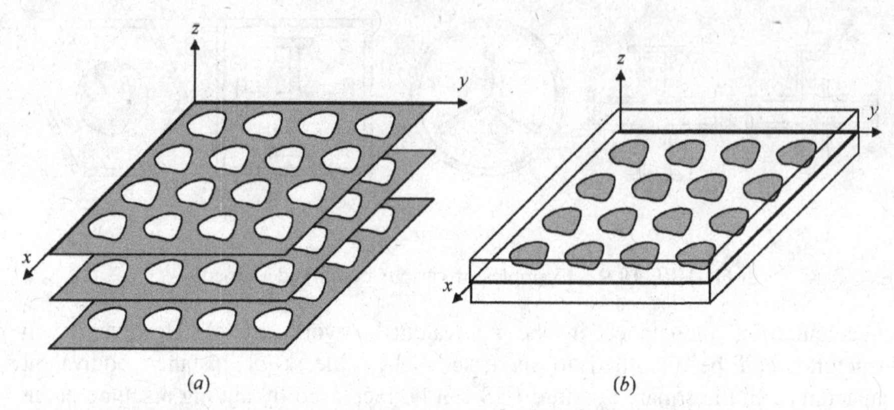

FIGURE 10.10 Cascade of FSSs (*a*) and FSSs loaded by dielectric slabs (*b*).

from a standard FSS, this effect can be achieved in different ways. For instance, it can be obtained by using more than one FSS (i.e., a cascade of FSSs with a suitable spacing, as sketched in Figure 10.10*a*) with equal [27–29] or different periodicities and elements [30]; alternatively, the single FSS can be loaded on each side with a dielectric slab of suitable thickness, as shown in Figure 10.10*b*, [31]. The first choice (cascading FSSs) shows a greater sensitivity of the bandwidth to the incidence angle and polarization, while the second (loaded FSSs) can furnish also a larger bandwidth (but with a less flat resonant curve near the resonant frequency). However, it should be mentioned that the presence of a dielectric loading can significantly affect the performance of an FSS, especially in terms of resonant frequencies, resonant curve shape, and sensitivity to incidence angle [3,32]. Finally a multilayer FSS can be considered that combines the advantages of the two approaches, whereby several cascading FSSs having two or more dielectric slabs are sandwiched between two surfaces [33–36].

Besides all the above-mentioned parameters (element type, type of periodic arrangement, spatial periods, number of cascading FSSs, their spacing, and slab thicknesses), there are some other important factors that determine the characteristic of an FSS. These are the conductivity of the conductors (important especially for the design of EM absorbers), their thickness, and the permittivity of the slabs. Because of the numbers of degrees of freedom, recently many optimization procedures have been proposed to make the design of the FSSs as effective as possible for the desired application [37–40].

10.5 RECONFIGURABLE AND ACTIVE FSSs

Over the last few years much of the research on FSSs has been devoted to tuning or reconfiguring an FSS, that is, to identifying efficient ways to shift or change its frequency behavior during operation. Basically this reconfigurable screen can be obtained in three ways: by changing the EM properties of the substrate of the FSS, by altering the geometry of the FSS, or by using circuit components with the possibility of varying the current distribution on the conducting elements of the FSS.

Some proposed reconfigurable FSSs (RFSSs) have at the base of their reconfigurable operation the use of a ferrite substrate [41,42]. It is well known that the constitutive parameters (in particular, the permeability) of ferrite can be changed by applying an external dc magnetic bias field. As already mentioned in the previous section, the spectral behavior of an FSS loaded with a substrate can strongly depend on the values of such constitutive parameters [32]; in particular, by changing the dc magnetic bias, the frequency response of a ferrite FSS can be shifted to higher or lower frequencies. However, because of some practical problems associated with the use of ferrite substrates (power requirements, complex bias network, etc.), other RFSSs based on substrates capable of changing their EM properties under an external control have also been proposed, such as liquid dielectric FSSs [43], ferroelectric FSSs [44], and silicon FSSs [45]. Indeed, in [43] and [44], the permittivity of the substrate is electronically controlled, while in [45], the silicon substrate behaves like a conductor when illuminated by an optical source.

The second type of RFSSs has at its operational core a change of geometry of the considered FSS. In [46] and [47], two cascaded periodic surfaces are suitably shifted horizontally or vertically to change both the resonant frequency and the bandwidth. Other structures based on microelectromechanical systems (MEMS) technology have been proposed as well [48,49].

Finally, the use of lumped-circuit elements on the periodic screen to vary the current distribution along the conducting elements of the FSS seems to provide the most promising candidate for the RFSSs. In particular, the use of variable reactive components [50], varactor diodes [51], or PIN diodes [52,53] has been proposed. In particular, when PIN diodes are used as switches among the conducting elements of an FSS (see Figure 10.11), the frequency response of the screen can dramatically change, depending on the voltage across the diodes. Additionally this kind of RFSS,

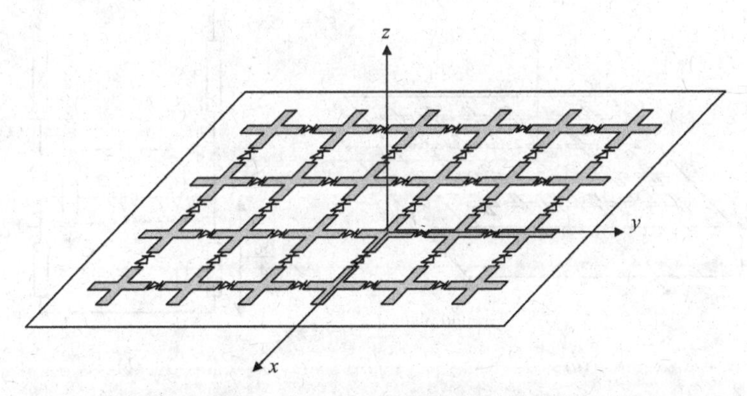

FIGURE 10.11 Reconfigurable FSS with PIN diodes.

along with same other FSSs, has been proposed [54,55] as walls to provide a selective shielding of enclosures.

10.6 FSSs AND CIRCUIT ANALOG ABSORBERS

Lossy FSSs have been used as well in the improvement of classical resonant absorbers, such as Jaumann layers and Salisbury screens. As is well known, such classical resonant absorbers mainly consist of purely resistive sheets and are very narrowband. Nevertheless with the addition of some capacitance and inductance they can become broadband absorbers (also called circuit analog, CA, absorbers) required in many applications. Although there are some bulk artificial materials that present filtering behavior [56], periodic surfaces made of lossy materials can also effectively act as broadband absorbers. Depicted in Figure 10.12*a* is a classic design that has an equivalent circuit representation reported in Figure 10.12*b*. In these representations the inductance is related to the straight part of the conductive elements, and the capacitance is associated with the gaps between the conductive elements. The resistance represents the lossy contribution of the conductive elements (characterized by a finite conductivity). Although the circuit representation is only a rough approximation, it provides useful insight into the working principle [3].

The CA absorber is in fact designed in such a way that at the resonant frequency f_R of the RLC circuit the distance between the ground plane and the FSS is a quarter wavelength (as for the Salisbury screen [57], the quarter wavelength transmission line transforms the short circuit corresponding to the ground plane into an open circuit at the FSS position). For frequencies lower than f_R, the input admittance Y_G is inductive, but the FSS admittance Y_{FSS} is capacitive, so their imaginary parts tend to cancel out each other in the total admittance of the CA absorber $Y_{CA} = Y_G + Y_{FSS}$. The same reasoning holds for frequencies higher than f_R, for which broadband behavior can be obtained with respect to the classic Salisbury screen. It should be noted that in the case of CA absorbers, the choice of the real part of Y_{CA} slightly

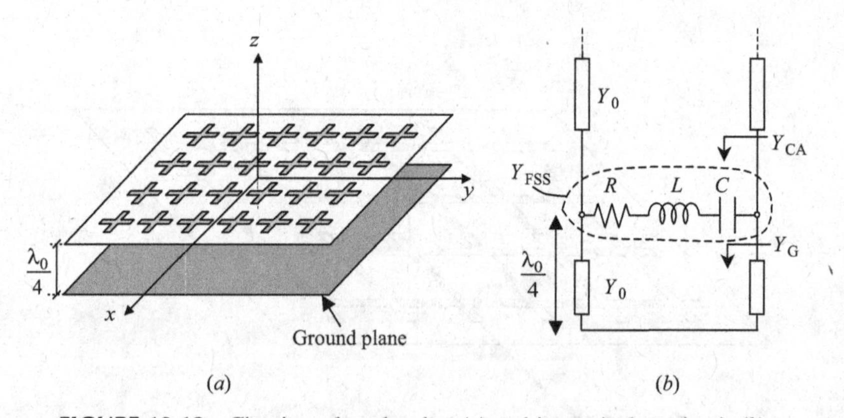

FIGURE 10.12 Circuit analog absorber (*a*) and its equivalent circuit (*b*).

larger than $Y_0 = \sqrt{\varepsilon_0/\mu_0}$ gives rise to a larger bandwidth with respect to the $\text{Re}\{Y_{CA}\} = Y_0$ case [3]. Like the Jaumann layers, to further improve the bandwidth performance, a CA absorber can be designed with more than one FSS [58]. Other designs of FSSs as EM absorbers, based on optimization techniques, can be found in [59–61].

10.7 MODELING AND DESIGN OF FSSs

The techniques adopted to model and analyze FSSs have their roots in past research on phased array antennas [62]. Many numerical approaches have been used to perform full-wave analyses of FSSs, based on finite-difference methods (FDTD) [63], finite-element methods (FEM) [64], the mode-matching method (MMM) [65], and hybrid methods combining the previous ones [66]. The FDTD and the FEM methods can handle arbitrary structures, but they are typically slow, and take up too much CPU time and memory. The circuit theory approach can be used under quasi-static assumptions by which to obtain the equivalent network of the considered FSS. Iterative techniques are being used to study realistic truncated structures with a finite number of elements (in this case the problem is the thousands of unknowns) [67]. However, the method of moments (MoM) is the most common approach, with infinite planar FSSs. Particularly efficient is the hybrid MoM approach presented in [68], which is based on the numerical determination of problem-matched entire-domain basis functions through the boundary–integral resonant–mode expansion method (BI-RME).

In general, most of the methods are based on the Floquet representation of the fields in the unit cell. Below we briefly summarize the method proposed by Chen in [20,21] and reviewed also in [2].

Basically the incident plane wave is decomposed into a sum of zeroth order TE and TM Floquet waves, and the scattered fields (reflected and transmitted) are expanded in a set of Floquet waves with unknown (reflection and transmission) coefficients. Then the components of the electric and magnetic fields tangential to the FSS are matched on a unit cell of the periodic surface, and an integral representation for the reflection and transmission coefficients is derived. At this stage, two different procedures can be followed, depending on the nature of the elements forming the periodic structure. In particular, by enforcing the tangential electric field to vanish on the conducting elements an electric field integral equation (EFIE) is obtained, where the unknown is the electric current on the conducting elements of the unit cell. Alternatively, by enforcing the continuity of the tangential magnetic field on the aperture elements, a magnetic field integral equation (MFIE) is derived, where the unknown is the aperture electric field on the aperture elements of the unit cell.

In both cases the EFIE and the MFIE can be solved by expanding the unknowns (electric current or aperture field, respectively) in a complete set of basis functions and performing a Galerkin testing procedure. When possible, the set of basis functions is chosen of the entire-domain type [69], and this set coincides with the set of modal functions of the waveguide having the aperture element (or the conducting element) as cross section [20,21]. The subdomain basis functions can also be used, but many hundreds may be needed [69]. For this reason the MoM-BI-RME method mentioned above [68] can be really effective.

Alternatively to the field approach described above, a modal transmission-line approach is used, as described by Orta and Tascone in [2]. In this approach an FSS with a finite number of elements is viewed as a planar discontinuity in a waveguide of infinite cross section, and an FSS with an infinite number of elements is viewed as a planar discontinuity in a waveguide having the unit cell as cross section and periodic boundary conditions. In the scattering problem the fields are expanded in terms of the modes of such a waveguide, and an equivalent transmission line is associated with each of these modes (in particular, voltages and currents are related to the Fourier transforms of the transverse electric and magnetic field, respectively, while the sources are represented by voltage and current generators). This way a generalized scattering matrix can be introduced, which is the basic tool to study the discontinuities in a waveguide and, for the present problem, allows to derive a functional equation solved by the method of moments.

Once an efficient and accurate numerical method is available for the study of an FSS, the problem of designing the periodic surface to provide a given frequency response is usually solved by means of optimization techniques; among them, evolutionary algorithms (e.g., genetic algorithms [37–39] and particle swarm optimization schemes [40]) are the most used.

REFERENCES

[1] K. Wu. *Frequency Selective Surface and Grid Array.* New York: Wiley, 1995.

[2] J. C. Vardaxoglou. *Frequency Selective Surfaces: Analysis and Design.* Taunton, Somerset: Research Studies Press, 1997.

[3] B. A. Munk. *Frequency Selective Surfaces: Theory and Design.* New York: Wiley, 2000.

[4] A. F. Peterson, S. L. Ray, and R. Mittra. *Computational Methods for Electromagnetics.* Piscataway, NJ: IEEE Press, 1998, ch. 7.

[5] S. Tretyakov. *Analitical Modeling in Applied Electromagnetics.* New York: Artech House, 2003.

[6] S. W. Schneider and B. A. Munk. "Scattering properties of 'super dense' arrays of dipoles." *IEEE Trans. Antennas Propagat.*, vol. 42, no. 4, pp. 463–472, Apr. 1994.

[7] J. C. Vardaxoglou and E. A. Parker. "Performance of two tripole arrays as frequency selective surfaces." *Electron. Lett.*, vol. 19, no. 18, pp. 709–710, Sep. 1983.

[8] P. W. B. Au, L. S. Musa, E. A. Parker, and R. J. Langley. "Parametric study of tripole and tripole loop arrays as frequency selective surfaces." *IEE Proc., Pt. H: Microw. Antennas Propagat.*, vol. 137, no. 5, pp. 263–268, Oct. 1990.

[9] E. A. Parker, S. M. A. Hamdy, and R. J. Langley. "Modes of resonance of the Jerusalem cross in frequency selective surfaces." *IEE Proc., Pt. H: Microw. Antennas Propagat.* vol. 130, no. 3, pp. 203–208, Apr. 1983.

[10] C. H. Tsao and R. Mittra. "Spectral-domain analysis of frequency selective surfaces comprised of periodic arrays of cross dipoles and Jerusalem crosses." *IEEE Trans. Antennas Propagat.*, vol. 32, no. 5, pp. 478–486, May 1984.

[11] R. J. Langley and E. A. Parker. "Equivalent circuit model for arrays of square loops." *Electron. Lett.*, vol. 18, no. 7, pp. 294–296, Apr. 1982.

[12] R. J. Langley and E. A. Parker. "Double-square frequency-selective surfaces and their equivalent circuit." *Electron. Lett.*, vol. 19, no. 17, pp. 675–677, Aug. 1983.

[13] E. A. Parker, S. M. A. Hamdy, and R. J. Langley. "Arrays of concentric rings as a frequency selective surface." *Electron. Lett.*, vol. 17, no. 23, pp. 880–881, Nov. 1981.

[14] E. A. Parker and J. C. Vardaxoglou. "Plane wave illumination of concentric ring frequency selective surfaces." *IEE Proc., Pt. H: Microw. Antennas Propagat.*, vol. 132, no. 3, pp. 176–180, Jun. 1985.

[15] A. Kondo. "Design and characteristics of ring slot type FSS." *Electron. Lett.*, vol. 27, no. 3, pp. 240–241, Jan. 1991.

[16] J. Huang, T. K. Wu, and S. W. Lee. "Tri-band frequency selective surface with circular ring elements." *IEEE Trans. Antennas Propagat.*, vol. 42, no. 2, pp. 166–175, Feb. 1994.

[17] T. K. Wu and S. W. Lee. "Multiband frequency selective surface with multiring elements." *IEEE Trans. Antennas Propagat.*, vol. 42, no. 11, pp. 1484–1490, Nov. 1994.

[18] E. A. Parker, C. Antonopoulos, and N. E. Simpson. "Microwave band FSS in optically transparent conducting layers: Performance of ring element arrays." *Microw. Opt. Tech. Lett.*, vol. 16, no. 2, pp. 61–63, Oct. 1997.

[19] R. B. Kieburtz and A. Ishimaru. "Scattering by a periodically apertured conducting screen." *IEEE Trans. Antennas Propagat.*, vol. AP-9, no. 6, pp. 506–514, Nov. 1961.

[20] C. C. Chen. "Scattering by a two-dimensional periodic array of conducting plates." *IEEE Trans. Antennas Propagat.*, vol. AP-18, no. 5, pp. 660–665, Sep. 1970.

[21] C. C. Chen. "Transmission through a conducting screen perforated periodically with apertures." *IEEE Trans. Microwave Theory Tech.*, vol. MTT-18, no. 9, pp. 627–632, Sep. 1970.

[22] E. A. Parker and A. N. A. El Sheikh. "Convoluted array elements and reduced size unit cells for frequency selective surfaces." *IEE Proc., Pt. H: Microw. Antennas Propagat.*, vol. 138, no. 1, pp. 19–22, Feb. 1991.

[23] A. D. Churpin, E. A. Parker, and J. C. Batchelor. "Convoluted double square: single layer FSS with close band spacings." *Electron. Lett.*, vol. 36, no. 22, pp. 1830–1831, Oct. 2000.

[24] J. Romeu and Y. Rahmat-Samii. "Dual band FSS with fractal elements." *Electron. Lett.*, vol. 35, no. 9, pp. 702–703, Apr. 1999.

[25] D. H. Werner and D. Lee. "Design of dual polarised multiband frequency selective surfaces using fractal elements." *Electron. Lett.*, vol. 36, no. 6, pp. 487–488, Mar. 2000.

[26] J. Romeu and Y. Rahmat-Samii. "Fractal FSS: A novel dual-band frequency selective surface." *IEEE Trans. Antennas Propagat.*, vol. 48, no. 7, pp. 1097–1105, Jul. 2000.

[27] B. A. Munk and R. J. Luebbers. "Reflection properties of two-layer dipole arrays." *IEEE Trans. Antennas Propagat.*, vol. AP-22, no. 6, pp. 766–773, Nov. 1974.

[28] B. A. Munk, R. J. Luebbers, and R. D. Fulton. "Transmission through a two-layer array of loaded slots." *IEEE Trans. Antennas Propagat.*, vol. AP-22, no. 6, pp. 804–809, Nov. 1974.

[29] J. C. Vardaxoglou and E. A. Parker. "Modal analysis of scattering from two-layer frequency-selective surfaces." *Int. J. Electr.*, vol. 58, no. 5, pp. 827–830, 1985.

[30] J. C. Vardaxoglou, A. Hossainzadeh, and A. Stylianou. "Scattering from two-layer FSS with dissimilar lattice geometries." *IEE Proc., Pt. H: Microw. Antennas Propagat.*, vol. 140, no. 1, pp. 59–61, Feb. 1993.

[31] P. Callaghan, E. A. Parker, and R. J. Langley. "Influence of supporting dielectric layers on the transmission properties of frequency selective surfaces." *IEE Proc., Pt. H: Microw. Antennas Propagat.*, vol. 138, no. 5, pp. 448–454, Oct. 1991.

[32] R. J. Luebbers and B. A. Munk. "Some effects of dielectric loading on periodic slot arrays." *IEEE Trans. Antennas Propagat.*, vol. AP-26, no. 4, pp. 536–542, Jul. 1978.

[33] R. C. Hall, R. Mittra, and K. M. Mitzner. "Analysis of multilayered periodic structures using generalized scattering matrix theory." *IEEE Trans. Antennas Propagat.*, vol. 36, no. 4, pp. 511–517, Apr. 1988.

[34] R. Orta, R. Tascone, and R. Zich. "Mutiple dielectric loaded perforated screens as frequency selective surfaces." *IEE Proc., Pt. H: Microw. Antennas Propagat.*, vol. 135, no. 2, pp. 75–82, Apr. 1988.

[35] A. K. Bhattacharyaa. "Analysis of multilayer infinite periodic array structures with different periodicities and axes orientations." *IEEE Trans. Antennas Propagat.*, vol. 48, no. 3, pp. 357–369, Mar. 2000.

[36] J.-F. Ma, R. Mittra, and N. T. Huang. "Analysis of multiple FSS screens of unequal periodicity using an efficient cascading technique." *IEEE Trans. Antennas Propagat.*, vol. 53, no. 4, pp. 1401–1414, Apr. 2005.

[37] G. Manara, A. Monorchio, and R. Mittra. "Frequency selective surface design based on genetic algorithm." *Electron. Lett.*, vol. 35, no. 17, pp. 1400–1401, Aug. 1999.

[38] S. Chakravarty and R. Mittra. "Application of the micro-genetic algorithm to the design of spatial filters with frequency-selective surfaces embedded in dielectric media." *IEEE Trans. Electromagn. Compat.*, vol. 44, no. 2, pp. 338–346, May 2002.

[39] M. Bozzi, G. Manara, A. Monorchio, and L. Perregrini. "Automatic design of inductive FSSs using the genetic algorithm and the MoM/BIR-ME analysis." *IEEE Antennas Wireless Propagat. Lett.*, vol. 1, pp. 91–93, 2002.

[40] S. Genovesi, R. Mittra, A. Monorchio, and G. Manara. "Particle swarm optimization for the design of frequency selective surfaces." *IEEE Antennas Wireless Propagat. Lett.*, vol. 5, pp. 277–279, 2006.

[41] T. K. Chang, R. J. Langley, and E. A. Parker. "Frequency selective surfaces on biased ferrite substrates." *Electron. Lett.*, vol. 30, no. 15, pp. 1193–1194, Jul. 1994.

[42] Y. C. Chan, G. Y. Li, T. S. Mok, and J. C. Vardaxoglou. "Analysis of a tunable frequency-selective surface on an in-plane biased ferrite substrate." *Microw. Opt. Tech. Lett.*, vol. 13, no. 2, pp. 59–63, Oct. 1996.

[43] A. C. de C. Lima, E. A. Parker, and R. J. Langley. "Tunable frequency selective surface using liquid substrates." *Electron. Lett.*, vol. 30, no. 4, pp. 281–282, Feb. 1994.

[44] E. A. Parker and S. B. Savia. "Active frequency selective surfaces with ferroelectric substrates." *IEE Proc. Microw. Antennas Propagat.*, vol. 148, no. 2, pp. 103–108, Apr. 2001.

[45] J. C. Vardaxoglou. "Optical switching of frequency selective surface bandpass response." *Electron. Lett.*, vol. 32, no. 25, pp. 2345–2346, Dec. 1996.

[46] D. S. Lockyer and J. C. Vardaxoglou. "Reconfigurable FSS response from two layers of slotted dipole arrays." *Electron. Lett.*, vol. 32, no. 6, pp. 512–513, Mar. 1996.

[47] D. S. Lockyer, J. C. Vardaxoglou, and R. A. Simpkin. "Complementary frequency selective surfaces." *IEE Proc. Microw. Antennas Propagat.*, vol. 147, no. 6, pp. 501–507, Dec. 2000.

[48] J. P. Gianvittorio, J. Zendejas, Y. Rahmat-Samii, and J. Judy. "Reconfigurable MEMS-enabled frequency selective surfaces." *Electron. Lett.*, vol. 38, no. 25, pp. 1627–1628, Dec. 2002.

[49] J. M. Zendejas, J. P. Gianvittorio, Y. Rahmat-Samii, and J. W. Judy. "Magnetic MEMS reconfigurable frequency-selective surfaces." *IEEE J. Microelectromech. Syst.*, vol.15, no. 3, pp. 613–623, Jun. 2006.

[50] C. Mias. "Frequency selective surfaces loaded with surface-mount reactive components." *Electron. Lett.*, vol. 39, no. 9, pp. 724–726, May 2003.

[51] C. Mias. "Varactor-tunable and dipole-grid-based frequency-selective surface." *Microw. Opt. Tech. Lett.*, vol. 43, no. 6, pp. 508–511, Dec. 2004.

[52] T. K. Chang, R. J. Langley, and E. Parker. "An active square loop frequency selective surface." *IEEE Microwave Guided Wave Lett.*, vol. 3, no. 10, pp. 387–388, Oct. 1993.

[53] T. K. Chang, R. J. Langley, and E. A. Parker. "Active frequency-selective surfaces." *IEE Proc. Microw. Antennas Propagat.*, vol. 143, no. 1, pp. 62–66, Feb. 1996.

[54] B. M. Cahill and E. A. Parker. "Field switching in an enclosure with active FSS screen." *Electron. Lett.*, vol. 37, no. 4, pp. 244–245, Feb. 2001.

[55] E. A. Parker and S. B. Savia. "Fields in an FSS screened enclosure." *IEE Proc. Microw. Antennas Propagat.*, vol. 151, no. 1, pp. 77–80, Feb. 2004.

[56] C. A. Kyriazidou, R. E. Diaz, and N. G. Alexopoulos. "Novel material with narrow-band transparency window in the bulk." *IEEE Trans. Antennas Propagat.*, vol. 48, no. 1, pp. 107–116, Jan. 2000.

[57] W. W. Salisbury. "Absorbent body for electromagnetic waves." U. S. Patent 2 599 944, Jun. 10, 1952.

[58] B. A. Munk, P. Munk, and J. Pryor. "On designing Jaumann and circuit analog absorbers (CA absorbers) for oblique angle of incidence." *IEEE Trans. Antennas Propagat.*, vol. 55, no. 1, pp. 186–193, Jan. 2007.

[59] S. Chakravarty, R. Mittra, and N. R. Williams. "Application of a microgenetic algorithm (MGA) to the design of broad-band microwave absorbers using multiple frequency selective surface screens buried in dielectrics." *IEEE Trans. Antennas Propagat.*, vol. 50, no. 3, pp. 284–296, Mar. 2002.

[60] D. J. Kern and D. H. Werner. "A genetic algorithm approach to the design of ultra-thin electromagnetic band-gap absorbers." *Microw. Opt. Tech. Lett.*, vol. 38, no. 1, pp. 61–64, Jul. 2003.

[61] A. Tennant and B. Chambers. "A single-layer tuneable microwave absorber using an active FSS." *IEEE Microwave Wireless Comp. Lett.*, vol. 14, no. 1, pp. 46–47, Jan. 2004.

[62] N. Amitay, W. Galindo, and C. P. Wu. *Theory and Analysis of Phased Array Antennas.* New York: Wiley, 1972.

[63] P. Harms, R. Mittra, and K. Wai. "Implementation of the periodic boundary condition in the finite-difference time-domain algorithm for FSS structures." *IEEE Trans. Antennas Propagat.*, vol. 42, no. 9, pp. 1317–1324, Sep. 1994.

[64] E. A. Parker, A. D. Chuprin, and R. J. Langley. "Finite element analysis of electromagnetic wave diffraction from buildings incorporating frequency selective walls." *IEE Proc. Microw. Antennas Propagat.*, vol. 146, no. 5, pp. 319–323, Oct. 1999.

[65] D. C. Love and E. J. Rothwell. "A mode-matching approach to determine the shielding properties of a doubly periodic array of rectangular apertures in a thick conducting screen." *IEEE Trans. Electromagn. Compat.*, vol. 48, no. 1, pp. 121–133, Feb. 2006.

[66] A. Monorchio, P. Grassi, and G. Manara. "A hybrid mode-matching finite-elements approach to the analysis of thick dichroic screens with arbitrarily shaped apertures." *IEEE Antennas Wireless Propagat. Lett.*, vol. 1, pp. 120–123, 2002.

[67] A. Stylianou, P. Debono, and J. C. Vardaxoglou. "Iterative computation of current and field distributions in multilayer frequency selective surfaces." *IEE Proc., Pt. H: Microw. Antennas Propagat.*, vol. 139, no. 6, pp. 535–541, Dec. 1992.

[68] M. Bozzi and L. Perregrini. "Analysis of multilayered printed frequency selective surfaces by the MoM/BI-RME method." *IEEE Trans. Antennas Propagat.*, vol. 51, no. 10, pp. 2830–2836, Oct. 2003.

[69] R. Mittra, C. H. Chan, and T. Cwik. "Techniques for analyzing frequency selective surfaces-a review." *IEEE Proc.*, vol. 76, no. 12, pp. 1593–1615, Dec. 1988.

CHAPTER ELEVEN

Shielding Design Guidelines

As is usual in engineering, the design stage is much more difficult than the analysis stage, and shielding does not represent an exception for this rule. The reason is manifold: on one hand, the solution of a shielding design problem is generally not unique, and therefore a number of choices has to be made involving very different aspects, often very far from electromagnetics. On the other hand, the concurrence of several coupling paths calls for some wise assumptions in order to simplify the procedure. The uncertainty as to the figures of merit to be selected and met to guarantee the correct operation of the shielding structure (i.e., the shielded devices and systems) is another delicate point (as pointed out in Chapter 3) that is often by-passed when compliance with standards is assumed to be sufficient for the achievement of EMC and EMI protection or information security objectives.

In general, any procedure aimed at designing a shielding structure consists of the following main steps:

1. Establishment of the shielding requirements, in terms of SE values or other figures of merit.
2. Assessment of type and number of functional discontinuities.
3. Assessment of dimensional constraints and nonelectromagnetic characteristics of the materials.
4. Formulation of a hypothesis of shielding configuration with all the details defined.
5. Estimation of the shielding performance.
6. Check on specs fulfillment and, if necessary, modification of the characteristics of the most relevant coupling paths responsible for the failure in the design.

Electromagnetic Shielding by Salvatore Celozzi, Rodolfo Araneo and Giampiero Lovat

Of course, steps 4 to 6 should be repeated until the check guarantees that a reasonable probability of success is achieved by the shielding configuration, once it is realized. It is important to observe the core of the design procedure that is in steps 4 and 5: experience will be of fundamental guidance in the choice of the structure (e.g., single shield or multilevel), the components, and the materials. The first step is also critical because it may lead to an over- or underestimation of the required shielding levels.

The way in which the steps of the above-outlined procedure are dealt with can be very different in the two typical situations of susceptibility-oriented or emission-oriented design. In the former, the frequency-spectrum of the EM field is assumed to be known, the immunity levels of the various components (in-band and out-of-band) are often known by manufacturers, and attention is paid to two key points: coupling paths and internal emissions (possibly adding to external threats). On the other hand, in emission-driven design procedures the EM field levels of the various subsystems and components are strongly dependent on assembly characteristics, while the emission level masks not to be exceeded are usually known and established by standards.

11.1 ESTABLISHMENT OF THE SHIELDING REQUIREMENTS

The frequency mask of the EM requirements is usually obtained by taking into account three terms that are then combined in different ways in case of emission or susceptibility problems:

- Levels of or limits for the environmental EM field, *EL*
- Margin, *M*
- Intrinsic and installation-dependent levels of EM susceptibility, *S*, and radiated emission of the involved components and systems, *RE*.

Each of the three terms calls for specific analysis and will be briefly considered in the following discussion by leaving to references a deeper treatment.

The margin *M* is often fixed between 10 and 30 dB, depending on the known accuracy of the adopted prediction methods and on the uncertainty existing on external field or radiated-emission levels and possible evolution in the system (system upgrade, time evolution of performance, etc.). However, a margin $M = 20$ dB is often chosen [1].

When the susceptibility problem is considered, an upper bound has to be selected between the predetermined frequency masks and the possible application-oriented EM levels of external fields due to interfering sources. For instance, the mask may be that considered in MIL-Std. 461E [2] and the frequency spectrum that ensuing from an EM pulse [3]. Furthermore the presence of sources in the interior of the shielded volume should also be carefully considered; depending on the field levels and on the frequencies they may generate, this aspect will in fact be of guidance in the choice of the number of shield levels necessary to guarantee the correct operation of the system and the compliance with standards. In general, the most difficult task is that concerning the inventory of the susceptibility levels S of all the components or subsystems. A careful analysis must be conducted not only on the in-band and out-of-band susceptibility of each component but also on their functional connection and on the influence that an external threat may have on nonexposed components through the exposed adjacent ones. In susceptibility-oriented shielding design, the required level of reduction is

$$\mathrm{SE} = EL + M - S. \tag{11.1}$$

The design of shielding configurations aimed at limiting the unwanted emission from devices, apparatus, and systems requires a careful analysis of the characteristics of the sources, since the radiating elements and mechanisms are manifold, and moreover near-field problems are often encountered. Their identification may be not as trivial as it might be expected at a first glance. The listing of all the radiating components within the shielding structure will not circumvent the task; for each of them it is necessary to estimate the voltage and the current spectra, taking into account that, for most components, only the maximum time response is declared and guaranteed by manufacturers, leaving an important uncertainty on the maximum involved frequency. After this operation is completed, the geometry of the subsystem connections should be considered to account for the radiating components' additional radiation at prescribed distances where limits are fixed. The shielding requirements are evaluated as

$$\mathrm{SE} = RE + M - EL. \tag{11.2}$$

When both susceptibility and radiated emission are of concern, the selection between values ensuing from (11.1) and (11.2) is not automatically the largest. The source type involved in the two situations is generally different, and the a priori choice of the most critical SE is as difficult as the separate verification for the two cases. Thus the values ensuing from (11.1) and (11.2) must be compared with the shielding effectiveness (in a broad meaning) of the designed structure under both the test conditions.

11.2 ASSESSMENT OF THE NUMBER AND TYPES OF FUNCTIONAL DISCONTINUITIES

There are several types of functional discontinuities. The main and most frequently encountered discontinuities are junctions and seams, cable pass-throughs, visualization apertures, air vents, operational devices, and connectors.

As described in Chapter 9, some functional discontinuities are dealt with by means of solutions enabling an adequate level of SE, whereas others can be rather critical. Shielded windows are probably the components with the lowest level of SE because they may achieve performance levels on the order of 40 to 50 dB (typical values for windows whose dimensions are up to few tens of centimeters in the maximum dimension), while levels up to and even above 100 dB are achieved with correctly designed air vents and shielded connectors. Windows of large dimensions can be expensive, so the use of multiple apertures for visualization is often preferred with respect to a unique large aperture. Very serious concern is generally directed to nonshielded operational devices (e.g., fuses, lamps, and switches) whose apertures offer an additional coupling path of large impact on the shielding performance of the overall system. Also the signal or power lines that connect the shielded region to the source region can be an efficient vehicle of coupling. However, the adoption of shielded operational devices and of proper grounding [4,5] of cable sheaths is usually sufficient to prevent any appreciable performance deterioration or to reduce to a minimum the undesired effects.

Junctions between panels are often sealed by gaskets, as described in Chapter 9. When gaskets are correctly selected and maintained, they guarantee adequate levels of SE in almost all the applications. Much more critical situations occur in dealing with permanent or semipermanent joints, typically welded, screwed, or bolted. At the design stage noncontinuously sealed joints are worthy of a careful analysis. For instance, the improvement in the SE achieved by subdividing a large aperture, whose length and height are L_{tot} and h, respectively, into N_a smaller apertures of length L_a is [1]

$$\Delta \mathrm{SE} = 20 \log \frac{L_{tot}}{L_a} - 20 \log \left[\frac{1 + \ln(L_{tot}/h)}{1 + \ln(L_a/h)} \right]. \tag{11.3}$$

It should be noted that adding screws or other types of contact points will reduce the height of the apertures with respect to the original one, thus further improving the shielding performance.

11.3 ASSESSMENT OF DIMENSIONAL CONSTRAINTS AND NONELECTROMAGNETIC CHARACTERISTICS OF MATERIALS

Dimensional constraints and other nonelectromagnetic considerations (corrosion, weight, cost, etc.) will influence the choice of materials and the shape of the shielding structure, which is closely related to the content's dimensions and to its emission and immunity characteristics. On the other hand, the shape and the geometrical dimensions of the shielded volume determine the frequency and the number of internal resonances. A preliminary and approximate analysis aimed at verifying whether or not the enclosure will work in an overmoded region of the frequency spectrum is useful to have as guidance on the importance that some aspects (e.g., the internal position of the most sensitive or radiating elements) can assume in the design.

Unfortunately, loading and apertures dimensions and positions can considerably affect the value of the resonant frequencies, so a careful numerical analysis is needed for a wise arrangement of components and subsystems in the shielded volume. However, in most cases the arrangement of such components and subsystems is determined on the basis of nonelectromagnetic considerations, such as those due to accessibility, ventilation, or visualization.

When compatible with costs and weight considerations, the use of partial internal shields is always recommended to protect the most susceptible components or to prevent the most radiating from causing both internal problems and emission levels larger than those admissible. Such a double level of shielding structure can also alleviate the requirements for the external shield with an overall advantage in terms of functionality and costs.

11.4 ESTIMATION OF SHIELDING PERFORMANCE

It should be recognized that shielding performance of actual configurations is very difficult to accurately predict. A rough estimation can be achieved by use of numerical simulations. These are generally accurate enough for prototyping purposes, provided that the constructive details are described with sufficient accuracy. However, depending on both system complexity and software/hardware characteristics, the input of details of an actual configuration, even in user-friendly commercial software, may require a lot of work and the output may be available after a long computational time. Therefore a preliminary approximate estimation can be very useful. Such an estimation of the SE can be obtained from the "ideal" SE achievable by means of a barrier of homogeneous material corrected by two terms: a worst-case term accounting for leakages and a rule-of-thumb term representing resonance deterioration. The consequent estimation of the SE is therefore

$$\mathrm{SE}' \simeq \mathrm{SE}_{\mathrm{barrier}} - \Delta\mathrm{SE}_{\mathrm{tot\_leakage}} - \Delta\mathrm{SE}_{\mathrm{standing\_waves}}. \tag{11.4}$$

The terms $\mathrm{SE}_{\mathrm{barrier}}$ and $\Delta\mathrm{SE}_{\mathrm{leakage}}$ (due to one aperture) were presented in Chapters 4, 6, and 7. Often, in a first attempt of shielding design, the deterioration effects due to internal resonances are taken into account by means of a $6 \div 10$ factor (in dB) [1].

If the leakage is due to several effects, their worst-case combination (i.e., the sum of in-phase field contributions arising from different coupling paths) is expressed as

$$\Delta\mathrm{SE}_{\mathrm{tot\_leakage}} = -20\log\left[\sum_{i=1}^{n} 10^{-L_{i(\mathrm{dB})}/20}\right], \tag{11.5}$$

where $L_{i(\mathrm{dB})}$ is the leakage (in dB) of the ith coupling path. Thus the general expression for a first performance estimation could be

$$\mathrm{SE} = -20\log\left[10^{-\mathrm{SE}_{\mathrm{barrier}}/20} + \sum_{i=1}^{n} 10^{-L_{i(\mathrm{dB})}/20}\right] - (6 \div 10). \tag{11.6}$$

REFERENCES

[1] D. R. J. White and M. Mardiguian. *Electromagnetic Shielding, A Handbook Series on EMI and Compatibility*, vol. 3. Gainsville, VA: Interference Control Technology, 1988.

[2] MIL-STD 461E. "Department of defense interface standard. requirements for the control of electromagnetic interface characteristics of subsystems and equipments." US Government Printing Office, Washington, DC, 20 August 1999.

[3] K. S. H. Lee, ed. *EMP Interaction: Principles, Techniques, and Reference Data*. Washington, DC: Hemisphere, 1986.

[4] R. R. Morrison and W. H. Lewis. *Grounding and Shielding in Facilities*. New York: Wiley, 1990.

[5] H. W. Ott. *Noise Reduction Techniques in Electronic Systems*, 2nd ed. New York: Wiley, 1988.

[6] MIL-HDBK 1857. "Grounding, bonding, and shielding design practices." US Government Printing Office, Washington, DC, 27 March 1998.

CHAPTER TWELVE

Uncommon Ways of Shielding

Some rather uncommon ways of shielding are briefly described in this chapter. They are based on different principles that call for such means as resorting to an active compensation, applying partial shields, or exploiting special material properties.

12.1 ACTIVE SHIELDING

Active shielding is the process of reducing the EM field in a region of interest by driving an active circuit that generates an EM reaction field characterized by the same frequency and amplitude as the incident field but in the opposite direction [1]. If the incident field presents a wide bandwidth, the final aim is to generate a reaction field in the same frequency range, or at least in a range as large as possible. Active shielding is sometimes termed active "cancellation" but this term seems to be not very appropriate because complete elimination of the incident field is never achieved, but rather field-cancellation efficiencies of 65% to 90% are typical (SE ranging from 10 to 25 dB).

To understand the true nature of the active shielding, this technique must be regarded as an alternative method to the basic passive shielding practice that limits an EM field (in both emission and susceptibility problems) by means of a barrier made of materials with high conductivity or permeability. Active shielding has remained an interesting way to attenuate mainly low-frequency magnetic fields. As explained in Appendix B, magnetic fields at extremely low frequencies are in fact attenuated with much difficulty via passive shields made of common materials (obtaining SE typically lower than 5 dB); alternatively, high-performance and expensive ferromagnetic materials can be used, such as permalloy, mumetal, or factory custom alloys. In addition it should be noted that passive shields can be used

Electromagnetic Shielding by Salvatore Celozzi, Rodolfo Araneo and Giampiero Lovat

only for single rooms because their use at the level of whole buildings would be impractical.

In this framework it should be clear that active shielding represents a practical and cost-effective approach when the mitigation of a power-frequency magnetic field is required in wide areas, such as for buildings placed near overhead power transmission lines or buried cables [2,3]. Similar technologies are often used also at the room level for the mitigation of a field produced by many different sources, such as stray fields from power transformers [4], bus bars, service panels [1], and inductions heaters [5,6], or to protect sensitive electron-beam devices, such as scanning electron microscopes and computer monitors, from possible interference.

The mitigation of the power-frequency magnetic fields in a wide region of space, as a building floor or a whole building, is mainly motivated by health issues. In the last 30 years the health effects of low-intensity power-frequency magnetic fields have become a subject of controversy. Several in vitro testing as well as in vivo studies on animals have not been able to show a clear correlation between certain cancer forms and domestic or work exposure to magnetic fields at levels normally existing near power lines. However, it is still under debate whether an interrelation between magnetic fields and human health can be confirmed in some way by the International Agency for Research on Cancer (IARC). This debate has generated so much concern about possible adverse health effects of power-frequency magnetic fields that many countries have adopted precautionary principles aimed at reducing magnetic-field exposure in existing installations and also at the preliminary planning stage in upcoming buildings.

Even if the mitigation of the magnetic field through the active shielding may be approached in a number of ways, the opposite field is generally produced by currents injected into adequately designed (round or square) active coils. A field-controlled configuration consists mainly of three basic elements: a sensor to monitor the incident field, a control and power unit, and a network of driven coils [1]. The system tracks the incident field and instantaneously adjusts to compensate for changes that can occur in the emitting source. Despite the rather simplicity of the configuration, several issues are critical, and in the past they precluded effective deployment of active shielding on a large scale. The first challenge in obtaining good performance is the proper design of the driven coil network that surrounds the treated area and produces the counterfield. The active shielding technique is more effective on those fields that are uniform in direction, intensity, and polarity.

However, the fields emitted by a real source can be very complex, showing elliptical polarization, strong field gradients, inductive delay, and spatial phase variance. In order to make the active shielding really effective, an accurate design procedure must be carried out in each location, to which the design is completely unique, choosing the optimal values of geometrical parameters (number, position, and dimensions of coils) and currents. Depending on the incident field, the values of current required to obtain a significant value of SE may be large (from several to hundred of amperes), making the system complex from a technical and safety point of view. In addition active shielding in its basic form works only at a given frequency (usually at power frequencies) so that it cannot be effective in reducing higher order harmonic fields as those radiated by power electronic equipment or high-voltage direct-current systems. To overcome this problem, it is necessary to construct the waveform of the coil currents as a function of the incident magnetic induction $\mathbf{b}(t)$. An effective and simple idea consists in obtaining instant by instant a current proportional to the field measured by a field sensor through an appropriate analog or discrete time controller.

Finally, there is the problem of finding where to place the active coils given an adequate free space surrounding the area to be protected or the source to be confined. When dealing with the fields emitted by three-phase power lines, these coils are usually buried in proximity of the lines. However, for complex elliptically polarized fields, as those emitted by high-voltage overhead lines, auxiliary conductors become necessary in addition to coils and these may be difficult to be installed for space and aesthetical reasons.

To overcome these problems, the mitigation of the magnetic field near power lines is usually obtained by means of auxiliary conductors whose currents exert a compensating influence on the primary field [2,3]. These wires can be driven, or not, by an active source. In the first case, in order to form a loop, the number of wires (which are passive components) is at least two, and the mitigation current flowing along the conductors is driven by the voltage induced by the currents of the transmission line. This way the mitigating field is always proportional to the incident field of the line, and the unaltered mitigation is achieved for any value of the load carried by the line. The mitigating wires are usually located near the conductors of the transmission line, such as under each outer phase of the line, or above the two outer phases but at a height lower than the guard wires, or at ground level; alternatively, they could be even constituted by the two guard wires, if already present on the line. To increase the versatility of this approach, an active source can be included into the mitigating loop. Hence, in order to establish the optimal mitigation that causes the minimum resultant field in a specified area, and in addition to choose the best position for the auxiliary conductors, it is also possible to use suitable values for the current amplitudes and phases.

The final generalization of this approach is to make all the currents of the active wires independent of each other, allowing them to return through the earth. Losses associated with the return current may be tolerated, since usually the mitigation needs to be achieved only for short lengths, that is, where human exposure is foreseen. For simple high-voltage overhead lines, it is generally possible to create in a given region a magnetic field almost opposite to that produced by the primary source by driving the guard-wire, while for three-phase cable sources or complex balanced or unbalanced three-phase systems, additional buried active conductors are necessary.

Examples of power-line mitigation can be found in [2,3], and an application of a three-phase line near a building is shown in Figure 12.1. Mitigation is often required on only one side of a line, at distances between 10 and 50 meters. An adequate choice of the equivalent height of the line conductors needs to be made to account for their sag, so the magnetic induction can be assumed to lie in a plane transverse to

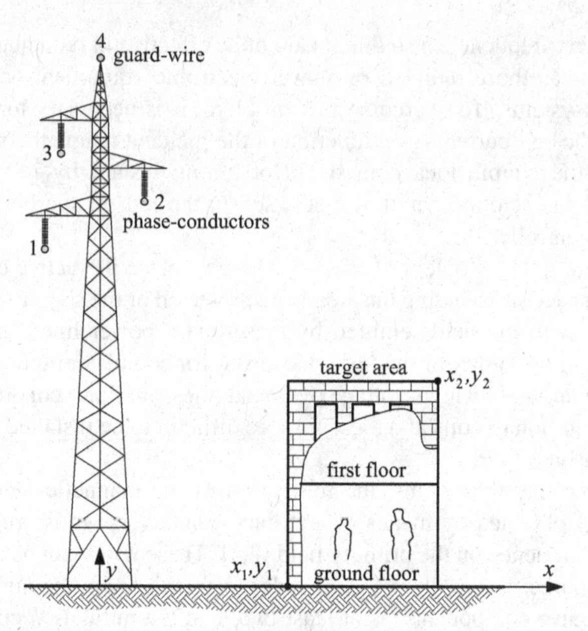

FIGURE 12.1 Three-phase line (conductor 1: $x = 2.3$ m, $y = 23$ m; conductor 2: $x = 2.3$, $y = 24.8$ m; conductor 3: $x = -2.3$ m, $y = 27.8$ m; guard-wire conductor 4: $x = 0$ m, $y = 32.9$ m) near a building with a target area defined by the coordinates $x_1 = 20$ m, $y_1 = 0$ m, $x_2 = 30$ m, and $y_2 = 6$ m.

the line direction, denoted as the z axis; the rms value of the magnetic induction is used as a measure of human exposure conditions.

The goal of the mitigation is the determination of the currents to be driven through the additional wires, whose position is also to be determined in such a way that a reduction of the magnetic induction is observed in the area of interest. In particular, the following inequality has to be satisfied:

$$B(x,y) = \sqrt{|B_x|^2 + |B_y|^2} \leq B_{\text{thres}} \quad \text{with} \quad (x,y) \in \{[x_1,x_2],[y_1,y_2]\}, \qquad (12.1)$$

where $B(x,y)$ is the total field due to the source conductors and to the compensation wires, B_{thres} is the threshold value of the magnetic flux density, and the point (x,y) belongs to the target area defined by the two intervals $[x_1,x_2]$ and $[y_1,y_2]$ along the x and y axis, respectively. The optimal solution may be sought by introducing a global parameter to be minimized, which is usually defined as

$$\Psi(x_1,x_2,y_1,y_2) = \int_{x_1}^{x_2}\int_{y_1}^{y_2} B(x,y)\,dx\,dy. \qquad (12.2)$$

The considered optimization problem is nonlinear, since the rms of the magnetic induction B has a nonlinear dependence on the distance R (which is also a parameter

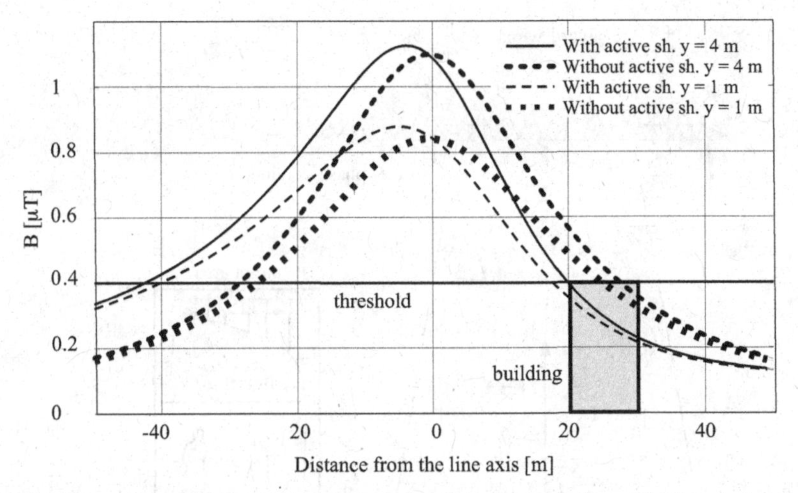

FIGURE 12.2 Magnetic-induction profiles with and without active shielding at 1 and 4 m above the ground for a system as in Figure 12.1.

to be optimized). This means that linear methods such as the standard Simplex are not suitable tools to solve the problem. In this case, genetic algorithms (GAs) find an optimal application, and they are usually chosen to search for the global minimum of the objective function (12.2). In fact GAs have been proved to be efficient and accurate in the solution of such a multidimensional optimization problem exhibiting several local minima [7].

The magnetic-induction profiles with and without active shielding for the configuration of Figure 12.1 are shown in Figure 12.2 with the line carrying a balanced system of currents at $I = 400$ A. As can be seen, the threshold of 0.4 μT is surpassed in a two-floor building located at 20 to 30 meters from the line axis. However, by means of an active compensation through a current driven in the guard-wire ($I = 60$ A), the threshold goal is attained.

Unfortunately, after the mitigation is achieved, an increase in the magnetic induction could be observed on the opposite side with respect to the line (as shown in Figure 12.2). To prevent this inconvenience, a passive partial shield is usually introduced to alter the field distribution on one side without affecting the other side. This technique, which combines active (wires and/or coils) and passive shielding, has general applicability and is also used to reduce stray fields from secondary transformer substations [4], induction heaters [5,6], and electric panels [1]. It is considered to be the most effective way to reduce extremely-low-frequency magnetic fields. An example of active/passive shielding is shown in Figure 12.3. To mitigate the magnetic field in the target, an active shielding is first applied; it drives a current of 212 A in guard wire 5 and a current of 98 A in the additional buried wire ($x = 0$ m, $y = -1$ m) as shown in Figure 12.4*a*. Next, since an increase of 12 μT above the buried wire is observed, a partial passive shield is introduced from $x = -1.5$ m to $x = 1.5$ m, buried at a depth of 0.5 m below the ground level. In

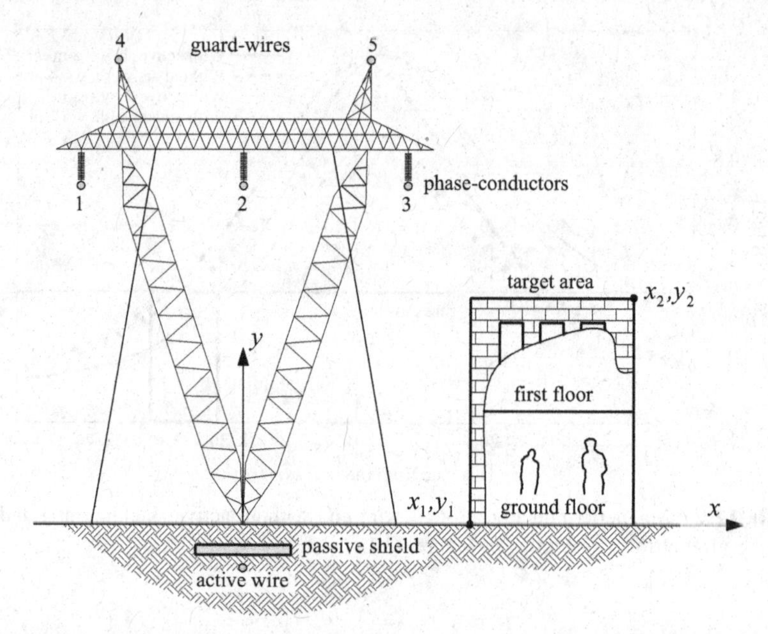

FIGURE 12.3 Three-phase line (conductor 1: $x = -7.2$ m, $y = 24.4$ m; conductor 2: $x = 0$ m, $y = 24.4$ m; conductor 3: $x = 7.2$ m, $y = 24.4$ m; guard-wire 4: $x = -4.9$ m, $y = 26.7$ m; guard-wire 5: $x = 4.9$ m, $y = 26.7$ m) near a building with a target area defined by the coordinates $x_1 = 20$ m, $y_1 = 0$ m, $x_2 = 30$ m, $y_2 = 6$ m.

Figure 12.4*b* the magnetic-induction profiles are reported for different thicknesses t of the shield and for different materials (e.g., aluminium with $\mu_r = 1$ and $\sigma = 3.5 \cdot 10^7$ S/m and steel with $\mu_r = 5 \cdot 10^4$ and $\sigma = 7.5 \cdot 10^5$ S/m); two-layer configurations are considered too.

12.2 PARTIAL SHIELDS

Open shielding topologies, namely, those where the shielded volume is not completely encircled by a shielding structure, allow a significant leakage at their edges, and are also called *partial shields*. They are realized by means of a finite-size barrier placed between the source and the observation point (victim). Such structures present worse shielding properties with respect to closed configurations (i.e., enclosures) and their performance is also strongly dependent on the source characteristics. Partial shields, however, are much simpler to install, as well as lighter, cheaper, and more effective as concerns heat dissipation. Because of their effectiveness in dissipating heat, they are frequently applied to shield electronic components and printed circuits.

Approximate analytical studies have been carried out on this subject in the presence of low-frequency sources, while a numerical analysis is appropriate in

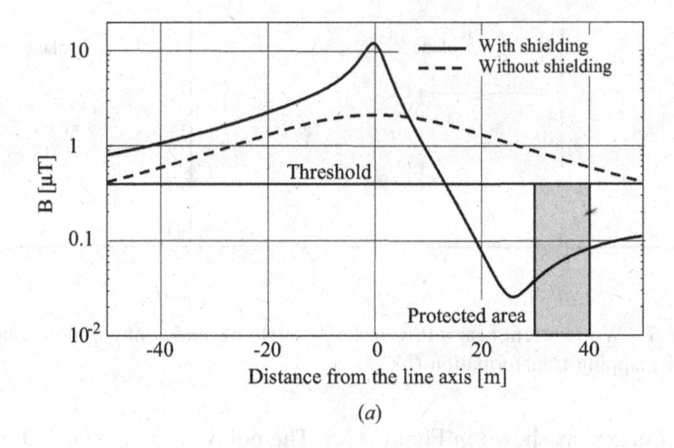

(*a*)

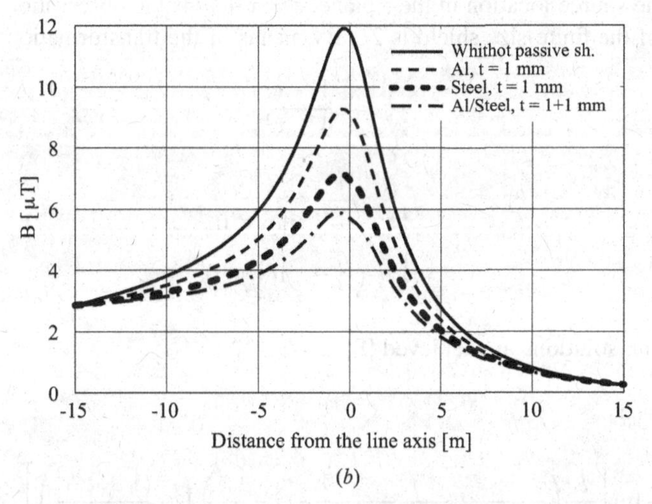

(*b*)

FIGURE 12.4 Magnetic-induction profiles with and without active shielding at 1 m above the ground plane (*a*); magnetic-induction profiles with and without passive shielding (*b*).

dealing with high-frequency problems. Nevertheless, some of the analytical considerations ensuing from low-frequency approximations may be valid when the distance between the source (or the victim) and the shield is much shorter than the wavelength.

Conformal mapping techniques have been used to evaluate the low-frequency magnetic field beyond a finite-width perfectly conducting (PEC) shield. Also problems concerning perfect magnetic conductor (PMC) shields and double-layer (PEC-PMC) shields have been studied [8]. Here only the main results are summarized.

In particular, the original problem is converted by means of a conformal transformation into a much simpler one, whose solution is achievable through the

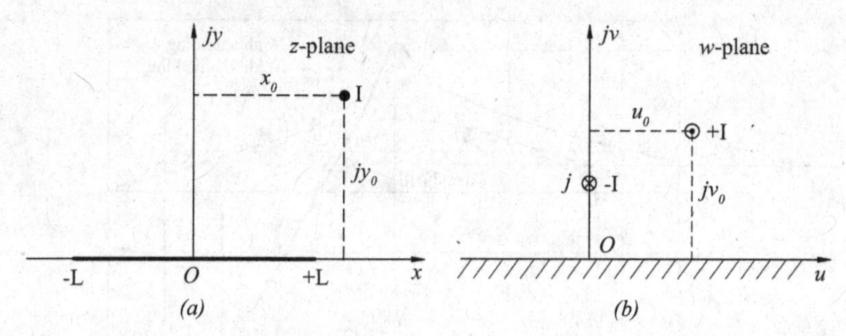

FIGURE 12.5 Wire current near a PEC or PMC shield (*a*) and configuration obtained after a conformal mapping transformation (*b*).

method of images, as shown in Figure 12.5. The point $z_0 = x_0 + jy_0$ in Figure 12.5 indicates the source location in the z plane; $z = x + jy$ is the observation point, and the width of the finite-size shield is $2L$. By means of the transformation

$$z = L\frac{1-\tau^2}{1+\tau^2} \tag{12.3}$$

and

$$\tau = \begin{cases} \sqrt{\frac{L-z}{L+z}} & \text{if } y < 0, \\ -\sqrt{\frac{L-z}{L+z}} & \text{if } y > 0, \end{cases} \tag{12.4}$$

the following solutions are achieved [8]:

PMC Case

$$H_x = -\mathrm{Im}\left[\frac{I}{2\pi}\left(\frac{1}{\tau-\tau_0} + \frac{1}{\tau-\tau_0^*} - \frac{1}{\tau-j} - \frac{1}{\tau+j}\right)\left(-\frac{(1+\tau^2)^2}{4L\tau}\right)\right], \tag{12.5a}$$

$$H_y = -\mathrm{Re}\left[\frac{I}{2\pi}\left(\frac{1}{\tau-\tau_0} + \frac{1}{\tau-\tau_0^*} - \frac{1}{\tau-j} - \frac{1}{\tau+j}\right)\left(-\frac{(1+\tau^2)^2}{4L\tau}\right)\right]. \tag{12.5b}$$

PEC Case

$$H_x = -\mathrm{Im}\left[\frac{I}{2\pi}\left(\frac{1}{\tau-\tau_0} - \frac{1}{\tau-\tau_0^*} - \frac{1}{\tau-j} + \frac{1}{\tau+j}\right)\left(-\frac{(1+\tau^2)^2}{4L\tau}\right)\right], \tag{12.6a}$$

$$H_y = -\mathrm{Re}\left[\frac{I}{2\pi}\left(\frac{1}{\tau-\tau_0} - \frac{1}{\tau-\tau_0^*} - \frac{1}{\tau-j} + \frac{1}{\tau+j}\right)\left(-\frac{(1+\tau^2)^2}{4L\tau}\right)\right]. \tag{12.6b}$$

This procedure may be extended to a double-layer shield [8]; a procedure that accounts for field penetration through the penetrable shield material is reported in [9].

12.3 CHIRAL SHIELDING

As described in Chapter 2, chiral media are natural or artificial materials whose constitutive relations relate both the electric displacement **D** and the magnetic induction **B** to the electric **E** and magnetic **H** fields, through (for simplicity, scalar) permittivity ε, permeability μ, and chirality parameter κ (see Section 2.7.1). These constitutive relations can also be cast in the form [10]

$$\begin{aligned} \mathbf{D} &= \varepsilon\mathbf{E} - j\xi_c\mathbf{B}, \\ \mathbf{H} &= j\xi_c\mathbf{E} + \frac{1}{\mu}\mathbf{B}, \end{aligned} \tag{12.7}$$

where ξ_c is the so-called chiral admittance that quantifies the strength of chirality and the handedness of the material. Chiral media have received considerable attention over the last few decades because, with respect to conventional materials, the chirality parameter provides an additional degree of freedom in the design of materials with specific EM characteristics.

In particular, chiral shields made of layered chiral structures have been studied to obtain efficient absorbers and reflectors. A layered chiral structure mainly consists of a layered structure (in a planar, cylindrical, or spherical fashion) in which one or more layers present chiral characteristics. In [11] a so-called chiroshield was proposed (and also patented) for shielding a region of space (or a scatterer) from EM radiation. In source-free chiral media the general time-harmonic field vector **V** satisfies the chiral wave equation

$$\nabla \times \nabla \times \mathbf{V} - 2\omega\mu\xi_c\nabla \times \mathbf{V} - k^2\mathbf{V} = \mathbf{0}. \tag{12.8}$$

Two eigenmodes exist, corresponding to a right-circular (RCP) and a left-circular polarized (LCP) plane wave, respectively. Their eigenvalues (wavenumbers) are

$$k_{\mathrm{RCP,LCP}} = k[\sqrt{1 + \eta^2\xi_c^2} \pm \eta\xi_c], \tag{12.9}$$

where $k = \omega\sqrt{\mu\varepsilon}$ and $\eta = \sqrt{\mu/\varepsilon}$. Moreover it turns out that

$$\frac{\mathbf{k}_{\mathrm{RCP,LCP}}}{k_{\mathrm{RCP,LCP}}} \times \mathbf{E} = \eta_{\mathrm{ch}}\mathbf{H}, \tag{12.10}$$

where $\mathbf{k}_{\mathrm{RCP,LCP}}$ is the wavevector associated with the wavenumber $k_{\mathrm{RCP,LCP}}$ and $\eta_{\mathrm{ch}} = \eta/\sqrt{1 + \eta^2\xi_c^2}$ is the so-called chiral impedance (which is independent of the handedness of the medium).

The planar chiroshield has been shown to be more efficient than the classical Dallenbach and Salisbury screens [12,13]. In particular, the magnetic chiroshield (consisting of a metal plate covered with a thin chiral layer of thickness *t*) has been proposed as an alternative to the Dallenbach screen, while the electric chiroshield (consisting of a thin chiral layer a quarter-wavelength spaced with respect to the metal plate) as an alternative to the Salisbury screen. The better performance of such chiral screens is likely due to enhanced impedance matching, larger absorptions, and larger bandwidths.

Similar advantages have been found for curved metallic surfaces coated with chiral materials. In particular, in [14] it has been shown how a dramatic decrease of the radar cross section (RCS) of a metallic sphere can be obtained with a chiral superstrate when the radius of curvature is larger than a half-wavelength.

The role of chirality in low-frequency shielding was investigated in [15], where it has been shown that chiral shields may offer a larger electric and magnetic SE with respect to conventional dielectric and magnetic coatings. In particular, the electric SE is affected also by the magnetic permeability of the medium, and a magnetic SE is present also with nonmagnetic coatings. Such properties are clearly a consequence of a crosscoupling of the electric and magnetic quantities.

More recently a state-equation approach has been presented in a study of the shielding properties of chiral-coated fiber-reinforced plastic composite cylinders. The influence of the chiral layer thickness and of the chiral admittance on possible invisible characteristics of the object was also investigated [16]. Bi-anisotropic structures (in particular, *omega media*) with additional degrees of freedom in order to design antireflection coverings were suggested in [17].

12.4 METAMATERIAL SHIELDING

The issue of invisibility by means of metamaterial coatings has continued to be studied, and a large number of papers are being published on this topic (e.g., see [18–23]). Invisibility means that an object is made nearly transparent to an external observer; that is, its scattering cross section is dramatically reduced, at least in a narrow frequency range. Much effort is directed toward the design of metamaterial structures that operate as cloaks of invisibility in the microwave and in the optical frequency range. Different ideas and techniques are being tried. On one hand, the use of anisotropic and/or inhomogeneous metamaterials has seemed to allow for a control of field distribution [19]. That is to roughly say, the EM field is swept around the coated scatterer, and it appears as if it had passed there through an empty volume of space; experimental results are already available [21]. On the other hand, similar effects have been theoretically predicted using isotropic and homogeneous metamaterials as coatings [18–23]. In any case, the progress on this topic is impressive but following it in the literature is beyond the scope of this section.

Another issue regarding metamaterial shielding is the design of metamaterial screens, which can present some advantages over conventional screens. In general,

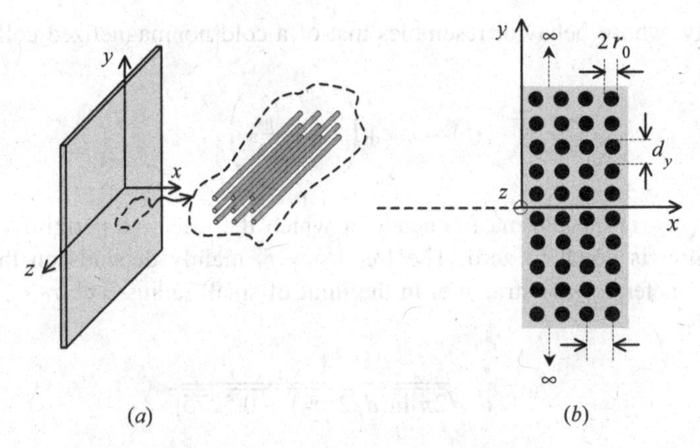

FIGURE 12.6 (*a*) Metamaterial wire-medium screen; (*b*) transverse view with geometrical parameters.

the metamaterial screen consists of a periodic arrangement of small dielectric and/or metallic inclusions in a host medium (with spatial period much smaller than the operating wavelength). The periodic structure can thus be homogenized and described by effective constitutive parameters.

In [24] the shielding performance of a planar metamaterial wire-medium (WM) screen under plane-wave illumination is studied. Such a screen consists of a finite number N of periodic layers of thin lossy metallic wires embedded in a dielectric host medium of finite thickness with relative dielectric permittivity $\varepsilon_{\mathrm{rh}}$. The structure is sketched in Figure 12.6.

With respect to other one- and two-dimensional periodic structures studied in the past, such as wire grids [25], the proposed structure contains more than a single row of conducting wires, so the usually applied equivalent shunt-impedance model cannot be employed. The cylinders are assumed to be infinitely long in the z direction; the spatial period along the y direction is d_y, and the diameter of the cylinders is $2r_0$. The distance d_x between each row of cylinders is assumed to be equal to the spatial period, meaning $d_x = d_y = d$. The main assumption that has to be made in order to correctly perform a homogenization of the periodic structure consists in considering the spatial period d suitably smaller than the operating wavelength λ_0. In such a case it can be shown that the periodic structure can be represented from the effective-medium-theory viewpoint as a homogeneous nonmagnetic medium characterized by an effective diagonal permittivity tensor. The two elements of such a tensor corresponding to directions orthogonal to the wires (ε_{xx} and ε_{yy}) are simply equal to the dielectric permittivity of the host medium, whereas the remaining element ε_{zz} is characterized by both temporal and spatial dispersion [26].

However, the propagation of EM waves with an electric field polarized along the wire direction z is unaffected by the anisotropy and the spatial dispersion. The medium can thus be represented by a simple scalar frequency-dependent permittivity, whose behavior resembles that of a cold nonmagnetized collisionless plasma:

$$\varepsilon(f) = \varepsilon_0 \varepsilon_{rh} \left(1 - \frac{f_p^2}{\varepsilon_{rh} f^2} \right), \tag{12.11}$$

where $f_p/\sqrt{\varepsilon_{rh}}$ is the plasma frequency at which the effective permittivity of the wire medium is equal to zero. The frequency f_p mainly depends on the geometrical parameters of the structure. In the limit of small radius (i.e., $r_0 \ll d$), there results [27]

$$f_p = \frac{c}{d} \frac{1}{\sqrt{2\pi[\ln(d/2\pi r_0) + 0.5275]}}, \tag{12.12}$$

where c is the speed of light in vacuum. Clearly, for frequencies smaller than $f_p/\sqrt{\varepsilon_{rh}}$, the effective permittivity is negative and increases in absolute value by lowering frequency. To derive (12.11), perfectly conducting wires were assumed. In general, a finite conductivity σ gives rise to a nonzero imaginary part of the effective permittivity, as it will be shown shortly.

In general, for arbitrarily polarized waves, the effective permittivity tensor is given by [26]

$$\underline{\boldsymbol{\varepsilon}} = \varepsilon_0 \varepsilon_{rh} \left[\mathbf{u}_x \mathbf{u}_x + \mathbf{u}_y \mathbf{u}_y + \left[1 - \frac{f_p^2}{\varepsilon_{rh} f^2 - (c^2 k_z^2 / 4\pi^2)} \right] \mathbf{u}_z \mathbf{u}_z \right], \tag{12.13}$$

where k_z is the wavenumber along the wire axis. It is immediate to see that (12.13) reduces to (12.11) for E-polarized waves propagating orthogonally to the wires. In what follows, we will limit ourselves to the study of normal incidence of E-polarized waves so that we can simply adopt the model (12.11) for the effective permittivity of the medium.

As mentioned above, when considering *lossy* wires (i.e., when a finite conductivity σ has to be taken into account), the expression (12.11) for the effective permittivity has to be suitably modified. Following a reasoning similar to that proposed in [28], it is simple to show that a correct model for the scalar effective permittivity is

$$\varepsilon_r(f) = \varepsilon_{rh} - \frac{c}{2\pi f d^2 [f \ln(d/4r_0) + [(1-j)/2\pi r_0]\sqrt{(\pi f / c\eta_0\sigma)}(I_0(\xi)/I_1(\xi))]}, \tag{12.14}$$

where η_0 is the free-space impedance, $I_0(\cdot)$ and $I_1(\cdot)$ are the zero- and first-order modified Bessel functions of the first kind, respectively, and

$$\xi = (1+j) r_0 \sqrt{\frac{\pi \eta_0 \sigma f}{c}}. \tag{12.15}$$

Finally, in the homogenization process a finite thickness h_{eff} has to be associated to the homogeneous slab equivalent to the periodic structure, which takes into account the fringing fields at the top and bottom layers of the structure. In particular, an equivalent thickness equal to Nd has been adopted in [24], in order to best match the reflection performance of the actual periodic structure and that of its homogenized model. At this point the calculation of the SE can easily be performed by means of the usual TL analogy illustrated in Chapter 4. It has been shown in [24] that the results obtained by means of the homogeneous model are in perfect agreement with those obtained through full-wave simulations of the actual periodic structure. An example is shown in Figure 12.7, where a comparison is reported between full-wave and homogenized results for the SE of a WM screen in air, constituted by $N=4$ layers of perfectly conducting wires with spacing $d=100$ mm and radius $r_0=0.1$ mm. The operating frequency is $f=100$ MHz, and the SE is reported as a function of the normalized abscissa x/d in the plane $y=0$ (see Figure 12.6*b*). A remarkable agreement is found between approximate and full-wave results also in the proximity of the air-screen interfaces and inside the WM screen. Such an agreement is maintained for all the frequencies below the plasma frequency.

Also in [24] the performance of the metamaterial WM screen is compared with the performance of a lossy *solid* metal screen to seek out any advantages of the metamaterial structure over the conventional one. To compare the performances of the two structures, the screens had to have the same volume occupancy of their metal constituents. Therefore, for a WM screen with N periodic layers of lossy wires with radius r_0 and spatial period d, the equivalent solid metal screen has a thickness h_{m} given by

$$h_{\text{m}} = \frac{N\pi r_0^2}{d}. \tag{12.16}$$

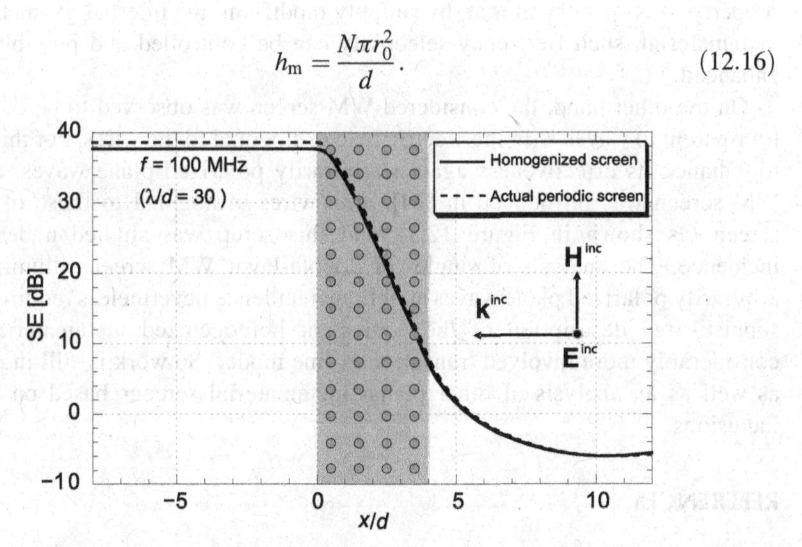

FIGURE 12.7 Comparison between homogenized and full-wave MoM results for the SE of a lossless WM screen in vacuum as a function of the normalized abscissa x/d in the plane $y = 0$, at the frequency $f = 100$ MHz. The WM screen has the following parameters: $N = 4$, $d = 100$ mm, and $r_0 = 0.1$ mm.

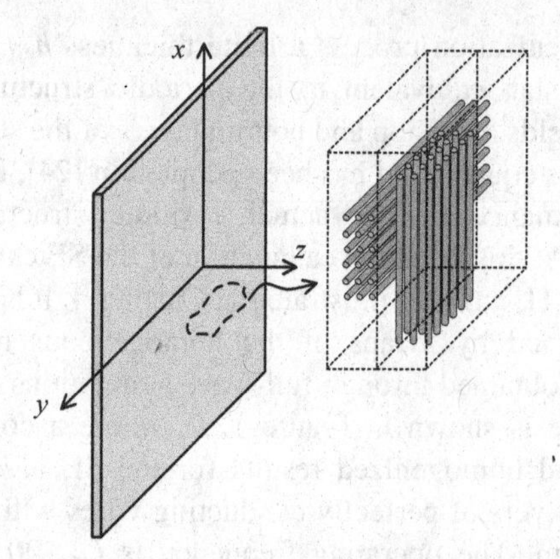

FIGURE 12.8 Sketch of a metamaterial double WM screen.

It is thus shown that there exists a frequency below which the performance of the lossy WM screen is superior to that of the solid metal screen. Such a frequency depends on the relevant physical and geometrical parameters of the actual periodic structure, and it can be estimated in planning an effective design for the WM screen. Furthermore a dramatically different behavior of the SE is observed as a function of frequency in the solid and WM screens. In particular, while the SE monotonically increases with the frequency for the solid screen, it first increases and then decreases in the WM case, thus showing a possibly desirable selective property. It is speculated that, by suitably modifying the internal geometry of the metamaterial, such frequency selectivity can be controlled and possibly further enhanced.

On the other hand, the considered WM screen was observed to be completely transparent to waves with the electric field orthogonal to the wires. For this reason, to enhance its effectiveness against arbitrarily polarized plane waves, a second WM screen was introduced in [24], with wires orthogonal to those of the first screen (as shown in Figure 12.8), and this setup was studied under normal incidence. The analysis of single- or double-layer WM screens illuminated by arbitrarily polarized plane waves at oblique incidence nevertheless requires a more sophisticated description of the equivalent homogenized metamaterial and a considerably more involved transmission-line model. So work is still in progress, as well as an analysis of other planar metamaterial screens based on different inclusions.

REFERENCES

[1] M. L. Hiles., R. G. Olsen, K. C. Holte, D. R. Jensen, and K. L. Griffing. "Power-frequency magnetic field management using a combination of active and passive shielding technology." *IEEE Trans. Power Deliv.*, vol. 13, no. 1, pp. 171–179, Jan. 1998.

[2] A. R. Memari and W. Janischewskyj. "Mitigation of magnetic field near power lines." *IEEE Trans. Power Deliv.*, vol. 11, no. 3, pp. 1577–1586, Jul. 1996.

[3] R. G. Olsen and P. Moreno. "Some observations about shielding extremely low-frequency magnetic fields by finite width shields." *IEEE Trans. Electromagn. Compat.*, vol. 30, no. 3, pp. 187–201, Aug. 1996.

[4] E. Salinas. "Passive and active shielding of power-frequency magnetic fields from secondary substation components." *Proc. 2000 IEEE Int. Symp. Electromagnetic Compatibility*, Washington, DC, 21–25 Aug. 2000, pp. 855–860.

[5] P. L. Sergeant, L. R. Dupre, M. De Wulf, and J. A. A. Melkebeek. "Optimizing active and passive magnetic shields in induction heating by a genetic algorithm." *IEEE Trans. Magn.*, vol. 39, no. 6, pp. 3486–3496, Nov. 2003.

[6] P. Sergeant, L. Dupre, and J. Melkebeek. "Active and passive magnetic shielding for stray field reduction of an induction heater with axial flux." *IEE Proc. Electr. Power Appl.*, vol. 152, no. 5, pp. 1359–1364, Sep. 2005.

[7] S. Celozzi and F. Garzia. "Active shielding for power-frequency magnetic field reduction using genetic algorithms optimisation." *IEE Proc. Sci. Meas. Technol.*, vol. 151, no. 1, pp. 2–7, Jan. 2004.

[8] P. Moreno and R. G. Olsen. "A simple theory for optimizing finite width elf magnetic field shields for minimum dependence on source orientation." *IEEE Trans. Elecromagn. Compat.*, vol. 39, no. 4, pp. 340–348, Nov. 1997.

[9] M. Istenic and R. G. Olsen. "A simple hybrid method for elf shielding by imperfect finite planar shields." *IEEE Trans. Elecromagn. Compat.*, vol. 46, no. 2, pp. 199–207, May 2004.

[10] D. L. Jaggard and X. Sun. "Theory of chiral multilayers." *J. Opt. Soc. Am. A*, vol. 9, no. 5, pp. 804–813, May 1992.

[11] D. L. Jaggard, N. Engheta, and J. Liu. "Chiroshield: A Salisbury/Dallenbach shield alternative." *Electron. Lett.*, vol. 26, no. 17, pp. 1332–1334, Aug. 1990.

[12] L. Dallenbach and W. Kleinsteuber. "Reflection and absorption of decimeter-waves by plane dielectric layers." *Hochfreq. u Elektroak*, vol. 51, pp. 152–156, 1938.

[13] W. W. Salisbury. "Absorbent body for electromagnetic waves." U.S. Patent 2 599 944, Jun. 10, 1952.

[14] D. L. Jaggard, J. Liu, and X. Sun. "Spherical chiroshield." *Electron. Lett.*, vol. 27, no. 1, pp. 77–79, Jan. 1991.

[15] A. H. Sihvola and M. E. Ermutlu. "Shielding effect of hollow chiral sphere." *IEEE Trans. Electromagn. Compat.*, vol. 39, no. 3, pp. 219–224, Aug. 1997.

[16] C.-N. Chiu and C.-I. G. Hsu. "Scattering and shielding properties of a chiral-coated fiber-reinforced plastic composite cylinder." *IEEE Trans. Electromagn. Compat.*, vol. 47, no. 1, pp. 123–130, Feb. 2005.

[17] S. A. Tretyakov and A. A. Sochava. "Proposed composite material for non-reflecting shields and antenna radomes." *Electron. Lett.*, vol. 29, no. 12, pp. 1048–1049, Jun. 1993.

[18] A. Alù and N. Engheta. "Achieving transparency with plasmonic and metamaterial coatings." *Phys. Rev. E*, vol. 72, pp. 016623-1–016623-9, Jul. 2005 (erratum in *Phys. Rev. E*, vol. 73, 019906, Jan. 2006).

[19] J. B. Pendry, D. Schurig, and D. R. Smith. "Controlling electromagnetic fields." *Science*, vol. 312, pp. 1780–1782, Jun. 2006.

[20] G. W. Milton and N. A. Nicorovici. "On the cloaking effects associated with anomalous localized resonance." *Proc. R. Soc. Lond. A: Math. Phys. Sci.*, vol. 462, no. 2074, pp. 3027–3059, Oct. 2006.

[21] D. Schurig, J. J. Mock, B. J. Justice, S. A. Cummer, J. B. Pendry, A. F. Starr, and D. R. Smith. "Metamaterial electromagnetic cloak at microwave frequencies." *Science*, vol. 314, pp. 977–980, Nov. 2006.

[22] M. G. Silveirinha, A. Alù, and N. Engheta. "Parallel plate metamaterials for cloaking structures." Phys. Rev. E, vol. 75, pp. 036603-1–036603-16, Mar. 2007.

[23] A. Alù and N. Engheta. "Plasmonic materials in transparency and cloaking problems: Mechanism, robustness, and physical insights." *Opt. Express*, vol. 15, no. 6, pp. 3318–3332, Mar. 2007.

[24] G. Lovat, P. Burghignoli, and S. Celozzi. "Shielding properties of a metamaterial wire-medium screen." *IEEE Trans. Electromagn. Compat.*, vol. 50, no. 1, pp. 80–88, Feb. 2008.

[25] J. L. Young and J. R. Wait. "Shielding properties of an ensemble of thin, infinitely long, parallel wires over a lossy half space." *IEEE Trans. Electromagn. Compat.*, vol. 31, pp. 238–244, Aug. 1989.

[26] P. A. Belov, R. Marqués, S. I. Maslovski, I. S. Nefedov, M. Silveirinha, C. R. Simovski, and S. A. Tretyakov. "Strong spatial dispersion in wire media in the very large wavelength limit." *Phys. Rev. B*, vol. 67, pp. 113103-1–113103-4, 2003.

[27] P. A. Belov, S. A. Tretyakov, and A. J. Viitanen. "Dispersion and reflection properties of artificial media formed by regular lattices of ideally conducting wires." *J. Electromag. Waves Appl.*, vol. 16, pp. 1153–1170, Sep. 2002.

[28] S. I. Maslovski, S. A. Tretyakov, and P. A. Belov. "Wire media with negative effective permittivity: A quasi static model." *Microw. Opt. Tech. Lett.*, vol. 35, pp. 47–51, Oct. 2002.

APPENDIX A

Electrostatic Shielding

Although most of the shielding techniques refer to radio-frequency (RF) and microwave ranges, also static and stationary applications deserve special attention. In this appendix, we focus on electrostatic shielding. Several electrostatic effects can in fact damage or cause failure in electronic components and assemblies. High-speed and high-density electronic devices are especially vulnerable because of their high static sensitivity. In general, the presence of even one very static sensitive device on a printed-circuit board requires that the entire assembly be handled with static protective precaution. The usual low-level static damages (such as component assembly) can increase leakages, alter functional characteristics, and, in general, weaken the devices performance.

The most typical example of electrostatic damage consists of a charged object that comes into contact with an electronic device. This causes a transient discharge to pass directly through the device (electrostatic discharge, ESD). Furthermore many devices can be damaged even without direct contact. For instance, one of the most basic semiconductor structures—the dielectric layer between the two conductive layers of a capacitor—can be completely spoiled by the presence of an electrostatic field. The dielectric breakdown in fact occurs when the field across the structure exceeds its dielectric strength. As is well known, such a problem is common to all the MOS devices.

Electrostatic shielding protects components and assemblies from damage and failure caused by external electrostatic fields. Clearly, the level of the required shielding is determined by the level of electric field that causes the failure.

Since the earliest age of electricity, the electrostatic effects that could be created within a volume by an external electrostatic field were known to be prevented by using a highly conductive enclosure that would act as an electrostatic shield. The fundamental principles of electrostatic shielding can easily be derived from basic

Electromagnetic Shielding by Salvatore Celozzi, Rodolfo Araneo and Giampiero Lovat

electrostatic knowledge. In Section A.1, the laws of electrostatics are briefly recalled, and in Section A.2, the basic concepts and tools are introduced. Finally, in Section A.3, several electrostatic shields are quantitatively analyzed and discussed. For a more extensive discussion of electrostatic problems, the interested reader is encouraged to consult the classic textbooks (e.g., [1,2]).

A.1 BASICS LAWS OF ELECTROSTATICS

In statics the field equations are decoupled into two independent sets of equations in terms of two independent sets of fields. In particular, the static electric field is described by the following equations written in differential form:

$$\begin{aligned} \nabla \times \mathbf{E}(\mathbf{r}) &= \mathbf{0}, \\ \nabla \cdot \mathbf{D}(\mathbf{r}) &= \rho_e(\mathbf{r}), \\ \mathbf{D} &= \mathbf{D}(\mathbf{E}). \end{aligned} \tag{A.1}$$

The second equation in (A.1) is the differential form of the Gauss law, and the third equation expresses the constitutive relationship in an operatorial form. For simplicity, in what follows we will assume that the dielectric media are linear, homogeneous, and isotropic so that

$$\mathbf{D}(\mathbf{r}) = \varepsilon \mathbf{E}(\mathbf{r}). \tag{A.2}$$

As is well known, the connection between electromagnetics and mechanics is established by Lorentz's force equation, for which a charge q moving at velocity $\mathbf{v}$ in the presence of an EM field experiences a force given by $\mathbf{F} = q\mathbf{E} + q\mathbf{v} \times \mathbf{B}$. Clearly, in electrostatics the Lorentz force is due only to the electric field and is thus given by $\mathbf{F} = q\mathbf{E}$ (which also expresses Coulomb's law).

Based on the above formulation, it is possible to derive the behavior of conductors under the influence of an external electric field. By definition, a conductor is a material having charges free to move under external influences, both electric and nonelectric. In terms of a rough atomic model, the electrons can migrate easily from one atom to another, moving through a background lattice. By definition, an uncharged conductor is neutral: this means that the amount of negative charge (associated with electrons) is equal to the positive one (associated with the background lattice). In an uncharged conductor at equilibrium, all the charges are distributed in such a way that the macroscopic electric field is zero inside and outside the conductor (a macroscopic field is that obtained by averaging over macroscopic dimensions). If it were not, a nonzero electric field would cause a Lorentz force to act on the free electrons, and therefore would give rise to a macroscopic motion of charges (in contrast with the assumed static equilibrium).

Similar reasoning can be applied to describe the state of equilibrium that is reached in a conductor under the influence of an external field, as illustrated in

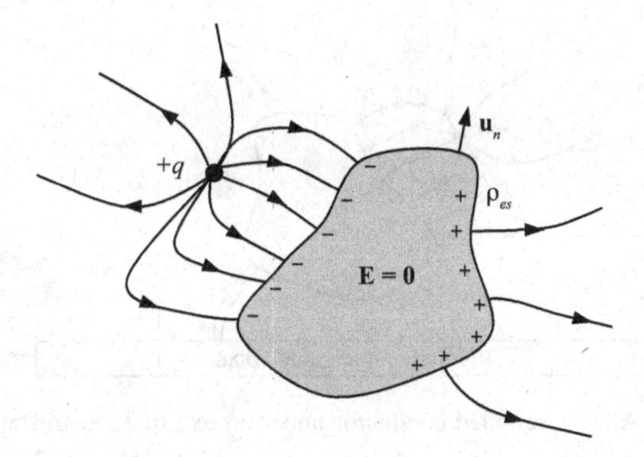

FIGURE A.1 Neutral conductor under an external electrostatic field.

Figure A.1. When a positive charge is placed in the proximity of a neutral conductor, the created electric field pushes the electrons of the conductor toward the positive charge, thus creating a conduction current. The relationship between the electric field and the resulting current density is described by Ohm's law, for which $\mathbf{J} = \sigma\mathbf{E}$ (the inverse of the conductivity σ thus describes a sort of obstacle for the charge motion through the background lattice). Therefore the negative charges continue to accumulate near the surface of the conductor until the field that these charges are creating becomes equal and opposite to the original one produced by the external positive charge. (Clearly, a reaction force also attracts the external positive charge to the conductor, but in this example we assume that both the outer charge and the conductor body are fixed in space through an external mechanical force.)

The rate of the whole process mainly depends on the value of the conductivity σ: for conductors such as copper or aluminium the relaxation time of the volume charge density is approximately 10^{-19} s. Therefore, when a conductor is subject to an applied electric field, within a very short time the freely moving charges inside the medium will rearrange themselves in such a way that they neutralize the effect of the original electric field inside the conductive medium. *Inside an isolated conductor at equilibrium no electrostatic field can exist.* As a consequence, inside a conductor, the macroscopic charge density ($\rho_e = \nabla \cdot \mathbf{D}$) is zero, so a net charge can exist only at the conductor surface (actually the charge will exist in a region near the surface of the conductor, but this comes out only from a microscopic description of the phenomenon). Finally, it can easily be shown that the electrostatic field at the conductor surface is perpendicular to the surface, and by a simple application of Gauss's law, it turns out that its magnitude is equal to $\mathbf{u}_n \cdot \mathbf{E} = \rho_{eS}/\varepsilon$, where $\mathbf{u}_n$ is the unit vector normal to the surface pointing outward from the conductor and ρ_{eS} is the electric surface charge density.

Another situation that is important to describe in discussing electrostatic shielding is the positively charged particle in the proximity of a conductor with the conductor *grounded*, as shown in Figure A.2. A conductor is grounded when it is in some way connected (e.g., via a grounding strap) to an ideal reservoir of charges (in practical applications the earth acts as this charge reservoir, also known as *ground*).

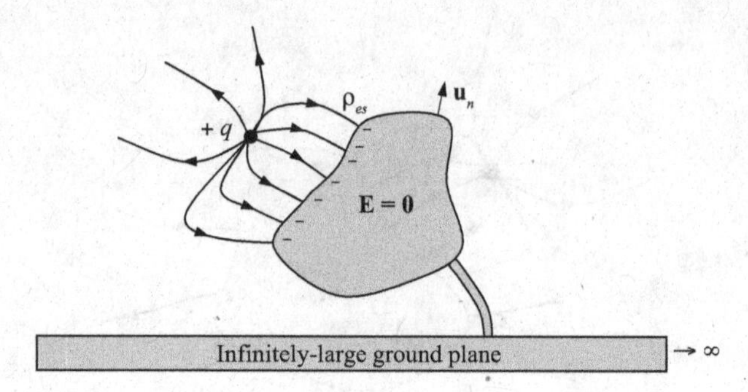

FIGURE A.2 Grounded conductor under an external electrostatic field.

Because of the presence of an outer positive charge (and of its electric field), the negative charges of the conductor move toward its surface (on the side of the outer positive charge). However, in this situation the positive charges of the lattice can be compensated by other negative charges coming from the ground. Unlike the nongrounded conductor, the grounded conductor is now negatively charged, but *in both cases the total electric field inside the conductor is zero.*

A.2 ELECTROSTATIC TOOLS: ELECTROSTATIC POTENTIAL AND GREEN'S FUNCTION

From the first equation of (A.1), it follows that a scalar field $V(\mathbf{r})$ (called the *electrostatic potential*) can be introduced such that

$$\mathbf{E}(\mathbf{r}) = -\nabla V(\mathbf{r}). \tag{A.3}$$

The potential difference $V(\mathbf{r}_2) - V(\mathbf{r}_1)$ represents the work per unit charge required to move a particle from $\mathbf{r}_1$ to $\mathbf{r}_2$ against a given electric field. Actually the scalar field $V(\mathbf{r})$ in (A.3) is defined up to an arbitrary fixed constant. By arbitrarily choosing a reference point $\mathbf{r}_0$, such a constant can be chosen as $V(\mathbf{r}_0)$ (usually $\mathbf{r}_0$ is chosen in such a way that $V(\mathbf{r}_0) = 0$). In any case, it can be seen that the potential difference between two points is unaffected by this choice.

At the interface between two different media (1 and 2) with dielectric permittivity ε_1 and ε_2, respectively, the following boundary conditions hold:

$$\mathbf{u}_n \times (\mathbf{E}_1 - \mathbf{E}_2) = \mathbf{0}, \tag{A.4}$$

$$\mathbf{u}_n \cdot (\varepsilon_1\mathbf{E}_1 - \varepsilon_2\mathbf{E}_2) = \rho_{eS}, \tag{A.5}$$

where $\mathbf{u}_n$ is the unit vector normal to the interface pointing into region 1 from region 2, the subscripts 1 and 2 label the fields just above and just below the interface, respectively, and ρ_{eS} is still the electric surface charge density. The boundary conditions (A.4) and (A.5) can also be expressed in terms of the electrostatic potential as

$$\begin{aligned} V_1(\mathbf{r}) &= V_2(\mathbf{r}), \\ \varepsilon_1 \frac{\partial V_1}{\partial n} - \varepsilon_2 \frac{\partial V_2}{\partial n} &= -\rho_{eS}, \end{aligned} \tag{A.6}$$

where the normal derivative is taken in the $\mathbf{u}_n$ direction. Other properties of the electrostatic potential can immediately be derived, based on the considerations of the previous section. We already know, for instance, that at equilibrium, the electrostatic field in the interior of a conductor is zero. From (A.3), it follows that the electrostatic potential inside the conductor must have the same value at all the points, meaning *the conductor is equipotential*, and in particular, its surface is an equipotential surface. For the same reason also the ground (introduced in the previous section) has to be equipotential.

In any problem of mathematical physics it is most important to identify conditions under which the considered problem admits a unique solution. As concerns electrostatics, it can be shown that the field within a region Ω enclosed by a surface S_Ω is unique provided that either V, $\partial V/\partial n$, or some combination of the two is specified over S_Ω. This specifying of the normal derivative of the potential over a conducting surface is equivalent to specifying the surface charge density. For problems defined in an infinite region of space, a suitable condition has to be imposed at infinity to guarantee the uniqueness of the solution. In particular, in electrostatics the electrostatic potential has to behave as $V \sim 1/r$ as $r \to \infty$.

By combining (A.1) through (A.3), we immediately derive the differential equation for the electrostatic potential, which turns out to be

$$\nabla^2 V(\mathbf{r}) = -\frac{\rho_{eS}(\mathbf{r})}{\varepsilon}, \tag{A.7}$$

and this is known as *Poisson's equation*. The corresponding homogeneous equation is

$$\nabla^2 V(\mathbf{r}) = 0, \tag{A.8}$$

known as *Laplace's equation*. It can be shown that solving (A.7) (or (A.8)) or directly the corresponding equation for $\mathbf{E}$ furnishes the same solution for the electrostatic field $\mathbf{E}$, provided that the scalar field $V(\mathbf{r})$ is a twice-differentiable function (actually this is a sufficient condition). Moreover the conditions for the uniqueness of the solution of (A.7) (or (A.8)) are the same as those provided above.

In general, the Poisson and Laplace equations can be solved by several methods, both analytical and numerical. Among the commonly used analytical techniques are separation of variables, Fourier transform, and conformal mapping; on the other hand, finite difference, finite elements, and moment methods are the most used numerical approaches. However, the so-called method of Green's function is probably the most useful technique to solve Poisson's equation and to gain physical insight into the electrostatic problem. Basically the solution for a single point source is determined and then, by using the second Green identity, the solution for an arbitrary charge distribution is expressed as a superposition integral.

Let us consider a region of space Ω. For simplicity, the medium filling the region is assumed to be linear, homogeneous, and isotropic with permittivity ε. In general, the region Ω can be multiply connected, meaning it can be bounded by an outer closed surface S_Ω and a number m of closed surfaces $S_1, \ldots, S_m$ inside the volume enclosed by S_Ω (such inner surfaces can be used to exclude material bodies). The potential $V(\mathbf{r})$ within Ω produced by a point source located at $\mathbf{r}'$ is called *Green's function* of the problem and is denoted $G(\mathbf{r}, \mathbf{r}')$. By definition, the static Green function $G(\mathbf{r}, \mathbf{r}')$ is a solution of the Poisson equation

$$\nabla^2 G(\mathbf{r}, \mathbf{r}') = -\delta(\mathbf{r} - \mathbf{r}'). \tag{A.9}$$

As is well known, the second Green identity establishes that given two twice-differentiable scalar functions $f(\mathbf{r})$ and $g(\mathbf{r})$ defined in a region Ω enclosed by a surface S, there results

$$\iiint_{\Omega} [f(\mathbf{r}')\nabla'^2 g(\mathbf{r}') - g(\mathbf{r}')\nabla'^2 f(\mathbf{r}')]\, \mathrm{d}\Omega' = -\oint\!\!\!\oint_{S} \left[f(\mathbf{r}') \frac{\partial g(\mathbf{r}')}{\partial n'} - g(\mathbf{r}') \frac{\partial f(\mathbf{r}')}{\partial n'} \right] \mathrm{d}S' \tag{A.10}$$

where the normal derivative is taken in the normal direction to S, inward to Ω. By using identity (A.10), with $f(\mathbf{r}') = V(\mathbf{r}')$, $g(\mathbf{r}') = G(\mathbf{r}, \mathbf{r}')$, and $S = S_\Omega \cup S_1 \cup \cdots \cup S_m$, we obtain

$$\begin{aligned} V(\mathbf{r}) = & \iiint_{\Omega} G(\mathbf{r}, \mathbf{r}') \frac{\rho_e(\mathbf{r}')}{\varepsilon} \mathrm{d}\Omega' + \oint\!\!\!\oint_{S_\Omega} \left[V(\mathbf{r}') \frac{\partial G(\mathbf{r}, \mathbf{r}')}{\partial n'} - G(\mathbf{r}, \mathbf{r}') \frac{\partial V(\mathbf{r}')}{\partial n'} \right] \mathrm{d}S' \\ & + \sum_{i=1}^{m} \oint\!\!\!\oint_{S_i} \left[V(\mathbf{r}') \frac{\partial G(\mathbf{r}, \mathbf{r}')}{\partial n'} - G(\mathbf{r}, \mathbf{r}') \frac{\partial V(\mathbf{r}')}{\partial n'} \right] \mathrm{d}S'. \end{aligned} \tag{A.11}$$

In the case of an unbounded region Ω, we choose points at infinity as zero-potential points. Then, provided that the sources are of finite extent, (A.11) reduces to

$$V(\mathbf{r}) = \iiint_{\Omega} G(\mathbf{r}, \mathbf{r}') \frac{\rho_e(\mathbf{r}')}{\varepsilon} \mathrm{d}\Omega' + \sum_{i=1}^{m} \oint\!\!\!\oint_{S_i} \left[V(\mathbf{r}') \frac{\partial G(\mathbf{r}, \mathbf{r}')}{\partial n'} - G(\mathbf{r}, \mathbf{r}') \frac{\partial V(\mathbf{r}')}{\partial n'} \right] \mathrm{d}S'. \tag{A.12}$$

It is well known that a solution of (A.9) is

$$G(\mathbf{r}, \mathbf{r}') = \frac{1}{4\pi|\mathbf{r} - \mathbf{r}'|}, \tag{A.13}$$

which is known as the *three-dimensional (3D) static free-space Green function*. Such a function is used to determine the potential due to a charge density in an unbounded space. From (A.12) with $m = 0$ there in fact follows

$$V(\mathbf{r}) = \frac{1}{4\pi\varepsilon} \iiint_{\Omega} \frac{\rho_e(\mathbf{r}')}{|\mathbf{r} - \mathbf{r}'|} \mathrm{d}\Omega'. \tag{A.14}$$

It is immediate to see that the static free-space Green function is reciprocal, meaning

$$G(\mathbf{r}_1, \mathbf{r}_2) = G(\mathbf{r}_2, \mathbf{r}_1). \tag{A.15}$$

Besides, the two-dimensional (2D) static free-space Green function (i.e., the potential at a point $\mathbf{r} = \mathbf{u}_\rho \rho + \mathbf{u}_z z$ due to a constant distribution of charge along the z direction at $\mathbf{r}' = \boldsymbol{\rho}'$) is

$$G(\boldsymbol{\rho}, \boldsymbol{\rho}') = \frac{1}{2\pi} \ln \frac{\rho_0}{|\boldsymbol{\rho} - \boldsymbol{\rho}'|}, \tag{A.16}$$

where ρ_0 is the reference point for the potential (with $\rho_0 \neq 0$ and $\rho_0 \neq \infty$).

In general, (A.9) has an infinite number of solutions that can be expressed as

$$G(\mathbf{r}, \mathbf{r}') = \frac{1}{4\pi|\mathbf{r} - \mathbf{r}'|} + F(\mathbf{r}, \mathbf{r}') \tag{A.17}$$

where $F(\mathbf{r}, \mathbf{r}')$ is an harmonic function, that is, by definition, it is a solution of the Laplace equation

$$\nabla^2 F(\mathbf{r}, \mathbf{r}') = 0. \tag{A.18}$$

The free-space static the Green function is a particular case of (A.17) with $F(\mathbf{r}, \mathbf{r}') = 0$. However, it is not the more convenient the Green function to use to express the potential $V(\mathbf{r})$ within a region containing internal surfaces, since, in force of (A.11) (or (A.12)), it requires the assignment of both $V(\mathbf{r})$ and its normal derivative over the boundaries. Based on the representation (A.11) (or (A.12)), the best choice for $G(\mathbf{r}, \mathbf{r}')$ is the one that requires less information about the values of $V(\mathbf{r})$ and its normal derivative over the boundaries. One possibility is to consider a version of the Green function that satisfies

$$G_D(\mathbf{r}, \mathbf{r}') = 0, \quad \forall \mathbf{r}' \in S, \tag{A.19}$$

which is known as the Dirichlet Green function (since it satisfies a Dirichlet-type boundary condition). This way (A.11) reduces to

$$V(\mathbf{r}) = \iiint_{\Omega} G_D(\mathbf{r},\mathbf{r}')\frac{\rho_e(\mathbf{r}')}{\varepsilon}\mathrm{d}\Omega' + \oiint_{S_\Omega} V(\mathbf{r}')\frac{\partial G_D(\mathbf{r},\mathbf{r}')}{\partial n'}\mathrm{d}S' + \sum_{i=1}^{m}\oiint_{S_i} V(\mathbf{r}')\frac{\partial G_D(\mathbf{r},\mathbf{r}')}{\partial n'}\mathrm{d}S'. \tag{A.20}$$

Representation (A.20) requires only the assignment of $V(\mathbf{r})$ over the boundary (but not its normal derivative). It is worth noting that if $S_\Omega, S_1, \ldots, S_m$ are the boundaries of conductors, the Dirichlet Green function corresponds to the potential due to a point source in the presence of conductors when these are all kept at zero potential. In such a case the specification of the values (constant) of the potential $V(\mathbf{r})$ on the conductors allows for the determination of $V(\mathbf{r})$ everywhere in Ω (the potential $F(\mathbf{r}, \mathbf{r}')$ thus represents the potential due to the surface charges on the conductors).

Besides, it should be mentioned that another good choice for the static Green function is one that satisfies the boundary condition

$$\frac{\partial G_N(\mathbf{r}, \mathbf{r}')}{\partial n'} = -\frac{1}{\mathrm{area}(S)}, \quad \forall \mathbf{r}' \in S, \tag{A.21}$$

which is known as the Neumann Green function (since it satisfies a Neumann-type boundary condition). Basically such a choice does not annul any term in (A.11) (or A.12) but reduces the surface integrals involving $V(\mathbf{r})$ to constants that leave the values of the electrostatic field $\mathbf{E}(\mathbf{r}) = -\nabla V(\mathbf{r})$ unaffected. Finally, by means of the second Green identity, it can easily be shown that both the Dirichlet and the Neumann Green functions are reciprocal (that is, they satisfy (A.15)).

A.3 ELECTROSTATIC SHIELDS

The goal of electrostatic shielding is *the creation of a region of space in which the electric field is independent of what happens outside this region*. In practice, very often a vanishingly small electric field inside the region of interest is required. As discussed at the beginning of this appendix, such "field-free" regions are often needed in experiments or for reliable operation of electronic devices.

A.3.1 Conductive Electrostatic Shields

As a first example of electrostatic shield, let us consider a conductive shell with no free charges inside, as shown in Figure A.3*a*. The shell could be an empty metallic rectangular box with finite-thickness walls, as shown in Figure A.3*b*.

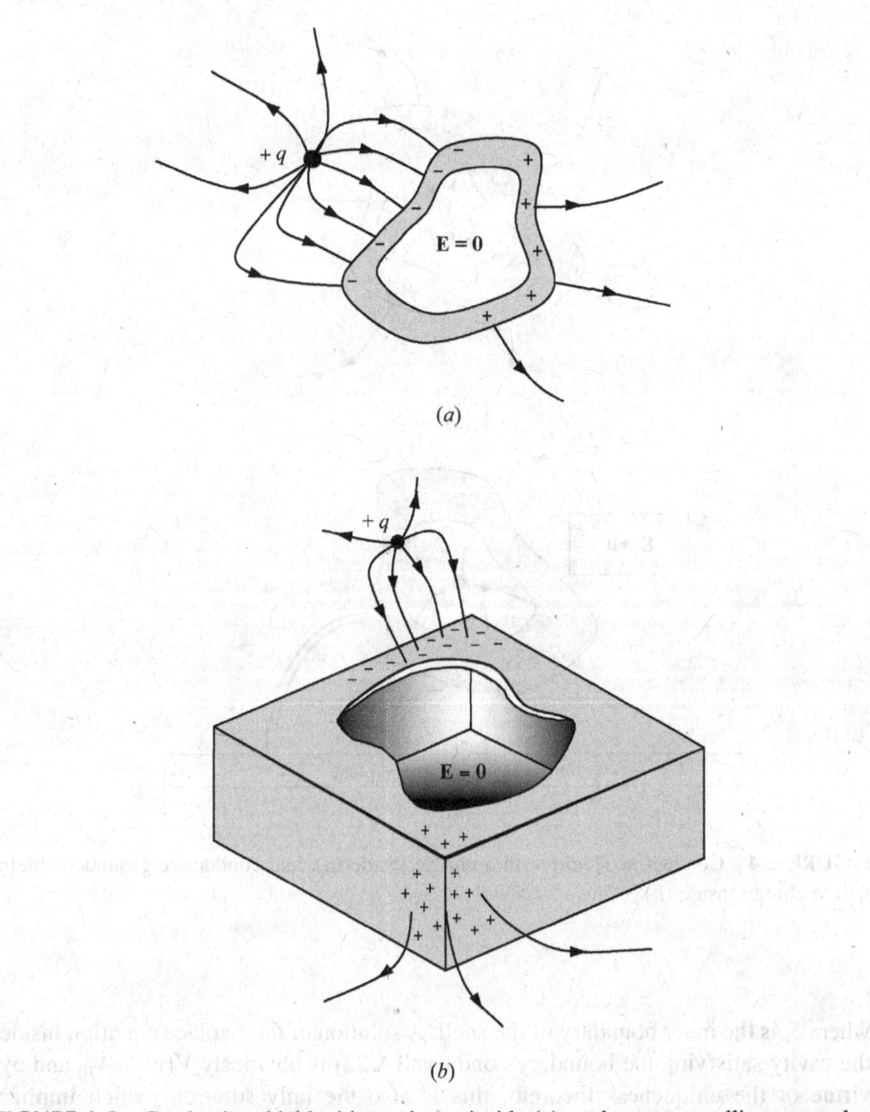

FIGURE A.3 Conductive shield with no charge inside (*a*), and empty metallic rectangular enclosure (*b*). (Please note that the hole in Figure A.3*b* is present only to give a picture of the enclosure's inside.)

Let us assume that a certain number of charges producing an impressed electrostatic field are present outside and close to the shell as in Figure A.3. Whatever the electrostatic field outside the cavity, the electric field inside is zero. This can easily be proved by the uniqueness theorem. Since the shell (i.e., the cavity walls) is conductive, the electrostatic potential has to be constant over the shell. In particular, it has to be

$$V(\mathbf{r}) = V_0, \quad \forall \mathbf{r} \in S_{\mathrm{i}} \tag{A.22}$$

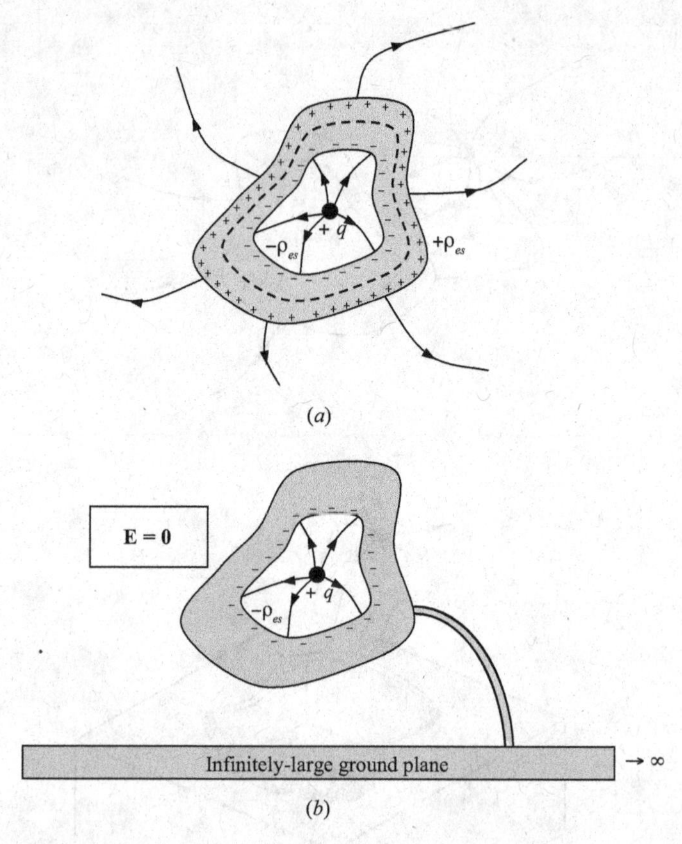

FIGURE A.4 Conductive shield with a charge inside (*a*), and conductive grounded shield with a charge inside (*b*).

where S_i is the inner boundary of the shell. A solution of the Laplace equation inside the cavity satisfying the boundary condition (A.22) is obviously $V(\mathbf{r}) = V_0$, and by virtue of the uniqueness theorem, this is also the only solution, which implies $\mathbf{E}(\mathbf{r}) = \mathbf{0}$ inside the cavity, as claimed above.

The reasoning above also allows us to conclude that whatever happens outside the closed conductive cavity has no influence on what happens inside. This means that the inner region is shielded from the effects of the external electrostatic field.

An alternative way to prove the result above is to use the representation (A.20) for the electrostatic potential: the volume Ω is the region enclosed by the shell (i.e., the cavity), $m = 0$, and S_Ω is the inner boundary of the shell. From (A.20) we thus obtain

$$V(\mathbf{r}) = \oiint_{S_\Omega} V(\mathbf{r}') \frac{\partial G_D(\mathbf{r}, \mathbf{r}')}{\partial n'} dS', \quad \mathbf{r} \in \Omega. \tag{A.23}$$

As before, the surface S_Ω is an equipotential surface, so we are free to choose the potential over this surface as the reference potential, which is $V(\mathbf{r}) = 0$ over S_Ω (clearly, according to this choice $V(\infty) \neq 0$, but this is of no importance for the present discussion). From (A.23), it immediately follows that $V(\mathbf{r}) = 0$ inside the cavity, and therefore $\mathbf{E}(\mathbf{r}) = \mathbf{0}$.

Let us consider now what happens in a symmetric situation, namely when a charge q is present inside the metallic cavity. The electric field inside the conductive shell is zero. Therefore, if we apply Gauss's law to a closed surface inside the shell (see Figure A.4*a*), it follows that the net charge inside the volume enclosed by such a surface is zero. Because the charge q is present inside the conductive cavity, an equal and opposite charge is induced on the inner surface S_i of the shell. Such an induced charge is distributed in such a way that it cancels the electric field produced by the original charge q at points outside the region enclosed by the inner surface S_i. Moreover, because the shell was originally neutral and is not grounded, an equal charge has to be present over its outer surface S_e. However, it can easily be shown that the distribution of this charge does depend only on external conditions, not on what happens inside the cavity. For instance, if the enclosed charge q is moved inside the cavity, then the charge distributed over the inner surface S_i will change, but the induced charge over the outer surface S_e is unaffected. Its distribution has only to ensure that its field inside S_e is zero. In this sense, also the region outside the shell is shielded. Clearly, this does not imply an *absence* of electric field in the region outside the shell but rather that the fields in the inner and outer regions are independent of each other (i.e., the redistribution of charge inside the shell does not influence the field outside the shell).

Still there is a way to cancel out the field in the region outside the shell. This is what happens if the conductive shell is grounded (as in Figure A.4*b*), and it can be proved by using the representation (A.20) for the electrostatic potential. The volume Ω is the region outside the shell, $m = 0$, and S_Ω is the outer boundary of the shell. From (A.20), since there are no charges inside Ω, we have

$$V(\mathbf{r}) = \oiint_{S_\Omega} V(\mathbf{r}') \frac{\partial G_\mathrm{D}(\mathbf{r}, \mathbf{r}')}{\partial n'} \mathrm{d}S', \quad \mathbf{r} \in \Omega. \tag{A.24}$$

The surface S_Ω is an equipotential surface, and we are free to choose the potential over this surface as the reference potential, meaning $V(\mathbf{r}) = 0$ over S_Ω (because of the grounding it is also $V(\infty) = 0$). From (A.24) it immediately follows that $V(\mathbf{r}) = 0$ in the region outside the shell, and therefore $\mathbf{E}(\mathbf{r}) = \mathbf{0}$.

It is to be noted that in the presence of the grounded shell, the two cases (charge outside or inside to the shell) are perfectly symmetric. This is because the outer and the inner regions are "enclosed" by surfaces at the same potential (considered also the ideal surface at infinity).

The assumption of an ideal ground is mandatory for the considerations above. In particular, if the ground is of finite size, the charges will be distributed over both the outer surface of the shell and the ground surface. The shell and the ground simply act like two connected conductors. With such a connection the initial charge on the outer surface of the shell distributes also over the surface of the ground. Only in the case of a very large ground will the surface charge density tend to zero (because of the very large surface over which it has to be distributed).

The absence of electrostatic field inside a closed conductive cavity is of fundamental importance for many practical applications of electric shielding. It is not really necessary that the conductive shield be perfectly closed, and in general, the shield can even have some small holes or apertures (which can be useful for many purposes, such as access to interior or ventilation). For such a shield, the electric field will not be exactly zero inside the cavity, but it can be made arbitrarily small for suitably small apertures. This is the basic principle of the well-known Faraday cage, usually constituted by a box whose walls consist of metallic grids.

A.3.2 Dielectric Electrostatic Shields

Another way to effectively shield a region of space consists of using screens made with high-permittivity materials. In general, the analysis of dielectric enclosures of arbitrary shape requires the use of numerical techniques. However, to gain some physical insight, in what follows we will consider the cases of spherical and cylindrical enclosures. This way, thanks to the particular geometry, a closed-form expression for the electrostatic SE can be obtained.

Let us thus consider a spherical shell of relative permittivity ε_r, exposed to an impressed uniform electrostatic field $\mathbf{E}^i = \mathbf{u}_z E^i$, as shown in Figure A.5, where the three distinct regions are the cavity region 1 ($r < r_i$), the shell region 2 ($r_i < r < r_e$), and the outer region 3 ($r > r_e$). Our goal is to determine the electric field inside the cavity bounded by the shell. As happens in scattering problems, the solution in the outer region is expressed as the sum of the impressed field and a *scattered field*: such

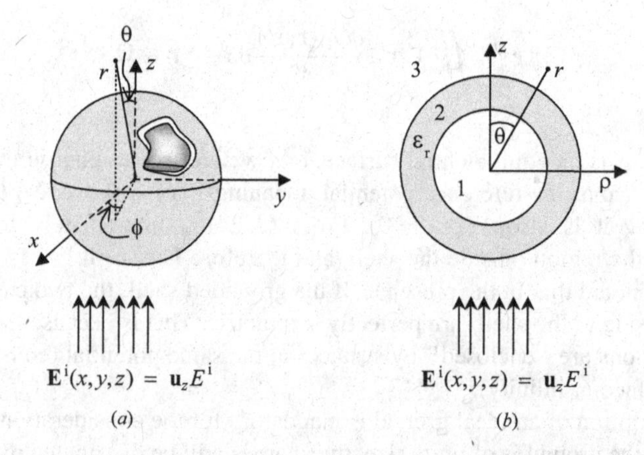

FIGURE A.5 Spherical dielectric shield under an external uniform electrostatic field. 3D view (*a*); 2D view (*b*).

a sum is known as the *total field.* While the potential corresponding to the scattered field has to satisfy Laplace's equation, the total field has to satisfy the boundary conditions due to the presence of the shell, namely the continuity of the total potential and of the product between its normal derivative and the relative permittivity across the inner and outer surfaces of the shell. Because of the spherical symmetry of the problem, it is convenient to use spherical coordinates. The impressed electrostatic potential is therefore $V^{\mathrm{i}}(\mathbf{r}) = -E^{\mathrm{i}}z = -E^{\mathrm{i}}r\cos\theta$ (up to a fixed constant). In the outer region 3 ($r > r_{\mathrm{e}}$) the total electrostatic potential is thus expressed as

$$V_3(\mathbf{r}) = V^{\mathrm{i}}(\mathbf{r}) + V^{\mathrm{s}}(\mathbf{r}), \tag{A.25}$$

where $V^{\mathrm{s}}(\mathbf{r})$ represents the scattered potential. The $V^{\mathrm{s}}(\mathbf{r})$ function is a solution of Laplace's equation, and because of the azimuthal symmetry, it can be expressed in spherical harmonics as

$$V^{\mathrm{s}}(\mathbf{r}) = V^{\mathrm{s}}(r, \theta) = \sum_{n=0}^{\infty} V_n^{\mathrm{s}}(r, \theta) = \sum_{n=0}^{\infty} [A_n r^n + B_n r^{-(n+1)}] P_n(\cos\theta), \tag{A.26}$$

where $P_n(\cdot)$ is the Legendre polynomial of order n. For the potential to be zero at infinity, the coefficients of the r^n terms have to be zero when region 3 is considered, meaning $A_n^{(3)} = 0$. So

$$V^{\mathrm{s}}(\mathbf{r}) = \sum_{n=0}^{\infty} B_n^{(3)} r^{-(n+1)} P_n(\cos\theta), \quad r > r_{\mathrm{e}}. \tag{A.27}$$

The potential in region 1 also satisfies Laplace's equation, and it can be expressed as in (A.26). However, in order to be bounded at the origin, the coefficients of the $r^{-(n+1)}$ terms have to be zero, meaning $B_n^{(1)} = 0$. So

$$V(\mathbf{r}) = \sum_{n=0}^{\infty} A_n^{(1)} r^n P_n(\cos\theta), \quad r < r_{\mathrm{i}}. \tag{A.28}$$

Finally, because in region 2 there are no restrictions, for $r_{\mathrm{i}} < r < r_{\mathrm{e}}$ the potential is

$$V(\mathbf{r}) = \sum_{n=0}^{\infty} [A_n^{(2)} r^n + B_n^{(2)} r^{-(n+1)}] P_n(\cos\theta), \quad r_{\mathrm{i}} < r < r_{\mathrm{e}}. \tag{A.29}$$

The constants $A_n^{(1)}, A_n^{(2)}, B_n^{(2)}$, and $B_n^{(3)}$ can be found by enforcing the continuity of the electrostatic potential and of the product between its normal derivative and the relative permittivity at the surfaces $r = r_{\mathrm{i}}$ and $r = r_{\mathrm{e}}$. Therefore, from (A.27) through (A.29), by taking into account that $\partial/\partial n = \partial/\partial r$, we have

$$\begin{cases} \sum_{n=0}^{\infty} A_n^{(1)} r_i^n P_n(\cos\theta) = \sum_{n=0}^{\infty} [A_n^{(2)} r_i^n + B_n^{(2)} r_i^{-(n+1)}] P_n(\cos\theta), \\ \varepsilon_0 \sum_{n=0}^{\infty} A_n^{(1)} n r_i^{n-1} P_n(\cos\theta) = \varepsilon_0 \varepsilon_r \sum_{n=0}^{\infty} [A_n^{(2)} n r_i^{n-1} - (n+1) B_n^{(2)} r_i^{-(n+2)}] P_n(\cos\theta), \end{cases} \tag{A.30}$$

$$\begin{cases} \sum_{n=0}^{\infty} [A_n^{(2)} r_e^n + B_n^{(2)} r_e^{-(n+1)}] P_n(\cos\theta) = -E^i r_e \cos\theta + \sum_{n=0}^{\infty} B_n^{(3)} r_e^{-(n+1)} P_n(\cos\theta), \\ \varepsilon_0 \varepsilon_r \sum_{n=0}^{\infty} [n A_n^{(2)} r_e^{n-1} - (n+1) B_n^{(2)} r_e^{-(n+2)}] P_n(\cos\theta) = \\ \varepsilon_0 \left[-E^i \cos\theta - \sum_{n=0}^{\infty} (n+1) B_n^{(3)} r_e^{-(n+2)} P_n(\cos\theta) \right]. \end{cases} \tag{A.31}$$

After multiplying (A.30) and (A.31) by $\sin\theta\, P_m(\cos\theta)$, integrating from $\theta = 0$ to $\theta = \pi$, and using the orthogonality relationship

$$\int_0^{\pi} P_m(\cos\theta) P_n(\cos\theta) \sin\theta\, d\theta = \frac{2}{2n+1} \delta_{mn}, \tag{A.32}$$

where δ_{mn} is the Kronecker symbol, we obtain

$$\begin{cases} A_m^{(1)} r_i^m = A_m^{(2)} r_i^m + B_m^{(2)} r_i^{-(m+1)}, \\ m A_m^{(1)} r_i^{m-1} = \varepsilon_r [m A_m^{(2)} r_i^{m-1} - (m+1) B_m^{(2)} r_i^{-(m+2)}], \end{cases} \tag{A.33}$$

$$\begin{cases} \frac{2}{2m+1} [A_m^{(2)} r_e^m + B_m^{(2)} r_e^{-(m+1)}] = -E^i r_e \int_0^{\pi} P_m(\cos\theta) \sin\theta \cos\theta\, d\theta + \\ \frac{2}{2m+1} [B_m^{(3)} r_e^{-(m+1)}], \\ \varepsilon_r \left[\frac{2m}{2m+1} A_m^{(2)} r_e^{m-1} - \frac{2(m+1)}{2m+1} B_m^{(2)} r_e^{-(m+2)} \right] = -E^i \int_0^{\pi} P_m(\cos\theta) \sin\theta \cos\theta\, d\theta \\ - \frac{2(m+1)}{2m+1} B_m^{(3)} r_e^{-(m+2)}. \end{cases} \tag{A.34}$$

Since

$$\int_0^{\pi} P_m(\cos\theta) \cos\theta \sin\theta\, d\theta = \begin{cases} 2/3 & m = 1 \\ 0 & m \neq 1 \end{cases} \tag{A.35}$$

then (A.34) can be rewritten as

$$\begin{cases} A_1^{(2)} r_e + B_1^{(2)} r_e^{-2} = -E^i r_e + B_1^{(3)} r_e^{-2}, \\ \varepsilon_r \left[A_1^{(2)} - 2 B_1^{(2)} r_e^{-3} \right] = -E^i - 2 B_1^{(3)} r_e^{-3}, \\ A_m^{(2)} r_e^m + B_m^{(2)} r_e^{-(m+1)} = B_m^{(3)} r_e^{-(m+1)}, \quad m \neq 1, \\ \varepsilon_r [m A_m^{(2)} r_e^{m-1} - (m+1) B_m^{(2)} r_e^{-(m+2)}] = -(m+1) B_m^{(3)} r_e^{-(m+2)}, \quad m \neq 1. \end{cases} \tag{A.36}$$

By grouping (A.33) and (A.36), for $m \neq 1$ we obtain the system

$$\begin{cases} r_i^m A_m^{(1)} - r_i^m A_m^{(2)} - r_i^{-(m+1)} B_m^{(2)} = 0, \\ m r_i^{m-1} A_m^{(1)} - \varepsilon_r m r_i^{m-1} A_m^{(2)} + \varepsilon_r (m+1) r_i^{-(m+2)} B_m^{(2)} = 0, \\ A_m^{(2)} r_e^m + B_m^{(2)} r_e^{-(m+1)} - B_m^{(3)} r_e^{-(m+1)} = 0, \\ \varepsilon_r m A_m^{(2)} r_e^{m-1} - \varepsilon_r (m+1) B_m^{(2)} r_e^{-(m+2)} + (m+1) B_m^{(3)} r_e^{-(m+2)} = 0, \end{cases} \tag{A.37}$$

while for $m = 1$ we have

$$\begin{cases} r_i A_1^{(1)} - r_i A_1^{(2)} - r_i^{-2} B_1^{(2)} = 0, \\ A_1^{(1)} - \varepsilon_r A_1^{(2)} + 2\varepsilon_r r_i^{-3} B_1^{(2)} = 0, \\ r_e A_1^{(2)} + r_e^{-2} B_1^{(2)} - r_e^{-2} B_1^{(3)} = -E^i r_e, \\ \varepsilon_r A_1^{(2)} - 2\varepsilon_r r_e^{-3} B_1^{(2)} + 2 r_e^{-3} B_1^{(3)} = -E^i. \end{cases} \tag{A.38}$$

It is easy to see that (A.37) cannot hold simultaneously unless

$$A_m^{(1)} = A_m^{(2)} = B_m^{(2)} = B_m^{(3)} = 0, \tag{A.39}$$

while solving for $A_1^{(1)}$ the system in (A.38) obtains

$$A_1^{(1)} = -\frac{9\varepsilon_r E^i}{D_S}, \tag{A.40}$$

where

$$D_S = (2 + \varepsilon_r)(1 + 2\varepsilon_r) - \left(\frac{r_i}{r_e}\right)^3 (\varepsilon_r - 1)^2. \tag{A.41}$$

Thus we have from (A.28), (A.39), and (A.40), and by taking into account that $P_1(\cos\theta) = \cos\theta$, the electrostatic potential within the enclosure:

$$V(\mathbf{r}) = -\frac{9\varepsilon_r E^i}{D_S} r \cos\theta, \quad r < r_i. \tag{A.42}$$

The electrostatic field is

$$\mathbf{E}(\mathbf{r}) = -\nabla V(\mathbf{r}) = \alpha E^i \mathbf{u}_z, \quad r < r_i, \tag{A.43}$$

where $\alpha = 9\varepsilon_r / D_S$ coincides with the inverse of the electrostatic SE. It is interesting in this case that the electric field inside the enclosure is uniform and that, since $\alpha < 1$ for $\varepsilon_r > 1$, it is certainly weaker than the impressed field. In particular, for $\varepsilon_r \gg 1$, (A.41) can be approximated as

$$D_S \simeq 2\varepsilon_r^2 \left[1 - \left(\frac{r_i}{r_e}\right)^3\right]. \tag{A.44}$$

By indicating the shell thickness as $d = r_e - r_i$ and assuming $d \ll r_i$, we can write the electrostatic SE approximately as

$$\mathrm{SE} = 20 \log\left|\frac{1}{\alpha}\right| \simeq 20 \log\left(\frac{2}{3}\frac{\varepsilon_r d}{r_i}\right). \tag{A.45}$$

It can thus be concluded that a good dielectric electrostatic shield has to be characterized by a very large dielectric permittivity.

The cylindrical dielectric shield consists of an ideal uniform infinitely long cylindrical shell made of a dielectric material with relative permittivity ε_r. The cross sections of the surface boundaries of the shell are assumed to be circular and concentric (with inner radius ρ_i and outer radius ρ_e), with the axis coincident with the z axis of the adopted reference system, as shown in Figure A.6. A uniform impressed electric field of amplitude E^i and directed along $\mathbf{u}_t$ (unit vector orthogonal to the z axis) is assumed. The problem of determining the electric field inside the cylindrical cavity bounded by the inner surface of the dielectric shell can be treated in a way similar to that adopted for the spherical shell, except that the Laplace equation has now to be solved in cylindrical coordinates (this implies that cylindrical harmonics, instead of spherical harmonics, are involved). If we omit the details, the electric field inside the cylindrical cavity turns out to be

$$\mathbf{E}(\mathbf{r}) = -\nabla V(\mathbf{r}) = \alpha E^i \mathbf{u}_t, \quad \rho < \rho_i, \tag{A.46}$$

where $\alpha = 4\varepsilon_r / D_C$ and

$$D_C = (\varepsilon_r + 1)^2 - \left(\frac{\rho_i}{\rho_e}\right)^2 (\varepsilon_r - 1)^2. \tag{A.47}$$

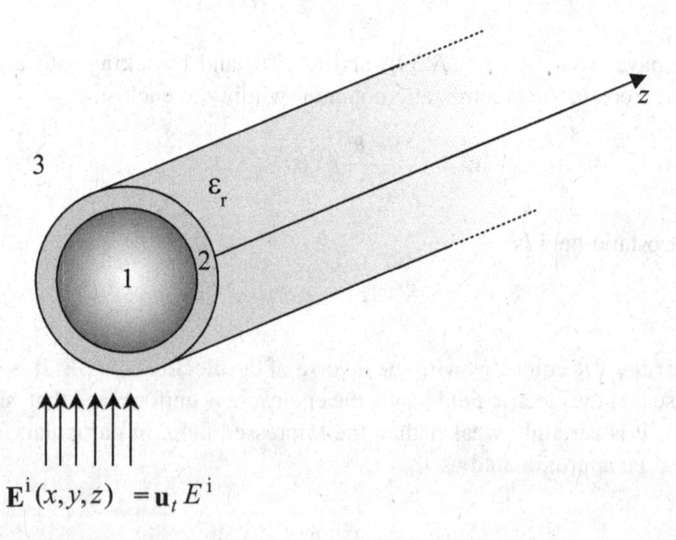

FIGURE A.6 Cylindrical dielectric shield under an external uniform electrostatic field transverse to the cylinder axis.

As in the spherical case, by indicating the shell thickness as $d = \rho_e - \rho_i$ and assuming both $d \ll \rho_i$ and $\varepsilon_r \gg 1$, we have from (A.46) and (A.47) the electrostatic SE approximately as

$$\mathrm{SE} = 20 \log\left|\frac{1}{\alpha}\right| \simeq 20 \log\left(\frac{1}{2}\frac{\varepsilon_r d}{\rho_i}\right). \tag{A.48}$$

Again, in order to have a sufficiently high value of the electrostatic SE, dielectric cylindrical shields with a large dielectric permittivity are required.

Last, it is important to note that when the cylindrical enclosure is exposed to a uniform electric field directed along the axis of the cylinder, no shielding effect occurs; that is, the dielectric shield is completely transparent to the external field. The potential $V(\mathbf{r}) = V^i(\mathbf{r}) = -E^i z$ is in fact the solution of Laplace's equation and of the boundary conditions: by the uniqueness theorem this is also the only solution.

A.3.3 Aperture Effects in Conductive Shields

Recall from the previous discussion that a closed conductive box can offer a perfect shield against electrostatic fields. Nevertheless, practical shields need some openings in their walls for many purposes, so electrostatic penetration can occur. In this section we illustrate the effects of such openings on the electrostatic SE of conductive enclosures, by considering, for simplicity, canonical geometries.

The first considered case consists of a infinitely thin, grounded conductive plane (i.e., a planar sheet) with a circular hole subject to a uniform electric field directed orthogonally to the hole, as shown in Figure A.7. In particular, the plane is placed at $z = 0$, the hole of radius ρ_0 is centered on the origin of the coordinates, and in the absence of the hole it is assumed:

$$\mathbf{E}^{\mathrm{inc}}(x, y, z) = \begin{cases} E^+\mathbf{u}_z, & z > 0, \\ E^-\mathbf{u}_z, & z < 0\,. \end{cases} \tag{A.49}$$

The analytical solution of such a problem requires the solution of Laplace's equation in cylindrical coordinates and the use of mixed boundary conditions [2]. Since

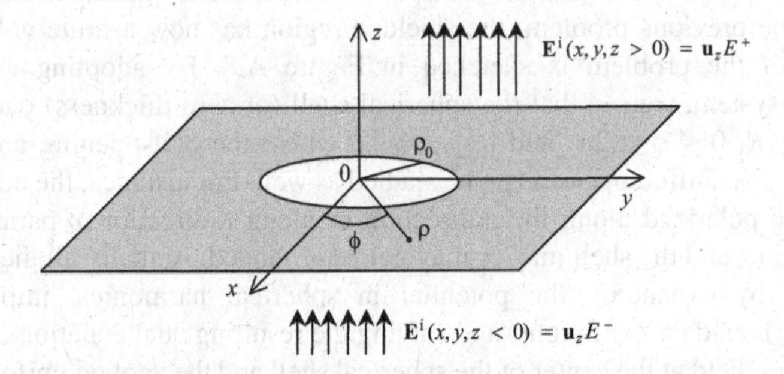

FIGURE A.7 Infinitesimally thin, grounded conductive plane with a circular hole under an external uniform electrostatic field that is orthogonal to the hole.

mixed boundary conditions result in analytical challenging problems, an alternative way to approach the problem is to separate Laplace's equation in oblate spheroidal coordinates and find the solution for the potential by an expansion in oblate spheroidal harmonics. The circular hole is thus seen as the limiting form of an oblate spheroidal surface.

By omitting the details (which can be found, e.g., in [1]), we find the solution for the potential to be

$$V(x, y, z) = \begin{cases} -E^{+}z + V_{\mathrm{p}}(\rho, z) & z > 0, \\ -E^{-}z + V_{\mathrm{p}}(\rho, z) & z < 0, \end{cases} \tag{A.50}$$

where

$$V_{\mathrm{p}}(\rho, z) = \frac{(E^{-} - E^{+})\rho_0}{\pi}\left[\sqrt{\frac{b-a}{2}} - \frac{|z|}{\rho_0}\tan^{-1}\left(\sqrt{\frac{2}{a+b}}\right)\right] \tag{A.51}$$

with

$$\rho = \sqrt{x^2 + y^2}, \quad a = \frac{\rho^2 + z^2 - \rho_0^2}{\rho_0^2}, \quad b = \sqrt{a^2 + \frac{4z^2}{\rho_0^2}}. \tag{A.52}$$

As can be seen, the perturbation potential $V_{\mathrm{p}}(\rho, z)$ gives rise also to a radial component of the electrostatic field, which nevertheless remains independent of the azimuthal coordinate ϕ. In particular, in correspondence of the hole ($z = 0$ and $\rho < \rho_0$) there results

$$\mathbf{E}(\rho, 0) = \frac{(E^{-} - E^{+})}{\pi}\frac{\rho}{\sqrt{\rho_0^2 - \rho^2}}\mathbf{u}_{\rho} + \frac{(E^{-} + E^{+})}{2}\mathbf{u}_z. \tag{A.53}$$

It can also be numerically verified that when $E^{-} = 0$, the electric field in the half-space $z < 0$ is less than 1% at distances from the hole larger than twice the radius.

Another interesting problem concerns the penetration of an impressed uniform electrostatic field $\mathbf{E}^{\mathrm{i}}$ through a circular aperture in a spherical conductive shell. With regard to the previous problem, the shielded region has now a finite volume. The geometry of the problem is sketched in Figure A.8. By adopting a spherical coordinate system, we see that the spherical shell (of zero thickness) occupies the surface $r = R$, $0 \le \phi < 2\pi$, and $0 \le \theta < \theta_0$, where the half-opening angle θ_{op} is equal to $\pi - \theta_0$. Different cases can be studied as well. For instance, the electrostatic field can be polarized along the z direction or along a direction ρ parallel to the aperture plane, and the shell may or may not be grounded. Actually all the problems are solved by expanding the potential in spherical harmonics, imposing the appropriate boundary conditions, and solving the resulting dual equations. The ratio of the electric field at the center of the spherical shell and the applied uniform field is usually taken as a measure of the electrostatic field penetration inside the shielded

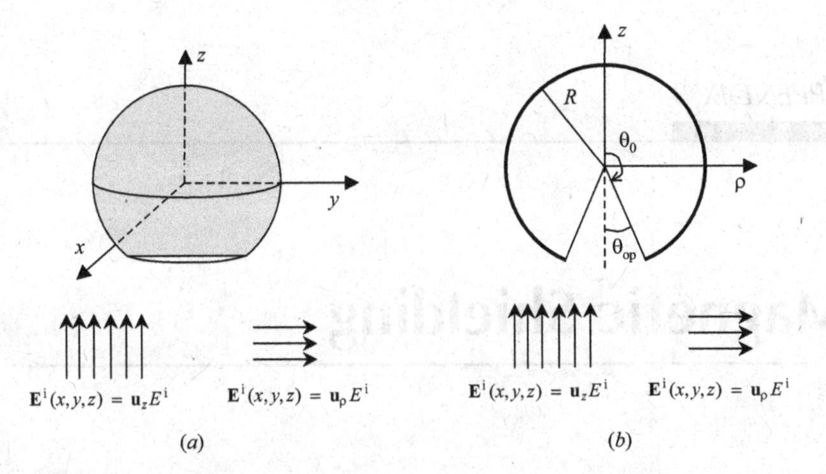

FIGURE A.8 Spherical conductive shield with a circular aperture under an external uniform electrostatic field.

region. In particular, it turns out that for the grounded shell with a z-polarized field the ratio is

$$\frac{E_z(0)}{E^i} = \frac{1}{\pi}\left(\theta_{op} - \frac{1}{3}\sin\theta_{op}\right), \tag{A.54}$$

while for the ungrounded case

$$\frac{E_z(0)}{E_z^i} = \frac{1}{\pi}\left(\theta_{op} - \frac{1}{3}\sin 3\theta_{op} + \frac{\sin\theta_{op} - \frac{1}{2}\sin 2\theta_{op}}{\pi - \theta_{op} + \sin\theta_{op}}\right). \tag{A.55}$$

Finally, the result for the transverse (radially) polarized impressed-field case is

$$\frac{E_\rho(0)}{E^i} = \frac{1}{\pi}\left(\theta_{op} - \frac{1}{2}\sin\theta_{op} - \frac{1}{2}\sin 2\theta_{op} + \frac{1}{6}\sin 3\theta_{op}\right). \tag{A.56}$$

REFERENCES

[1] J. D. Jackson. *Classical Electrodynamics*, 3rd ed. New York: Wiley, 1998.

[2] W. R. Smythe. *Static and Dynamic Electricity*, 2nd ed. New York: McGraw-Hill, 1950.

APPENDIX B

Magnetic Shielding

For almost a century the subject of shielding extremely low-frequency (ELF) and very low-frequency (VLF) magnetic fields has been of interest [1]. The interest originated from the design necessity to protect part of the circuit in radio-receiving apparatus from the disturbing effect of the radiated field in its neighborhood [2–4]. Over the decades the awareness of possible interferences from radiated fields rapidly increased. Apart from special applications, interference on electronic systems became more and more evident with the widespread use of computers. Such interference occurred mainly as frame disturbances on CRT displays when the external magnetic flux density exceeded fixed values [5]. In the past decade a more serious issue about the possible health hazards for persons being exposed to magnetic fields at low frequencies has led to a renewed interest in the subject [6]. Nowadays, shielding of low-frequency magnetic fields is a topic of interest for a number of applications, ranging from the mitigation of power-line sources to protection of sensitive equipment, as schematically depicted in Figure B.1.

At low frequencies the magnetic field is due either to the electric current flowing in conductors of various geometries or to the magnetization of surrounding ferromagnetic materials. The classical strategy for reducing quasi-static magnetic fields in a specific region consists in inserting a shield of appropriate material, whose properties are used to alter the spatial distribution of the magnetic field emitted by the source. The shield in fact causes a change in the behavior of the field, diverting the lines of the magnetic induction away from the shielded region.

A quantitative measure of the effectiveness of a shield in reducing the magnetic-field magnitude at a given point is the shielding effectiveness SE_B, defined as

$$\mathrm{SE}_B = \frac{|\mathbf{B}_0(\mathbf{r})|}{|\mathbf{B}_S(\mathbf{r})|}, \tag{B.1}$$

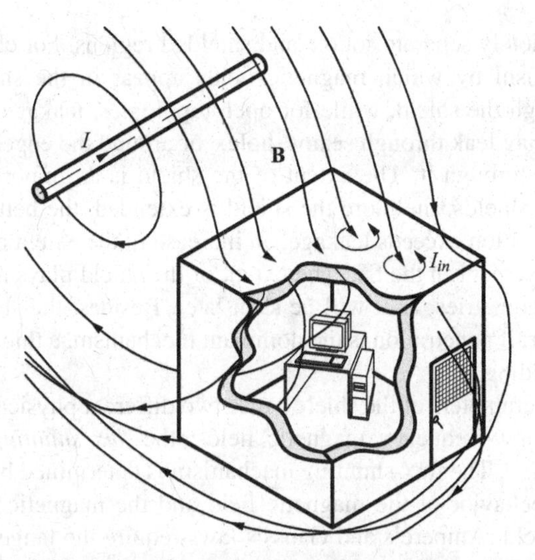

FIGURE B.1 Low-frequency shielding scenario.

where $\mathbf{B}_0$ is the magnetic induction at the observation point $\mathbf{r}$ when the shield is absent and $\mathbf{B}_S$ is the magnetic induction at the same point with the shield applied. In general, the SE is a function of the position $\mathbf{r}$ at which it is calculated (or measured). If the properties of a shield material are independent of field magnitudes, as it happens for good conductors such as copper and aluminum, the SE is correspondingly independent of the excitation amplitude. However, if the magnetic permeability of a shield material depends on the magnetic induction within the material (as it happens for ferromagnetic materials such as nickel alloys like Mumetal and Ultraperm or low-carbon steels), the SE is also dependent on the excitation amplitude.

In addition to (B.1) the shielding effectiveness $\mathrm{SE}_B^{\mathrm{dB}}$, defined as

$$\mathrm{SE}_B^{\mathrm{dB}} = 20 \log \mathrm{SE}_B = 20 \log \frac{|\mathbf{B}_0(\mathbf{r})|}{|\mathbf{B}_S(\mathbf{r})|}, \tag{B.2}$$

is also used.

B.1 MAGNETIC SHIELDING MECHANISM

When the shield is inserted between the source and the region where a reduction of the field magnitude is desired, the resulting shape of the field is generally dependent on the shield geometry, the material parameters, and the frequency of the emitted field [7–9].

Shield geometries that completely divide the space into source and shielded regions (e.g., infinite planar shields, infinite cylindrical shields, and spherical shields) define closed topologies. Open topologies are defined as shield geometries that do not completely separate source and shielded regions. For closed topologies, the only mechanism by which magnetic fields appear in the shielded region is penetration through the shield, while for open topologies, leakage may also occur. Magnetic fields may leak through seams, holes, or around the edges of the shield as well as penetrate through it. The extent of the shield is an important factor when considering open shields: the more the shield is extended, the better the shielding. However, if penetration exceeds leakage, an increase in the extent of the shield may bring little improvement in the SE. The extent of the shield plays an important role also for closed geometries, as it will be seen later. Besides, the shield thickness is another key factor; if penetration is the dominant mechanism, a thicker shield results in improved shielding.

The material parameters of the shield cause two different physical mechanisms in the shielding of low-frequency magnetic fields: the *flux shunting* and the *eddy-current cancellation.* The flux-shunting mechanism is determined by two conditions that govern the behavior of the magnetic field and the magnetic induction at the surface of the shield: Ampere's and Gauss's laws require the tangential component of the magnetic field and the normal component of the magnetic induction to be continuous across material discontinuities. Hence, in order to simultaneously satisfy both conditions, the magnetic field and the magnetic induction can abruptly change direction when crossing the interface between two different media. At the interface between air and a ferromagnetic shield material having a large relative permeability, the field and the induction on the air side of the interface are pulled toward the ferromagnetic material nearly perpendicular to the surface, whereas on the ferromagnetic side of the interface, they are led along the shield nearly tangential to the surface. The resulting overall effect of the shielding structure is that the magnetic induction produced by a source is diverted into the shield, then shunted within the material in a direction nearly parallel to its surface, and finally released back into the air. In Figure B.2*a*, the typical behavior of a cylindrical shield placed in an external uniform magnetic field is reported.

The field map refers to a structure with internal radius $a = 0.1$ m, thickness $\Delta = 1.5$ cm, and $\mu_r = 50$ at dc ($f = 0$ Hz). The SE is determined by the material permeability and the geometry of the shield. The shield in fact gathers the flux over a

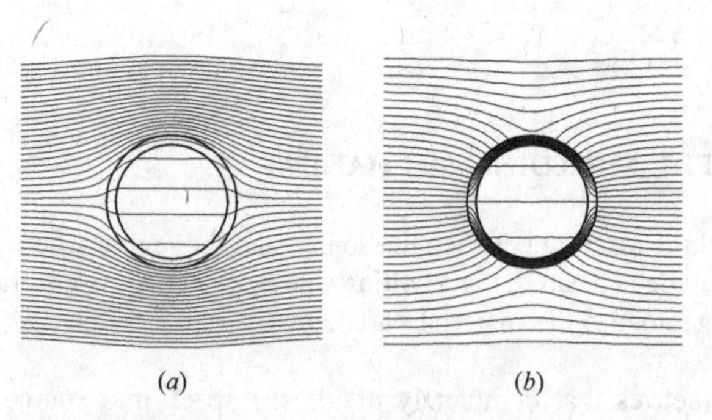

FIGURE B.2 Magnetic-field distribution for cylindrical shields subjected to a uniform impressed field: (*a*) ferromagnetic shield; (*b*) highly conductive shield.

region whose size is determined by the large-scale dimension of the shield (i.e., the diameter) and shunts it through the thickness of the ferromagnetic material. Therefore the magnetic induction within the shield material is amplified by a factor determined by the shield diameter–thickness ratio. Besides, the amount of the magnetic induction pulled into the shield and the reduction of leakage into the shielded region are determined by the relative permeability of the ferromagnetic material. These effects combine to produce a SE that can be improved either by increasing the material relative permeability or by increasing the material thickness with respect to the shield diameter.

The eddy-current cancellation mechanism is determined by the eddy currents that arise in the shield material due to the presence of a time-varying incident magnetic field. When the shield is exposed to a time-varying magnetic field, an electric field is induced within the material, as described by Faraday's law, and when the shield is highly conductive, an induced electric current density is driven as described by the Ohm law. The induced current density gives rise to a magnetic field opposing the incident one, which is therefore repulsed by the metal and forced to run parallel to the surface of the shield, yielding a small magnetic induction inside the metal. In Figure B.2*b* the typical SE behavior of the previously considered cylindrical shield (but with $\sigma = 5 \cdot 10^8$ S/m and $f = 50$ Hz) placed in an external uniform ac magnetic field is reported. Unlike the flux-shunting mechanism, the eddy-current cancellation occurs only when the incident field is time-varying; it occurs in any electrically conducting material, regardless of the values of the relative permeability.

A fundamental shielding parameter in determining the mechanism of the eddy-current cancellation, but also in the flux-shunting mechanism when dealing with ac fields, is the value of the shield thickness compared with the skin depth. The propagation constant of a uniform plane wave in a conductive material is (see Chapter 4)

$$\gamma = \sqrt{j\omega\mu_0\mu_r(\sigma + j\omega\varepsilon_0\varepsilon_r)}. \tag{B.3}$$

If the shield is a good conductor (i.e., the product of the frequency and the dielectric permittivity can be assumed to be small enough with respect to the conductivity, $\omega\varepsilon_0\varepsilon_r \ll \sigma$), then the propagation constant γ can be expressed as

$$\gamma \simeq \sqrt{j\omega\mu_0\mu_r\sigma} = (1+j)\sqrt{\frac{\omega\mu_0\mu_r\sigma}{2}} = \frac{(1+j)}{\delta}, \tag{B.4}$$

where δ is the frequency-dependent skin depth

$$\delta = \sqrt{\frac{2}{\omega\mu_0\mu_r\sigma}}. \tag{B.5}$$

For an ac field the total magnetic induction decays exponentially into the material, away from the air–shield interface, with a characteristic decay length equal to the skin depth δ. Therefore, when the shield thickness Δ is much larger than the skin depth δ, large SE values can readily be obtained, whereas when $\delta \gg \Delta$, the induced currents flow uniformly over the shield thickness and the resultant shielding is not as effective as in the previous case. Nevertheless, significant shielding can be accomplished with highly conductive materials (e.g., copper and aluminum) when the geometrical dimensions are suitably chosen. As concerns the induced-current mechanism, improved shielding can be obtained by increasing the large-scale dimension of the structure, for a fixed thickness. The inductive coupling with the source is in fact proportional to the area that intercepts the source flux, while the resistance presented to the induced current is proportional to the circumference of the shield. Since the shielding mechanism is also governed by the ratio between the inductive induced voltage and the resistance, highly conductive materials give larger attenuations when the shield is large, even if the thickness is electrically small. Interestingly an opposite behavior is obtained in the case of the flux-shunting mechanism, where an increase in radius produces a poorer shielding because a larger total flux has to be shunted through a fixed thickness.

It is now clear that the geometry of the shield (and of the source) and the material parameters, together with the frequency of the source field, determine which shielding mechanism is dominant (the flow of the eddy currents or the channeling of the magnetic induction) and consequently determine the SE. Furthermore these two shielding mechanisms are characterized by two different boundary conditions: in the flux-shunting case, the tangential component of the magnetic field is nearly zero, whereas in the eddy-current cancellation case, the normal component of the magnetic field is nearly zero [9].

B.2 CALCULATION METHODS

Generally, theoretical analysis of low-frequency magnetic-field shielding can be very trying to carry out. The major difficulties are the geometric complexity of both shield and sources, as well as the possible nonlinearity of the materials (e.g., saturation and hysteresis) that call for more complex permeability modeling, thus making the magnetic-field analysis less tractable. Although an analytic solution can be achieved for simple idealized geometries and linear-medium configurations, the solution might not be capable of giving practical shielding design guidelines. Presently it is possible to identify several methods of analysis each involving various degrees of approximation.

General complex geometries can be treated only by means of numerical methods. However, the solution of complex shielding configurations is not a simple task. Since the geometrical dimensions of the model vary from the shield thickness (order of millimeters) up to the size of the ambient to be protected (order of meters), the FEM in the frequency domain (see Section 5.1) has been chosen as the main tool of calculation [10]. Because of its unstructured nature, the FEM is able to accurately model all the geometrical features without additional cost. Furthermore it has been widely used with good results for the calculation of eddy currents at low frequencies [11]. However, nonlinear effects, such as saturation and hysteresis, naturally call for a time-domain analysis where they can be treated instead with plainness. From this point of view, the FDTD technique (see Section 5.3) could be an alternative choice. However, one area where FDTD encounters serious difficulties is in problems where the object under analysis is electrically very small. In this case the CFL stability condition on the maximum allowable time step can make direct application of the explicit Yee algorithm infeasible. Implicit methods, subgridding techniques, or impedance type boundary conditions can partially alleviate the problem, but they give arise to additional issues concerning numerical accuracy and stability (e.g., see Chapter 5).

Beside numerical approaches, three analytical techniques are available. They are the more common methods of analysis based on the exact solution of the boundary-value problem, on the transmission-line model, and on the equivalent lumped-circuit model.

The classical boundary-value problem method has been applied to very basic shield structures such as uniform plates, cylinders, and spheres [1–4]. An exact treatment of nonuniform shields is not available even for these idealized shapes. The disadvantages of the analytical methods are usually the complexity of the expressions involved and the oversimplification of the real structure. Nevertheless, such disadvantages depend somewhat on the viewpoints, since it can be argued that the analytical solutions are exact only for idealized geometries. Furthermore, although most of the actual shields are not spherical, cylindrical, or planar with infinite length or infinite extent, analytical expressions can provide some reasonable indications and often a useful upper limit of what can be expected for shields of other shapes.

The uniform transmission-line method reduces the problem of predicting the SE to computing the transmission of a plane wave through a uniform material of the same thickness of the enclosure. The geometrical features of the source and enclosure are simply estimated by considering a wave of appropriate impedance incident on the structure (usually low-impedance field sources are considered). This method can be seen as a natural extension of the Schelkunoff approach [12] (originally developed for high-frequency traveling plane waves) to the near-field low-frequency limit. Results show a reasonable accuracy of the method only when the appropriate wave impedance for the source and enclosure orientation can be determined.

An alternative approach is the lumped-circuit approach based on circuit theory [13,14]. An equivalent circuit model is constructed for both the enclosure and the source and from its analysis the SE is determined. The circuit usually includes an independent source and an equivalent resistance and inductance to model the shield. The method gives accurate results only at low frequencies where the lumped-circuit theory is reliable and is, in any case, limited by the need of closed-form expressions for the equivalent parameters of the screen, which are available only for simple geometries. An important feature is that it can provide some insight into the problem of imperfect seams, provided that suitable mutual inductances can be determined.

In the following sections some important boundary-value problems will be revised and important details on some peculiar aspects will be provided. In representing the shielding materials, it will be assumed that their constitutive relations are linear; this means that hysteresis effects will be neglected and magnetic materials will be represented by their initial permeability. Only in the last section some details will be given on shields with hysteresis.

B.3 BOUNDARY-VALUE PROBLEMS

Expressions for the magnetic SE of spherical and cylindrical shields were originally derived by King [4]. Next an extensive treatment based on an approximated approach was presented by Kaden [15]. Later accurate analytical expressions were derived by Wait [16] and Hoburg [7]. The SE of a planar shield, on the other hand, has been studied by many authors considering as transmitting and receiving sources two coaxial wire loops [3,17–27], arbitrarily oriented magnetic dipoles [28,29], or parallel long straight current conductors [9,30,31]. To be consistent with the published literature, in the following sections the inner and outer radii of the spherical and cylindrical shields will be denoted by a and b, respectively, and the thickness of the shield will be denoted by Δ.

B.3.1 Spherical Magnetic Conducting Shield

Let us consider a spherical shell made of a magnetic conducting material with inner radius a, outer radius b, and wall thickness Δ (i.e., $\Delta = b - a$) placed in a uniform incident magnetic field with amplitude H_0, as shown in Figure B.3. The boundary-value problem for the quasi-stationary EM field can be solved in a closed form in terms of the magnetic vector potential **A** in a spherical reference coordinate system. Because of the symmetry of the sphere's geometry, the vector potential is in fact ϕ-directed and independent of the ϕ coordinate. Under the magnetic quasi-stationary approximation the vector potential in the free-space outside and inside the spherical shield satisfies the Laplace equation, while inside the conducting magnetic material it satisfies the diffusion equation [32].

$$\nabla^2 A_\phi^{\mathrm{e}} = 0, \qquad r \geq b, \tag{B.6a}$$

$$\nabla^2 A_\phi^{\mathrm{sh}} - \gamma^2 A_\phi = 0, \qquad a < r < b, \tag{B.6b}$$

$$\nabla^2 A_\phi^{\mathrm{i}} = 0, \qquad r \leq a, \tag{B.6c}$$

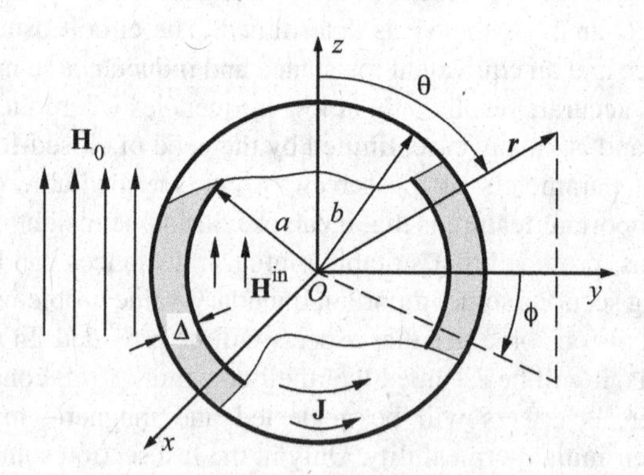

FIGURE B.3 Spherical shell placed in a uniform external magnetic field.

where $\nabla^2 A_\phi$ can be written in spherical coordinates as

$$\nabla^2 A_\phi = \left[\frac{1}{r^2}\frac{\partial}{\partial r}\left(r^2\frac{\partial}{\partial r}\right) + \frac{1}{r^2 \sin\theta}\frac{\partial}{\partial\theta}\left(\sin\theta\frac{\partial}{\partial\theta}\right) - \frac{1}{r^2\sin^2\theta}\right]A_\phi. \tag{B.7}$$

Since the current density is proportional to $\sin\theta$, it is reasonable to express A_ϕ as a product of a radial function and $\sin\theta$, that is,

$$A_\phi^{\mathrm{e}}(r,\theta) = \mu_0 H_0 \sin\theta\left(\frac{r}{2} - \frac{c_1}{r^2}\right), \qquad r \geq b, \tag{B.8a}$$

$$A_\phi^{\mathrm{sh}}(r,\theta) = \mu_0 H_0 \sin\theta[c_2 i_1(\gamma r) + c_3 k_1(\gamma r)], \qquad a < r < b, \tag{B.8b}$$

$$A_\phi^{\mathrm{i}}(r,\theta) = \mu_0 H_0 c_4 r \sin\theta, \qquad r \leq a, \tag{B.8c}$$

where $i_1(\cdot)$ and $k_1(\cdot)$ in (B.8b) are the first-order modified spherical Bessel functions of the first and second kind, respectively [33]. The unknown constants c_1, c_2, c_3, and c_4 can be determined by enforcing the boundary conditions on the outer and inner surfaces of the shell, namely the continuity of the tangential component of the magnetic field and the continuity of the normal component of the magnetic induction:

$$B_r^{\mathrm{e}}\big|_{r=b} = B_r^{\mathrm{sh}}\big|_{r=b}, \quad H_\theta^{\mathrm{e}}\big|_{r=b} = H_\theta^{\mathrm{sh}}\big|_{r=b}, \tag{B.9a}$$

$$B_r^{\mathrm{i}}\big|_{r=a} = B_r^{\mathrm{sh}}\big|_{r=a}, \quad H_\theta^{\mathrm{i}}\big|_{r=a} = H_\theta^{\mathrm{sh}}\big|_{r=a}. \tag{B.9b}$$

Once the unknown constants have been determined, the final expressions of the magnetic vector potential in all the three regions of space are obtained. Hence the SE can be expressed as

$$\begin{aligned} \mathrm{SE} = \Bigg| & \frac{1}{3b^3\gamma^2\mu_\mathrm{r}}\{2\Delta\mu_\mathrm{r}^2 + \mu_\mathrm{r}[ab(3b-\Delta)\gamma^2 - \Delta] + \Delta(ab\gamma^2 - 1)\}\cosh(\gamma\Delta) \\ & + \frac{1}{3b^3\gamma^3\mu_\mathrm{r}}[\gamma^2(a^2b^2\gamma^2 + b^2 - a\Delta) + (3b-\Delta)\Delta\mu_\mathrm{r}\gamma^2 + 2(ab\gamma^2-1)\mu_\mathrm{r}^2 \\ & + \mu_\mathrm{r} + 1]\sinh(\gamma\Delta)\Bigg|, \end{aligned} \tag{B.10}$$

where the expressions for positive arguments and nonnegative integer indexes of the modified spherical Bessel functions of order zero, one, and two [33] have been used. It is possible to note that the magnetic field is uniform inside the shell regardless of the material and size. Expression (B.10) is the same as the one derived by Hoburg [7].

By assuming very thin shells (i.e., $a \simeq b = r_0$) and $a, b \gg \delta$, from (B.10) it is possible to obtain the approximate solution [4,15]

$$\mathrm{SE} \simeq \left|\cosh(\gamma\Delta) + \frac{1}{3}\left(\frac{\gamma r_0}{\mu_\mathrm{r}} + \frac{2\mu_\mathrm{r}}{\gamma r_0}\right)\sinh(\gamma\Delta)\right|. \tag{B.11}$$

If $a, b \gg \delta$, instead of solving the exact eddy-current problem inside the shell to find afterward the approximate expression, it is more convenient to neglect from the beginning the terms with coefficients proportional to $1/r$ and $1/r^2$ in the diffusion equation (B.6b), as done by Kaden [15]. In this case the approximate solution for the magnetic vector potential inside the magnetic conducting material is

$$A_\phi^{\mathrm{sh}}(r,\theta) \simeq \mu_0 H_0 \sin\theta(c_2 e^{\gamma r} + c_3 e^{-\gamma r}), \qquad a < r < b. \tag{B.12}$$

Expression (B.12) can be used instead of (B.8b) to solve the boundary-value problem; the resulting SE is

$$\mathrm{SE} = \left| \frac{1}{3b\gamma\mu_\mathrm{r}} \{\gamma[3\mu_\mathrm{r} a + (2\mu_\mathrm{r} - 1)\Delta]\cosh(\gamma\Delta) + (\gamma^2 ab + \mu_\mathrm{r} - 1 + 2\mu_\mathrm{r}^2)\sinh(\gamma\Delta)\} \right|, \tag{B.13}$$

which reduces to (B.11) for the very thin shell (i.e., $a \simeq b = r_0$ and $\Delta \to 0$) with $a, b \gg \delta$.

To obtain a low-frequency approximation of (B.11), the shell thickness is considered to be much thinner than the penetration depth ($\Delta \ll \delta$). By use of the approximations of the hyperbolic functions for small arguments with a first-order accuracy, the SE can be expressed as

$$\mathrm{SE} \simeq \left| 1 + \frac{2\mu_\mathrm{r}}{3r_0}\Delta + \frac{r_0}{3\mu_\mathrm{r}}\Delta\gamma^2 \right|. \tag{B.14}$$

On the other hand, in order to obtain a thick-shield approximation, the frequency is considered sufficiently high so that $|\gamma\Delta| \gg 1$; that is, the shell is considered electrically thick compared to the skin depth ($\Delta \gg \delta$), and likewise the skin depth is much smaller than the shell radius ($r_0 \gg \delta$). Furthermore the shell is considered geometrically thin so that $a \simeq b = r_0$. With these assumptions, the hyperbolic functions can be approximated as $\cosh\gamma\Delta \simeq \sinh\gamma\Delta \simeq e^{\gamma\Delta}/2$, and the SE is readily obtained from (B.11) as

$$\mathrm{SE} \simeq \left| \frac{r_0\gamma}{6\mu_\mathrm{r}} e^{\gamma\Delta} \right| = \frac{r_0}{3\sqrt{2}\mu_\mathrm{r}\delta} e^{\Delta/\delta}. \tag{B.15}$$

This expression clearly indicates that in the high-frequency range a conducting shell is effective in shielding magnetic fields due to the eddy currents induced on it (which appears as the usual material attenuation factor $e^{\Delta/\delta}$). Moreover, under these conditions, a larger shield will allow a larger radius for the eddy-current flow, yielding a larger field attenuation. Thus, for practical shield dimensions, a large closed highly conductive shield performs better than a small one.

The static problem of the dc shielding of a spherical shell in a uniform magnetic field can be solved in terms of spherical-harmonic functions for the magnetic vector potential **A** with a procedure similar to that adopted in the electrostatic case (see Appendix A). The magnetic-field problem in fact reduces to a potential problem described throughout all the space by the Laplace equation [32]:

$$\nabla^2 A_\phi = 0. \tag{B.16}$$

By using arguments similar to those of the electrostatic case, the dc SE is obtained as [4,34]

$$\mathrm{SE} = \frac{1}{9\mu_\mathrm{r}}\left[(2\mu_\mathrm{r}+1)(\mu_\mathrm{r}+2) - 2\frac{a^3}{b^3}(\mu_\mathrm{r}-1)^2\right]. \tag{B.17}$$

It can also be shown that (B.17) is merely the static limit of the exact solution of the eddy-current problem (B.10) in the limit $\gamma \to 0$.

If the shield thickness is small ($\Delta \to 0$), (B.17) reduces to

$$\mathrm{SE} \simeq 1 + \frac{2}{3}\frac{(\mu_\mathrm{r}-1)^2}{\mu_\mathrm{r} r_0}\Delta \tag{B.18}$$

which, if the relative permeability is large (i.e., $\mu_\mathrm{r} \gg 1$), further reduces to

$$\mathrm{SE} \simeq 1 + \frac{2\mu_\mathrm{r}}{3r_0}\Delta. \tag{B.19}$$

This is the static limit of expression (B.11), which is valid for a thin spherical shell with a large relative permeability. However, from (B.18) it is possible to obtain a new low-frequency approximation for the SE of a thin spherical shell as

$$\mathrm{SE} \simeq \left|1 + \frac{2}{3}\frac{(\mu_\mathrm{r}-1)^2}{\mu_\mathrm{r} r_0}\Delta + \frac{r_0}{3\mu_\mathrm{r}}\Delta\gamma^2\right|. \tag{B.20}$$

From the previous equations it can be seen that the magnetic bypass of the field is more efficient with a large $\mu_\mathrm{r}\Delta$ factor or with a small shell radius r_0. This result is consistent with the observation that for practical shield dimensions, a smaller closed ferromagnetic shield performs better than a larger one.

It is worth noting that under the assumption of a nonmagnetic spherical shell ($\mu_\mathrm{r} = 1$), (B.10) reduces to

$$\mathrm{SE} \simeq \left|\frac{(\gamma a)^2}{3\gamma b}\left\{\frac{3}{\gamma b}\cosh(\gamma\Delta) + \left[1 + \frac{3}{(\gamma b)^2}\right]\sinh(\gamma\Delta)\right\}\right| \tag{B.21}$$

which is exactly the same expression as the one derived by Harrison [35] from the expression given by King [4]

$$\mathrm{SE} = \left| \frac{(\gamma a)^{5/2}}{3\sqrt{\gamma b}} \left[I_{1/2}(\gamma b) K_{3/2}(\gamma a) - K_{1/2}(\gamma b) I_{3/2}(\gamma a) \right] \right|. \tag{B.22}$$

where I_q and K_q are the qth-order modified Bessel functions of the first and second kind, respectively. In the $\mu_r = 1$ case two different expressions could be obtained by comparing (B.21) and (B.11). However, it should be noted that (B.21) is exact (under the quasi-stationary approximation) but limited to nonmagnetic shells, whereas (B.11) is an approximation of the exact solution (B.10) that is accurate when $r_0 \gg \delta$.

When dealing with N spherical concentric shells with large permeability ($\mu_r \gg 1$) and small thickness ($|\gamma\Delta| \ll 1$), the total SE can be expressed as [36]

$$\mathrm{SE}_T = 1 + \sum_{i=1}^{N} S_i' + S_1' \prod_{i=2}^{N} S_i' \left(1 - \frac{r_{i-1}^3}{r_i^3} \right), \tag{B.23}$$

where S_i' is the individual modified SE of the ith sphere (with radius r_i) and is defined as $S_i' = S_i - 1$, (S_i is the relevant SE calculated with one of the previous expressions).

The frequency behavior of the SE of a spherical shell with radius $r_0 = 30$ cm is shown in Figure B.4. Different materials have been considered to investigate the effects of the two shielding mechanisms (i.e., the flux shunting and the eddy-current cancellation). The considered materials are Duranickel stainless steel ($\mu_r = 10.58$, $\sigma = 2.35 \cdot 10^6$ S/m) with thickness $\Delta = 2$ mm, a copper casting alloy ($\mu_r = 1.09$, $\sigma = 1.18 \cdot 10^7$ S/m) with thickness $\Delta = 2$ mm, and an iron-nickel alloy ($\mu_r = 75 \cdot 10^3$, $\sigma = 2 \cdot 10^6$ S/m) with thickness $\Delta = 0.15$ mm. Note in Figure B.4 the excellent agreement of the approximate expression (B.11) with the exact solution and in addition the low- and high-frequency approximations. To fully understand these curves, it is necessary to first evaluate the critical frequency f_0 at which the skin depth δ is equal to the thickness Δ; such a frequency is 2.7 kHz, 5.3 kHz, and 75 Hz, respectively, for the above-mentioned structures. Figure B.5 compares the skin depth versus frequency trends for the above-considered materials with respect to the two thicknesses of the shell.

In Figure B.4 it is possible to see that the low-frequency and high-frequency approximations are sufficiently accurate below and above f_0, respectively. Nevertheless, the high-frequency approximation accounts for only the eddy-current cancellation; the flux-shunting mechanism (the third term in (B.11)) has been neglected. Therefore, in the iron-nickel alloy, because of its large permeability, it is necessary to operate at frequencies higher than those for the first two materials in order to make the eddy-current cancellation mechanism prevail and have an accurate high-frequency approximation.

B.3.2 Cylindrical Magnetic Conducting Shield in a Transverse Magnetic Field

Let us consider an infinitely long cylindrical shell with inner radius a, outer radius b, and a wall thickness Δ (i.e., $\Delta = b - a$). The shell is placed in a transverse uniform ac magnetic field (i.e., perpendicular to the axis of the cylindrical shell) of amplitude H_0, as shown in Figure B.6.

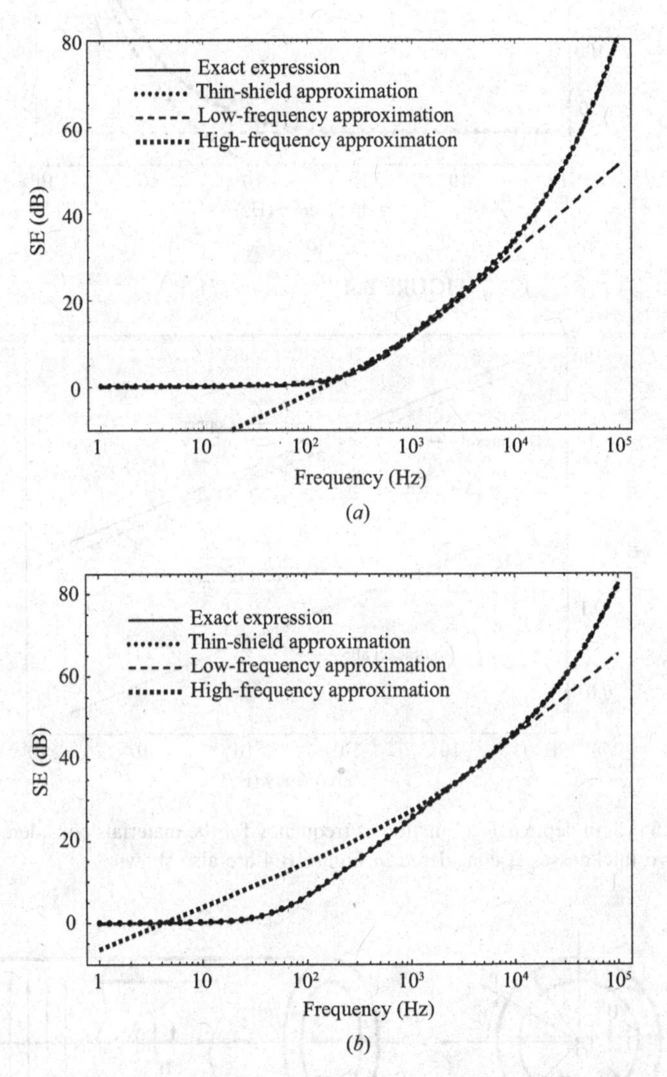

FIGURE B.4 SE of a spherical shell ($r_0 = 30$ cm) compared using different approximations and the exact expression for different materials: (*a*) Duranickel stainless steel ($\mu_r = 10.58$, $\sigma = 2.35 \cdot 10^6$ S/m, $\Delta = 2$ mm); (*b*) copper casting alloy ($\mu_r = 1.09$, $\sigma = 1.18 \cdot 10^7$ S/m, $\Delta = 2$ mm); (*c*) iron-nickel alloy ($\mu_r = 75 \cdot 10^3$, $\sigma = 2 \cdot 10^6$ S/m, $\Delta = 0.15$ mm).

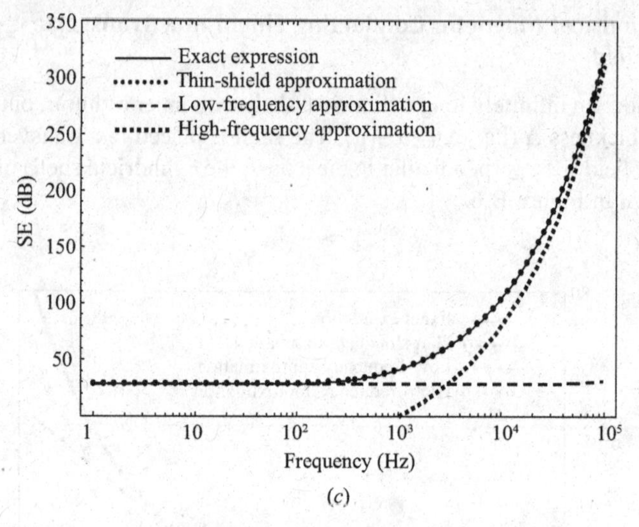

(*c*)

FIGURE B.4 (*Continued*)

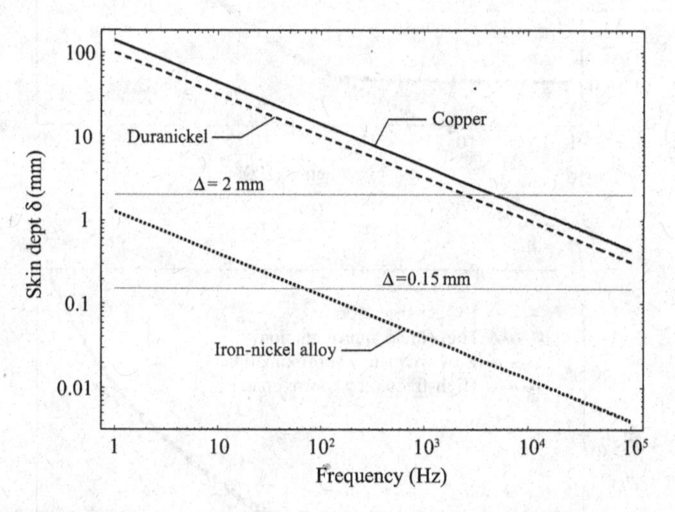

FIGURE B.5 Skin depth δ as a function of frequency for the materials considered in Figure B.4. The two thicknesses Δ considered in Figure B.4 are also shown.

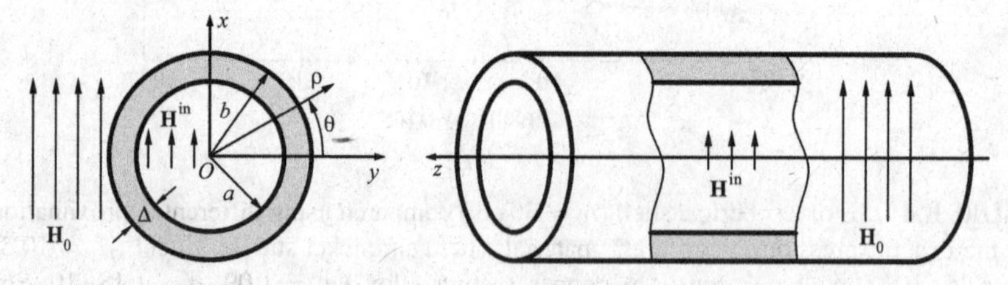

FIGURE B.6 Cylindrical shell placed in a uniform external "transverse" magnetic field.

The problem can be solved in terms of the magnetic vector potential **A** in a cylindrical reference coordinate system. Because of the symmetry of the geometry, the vector potential is z directed and independent of the z coordinate. Under the magnetic quasi-stationary approximation the vector potential in the free space outside and inside the cylindrical shield satisfies the Laplace equation, while inside the conducting magnetic material it satisfies the diffusion equation. Therefore (B.6) hold for the z component of **A**, and $\nabla^2 A_z$ can be written in cylindrical coordinates as

$$\nabla^2 A_z = \left(\frac{\partial^2}{\partial \rho^2} + \frac{1}{\rho}\frac{\partial}{\partial \rho} + \frac{1}{\rho^2}\frac{\partial^2}{\partial \phi^2} \right) A_z. \tag{B.24}$$

Since the current density is proportional to a $\cos\phi$ factor, A_z is expressed as a product of a radial function and $\cos\phi$, that is,

$$A_z^{\mathrm{e}}(\rho,\phi) = \mu_0 H_0 \cos\phi \left(\rho - \frac{c_1}{\rho} \right), \qquad \rho \geq b, \tag{B.25a}$$

$$A_z^{\mathrm{sh}}(\rho,\phi) = \mu_0 H_0 \cos\phi [c_2 I_1(\gamma\rho) + c_3 K_1(\gamma\rho)], \qquad a < \rho < b, \tag{B.25b}$$

$$A_z^{\mathrm{i}}(\rho,\phi) = \mu_0 H_0 c_4 \rho \cos\phi, \qquad \rho \leq a, \tag{B.25c}$$

where $I_1(\cdot)$ and $K_1(\cdot)$ in (B.25b) are the first-order modified Bessel functions of the first and second kind, respectively. The unknown coefficients c_1, c_2, c_3, and c_4 can be determined by enforcing the boundary conditions on the outer and inner surfaces of the cylindrical shell, namely the continuity of the tangential component of the magnetic field and the continuity of the normal component of the magnetic induction. The resulting linear system must be solved to obtain the unknown coefficients and to derive the final expressions of the magnetic vector potential in all the three regions of space. Then the SE is found to be [7,16]

$$\begin{aligned} \mathrm{SE} = \Bigg| \frac{a}{2b\mu_{\mathrm{r}}} \{ & [\mu_{\mathrm{r}} K_1(\gamma a) - \gamma a K_1'(\gamma a)][\mu_{\mathrm{r}} I_1(\gamma b) + \gamma b I_1'(\gamma b)] \\ & - [\mu_{\mathrm{r}} I_1(\gamma a) - \gamma a I_1'(\gamma a)][\mu_{\mathrm{r}} K_1(\gamma b) + \gamma b K_1'(\gamma b)] \} \Bigg|, \end{aligned} \tag{B.26}$$

where $I_1'(\cdot)$ and $K_1'(\cdot)$ are the first derivative of the first-order modified Bessel functions of the first and second kind, respectively. Inside the cylindrical shell the magnetic field is uniform, the same as for the spherical one.

A substantial simplification of (B.26) can be obtained when the radius of the shell is large compared to the skin depth (i.e., $a, b \gg \delta$), since approximations for the large argument of the Bessel functions can be used [33], resulting in

$$\begin{aligned} \mathrm{SE} \simeq \Bigg| \frac{\sqrt{a}}{8\mu_{\mathrm{r}} \gamma b \sqrt{b}} & [\gamma(b + 8\mu_{\mathrm{r}} a + 8\mu_{\mathrm{r}} \Delta) \cosh(\gamma\Delta) \\ & + (\gamma\Delta + 4\gamma^2 a^2 + \gamma a + 4\gamma^2 a \Delta + 4\mu_{\mathrm{r}}^2) \sinh(\gamma\Delta)] \Bigg|. \end{aligned} \tag{B.27}$$

Now, for a magnetic conducting thin shell with $a \simeq b = \rho_0$ (i.e., small thickness Δ), the following approximate expression can be obtained [4,15]:

$$\mathrm{SE} \simeq \left| \cosh(\gamma\Delta) + \frac{1}{2}\left(\frac{\gamma\rho_0}{\mu_r} + \frac{\mu_r}{\gamma\rho_0}\right)\sinh(\gamma\Delta) \right|. \tag{B.28}$$

The same for the spherical shell, under the assumption $a, b \gg \delta$ it is possible to derive an approximate expression for the SE by expressing the magnetic vector potential inside the conducting material as

$$A_z^{\mathrm{sh}}(\rho, \phi) \simeq \mu_0 H_0 \cos\phi (c_2 e^{\gamma\rho} + c_3 e^{-\gamma\rho}), \qquad a < \rho < b. \tag{B.29}$$

From (B.25a), (B.29), and (B.25c), the following expression for the SE is obtained:

$$\mathrm{SE} = \left| \frac{a+b}{2b}\cosh(\gamma\Delta) + \frac{1}{2}\left(\frac{\gamma a}{\mu_r} + \frac{\mu_r}{\gamma b}\right)\sinh(\gamma\Delta) \right|. \tag{B.30}$$

Expression (B.28) can be still derived by the assumption of a thin shell (i.e., $a \simeq b = \rho_0$).

The low-frequency approximation ($|\gamma\Delta| \ll 1$) of (B.28) is readily obtained by replacing the hyperbolic functions with their approximations for small arguments with a first-order accuracy, that is,

$$\mathrm{SE} \simeq \left| 1 + \frac{\mu_r}{2\rho_0}\Delta + \frac{\rho_0}{2\mu_r}\Delta\gamma^2 \right|. \tag{B.31}$$

The thick-shield approximation for the magnetic conducting cylindrical shell can be derived in a way similar to the spherical-shell case, after the usual approximations ($|\gamma\Delta| \gg 1$, $\rho_0 \gg \delta$, and $a \simeq b = \rho_0$) are made. After simple algebraic manipulations the SE is expressed as

$$\mathrm{SE} \simeq \left| \frac{\rho_0 \gamma}{4\mu_r} e^{\gamma\Delta} \right| = \frac{\rho_0}{2\sqrt{2}\mu_r \delta} e^{\Delta/\delta}. \tag{B.32}$$

The static problem of dc shielding of a cylindrical shell in a uniform magnetic field transverse to the axis can be solved in terms of cylindrical harmonic functions for the magnetic scalar potential Φ_m [32]. If the magnetic field is expressed as the gradient of a scalar potential ($\mathbf{H} = -\nabla\Phi_m$), the magnetic-field problem reduces to a potential problem described throughout all the space by the Laplace equation

$$\nabla^2 \Phi_m = 0. \tag{B.33}$$

By arguments similar to those of the electrostatic case, the SE is expressed as [4,34]

$$\mathrm{SE} = \frac{1}{4\mu_r}\left[(\mu_r+1)^2 - \frac{a^2}{b^2}(\mu_r-1)^2\right]. \tag{B.34}$$

When the shell is nonmagnetic ($\mu_r = 1$), the SE is 1, meaning there is no difference between the field inside and outside the shell. It is also possible to show that (B.34) is the static limit of the exact solution of the eddy-current problem (B.26) for $\gamma \to 0$.

If the shield thickness is small ($\Delta \to 0$), then (B.34) reduces to

$$\mathrm{SE} \simeq 1 + \frac{1}{2}\frac{(\mu_r-1)^2}{\mu_r\rho_0}\Delta, \tag{B.35}$$

which, if the relative permeability is large ($\mu_r \gg 1$), further reduces to

$$\mathrm{SE} \simeq 1 + \frac{\mu_r}{2\rho_0}\Delta. \tag{B.36}$$

This is the static limit of (B.28), which is valid for a thin cylindrical shell with a large relative permeability. From (B.35), it is possible to obtain a new low-frequency approximation for a thin shell as

$$\mathrm{SE} \simeq \left|1 + \frac{1}{2}\frac{(\mu_r-1)^2}{\mu_r\rho_0}\Delta + \frac{\rho_0}{2\mu_r}\Delta\gamma^2\right|, \tag{B.37}$$

which is more accurate than (B.31). Expression (B.37) can be obtained from the exact solution (B.26) by making a second-order Taylor series expansion with respect to γ around zero and then a new second-order Taylor expansion with respect to Δ around zero.

The SE of a cylindrical shell with radius $\rho_0 = 30$ cm under a uniform transverse magnetic field is shown in Figure B.7. The material of the shell is an iron-nickel alloy with $\mu_r = 75 \cdot 10^3$ and $\sigma = 2 \cdot 10^6$ S/m, and the shield thickness is $\Delta = 0.15$ mm. The same observations made for the spherical shell apply in this case.

B.3.3 Cylindrical Magnetic Conducting Shield in a Parallel Magnetic Field

For an infinitely long cylindrical shell placed in a uniform magnetic field of amplitude H_0 parallel to the axis, as shown in Figure B.8, the final approximate expression for the SE is close to that obtained for the spherical shield. The problem can be solved directly in terms of the magnetic field. Because of the symmetry of the problem, the magnetic field has only the z component, which depends only on the ρ

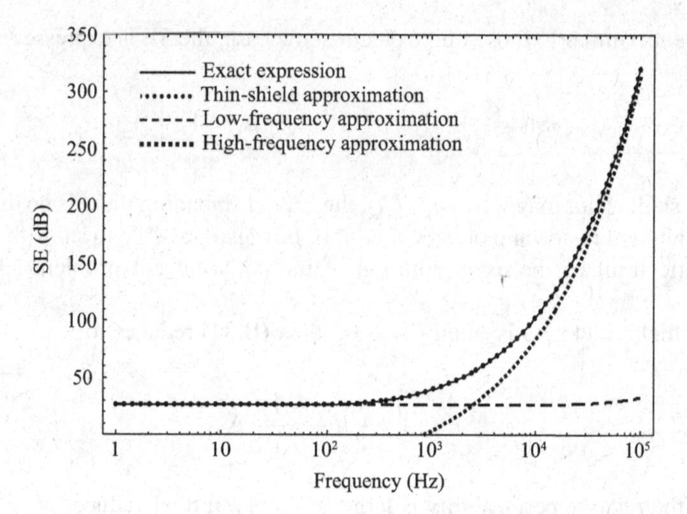

FIGURE B.7 SE of a cylindrical shell ($\rho_0 = 30$ cm, $\Delta = 0.15$ mm) under a uniform transverse magnetic field. Various approximations are compared with the exact result of a shell made of an iron-nickel alloy ($\mu_r = 75 \cdot 10^3$, $\sigma = 2 \cdot 10^6$ S/m).

coordinate in a cylindrical reference coordinate system. Under the magnetic quasi-stationary approximation, H_z satisfies the magnetic diffusion equation inside the conductor:

$$\nabla^2 H_z^{\mathrm{sh}} - \gamma^2 H_z^{\mathrm{sh}} = 0, \qquad a < \rho < b, \tag{B.38}$$

where

$$\nabla^2 H_z = \left(\frac{\partial^2}{\partial \rho^2} + \frac{1}{\rho}\frac{\partial}{\partial \rho}\right) H_z. \tag{B.39}$$

The general exact solution is

$$H_z^{\mathrm{sh}}(\rho) = H_0[c_1 I_0(\gamma\rho) + c_2 K_0(\gamma\rho)], \qquad a < \rho < b, \tag{B.40}$$

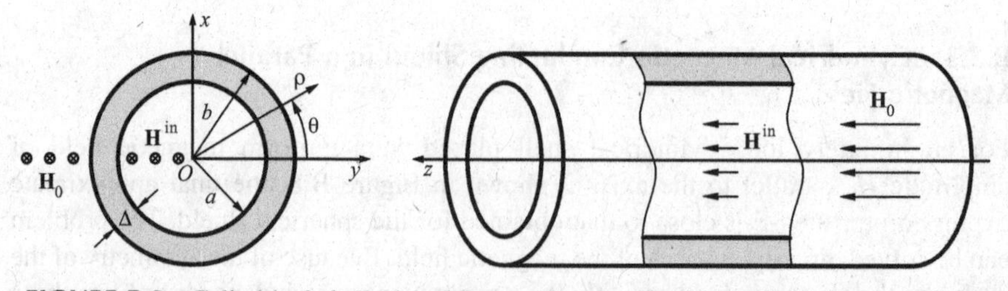

FIGURE B.8 Cylindrical shell placed in a uniform external "parallel" magnetic field.

where $I_0(\cdot)$ and $K_0(\cdot)$ are the zero-order modified Bessel functions of the first and second kind, respectively. The unknown coefficients c_1 and c_2 can be determined by enforcing the boundary conditions on the surfaces of the cylinder. On the outer surface the continuity of the tangential component of the magnetic field is enforced,

$$H_z^{\mathrm{sh}}\Big|_{\rho=b} = H_0, \tag{B.41}$$

while on the inner surface the continuity of the tangential component of the electric field is enforced, giving [32]

$$\frac{1}{2} j\omega a \mu_0 \sigma H_z^{\mathrm{sh}}\Big|_{\rho=a} = \frac{\mathrm{d}H_z^{\mathrm{sh}}}{\mathrm{d}\rho}\Big|_{\rho=a}. \tag{B.42}$$

By taking into account the relations

$$\frac{\mathrm{d}I_0(z)}{\mathrm{d}z} = I_1(z), \quad \frac{\mathrm{d}K_0(z)}{\mathrm{d}z} = -K_1(z), \tag{B.43}$$

from (B.41) and (B.42), the uniform internal magnetic field can be obtained so that the SE reads

$$\mathrm{SE} = \left| \frac{(a\gamma)^2}{2} \left[I_0(\gamma b) K_2(\gamma a) - K_0(\gamma b) I_2(\gamma a) \right] - \frac{(a\gamma)^2}{2} \frac{(\mu_\mathrm{r} - 1)}{\mu_\mathrm{r}} \left[I_0(\gamma b) K_0(\gamma a) - K_0(\gamma b) I_0(\gamma a) \right] \right|. \tag{B.44}$$

When $\mu_\mathrm{r} = 1$, the previous expression is exactly the same as derived by King [4]. By assuming $a, b \gg \delta$, the Bessel functions can be substituted with their approximations for large arguments yielding to

$$\mathrm{SE} \simeq \left| \frac{\sqrt{a}}{\sqrt{b}} \left[\cosh(\gamma\Delta) + \frac{1}{2} \frac{\gamma a}{\mu_\mathrm{r}} \sinh(\gamma\Delta) \right] \right|. \tag{B.45}$$

When the cylinder is thin (i.e., $a \simeq b = \rho_0$), (B.44) reduces to [4,15,33]

$$\mathrm{SE} = \left| \cosh(\gamma\Delta) + \frac{1}{2} \frac{\gamma \rho_0}{\mu_\mathrm{r}} \sinh(\gamma\Delta) \right|. \tag{B.46}$$

As was previously shown, instead of solving the exact eddy-current problem inside the shell and then deriving the approximate expression, under the assumption of $a, b \gg \delta$, it is more convenient to follow the approximate procedure proposed by Kaden [15], neglecting the first-order derivative with respect to ρ in the diffusion equation (B.38) from the beginning. The approximate solution for the magnetic field inside the conducting material is thus

$$A_z^{\mathrm{sh}}(\rho) \simeq H_0(c_1 e^{\gamma\rho} + c_2 e^{-\gamma\rho}), \qquad a < \rho < b. \tag{B.47}$$

By solving again (B.41) and (B.42) for the unknown coefficients c_1 and c_2, the following expression for the SE can easily be obtained:

$$\mathrm{SE} = \left| \cosh(\gamma\Delta) + \frac{1}{2}\frac{\gamma a}{\mu_\mathrm{r}} \sinh(\gamma\Delta) \right|, \tag{B.48}$$

which clearly reduces to (B.46) under the assumption of a thin shell (i.e., $a \simeq b = \rho_0$).

The low-frequency approximation (i.e., $|\gamma\Delta| \ll 1$) of (B.46) is

$$\mathrm{SE} \simeq \left| 1 + \frac{\rho_0}{2\mu_\mathrm{r}} \Delta\gamma^2 \right|, \tag{B.49}$$

which clearly shows that for magnetic fields parallel to the cylinder surface, a magnetic shell does not distort the field at dc ($f = 0$ Hz) and thus does not provide any shielding of static magnetic fields. It is also possible to show that the high-frequency approximation of (B.46) is the same as (B.32), after making the usual approximations ($|\gamma\Delta| \gg 1$, $\rho_0 \gg \delta$, and $a \simeq b = \rho_0$) and using the above-described approximations of the hyperbolic functions.

Interestingly, when $\mu_\mathrm{r} = 1$, the two expressions (B.26) and (B.44) are the same, giving [4]

$$\mathrm{SE} = \left| \frac{(a\gamma)^2}{2} [I_0(\gamma b) K_2(\gamma a) - K_0(\gamma b) I_2(\gamma a)] \right|. \tag{B.50}$$

By a superposition of these two solutions it is possible to conclude that this expression holds when the external magnetic field makes an arbitrary angle with the axis of the cylindrical shell.

It is worth noting that the expressions obtained for a cylindrical shell can be used for a first estimation of the SE of a rectangular enclosure with one predominant dimension, by way of an equivalent radius defined as $R = ab/(a + b)$, where a and b are the transversal dimensions of the rectangular cross section.

The SE of a cylindrical shell, with radius $\rho_0 = 30$ cm under a uniform parallel magnetic field, is shown in Figure B.9. The material of the shell is an iron-nickel alloy with $\mu_\mathrm{r} = 75 \cdot 10^3$, $\sigma = 2 \cdot 10^6$ S/m, and the shield thickness is $\Delta = 0.15$ mm. Also in this case the same observations made for the spherical shell apply. This result does not surprise because even if the geometry changes, the shielding mechanisms remain always the same.

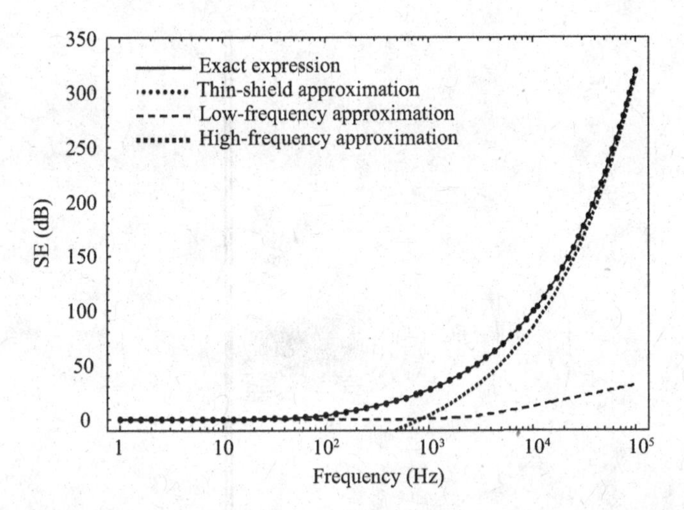

FIGURE B.9 SE of a cylindrical shell ($\rho_0 = 30$ cm, $\Delta = 0.15$ mm) in a transverse uniform magnetic field. Comparison among the exact expression and different approximations in the case of a shell of an iron-nickel alloy ($\mu_r = 75 \cdot 10^3$, $\sigma = 2 \cdot 10^6$ S/m).

B.3.4 Infinite Plane

The infinite planar shield has been studied as a canonical geometry for the design of EM shields. The shield consists of an infinite planar sheet with thickness Δ, with large values of the conductivity σ, and/or of the relative magnetic permeability μ_r. The sheet completely separates the shielded region from the source region where a low-frequency low-impedance field is produced by magnetic sources (e.g., a wire loop, a line current, or a magnetic dipole). Although the shield is of infinite extent, some field lines penetrate through the shield as shown in Figure B.10, depending on the material parameters and the operating frequency.

In the following discussion exact solutions of canonical problems will be briefly presented together with some approximate solutions that are wieldier but often give an insight into the shielding mechanisms.

Parallel Loop

The solution of the shielding problem when a current loop is parallel to the screen (considered of infinite extent) is carried out through the use of the magnetic vector potential **A**. The geometry of the problem, shown in Figure B.11, suggests that we look for the solution in a cylindrical reference coordinate system whose origin is placed at the center of the transmitting loop. Because of the symmetry of the problem, the magnetic vector potential has only the ϕ component, which is independent of the coordinate ϕ (at low frequencies the loop radius R is in fact assumed to be much smaller than the operating wavelength). Consequently any propagation of the current on the loop wire is neglected, and the current is approximated as uniform with a constant value I along the loop. Furthermore for the

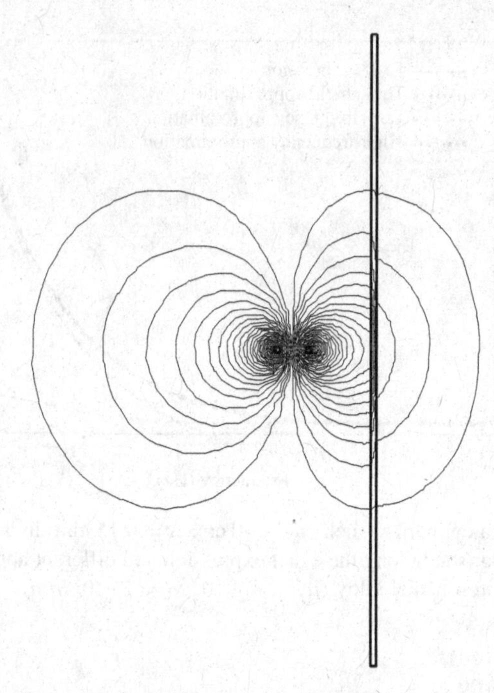

FIGURE B.10 Magnetic-induction distribution in the presence of an infinite planar metallic shield illuminated by the magnetic field of a current pair.

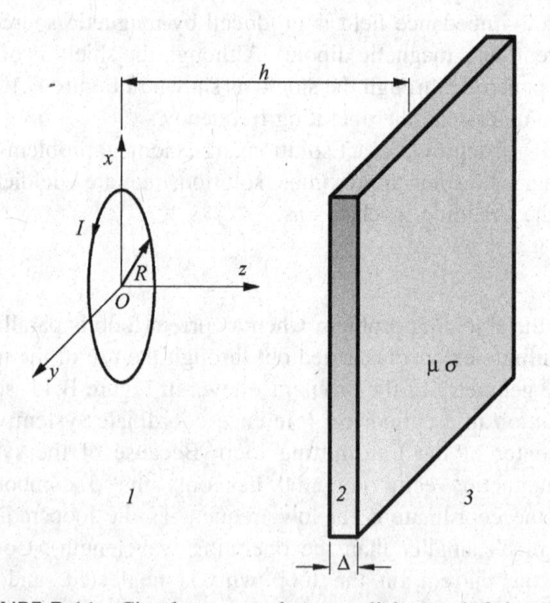

FIGURE B.11 Circular current loop parallel to an infinite plate.

magnetic quasi-stationary approximation the displacement currents are neglected in the highly conductive metallic shield.

By the foregoing assumptions, the wave equations that describe the problem in air and inside the shield can be written as

$$\begin{aligned}\nabla^2 A_\phi + k_0^2 A_\phi &= 0,\\ \nabla^2 A_\phi - \gamma^2 A_\phi &= 0,\end{aligned} \tag{B.51}$$

where k_0 is the free-space wavenumber, $\gamma \cong \sqrt{j\omega\mu_0\mu_r\sigma}$ is the propagation constant inside the conductive plane, and

$$\nabla^2 A_\phi = \left(\frac{\partial^2}{\partial\rho^2} + \frac{1}{\rho}\frac{\partial}{\partial\rho} + \frac{\partial^2}{\partial z^2}\right) A_\phi. \tag{B.52}$$

After the separation of variables is applied, the general solution of (B.51) in the three regions of space shown in Figure B.11 can be written as [17,18]

$$A_\phi^{(1)}(\rho, z) = \frac{\mu_0 RI}{2} \int_0^{+\infty} \frac{\lambda}{\tau_0} J_1(\lambda R) J_1(\lambda\rho)[e^{-\tau_0|z|} + k_1(\lambda)e^{+\tau_0 z}]\mathrm{d}\lambda, \tag{B.53a}$$

$$A_\phi^{(2)}(\rho, z) = \frac{\mu_0 RI}{2} \int_0^{+\infty} \frac{\lambda}{\tau} J_1(\lambda R) J_1(\lambda\rho)[k_2(\lambda)e^{-\tau z} + k_3(\lambda)e^{+\tau z}]\mathrm{d}\lambda, \tag{B.53b}$$

$$A_\phi^{(3)}(\rho, z) = \frac{\mu_0 RI}{2} \int_0^{+\infty} \frac{\lambda}{\tau_0} J_1(\lambda R) J_1(\lambda\rho) k_4(\lambda)e^{-\tau_0 z}\mathrm{d}\lambda, \tag{B.53c}$$

where $J_1(\cdot)$ is the first-order Bessel function of the first kind, while $\tau_0 = \sqrt{\lambda^2 - k_0^2}$ and $\tau = \sqrt{\lambda^2 - \gamma^2}$. In (B.53a) the first term is the vector potential impressed by the loop without the shield. By enforcing the boundary conditions on the two surfaces of the sheet (i.e., continuity of the tangential electric field (A_ϕ) and continuity of the tangential magnetic field ($\partial A_\phi/\partial z$)), the four unknown coefficients $k_i(\lambda)$ (which are functions of λ) can be obtained. Once the expressions of the magnetic vector potential beyond the screen (i.e., in region 3), with and without the shield, have been obtained, the z component of the magnetic induction can be obtained at the observation point to evaluate the SE as

$$\mathrm{SE} = \left|\frac{1}{4\mu_r} \frac{\int_0^{+\infty} \lambda^2 \tau_0^{-1} J_1(\lambda R) J_0(\lambda\rho) e^{-\tau_0 z}\mathrm{d}\lambda}{\int_0^{+\infty} K\lambda^2 \tau\tau_0^{-2} J_1(\lambda R) J_0(\lambda\rho) e^{-\tau_0 z - (\tau-\tau_0)\Delta}\mathrm{d}\lambda}\right|, \tag{B.54}$$

with

$$K = \left[\left(\frac{\tau}{\tau_0} + \mu_r\right)^2 - \left(\frac{\tau}{\tau_0} - \mu_r\right)^2 e^{-2\tau\Delta}\right]^{-1}. \tag{B.55}$$

Note the SE independence of the source-to-shield spacing. This is an interesting result, and it has been shown to be consistent with experimental results [17–20]. When $\rho = 0$, the numerator of (B.54) can be evaluated analytically, giving

$$\left| \left(jk_0 R + \frac{R}{\sqrt{R^2+z^2}} \right) \frac{e^{-jk_0\sqrt{R^2+z^2}}}{R^2+z^2} \right|, \tag{B.56}$$

which can be simplified at very low frequencies (i.e., in the static limit) as $R(R^2+z^2)^{-3/2}$.

Bannister [19,20] gave interesting approximate formulas for the SE of coaxial loops for electrically thin and thick walls. In the low-frequency case, two quasi–near approximations are introduced: when the measurement distance $L = \sqrt{R^2 + (z-\Delta)^2}$ is much smaller than the operating wavelength (i.e., $L \ll \lambda_0$), the propagation constant in air can be neglected in the computation of τ_0, which thus coincides with the integration variable λ; when the measurement distance is much greater than the skin depth in the shield (i.e., $L \gg \delta$) and the shield is thicker than twice the skin depth (i.e., $\Delta > 2\delta$), the integration variable λ can be neglected in the computation of τ, which becomes equal to the propagation constant γ. By these assumptions, (B.54) with $\rho = 0$ can be analytically evaluated, resulting in

$$\mathrm{SE_{dB}} = 8.686\frac{\Delta}{\delta} + 20\log\left[\frac{L}{8.485\mu_r\delta}\frac{L}{(z-\Delta)}\left(\frac{L}{\sqrt{R^2+z^2}}\right)^3\right] \tag{B.57}$$

under the additional restriction $L \gg \delta\mu_r$. On the other hand, by the assumption $L \ll \delta\mu_r$, it results in

$$\mathrm{SE_{dB}} = 8.686\frac{\Delta}{\delta} + 20\log\left[\frac{\mu_r\delta L(L+z-\Delta)}{5.657(R^2+z^2)^{3/2}}\right]. \tag{B.58}$$

Interestingly the first term in the previous equations is exactly the absorption term A in the TL theory of planar shields (see Chapter 4). Consequently the second term can be identified as the reflection coefficient R, which depends on the mismatch between the characteristic impedance of the impinging wave and the intrinsic impedance of the wave inside the conducting material.

In the high-frequency range, the measurement distance L is comparable with the operating wavelength, so the first quasi–near approximation is no longer valid. However, the second quasi–near approximation can still be applied (i.e., $L \gg \delta$). Furthermore the additional restriction $L \gg \delta\mu_r$ is assumed, which is valid in most cases at high frequencies. By these assumptions, (B.54) reduces to

$$\mathrm{SE_{dB}} = 8.686\frac{\Delta}{\delta} + 20\log\left|\frac{L}{2.828\mu_r\delta}\frac{L}{(z-\Delta)}\left(\frac{L}{\sqrt{R^2+z^2}}\right)^3 \frac{1+jk_0\sqrt{R^2+z^2}}{3+j3k_0L-k_0^2L^2}\right|. \tag{B.59}$$

Other approximate formulas have been obtained by Dahlberg [21,22]. Under the restriction $L \gg \delta\mu_r$ and $L \gg \delta^2\mu_r/\Delta$ (i.e., when the eddy-current cancellation mechanism is dominant), the SE is

$$\mathrm{SE} = \left| j\gamma \frac{\sinh(\gamma\Delta)}{2\mu_r} \frac{(R^2 + z^2)}{3z} \right|. \tag{B.60}$$

When instead $L \ll \delta\mu_r$ and $L \ll \delta^2\mu_r/\Delta$ (i.e., when the flux-shunting mechanism is dominant), the SE reduces to

$$\mathrm{SE} = \left| \mu_r \frac{\sinh(\gamma\Delta)}{4\gamma} \frac{\sqrt{1 + (z/R)^2} + z/R}{[1 + (z/R)^2]} \right|. \tag{B.61}$$

The SE in the parallel-loop configuration is reported in Figure B.12 for a geometry with $R = 17.25$ cm, $h = 30.5$ cm, and $z = 61$ cm. Three materials have been selected for the analysis: one having $\mu_r = 1$ and $\sigma = 54 \cdot 10^6$ S/m (annealed copper), one having $\mu_r = 1$ and $\sigma = 28 \cdot 10^6$ S/m (household foil aluminum), and one with $\mu_r = 200$ (independent of frequency) and $\sigma = 9 \cdot 10^6$ S/m (commercial 1010 low-carbon steel). All the shields are assumed to have a thickness $\Delta = 0.5$ mm. The exact formulation has been compared with several approximations, some of which will be presented next. First of all, there should be noted the excellent agreement of the small-dipole approximation of the source (see equation (B.67) below). The Bannister approximation is accurate only in the very high-frequency range, while Dahlberg approximation shows a better accuracy in the whole frequency window. The TL approach is accurate only if the correct wave impedance is used [23,24].

Perpendicular Loop

The perpendicular-loop shielding problem, shown in Figure B.13, is solved by means of the magnetic vector potential of the second-order $\mathbf{W}$, which is related to the magnetic vector potential $\mathbf{A}$ through $\mathbf{A} = \nabla \times \mathbf{W}$. The complete expression of $\mathbf{W}$ is

$$\mathbf{W} = \mathbf{u}W_1 + \mathbf{u} \times \nabla W_2, \tag{B.62}$$

where W_1 and W_2 are two scalar functions and $\mathbf{u}$ is an arbitrary vector that is set equal to the unit vector $\mathbf{u}_y$ (according to the adopted reference system, it is in fact easy to show that the magnetic induction in air does not depend on the x and z components of $\mathbf{W}$). Equation (B.62) describes a field that is the sum of two terms, one transverse electric and the other transverse magnetic with respect to the unit vector $\mathbf{u}$. Under the quasi-stationary approximation, with $\nabla \cdot \mathbf{A} = 0$ assumed, the two scalar functions W_i $(i = 1, 2)$ satisfy the following two differential equations:

$$\nabla^2 W_i = 0, \tag{B.63a}$$
$$\nabla^2 W_i = \gamma^2 W_i, \tag{B.63b}$$

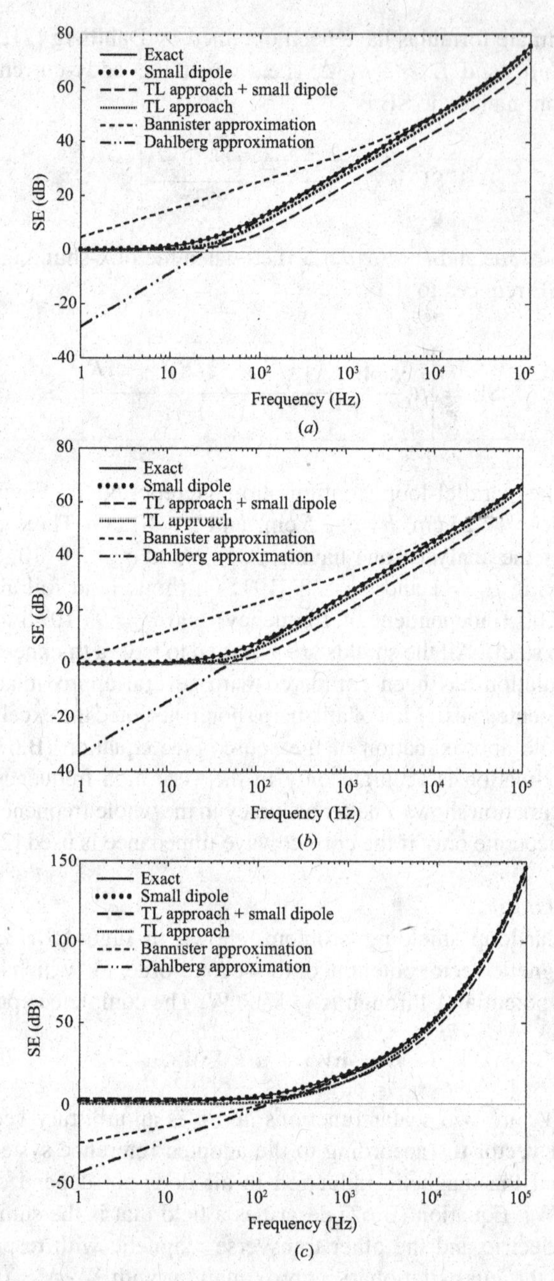

FIGURE B.12 Theoretical SE as a function of frequency of an infinite planar metallic shield when a current loop source parallel to the screen is considered. Comparison among exact results and different approximations for a screen of thickness $\Delta = 0.5$ mm: (*a*) copper screen; (*b*) aluminum screen; (*c*) low-carbon steel. Other parameters: $R = 17.25$ cm, $h = 30.5$ cm, $z = 61$ cm.

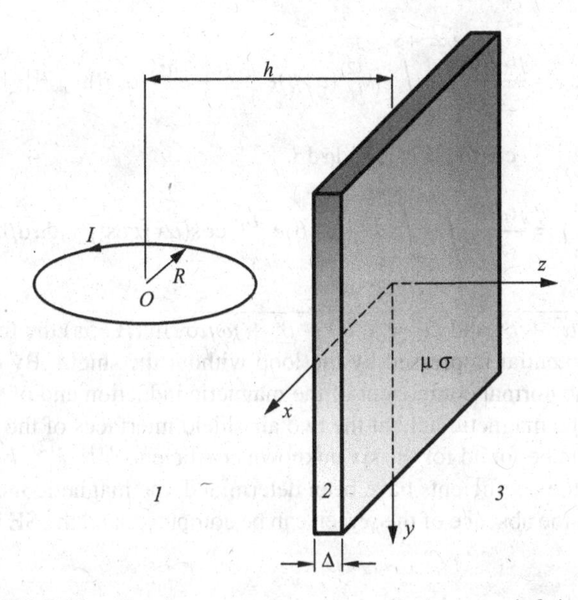

FIGURE B.13 Circular current loop perpendicular to an infinite plate.

in free space and in the conducting medium, respectively. As a consequence the magnetic induction is defined as

$$\mathbf{B} = \nabla \times (\nabla \times \mathbf{W}) = \left(\frac{\partial^2 W_1}{\partial x \partial y} + j\omega\mu_0\mu_r\sigma \frac{\partial W_2}{\partial z} \right) \mathbf{u}_x + \left(\frac{\partial^2 W_1}{\partial y^2} + j\omega\mu_0\mu_r\sigma W_2 \right) \mathbf{u}_y + \left(\frac{\partial^2 W_1}{\partial y \partial z} - j\omega\mu_0\mu_r\sigma \frac{\partial W_2}{\partial x} \right) \mathbf{u}_z, \quad \text{(B.64)}$$

which clearly shows that W_2 does not contribute to the magnetic induction in free space and therefore can be removed from there. The field radiated by a loop is in fact transverse electric with respect to its axis so that only W_1 is required, but W_2 must be accounted for in the conducting medium.

As for the coaxial shielding problem, the current I in the source antenna is assumed to be independent on the angular position and constant. Moreover an important simplification in the mathematical analysis is obtained in the very low-frequency range, where it can be assumed that $\exp(-jk_0r) \simeq 1$. By these assumptions and with use of the double Fourier integral [25], the following expressions for the second-order potential in the three regions of space can be obtained:

$$W_1^{(1)}(x, y, z) = \frac{\mu_0 IR}{\pi} \int_0^{+\infty} \int_0^{+\infty} \left[-\frac{I_1(\beta R)}{\beta D_1} e^{-D_1|z+d|} + k^{(1)}(\alpha, \beta) e^{+D_1 z} \right] \cdot \cos(\alpha x) \cos(\beta y) d\alpha d\beta, \quad \text{(B.65a)}$$

$$W_i^{(2)}(x,y,z) = \frac{\mu_0 IR}{\pi} \int_0^{+\infty}\int_0^{+\infty} [k_i^{(2)}(\alpha,\beta)e^{-D_2 z} + k_i'^{(2)}(\alpha,\beta)e^{+D_2 z}] \cdot \cos(\alpha x)\cos(\beta y) d\alpha d\beta, \tag{B.65b}$$

$$W_1^{(3)}(x,y,z) = \frac{\mu_0 IR}{\pi} \int_0^{+\infty}\int_0^{+\infty} k^{(3)}(\alpha,\beta)e^{-D_1 z}\cos(\alpha x)\cos(\beta y) d\alpha d\beta, \tag{B.65c}$$

where $D_1 = \sqrt{\alpha^2+\beta^2}$ and $D_2 = \sqrt{\alpha^2+\beta^2+j\omega\mu\sigma}$. In (B.65a) the first term is the second-order potential impressed by the loop without the shield. By enforcing the continuity of the normal component of the magnetic induction and of the tangential component of the magnetic field at the two air-shield interfaces of the plane screen, six equations can be found for the six unknown coefficients $k^{(1)}$, $k_1^{(2)}$, $k_2^{(2)}$, $k_1'^{(2)}$, $k_2'^{(2)}$ and $k^{(3)}$. Once the coefficients have been determined, the magnetic induction in the presence and in the absence of the screen can be computed, and the SE is obtained as [26,27]

$$\mathrm{SE} = \left| \frac{\int_0^{+\infty}\int_0^{+\infty} [\beta I_1(\beta R)/D_1]\cos(\alpha x)\cos(\beta y)e^{-D_1|z-h|} d\alpha d\beta}{\int_0^{+\infty}\int_0^{+\infty} [4\beta I_1(\beta R)D_3 e^{-(D_2-D_1)\Delta}/D_4]\cos(\alpha x)\cos(\beta y)e^{-D_1|z-h|} d\alpha d\beta} \right|, \tag{B.66}$$

where $D_3 = D_2/\mu_r$ and $D_4 = (D_3+D_1)^2 - (D_3-D_1)^2 e^{-2D_2\Delta}$. Note that unlike the SE of the coaxial-loop configuration, the SE is here a function of the shield distance d from the transmitting loop. Further details and discussions on the low-frequency coplanar-loop shielding configuration can be found in [26,27].

By the small-dipole approximation the transmitting loop can be modeled as a small dipole whose magnetic moment vector is $\mathbf{M} = \mu_0 \pi R^2 I \mathbf{u}$. By representing the fields produced by the elemental dipole source in the spectral domain [28,29], the SE in the parallel-loop configuration can be computed as

$$\mathrm{SE} = \left| \frac{\int_0^{+\infty} \xi^3 \zeta^{-1} e^{-\zeta k_0(z+d)} J_0(\xi k_0 \rho) d\xi}{\int_0^{+\infty} T_{\mathrm{TE}}(\xi)\xi^3\zeta^{-1} e^{-\zeta k_0(z+d)} J_0(\xi k_0\rho) d\xi} \right|, \tag{B.67}$$

while in the perpendicular-loop configuration there results

$$\mathrm{SE} = \left| \frac{\int_0^{+\infty} [\xi\zeta^{-1}J_+(\xi k_0\rho) - \xi\zeta J_-(\xi k_0\rho)]e^{-\zeta k_0(z+d)} d\xi}{\int_0^{+\infty} [T_{\mathrm{TM}}(\xi)\xi\zeta^{-1}J_+(\xi k_0\rho) - T_{\mathrm{TE}}(\xi)\xi\zeta J_-(\xi k_0\rho)]e^{-\zeta k_0(z+d)} d\xi} \right|, \tag{B.68}$$

where $J_\pm(\cdot) = [J_0(\cdot) \pm J_2(\cdot)]/2$. A cylindrical coordinate system has been adopted in both (B.67) and (B.68); in each case the z axis is perpendicular to the plane in which the loop lies and ρ indicates the distance of the observation point from the origin (placed at the center of the loop) on the plane of the loop. In the previous expressions the plane-wave TM and TE transmission coefficients were defined as

$$T_{\mathrm{TM}}(\xi) = \frac{e^{(\zeta-\zeta_{\mathrm{m}})k_0\Delta}}{1 - [(\dot{\varepsilon}_{\mathrm{r}}\zeta - \zeta_{\mathrm{m}})^2/4\dot{\varepsilon}_{\mathrm{r}}\zeta\zeta_{\mathrm{m}}](e^{-2\zeta_{\mathrm{m}}k_0\Delta} - 1)}, \tag{B.69a}$$

$$T_{\mathrm{TE}}(\xi) = \frac{e^{(\zeta-\zeta_{\mathrm{m}})k_0\Delta}}{1 - [(\mu_{\mathrm{r}}\zeta - \zeta_{\mathrm{m}})^2/4\mu_{\mathrm{r}}\zeta\zeta_{\mathrm{m}}](e^{-2\zeta_{\mathrm{m}}k_0\Delta} - 1)}, \tag{B.69b}$$

with $\zeta = \sqrt{\xi^2 - 1}$, $\zeta_{\mathrm{m}} = \sqrt{\xi^2 - \mu_{\mathrm{r}}\varepsilon_{\mathrm{r}}}$, and $\dot{\varepsilon}_{\mathrm{r}} \cong (-j\sigma)/(\omega\varepsilon_0)$ under the quasi-static assumption.

The SE in the perpendicular-loop configuration is reported in Figure B.14 for a geometry with $R = 17.25$ cm, $h = 30.5$ cm, and $z = h$. The same materials as those used for the parallel-loop configuration were considered. The exact formulation was compared with several approximations; the results obtained through the small-dipole approximation turned out to be in this case also in excellent agreement with the exact ones, whereas the TL approach is reasonably accurate only if the correct wave impedance is used.

Horizontal Current Line

The shield is illuminated by the magnetic field radiated by a single infinitely long line current that runs parallel to the shield, as shown in Figure B.15. The line carries a constant current I that is independent of the z coordinate because no propagation

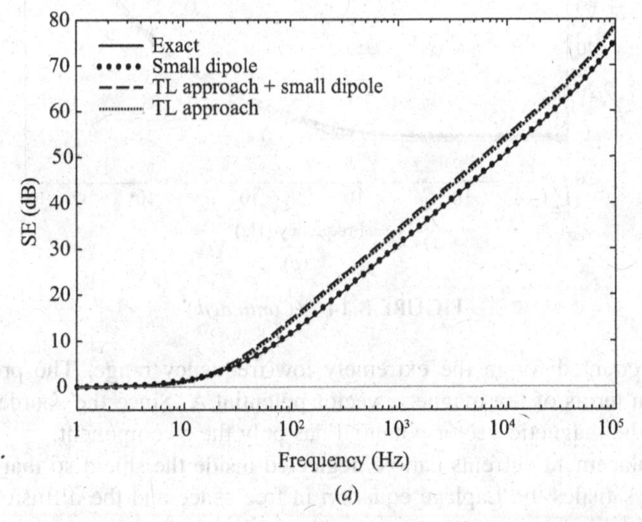

FIGURE B.14 Theoretical SE as a function of frequency of an infinite planar metallic shield when a current loop source perpendicular to the shield is considered. Comparison among the exact formulation and different approximations for a screen with thickness $\Delta = 0.5$ mm: (*a*) copper screen; (*b*) aluminum screen; (*c*) low-carbon steel. Other parameters: $R = 17.25$ cm, $h = 30.5$ cm, and $z = h$.

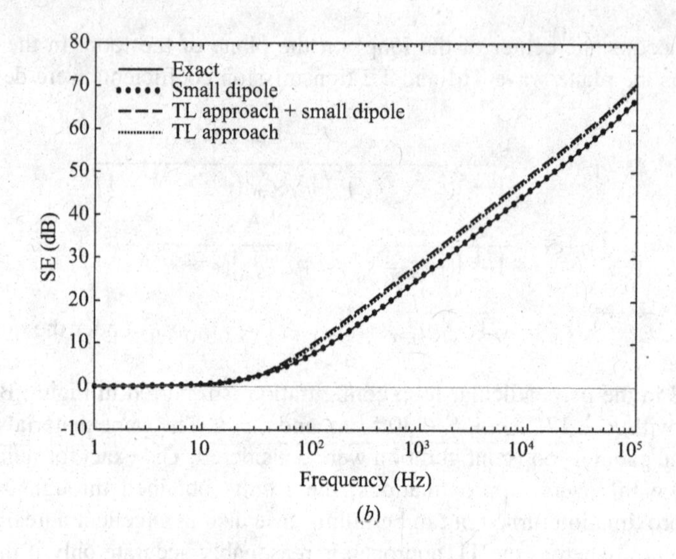

(*b*)

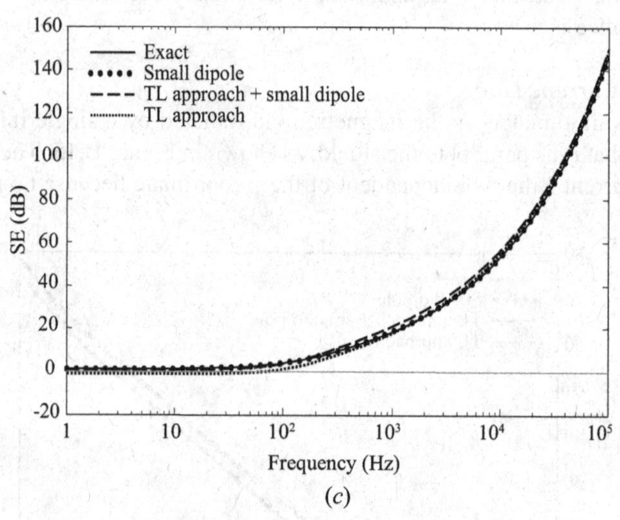

(*c*)

FIGURE B.14 (*Continued*)

effect is accounted for in the extremely low-frequency range. The problem can be solved in terms of the magnetic vector potential **A**. Since the source current is z-directed, the magnetic vector potential has only the z component.

The displacement currents can be neglected inside the shield so that the vector potential **A** satisfies the Laplace equation in free space and the diffusion equation inside the shield:

$$\nabla^2 A_z = 0, \qquad y < 0, y > \Delta, \tag{B.70a}$$

$$\nabla^2 A_z - \gamma^2 A_z = 0, \qquad 0 < y < \Delta. \tag{B.70b}$$

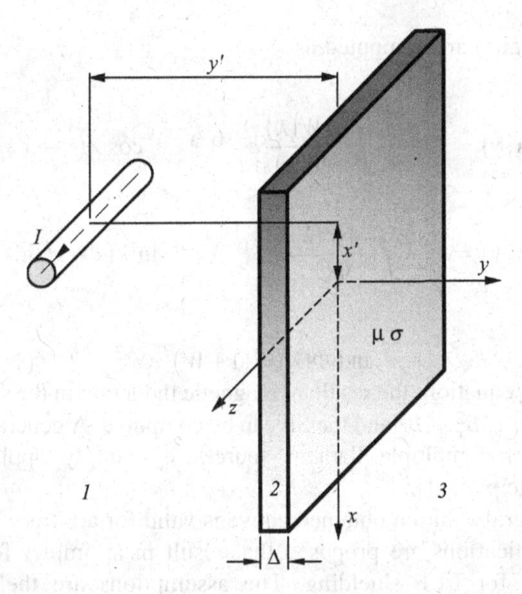

FIGURE B.15 Horizontal current wire parallel to an infinite plate.

By applying the separation of variables and by introducing the Fourier spatial transform with respect to the x variable [30], we obtain the following expressions for the magnetic vector potential in the three regions:

$$A_z^{(1)}(x,y) = \int_0^{+\infty} \left[\frac{\mu_0 I}{2\pi} e^{-\lambda(y-y')} + k_1(\lambda) e^{+\lambda y}\right] \cos[\lambda(x-x')] \frac{1}{\lambda} \mathrm{d}\lambda, \tag{B.71a}$$

$$A_z^{(2)}(x,y) = \int_0^{+\infty} [k_2(\lambda) e^{-\sqrt{\gamma^2+\lambda^2} y} - k_3(\lambda) e^{+\sqrt{\gamma^2+\lambda^2} y}] \cos[\lambda(x-x')] \frac{1}{\lambda} \mathrm{d}\lambda, \tag{B.71b}$$

$$A_z^{(3)}(x,y) = \int_0^{+\infty} k_4(\lambda) e^{-\lambda y} \cos[\lambda(x-x')] \frac{1}{\lambda} \mathrm{d}\lambda, \tag{B.71c}$$

where k_1, k_2, k_3, and k_4 are unknown coefficients. In (B.71a) the first term is the magnetic vector potential impressed by the line current without the shield which results in the well-know Biot-Savart's expression $\mathbf{B} = \mathbf{u}_\phi \mu_0 I/(2\pi\rho)$. The four unknown coefficients can be obtained by enforcing the boundary conditions on the two interfaces of the shield (i.e., the continuity of the tangent x component of the magnetic field vector and of the normal y component of the magnetic induction). Once the coefficients have been obtained, the components of the magnetic induction in the shielded region are computed as

$$B_x^{(3)}(x, y) = -\frac{\mu_0 I}{2\pi} \int_0^{+\infty} \frac{4W(\lambda)}{\Phi(\lambda)} e^{-\lambda(y-y'-\Delta)} \cos[\lambda(x-x')]\, d\lambda, \tag{B.72a}$$

$$B_y^{(3)}(x, y) = \frac{\mu_0 I}{2\pi} \int_0^{+\infty} \frac{4W(\lambda)}{\Phi(\lambda)} e^{-\lambda(y-y'-\Delta)} \sin[\lambda(x-x')]d\lambda, \tag{B.72b}$$

where $W(\lambda) = (\mu_r \lambda)/\sqrt{\lambda^2+\gamma^2}$ and $\Phi(\lambda) = (1+W)^2 e^{\sqrt{\lambda^2+\gamma^2}\Delta} - (1-W)^2 e^{-\sqrt{\lambda^2+\gamma^2}\Delta}$. From the previous equations the resultant magnetic induction in the shielded region is obtained as $|B_S| = \sqrt{B_x^2 + B_y^2}$ and the SE can be computed. A general expression can also be derived for multiple-filament sources by simply applying the linear superposition principle [31].

The exact integral solution obtained above is valid for arbitrary linear materials. In [30,31] simplifications are proposed that result in a simpler formula, without infinite integrals, for ELF shielding. The assumptions are the same as those previously described for the parallel loop: the displacement currents in the metallic shield are ignored, the distance L between the observation point and the wire is much smaller than the wavelength ($L \ll \lambda_0$), and the distance between the wire and the shield is much larger than the skin depth ($y' \gg \delta$, i.e., $\sqrt{\lambda^2 + \gamma^2} \simeq \gamma$).

For the case where the eddy-current shielding mechanism dominates (i.e., the screen is highly conductive) Olsen [31] reduced the resultant magnetic-induction expression to

$$B_\rho = \frac{\mu_0 I}{2\pi \rho_e} \frac{2\sqrt{2}\mu_r \delta}{(e^{\gamma\Delta} - e^{-\gamma\Delta})\rho_e}, \tag{B.73}$$

where $\rho_e = \sqrt{(\tilde{y} + \Delta_e)^2 - (x-x')^2}$ and $\Delta_e = -j\mu_r \delta^2/\Delta$ is a complex distance that becomes significant only when the distance $\tilde{y} = y + y' + \Delta$ between the source and the observation point is small. Interestingly we have in (B.73) a magnetic induction produced by a straight wire in the dc limit and corrected by a transmission coefficient that accounts for the eddy-current shielding mechanism.

When the flux-shunting mechanism dominates, no simple closed formula is available. A nearly exact formula that uses exponential integrals can be found in [31]. It is simpler than (B.72) from a computation point of view, but it does not lead to any physical interpretation of the result.

The SE of an infinite metallic plane exposed to a parallel current pair is reported in Figure B.16; the current is assumed along the y axis, the spacing between the wires is $s = 0.4$ m, the distance of both the wires from the shield is $y' = 1$ m, and $y = 0.3$ m. The same shield materials previously introduced for the loop configurations have been considered again. The exact formulation is compared with several approximations. The Olsen approximation is fairly accurate, except at

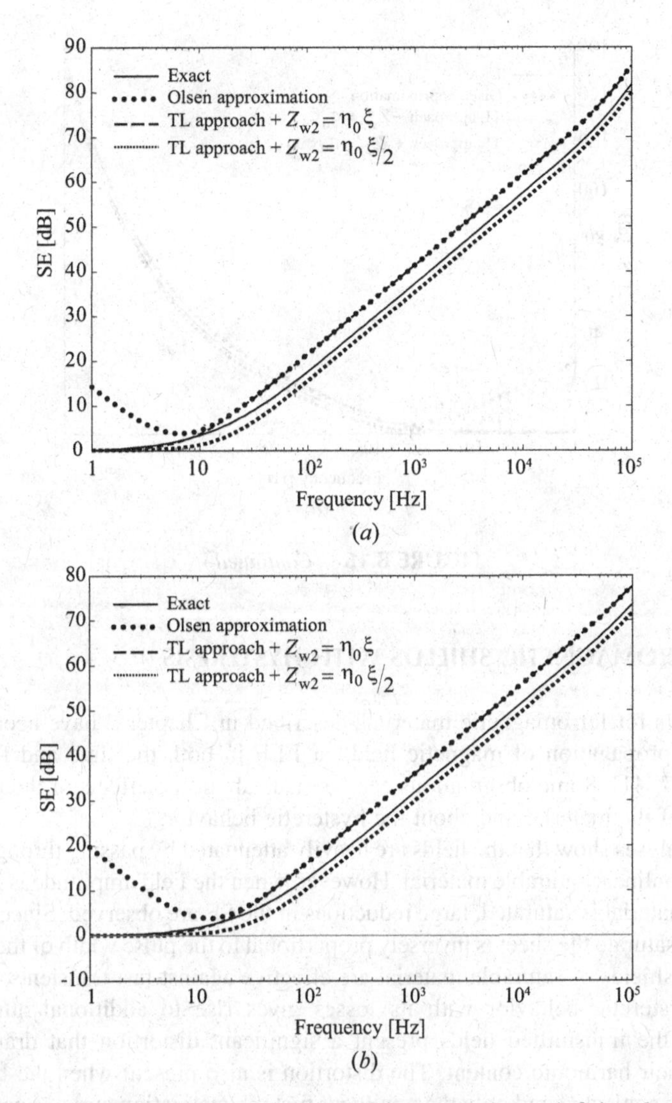

FIGURE B.16 Theoretical SE as a function of frequency for an infinite planar metallic shield in the presence of two infinitely long current-line sources placed along the y axis. Comparison among exact results and different approximations for a screen of thickness $\Delta = 0.5$ mm: (*a*) copper screen; (*b*) aluminum screen; (*c*) low-carbon steel. Other parameters: $s = 0.4$ m, $y' = 1$ m, and $y = 0.3$ m.

very low frequencies where the complex distance factor is no longer able to account for the effects induced in the screen due to the small electrical distance of the source. The TL approach is reasonably accurate only if the correct wave impedance is used. It is interesting that if the wave impedance $Z_{w2} = \eta_0 y'/2$ (proposed by Olsen [31] through an analogy between the exact solution and the TL theory, see Section 4.5) is used, the accuracy of the TL approach is greatly improved.

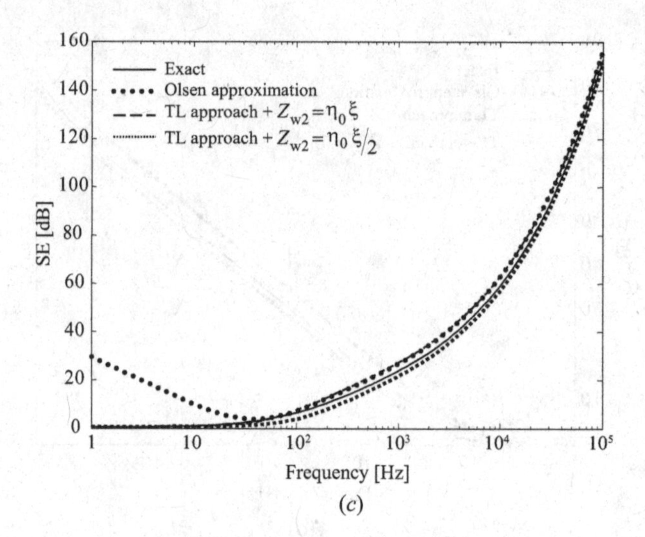

(*c*)

FIGURE B.16 (*Continued*)

B.4 FERROMAGNETIC SHIELDS WITH HYSTERESIS

The models for ferromagnetic materials described in Chapter 2 have been used to study the propagation of magnetic fields at ELF in both the time and frequency domain [37–41]. Some observations can be made about the effects of the nonlinear behavior of the material and about the hysteretic behavior.

The analyses show that the fields are heavily attenuated by passing through a sheet made of nonlinear saturable material. However, when the field amplitude is increased until the material is saturated, large reductions in the SE are observed. Since the level needed to saturate the sheet is inversely proportional to the pulse width of the incident field, thin shields of saturable material are effective against fast transients.

The hysteretic behavior with its losses gives rise to additional attenuation. However, the transmitted fields present a significant distortion that dramatically changes their harmonic content. The distortion is also present when the hysteretic behavior is neglected and only the nonlinear first magnetization curve is considered, but in this case it is much less prominent. The distortion effect generally gives rise to a marked third harmonic that must be properly accounted for when the rise of a high-frequency content in the transmitted spectrum can disturb protected equipments.

REFERENCES

[1] A. P. Wills. "On the magnetic shielding effect of tri-lamellar spherical and cylindrical shells." *Phys. Rev.*, vol. 9, pp. 193–213, Oct. 1899.

[2] T. E. Sterne. "Multi-lamellar cylindrical magnetic shields." *Rev. Sci. Inst.*, vol. 6, pp. 324–326, Oct. 1935.

[3] S. Levy. "Electromagnetic shielding of an infinite conducting plane conducting sheet placed between circular coaxial cables." *Proc. IRE*, vol. 21, pp. 923–941, Jun. 1936.

[4] L. V. King. "Electromagnetic shielding at radio frequencies." *Phil. Mag. J. Sci.*, *VII*, vol. 15, pp. 201–223, Feb. 1933.

[5] T. S. Perry. "Today's view of magnetic fields." *IEEE Spectrum*, pp. 14–23, Dec. 1994.

[6] World Health Organization. "Non-ionizing radiation, Part 1: Static and extremely-low-frequency (ELF) electric and magnetic fields." In *IARC Monograph on the Evaluation of Carcinogenic Risks to Humans*, vol. 80. Lyon, France: WHO/IARC, IARC Press, 2002.

[7] J. F. Hoburg. "Principles of quasistatic magnetic shielding with cylindrical and spherical shields." *IEEE Trans. Electromagn. Compat.*, vol. 37, no. 4, pp. 574–579, Nov. 1995.

[8] L. Hasselgren and J. Luomi. "Geometrical aspects of magnetic shielding at extremely low frequency." *IEEE Trans. Electromagn. Compat.*, vol. 37, no. 3, pp. 409–420, Aug. 1995.

[9] R. G. Olsen. "On low frequency shielding of electromagnetic fields." *Proc. 10th Int. Symp. High Voltage Engineering*, 25–29 August 1997. Montreal, Canada, pp. 1–12.

[10] I. Hasselgren and Y. Hamnerius. "Calculation of low-frequency magnetic shielding of a substation using a two-dimensional finite-element method." *European Trans. Elect. Power Eng.*, vol. 5, no. 2, pp. 81–90, Mar. 1995.

[11] O. Biro and K. Preis. "Finite element analysis of 3-D eddy currents." *IEEE Trans. Magn.*, vol. 26, no. 2, pp. 418–423, Mar. 1990.

[12] S. A. Schelkunoff. *Electromagnetic Waves.* New York: Van Nostrand, 1943.

[13] D. A. Miller and J. E. Bridges. "Review of circuit approach to calculate shielding effectiveness." *IEEE Trans. Electromagn. Compat.*, vol. 10, no. 1, pp. 52–62, Mar. 1968.

[14] J. E. Bridges. "An update of the circuit approach to calculate shielding effectiveness." *IEEE Trans. Electromagn. Compat.*, vol. 30, no. 3, pp. 211–221, Aug. 1988.

[15] H. Kaden. *Wirbelströmeund Schirmung in der Nachrichtentechnick.* Berlin: Springer, 1959.

[16] J. R. Wait and D. A. Hill. "Electromagnetic shielding of sources within a metal-cased bore hole." *IEEE Trans. Geosci. Electr.*, vol. 15, pp. 108–112, Apr. 1977.

[17] J. R. Moser. "Low-frequency shielding of a circular loop electromagnetic field source." *IEEE Trans. Electromag. Compat.*, vol. 9, no. 1, pp. 6–18, Jan. 1967.

[18] J. R. Moser. "Low-frequency low-impedance electromagnetic shielding." *IEEE Trans. Electromagn. Compat.*, vol. 30, no. 3, pp. 202–210, Aug. 1988.

[19] P. Bannister. "New theoretical expressions for predicting shielding effectiveness for the plane shield case." *IEEE Trans. Electromag. Compat.*, vol. 10, no. 1, pp. 2–7, Oct. 1968.

[20] P. Bannister. "Further notes for predicting shielding effectiveness for the plane shield case." *IEEE Trans. Electromagn. Compat.*, vol. 11, no. 5, pp. 50–53, May 1969.

[21] E. Dahlberg. "On shielding estimates for magnetic and non-magnetic shield materials." *Dept. of Plasma Physics, Royal Institute of Technology, Stockholm*, Tech. Rep. TRITA-EPP-75-05, pp. 21–26, 1975.

[22] E. Dahlberg. "Electromagnetic shielding, some simple formulae for closed uniform shields." *Dept. of Plasma Physics, Royal Institute of Technology, Stockholm*, Tech. Rep. TRITA-EPP-75-27, pp. 7–14, 1975.

[23] A. C. D. Whitehouse. "Screening: New wave impedance for the transmission line analogy." *IEE Proc.*, vol. 116, no. 7, pp. 1159–1164, Jul. 1969.

[24] S. Celozzi and M. D'Amore. "Shielding performance of ferromagnetic cylindrical cans." *Proc. 1996 IEEE Int. Symp. Electromagnetic Compatibility.* S. Clara, CA, 19–23 Aug. 1996, pp. 95–100.

[25] J. A. Tegopoulos and E. E. Kriezis. *Eddy Currents in Linear Conducting Media.* Amsterdam: Elsevier, 1985.

[26] S. Celozzi. "Shielding effectiveness prediction in MIL-STD 285 loop source configuration." *Proc. 1998 IEEE Int. Symp. Electromagnetic Compatibility.* Denver, CO, 24–28 Aug. 1998, pp. 1125–1130.

[27] R. Araneo and S. Celozzi. "Exact solution of low-frequency coplanar loops shielding configuration." *IEE Proc. Sci. Meas. Technol.*, vol. 149, no. 1, pp. 37–44, Jan. 2002.

[28] R. Yang and R. Mittra. "Coupling between two arbitrarily oriented dipoles through multilayered shields." *IEEE Trans. Electromagn. Compat.*, vol. 27, no. 3, pp. 131–136, Mar. 1985.

[29] A. Nishikata and A. Sugiura. "Analysis for electromagnetic leakage through a plane shield with an arbitrarily-oriented dipole source." *IEEE Trans. Electromagn. Compat.*, vol. 34, no. 8, pp. 284–291, Aug. 1992.

[30] Y. Du, T. C. Cheng, and A. S. Farag. "Principles of power frequency magnetic field shielding with flat sheets in a source of long conductors." *IEEE Trans. Electromagn. Compat.*, vol. 38, no. 8, pp. 450–459, Aug. 1996.

[31] R. G. Olsen, M. Istenic, and P. Zunko. "On simple methods for calculating ELF shielding of infinite planar shields." *IEEE Trans. Electromagn. Compat.*, vol. 45, no. 3, pp. 538–547, Aug. 2003.

[32] H. E. Knoepfel. *Magnetic Fields: A Comprehensive Theoretical Treatise for Practical Use.* New York: Wiley, 2000.

[33] M. Abramowitz and I. A. Stegun. *Handbook of Mathematical Functions with Formulas, Graphs, and Mathematical Tables*, 9th ed. New York: Dover, 1972.

[34] A. K. Thomas. "Magnetic shielded enclosure design in the DC and VLF region." *IEEE Trans. Electromagn. Compat.*, vol. 10, no. 1, pp. 142–152, Mar. 1968.

[35] C. W. Harrison. "Transient electromagnetic field propagation through infinite sheets, into spherical shells and into hollow cylinders." *IEEE Trans. Antennas Propagat.*, vol. 12, no. 5, pp. 319–334, May 1964.

[36] W. C. Cooley. "Low-frequency shielding effectiveness of nonuniform enclosures." *IEEE Trans. Electromagn. Compat.*, vol. 10, no. 1, pp. 34–43, Mar. 1968.

[37] D. E. Merewether. "Electromagnetic pulse transmission through a thin sheet of saturable ferromagnetic material of infinite surface area." *IEEE Trans. Electromagn. Compat.*, vol. 11, no. 4, pp. 139–143, Nov. 1969.

[38] S. Celozzi and M. D'Amore. "Magnetic field attenuation of nonlinear shields." *IEEE Trans. Electromagn. Compat.*, vol. 38, no. 3, pp. 318–326, Aug. 1996.

[39] R. Araneo and S. Celozzi. "On the evaluation of the shielding performance of ferromagnetic shields." *IEEE Trans. Magn.*, vol. 39, no. 2, pp. 1046–1052, Mar. 2003.

[40] T. Barbarics, A. Kost, D. Lederer, and P. Kis. "Electromagnetic field calculation for magnetic shielding with ferromagnetic material." *IEEE Trans. Magn.*, vol. 36, no. 4, pp. 986–989, Jul. 2000.

[41] F. Bertoncini, E. Cardelli, S. Di Fraia, and B. Tellini. "Analysis of shielding performance in nonlinear media." *IEEE Trans. Magn.*, vol. 38, no. 2, pp. 817–820, Mar. 2002.

APPENDIX C

Standards and Measurement Methods

Over the last few decades the proliferation of electrical and electronic products with their associated EMI problems have raised the necessity for the development and enforcement of mandatory EMC standards in many countries around the world. From an historical point of view, since World War II the defense sector pioneered in this area because the success of military missions greatly depended, and still depends, on the good performance of the adopted electronic/communication equipment, which must be free of EMI/EMC related problems. The civilian sector followed quickly, as many national governments understood the importance of protecting the RF spectrum from the unwanted EM noise emission of systems. Besides, the rapidly increasing use of radio services has led to dramatic increases in the proliferation of high-level EMIs in urban areas where, nowadays, electric/electronic equipment operate with levels of interfering fields of about 1 to 15 V/m or higher. Consequently standards have been promulgated throughout the years to ensure that the products marketed for use in residential, business, and industrial environments have a sufficient immunity to external EMI to enable them to operate as designed and have disturbing emissions sufficiently lower than a level that could prevent other apparatus from working as intended.

Making a comprehensive survey of the existing standards is a daunting task. Even if international organizations have brought out documents providing recommendations for implementation of EMC requirements, each country has its own recommendations in its national standards, with its set of test instruments, test procedures, and test limits, resulting in differences (sometimes unjustified).

The major international organizations are the International Standards Organization (ISO) and the International Electrotechnical Commission (IEC). Anyway,

Electromagnetic Shielding by Salvatore Celozzi, Rodolfo Araneo and Giampiero Lovat

besides their standards there are the standards published by the European Committee for Electrotechnical Standardization (CENELEC) in Europe, the Federal Communications Commission (FCC) in the United States, the Voluntary Control Council for Interference by Information Technology Equipment (VCCI) in Japan that are also considered unique international standards in the civilian sector. Added to these international standards are more or less international standards, civilian and military, that are published by several other organizations like the American National Standards Institute (ANSI), the Institute of Electric and Electronics Engineers (IEEE), the US Department of Defense (DoD), the National Security Agency (NSA), the International Telecommunication Union (ITU), the Society of Automotive Engineers (SAE), the European Telecommunications Standards Institute (ETSI), the Association for Electrical, Electronic and Information Technologies (VDE), the National Electrical Manufacturers Association (NEMA), the Energy Information Administration (EIA), the Radio Technical Commission for Aeronautics (RCTA), the American Society for Testing and Materials (ASTM), and the Society of Cable Telecommunications Engineers (SCTE). The list makes no claim to be exhaustive.

In addition, as mentioned above, each country has its own set of national standards. The position of the manufacturer becomes more difficult when a particular product is to be supplied to both civilian and military agencies. Different EMC standards specifying a variety of tests and standards, with similar tests but with different test instrumentation, have been bothering the industries by increasing the duration of the product development cycle and the testing costs. Today the harmonization of standards is being greatly demanded by many to reduce trade barriers among countries, especially as many of the major world economies like China, India, and Russia have started opening up.

Despite the plethora of standards it is possible to define some major classes of standards. Broadly speaking, the standards can be divided as basic standards and product standards.

Basic standards cover measurement methods for certain EMC phenomena and they should be referenced by product committees. These standards sometimes include a list of preferred test levels from which a product committee can choose the level most applicable to its product. *Product standards* define the test levels or the mandatory limits to be applied to a specific product in referring to the basic standards for the test methodologies. The distinction is clearly evident in the European Union (EU) where the norms are categorized as product/family product standards, generic standards (to be used for product types that have not product/ product family standards), and basic standards.

The major EMC standards on subjects related to shielding mainly deal with three subjects: shielding, radiated emission, and radiated immunity. On the whole, they provide measurement test procedures and methods as well as performance levels and maximum allowable limits. By the end of this appendix, the reader should have a clear sense of the serious problems to be met for a harmonization of the standards. There are differences among the large number of civilian standards from different organizations as well as often within the same organizations; there are differences between military environments and civilian environments, particularly as concerns prescribed levels and limits; there are differences in testing procedures and in the used equipment (e.g., a peak detector is used in military standards while quasi-peak or average detectors are used in civilian standards); and there are important differences in the considered frequency ranges (e.g., radiated electric field measurements start from 10 kHz in military standards while they usually start from 30 MHz in civilian standards).

C.1 MIL-STD 285 AND IEEE STD-299

The military standard MIL-STD 285 [1] was published in 1956 by the US Department of Defense for evaluating SE for military purposes. It remained unchanged until the Department canceled official support in 1997. Although the specification was originally a set of test methods for evaluating shielding enclosures of the mesh-screen variety, it has been quickly adopted for use on all types of facilities, becoming probably the most frequently referenced standard in the RF shielding industry. The document covers measurements within the frequency range from 100 kHz to 10 GHz. It defines the frequencies and the EM-field components that are subject to testing, and states the required antenna configuration and equipment. Although nowadays the standard is obsolete, the basic procedures for using the specified types of antennas and their separation distance are still adopted and are the standard in all testing. The figure of merit of an enclosure is the SE defined as the increase in the setting of the attenuator necessary to obtain the same reference reading level in the detector as when the shielding enclosure wall is removed. Thus the measurement consists of a reference and a shielding measurement, leaving the relative positions of the antennas unchanged.

Briefly, the standard calls for the signal source to be placed outside the tested enclosure, while the measurement device is located inside. The source may be driven with continuous wave (CW), modulated CW, or pulsed CW signals. The standard identifies three different field sources against which the SE may be defined, in accord with the following specifications and requirements:

1. Low-impedance magnetic fields. Loops 12 inches in diameter are to be used as transmitting and receiving antennas, spaced 12 inches from the shield walls. The test frequency must be one in the range 150 to 200 kHz. The attenuation provided by the enclosure must be at least 70 dB.
2. High-impedance electric fields. Monopole rod antennas, 41 inches long, are to be used as both transmitting and receiving antennas, spaced 12 inches from the shield walls. The test frequencies must be 200 kHz, 1 MHz, and 18 MHz. The attenuation provided by the enclosure must be at least 100 dB.
3. Plane waves. Dipoles tuned at 400 MHz are to be used as both transmitting and receiving antennas, and placed 72 and 2 inches from the shielding walls, respectively. The test frequency must be 400 MHz and the provided attenuation must be at least 100 dB.

The standard suffers of some defects. The location and orientation of the antennas are not defined and are left up to the tester. Although it can be assumed that the intent is to find the orientation and location that produce the largest leakage, the standard does not clearly state this, and it seems likely that different laboratories can obtain different results by testing the same enclosure. The loop and monopole antennas required by the standard do not appear suitable for measurements up to 10 GHz. Finally, during the reference-level measurement, the antennas are inductively coupled so that the antenna characteristics are quite different in the presence and in the absence of the shield.

The IEEE Std-299 (published by the Institute of Electrical and Electronics Engineers (IEEE)) originated in the late 1960s. It has served mainly as a detailed and thorough testing procedure for high-performance shielding enclosures. Some significant changes were made in 1991, but the standard still remains oriented toward high-performance enclosure testing. The revision approved in the late 1997 made several significant changes in the document, incorporating the basic concepts of MIL-STD 285 (which was withdrawn in that year). The 1997 version broadened the applicability beyond high-performance enclosures, provided a procedure for testing of smaller enclosures, and moved toward an economic testing. Nevertheless, the thoroughness and accuracy of the testing were maintained, together with a greater emphasis on assuring test-result repeatability. The latest revision of IEEE Std-299 [2] does not introduce any major changes to the measurement methodology but adds a section dealing with measurement uncertainty. In the future this standard (a new revision is foreseen for 2010) is expected to include methods allowing for the evaluation of small enclosures and shielding materials.

The present IEEE Std-299 describes uniform procedures for measuring SE for enclosures at frequencies from 9 kHz to 18 GHz (extendable down to 50 Hz and up to 100 GHz), although the smallest linear dimension of the enclosure is assumed to be at least 2 m; the selection of 2 m as the smallest linear dimension was originally based upon being able to fit a typical bicone-style antenna inside an enclosure to perform plane-wave testing down to the range from 30 to 50 MHz. The document does not give any limits for pass/fail, leaving to the owner of the shielding enclosure to provide these limits. In addition, although the standard suggests a range of test frequencies that can provide confidence in the effectiveness of the shield, it is clearly stated that the actual test frequencies must be chosen according to a test plan approved by the shield owner, the tester, and the shielding provider/vendor. The measurement range in this method is divided into three subranges, and in each subregion, some subgroups of testing frequencies are suggested:

1. A low-frequency range, from 9 kHz (50 Hz) to 20 MHz (a single frequency is suggested within 9–16 kHz, 140–160 kHz, and 14–16 MHz) where the SE is defined in terms of magnetic-field performance as

$$\mathrm{SE}_H = 20 \log \frac{|H_1|}{|H_2|}, \tag{C.1}$$

where the subscripts 1 and 2 indicate the field measurement in the absence and in the presence of the enclosure, respectively.

2. A resonant range, from 20 to 300 MHz, where the SE is expressed in terms of either electric field or power, according to

$$\mathrm{SE}_E = 20 \log \frac{|E_1|}{|E_2|}, \tag{C.2}$$

$$\mathrm{SE}_P = 10 \log \frac{|P_1|}{|P_2|}. \tag{C.3}$$

3. A high-frequency range from 300 MHz to 18 GHz (100 GHz) (a single frequency is suggested within 300–600 MHz, 600–1000 MHz, 1–2 GHz, 2–4 GHz, 4–8 GHz, and 8–18 GHz) where the shielding performance of the enclosure is expressed in power terms according to (C.3).

In the low-frequency range the standard utilizes a small electrostatically shielded loop, with a 0.3 m diameter, as the source of the magnetic field and as the receiving antenna. The source loop is driven by an ordinary audio frequency generator, plus an amplifier, that is usually adequate to supply the current if a suitable impedance matching device is used. The receiving loop is connected to a field-strength meter, a spectrum analyzer, or a similar device. A CW signal without modulation must be used to drive the transmitting antenna.

The measurement of H_2 is obtained as shown in Figure C.1*a*, with the transmitting and receiving loops spaced 0.3 m from the respective shielding barrier and in a coplanar position on a plane perpendicular to the surface being measured. The document clarifies that small loops are used because their sizes allow for the evaluation of the enclosure's performance when it is exposed to magnetic-field sources near the enclosure walls. Moreover the use of coplanar loops is advocated (as opposed to coaxial loops) because of their precision in locating defects and the facility in measuring their effects. The specification calls for the receiver to be located inside the enclosure, with an external transmitter. The ancillary equipment can remain in place during the test, while other equipment that is not a usual part of the enclosure must be removed.

The reference field H_1 is obtained by direct measurements with the receiving antenna spaced from the transmitting loop by 0.6 m edge to edge to which is added the thickness of the shielding barrier. This way the same total distance is maintained between the loops in the reference and shielding measurements. During the reference measurement the dynamic range (DR) of the equipment must be shown to be adequate to the measurement, being at least 6 dB greater than the SE to be measured. The DR of the receiving system is the range of amplitudes over which the system operates linearly. For a shielding-effectiveness measurement the important portion of the DR is from the reference level to the minimum discernable signal above the noise floor, defined as that with an amplitude of at least 3 dB above the test-system noise floor. The standard requires that during the demonstration the

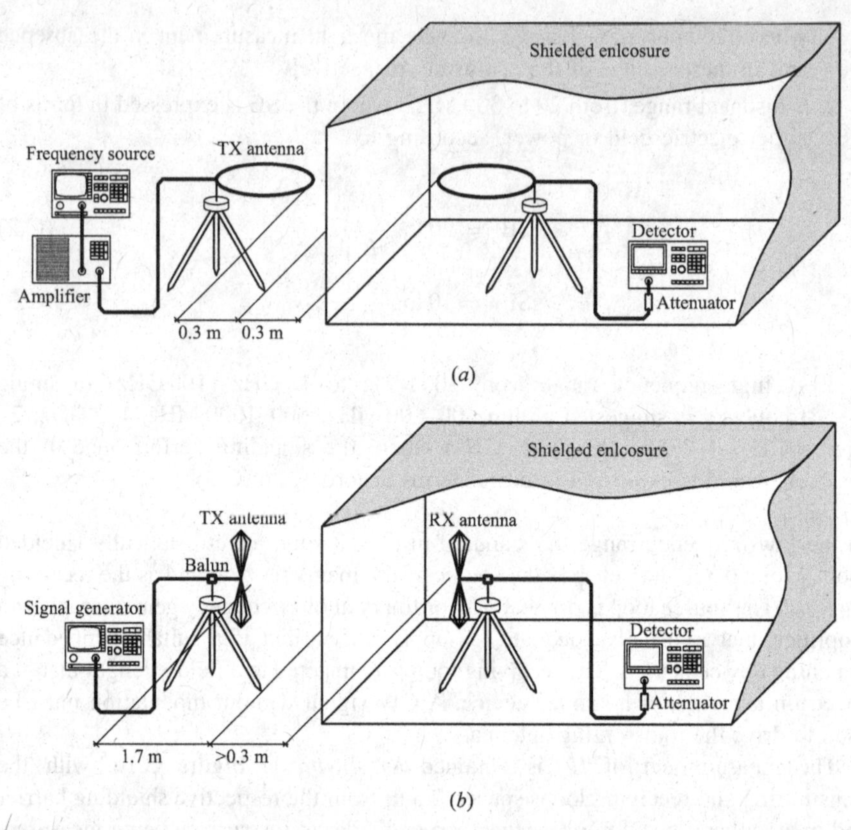

FIGURE C.1 IEEE Std-299 measurement setup: (*a*) low-frequency range; (*b*) resonant and high-frequency range.

receiving equipment remain calibrated for all the levels of received and transmitted signals being experienced and make use of an attenuator external to the receiver. It should be noted that in the MIL-STD 285 calibration setup, the receiving antenna was to be located outside the enclosure, leaving the receiving equipments inside, and the connecting cable was to be run through a feed-through connector in the enclosure wall. This setup had the advantage of protecting the sensitive receiver from the field of the transmitting antenna and, since it is difficult to provide over 100 dB of SE with the metal case of the receiver, of keeping it operating in the linear part of its DR.

The document recommends several loop positions to use in testing for the effects of common electrical nonuniformities that allow penetration of the magnetic field: single- and multiple-panel entry doors, seams, joints, accessible and not fully accessible corners, air vents, access panels, and connector panels. In addition the document indicates that the position be swept of the receiving loop, keeping fixed the transmitting one, and that the maximum value indicated by the field detector be looked for (i.e., the worst-case measurement). However, the final measurement must be always made with the coplanar configuration.

In the resonant range the standard calls for biconical antennas as the source and detector of the electric field in the range 20 to 100 MHz, and half-wavelength dipoles for frequencies above 100 MHz. The transmitting antenna is always placed outside the enclosure, with the receiving antenna inside. The receiving antenna is connected to the field-strength meter via a coaxial cable, through a balun transformer. To avoid false resonances in the measured SE, the cable must be perpendicular to the axis of the antenna for a distance of at least 1 m and must be loaded with ferrite jacketing or ferrite beads. A CW signal without modulation is still used to drive the transmitting antenna.

The document recommends that most enclosures have their fundamental resonances in this frequency range, and consequently to avoid testing the enclosure at these frequencies. In addition it states that SE measurements made at a single frequency in this range may not be representative of measurements made at other frequencies because there may be significant variations due to resonances. However, when measurement at a single frequency is necessary, specifications are given to verify if the measurement is acceptable or if additional tests are required because of the proximity of the resonance.

The test of the enclosure is rather extensive in this frequency range. Several positions of the transmitting antenna must be selected to cover the various parts of the shield. A distance of about 1.7 m from the shield surface must be maintained, and both horizontal and vertical polarizations are required. For each position/polarization of the transmitting antenna the receiving antenna must be swept in position throughout the interior of the enclosure, and in polarization, to search for the largest detector response (i.e., the minimum SE), with a minimum distance of 0.3 m always maintained from the enclosure walls. The reference level is measured by locating the receiving antenna outside the enclosure at a distance of 2 m from the transmitting antenna, leaving the attenuator and the receiver inside the enclosure. The connecting coaxial cable must be routed through the wall of the shield via a bulkhead coaxial connector or an open shield door far enough to pass the cable.

The basic reference and shielding measurements in the high-frequency range are similar to those in the resonant range. The noticeable differences are obviously in the antennas: dipoles, biconical antenna, horns, Yagi, log-periodic and other linear antennas can be used as sources; half-wavelength dipoles and horn antennas are used as receiving antennas from 300 MHz to 1 GHz and above 1 GHz, respectively. An accurate sweep in the position and polarization of both the transmitting and receiving antennas is still required to search for the worst-case measurement.

Further the IEEE Std-299 is not free from defects and shares some problems of MIL-STD 285. During the calibration procedure the measurement in the absence of the shield is made with a sensor having different characteristics from that used with the shield in place during the measurement procedure. In fact the presence of the shield undoubtedly alters the antenna characteristics. Three different SE ratios are used, one for each frequency range, however a direct comparison among these numbers is questionable. The low-frequency tests are not representative of the typical environment of a shielded enclosure because external conductors, which are usually the principal low-frequency magnetic-field sources, are only incidentally excited by the test loops.

The definitions, procedures, and methodologies described in the MIL-STD 285 / IEEE Std-299 can be found in a number of military handbooks published by the US Department of Defense. These handbooks provide basic information on applying shielding theory and on usual practices. The MIL-HDBK 1195 [3] gives a brief introduction to EMI shielding theory and presents basic criteria that are important to observe during the planning, design, and construction of a typical facility containing an EMI shielded enclosure. The MIL-HDBK 419A [4] is more general in its content. It provides basic information on grounding, bonding, and shielding practices for electronic equipment. The MIL-HDBK 1857 [5] is a version of the old MIL-STD-1857, dated 1976, that has been re-designated as a handbook. Its purpose is to provide guidance in the design of shielding.

C.2 NSA 65-6 AND NSA 94-106

The National Security Agency (NSA) is responsible for the analysis of foreign communications (cryptanalysis), and for protecting US government communications from similar agencies elsewhere (cryptography). Through the years the NSA has issued several standards concerning shielding performance of enclosures [6–9], among which the most famous was the NSA 65-6 [8] today superseded by the NSA 94-106 [9].

In its scope the NSA 94-106 covers the general requirements for the installation and performance of shielded enclosures that attenuate EM radiation. The requirements apply to all the associated and auxiliary facilities furnished as a part of the shielded enclosure as well. The document provides complete instructions on the purchase and testing of modular enclosures' housing and on protecting communications equipment used in the transmission of intelligence information.

The test methods are essentially the same as those in MIL-STD 285/IEEE Std-299, with two noticeable exceptions. In the magnetic test, the loop-antenna orientation is not collinear but planar and the receiver is located outside the enclosure. The outside placement was chosen to simulate the classic model of the emitter being inside the enclosure and the eavesdropping receiver being outside the enclosure. The number of specified test frequencies is quite large: magnetic field attenuation must be measured at 1 kHz, 10 kHz, 100 kHz, and 1 MHz; electric-field attenuation must be measured at 1 kHz, 10 kHz, 100 kHz, 1 MHz, and 10 MHz; the plane-wave attenuation must be measured at 100 MHz, 400 MHz, 1 GHz, and 10 GHz, with a transmitting antenna placed at least 6 m away from the shielded wall and the receiving antenna set no closer than 5.1 cm. The SE requirements are shown in Figure C.2.

The main difference between MIL-STD 285 and NSA 94-106 is in the low-frequency magnetic-field shielding performance above 100 kHz. The former is based on screen-wire performance at 150 kHz, whereas the latter is based on the values of magnetic-field attenuation attainable by means of solid shield of galvanized sheet metal.

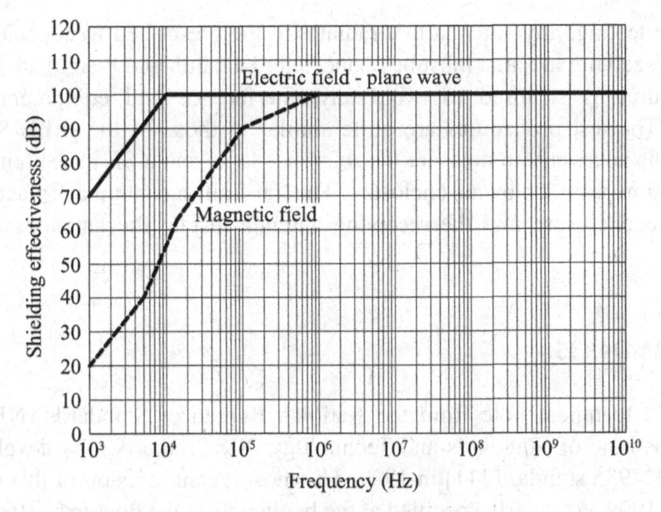

FIGURE C.2 NSA 94-106 performance requirements.

C.3 ASTM E1851

After a careful review of IEEE Std-299-1991 revealed that the standard developers tailored the standard for a shielded enclosure installed in a building or large facility and that the standard was not well suited for a transportable shielded enclosure, some members of the Department of Defense organization (called Joint Committee on Tactical Shelters (JOCOTAS)) asked the American Society for Testing and Materials (ASTM) to publish a standard specifically intended for a transportable enclosure or shelter. As a result ASTM Standard E1851 was published in February 1997.

In its latest version dated 2004 [10], the standard provides the test methods for the determination of the electromagnetic SE of durable relocatable shielded enclosures and shelters that do not have any equipment or equipment racks. The main text of the standard is written for a first-article testing that assesses the adequacy of an enclosure design and fabrication; the test requires a few days to be completed. An appendix is provided to verify the construction quality of the shielded enclosure in about half a day.

The standard is very similar to the IEEE Std-299. It requires the use of five specific frequencies for testing: magnetic SE measurements between 140 and 160 kHz and between 14 and 16 MHz; far-field shielding measurements between 300 and 500 MHz, 900 and 1000 MHz, and 8.5 and 10.5 GHz. Use of high-impedance electric fields is avoided because of the difficulties in making measurements and in detecting leaks. For specific applications, the frequency range may be extended from 50 Hz to 40 GHz. The test equipment has to provide a dynamic range of at least 10 dB above the SE requirement at a test frequency. For magnetic-field testing, circular-loop antennas that are 1 ft in diameter have to be used; the shielded circular receiving antenna can have multiple turns. For plane-wave testing, any antenna that radiates at the prescribed frequencies may be used. The receiving antenna must be connected to a balun and then to an attenuator. A CW source is required to avoid any interference and equipment-coupling problems. The test procedures are quite similar to those of the IEEE Std-299: a calibration is necessary to measure the reference level and a SE measurement gives the attenuation provided by the enclosure. The transmitting antenna is placed outside the shielded enclosure and the receiving antenna inside; the antennas should be coplanar.

C.4 ASTM D4935

Basing on a technical note from the National Bureau of Standards (NBS; today National Institute of Standards and Technology, NIST), the ASTM developed and issued the D4935 standard [11] in 1989. The most recent revision of this document dates from 1999. As clearly specified at the beginning of the document, the scope of the standard is to provide a test procedure for measuring the SE of planar materials due to a plane EM wave. The claim is that from the measured data, the shielding performance against near-field magnetic and electric sources can be computed, but the validity of the results has never been established.

ASTM regulations lay claim to technical expertise in evaluating the current standard every five years, in order to decide whether the standard should remain in force or be withdrawn. Formerly under the jurisdiction of Committee D09 on Electrical and Electronic Insulating Materials, the standard did not receive acceptance in September 2005 and was withdrawn without replacement. The rational given was that "Committee D09 cannot maintain a standard for which the expertise may not lie within the current committee membership, or for which the utilization of the standard is questionable." Although the D4935 document is no longer supported by the ASTM, it is still being supplied for information purposes. In addition the method described in the standard is still widely used for measuring the SE of planar materials against a plane wave.

The test procedure is based on the use of a specimen holder, shown in Figure C.3, that is constructed with an enlarged coaxial transmission line (having an external-to-internal diameter ratio of 76 mm to 33 mm) with special taper sections and notched matching grooves to maintain a characteristic impedance of 50 Ω throughout the entire length of the holder. The measurement method is valid over a frequency range from 30 MHz to 1.5 GHz. The limits are not exact, but certain limitations arise outside this range. At frequencies lower than 30 MHz, the capacitive coupling of energy into the specimen through displacement currents decreases, and the dynamic range of the measurement devices (or to be more precise, that of the network analyzer used for these measurement) is not adequate. For frequencies above 1.5 GHz, the field inside the test adapter is no longer a TEM wave because of the onset of higher order modes, the first of which is the TE_{11} mode whose cutoff frequency is around 1.7 GHz. The specimen holder is equipped with a 133-mm flange, which increases the capacitive coupling between its two halves. The

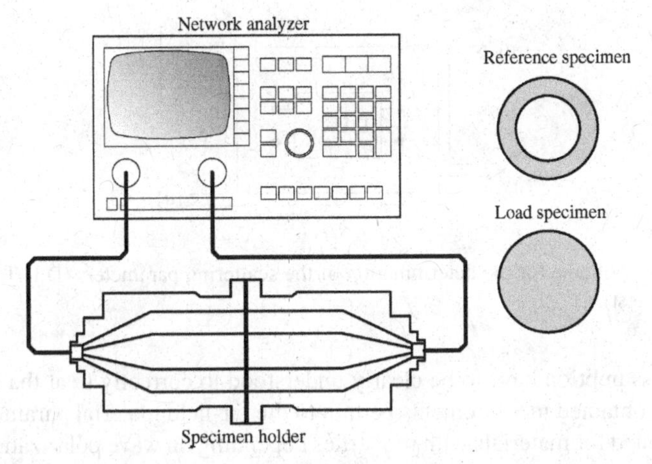

FIGURE C.3 ASTM D4935 measurement setup.

measurement uncertainty usually does not exceed ±2 dB with a properly established test setup that generally provides a dynamic range of 100 dB. The measurement device may consist of a network analyzer, which is capable of measuring both insertion and return loss.

The SE is determined by comparing the difference in attenuation of a reference sample to the test sample, taking into account the insertion and return power losses. It is defined on a power ratio basis according to equation (C.3). With reference to Figure C.4, the total power absorbed by the system can be computed, normalized with respect to the power available from the source, as

$$P_{\text{abs}} = 1 - |S_{11}|^2 - |S_{21}|^2, \tag{C.4}$$

where S_{11} and S_{21} are the scattering parameters measured by means of the network analyzer (as depicted in Figure C.4). The measurement procedure consists of two stages: in the first stage, a reference sample is placed in the test adapter to compensate for the coupling capacitance; the second stage uses the actual test specimen. The sample is in the form of a 33-mm circle inside a 133/76-mm ring.

Several prerequisites and cautions are necessary to correctly apply the procedure described in the standard. The thickness of the tested materials cannot exceed $\lambda_0/100$, being λ_0 the free-space wavelength of the EM wave (2 mm for a test frequency of 1500 MHz). It is necessary to guarantee a fixed distance between the adapter elements to ensure identical pressure on the surface of the sample both for the test and reference samples. For frequencies above 200 MHz, a calibration procedure must be performed to compensate for any capacitive coupling between the elements of the measuring adapter. In homogeneous materials with frequency-independent permittivity and permeability, it is sufficient to perform the measurements for just a few selected frequencies, whereas for frequency-dependent materials or just thick materials (skin depth less than the specimen thickness), the measurements must be performed for the entire frequency band.

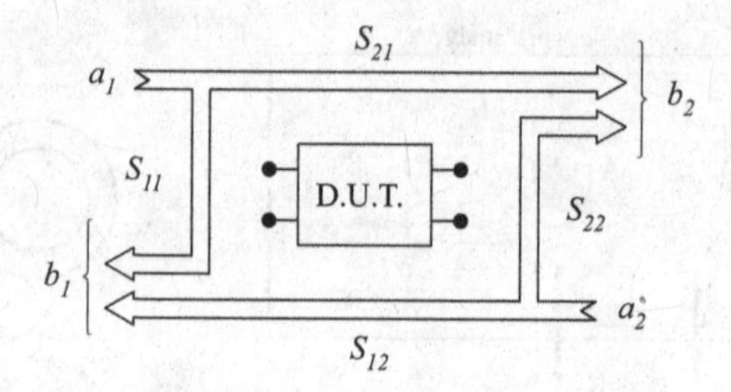

FIGURE C.4 Scheme for the determination of the scattering parameters (D.U.T. stands for device under test).

All the assumption have to be clearly understood to correctly treat the measured results. The obtained measurements pertain to the far-field material parameters. The results obtained for materials with properties depending on wave polarization will be average results. The measurement uncertainty usually falls within ±5 dB.

C.5 MIL-STD 461E

The military's concern for EMI began with the installation of the first radio in a vehicle before World War I. However, it was only in 1934 that the US Army Signal Corps published its first EMI standard (SCL-49—*Electrical Shielding and Radio Power Supply in Vehicles*). From this simple beginning, military EMIs evolved and changed as the complexity of the systems increased and the threats from electromagnetic pulses (EMP) were documented. Eventually each branch of the service defined specific requirements for their departments or platforms, and thus forced manufacturers to comply with significantly different specifications. As a result the Department of Defense formed a working group to consolidate and replace approximately 20 requirements into three initial fundamental standards that were published in 1967: MIL-STD 461 about the requirements, MIL-STD 462 about the measurement methodology, and MIL-STD 463 concerning definitions and acronyms. Through the years, revisions were required, resulting in MIL-STD 461A and MIL-STD 462A being issued and then revised three times (from revision A to D). In 1993, MIL-STD 463 was dropped, and its definitions referenced to the standard ANSI C63.14, *Standard Dictionary for Technologies of Electromagnetic Compatibility (EMC), Electromagnetic Pulse (EMP) and Electrostatic Discharge (ESD)*, developed by the Accredited Standards Committee C63 of the American National Standards Institute (ANSI), which works on EMC standards. In 1999, 461 and 462 [12] were combined in the currently enforced MIL-STD 461E [13].

The new document establishes the design requirements for the control of the EM emission and susceptibility characteristics of equipments and subsystems used by activities and agencies of the US Department of Defense, updating the previous mandatory requirements. At the same time, in referring to the superseded MIL-STD 462D, it provides general guidance on the measurement and determination of such characteristics. The requirements about conducted and radiated emissions and susceptibility are intended to serve a wide range of applications, from trucks to ships, to aircraft, to fixed installations. In addition the document provides the opportunity to tailor requirements to an application without having to issue exceptions to the standard.

As concerns the radiated emissions and susceptibility, five different requirements are specified: RE101 for radiated magnetic-field emissions in the frequency range from 30 Hz to 100 kHz; RE102 for radiated electric-field emissions in the frequency range from 10 kHz to 18 GHz; RS101 for radiated magnetic-field susceptibility in the frequency range from 30 Hz to 100 kHz; RS103 for radiated electric-field susceptibility in the frequency range from 2 MHz to 40 GHz; RS105 for radiated susceptibility to transient EM fields, such as a pulsed electromagnetic interference (EMI) or an electromagnetic pulse (EMP). Requirements are also presented for a radiated-emission test RE103 that can be used to measure the spurious and harmonic outputs from transmitters with their antennas in the range 10 kHz to 40 GHz as an alternative to conducting emission tests.

Applicable sections are summarized in Table C.1 and cross referenced as to how and where the equipments are intended to be installed in or on. If the equipment or subsystem can be installed on more than one platform, the standard requires that it comply with the most stringent requirement. An "A" entry in the table indicates that the requirement is applicable and must be followed, and an "L" means that the applicability of the requirement is limited, as specified in the relevant requirement paragraphs of the standard. Absence of an entry means that the requirement is not applicable for that application.

The standard provides four basic test setups (shown in Figure C.5) for the equipment under test (EUT). These setups are representative of typical system configurations: a general test setup (Figure C.5*a*), a test setup for nonconductive surfaces EUTs (Figure C.5*b*), a test setup for free-standing EUTs in shielded enclosures (Figure C.5*c*), and a test setup for free-standing EUTs (Figure C.5*d*). The test environment has to make a trade-off between real-world evaluation and laboratory repeatability. Except for the last setup, all the tests are performed in an ordinary metal-box shielded enclosure to prevent external environment signals from contaminating emission measurements and susceptibility test signals from interfering with electrical and electronic items in the vicinity of the test facility.

TABLE C.1 MIL-STD 461E Applicable Sections

Equipment	RE101	RE102	RE103	RS101	RS103	RS105
Surface ships	A	A	L	A	A	L
Submarines	A	A	L	A	A	L
Aircraft, Army	A	A	L	A	A	L
Aircraft, Navy	L	A	L	L	A	L
Aircraft, Air Force		A	L		A	
Space systems		A	L		A	
Ground, Army		A	L	L	A	
Ground, Navy		A	L	L	A	L
Ground, Air Force		A	L		A	

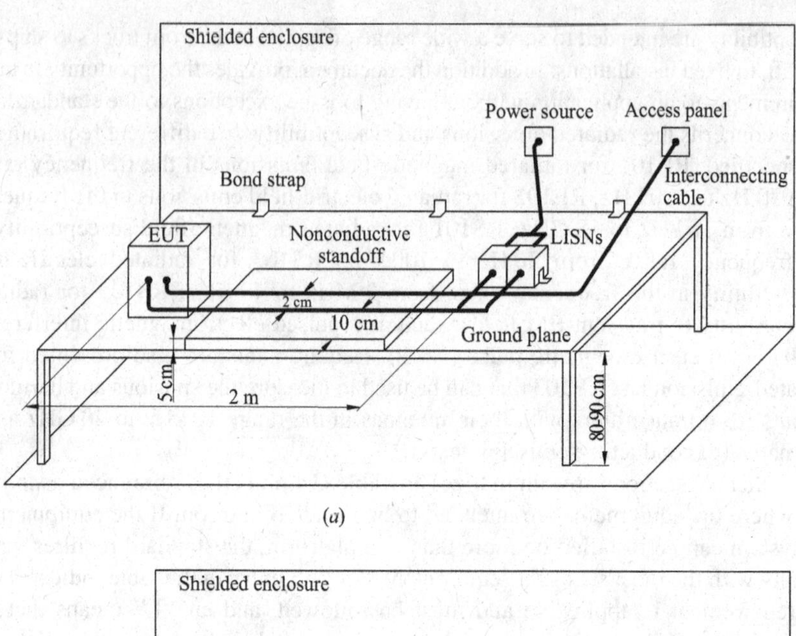

(*a*)

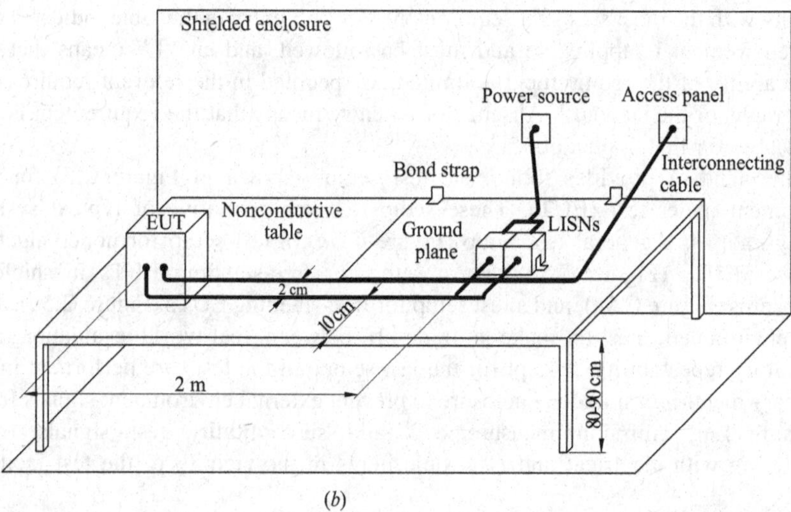

(*b*)

FIGURE C.5 MIL-STD 461E basic measurement setups: (*a*) general test setup; (*b*) test setup for nonconductive surfaces mounted EUTs; (*c*) test setup for free-standing EUTs in shielded enclosures; (*d*) test setup for free-standing EUTs.

Regardless of which setup is used, the test environment must have always conducted and radiated RF ambient levels below the specification. When the EUT is not free-standing, the EUT is placed on a table with a stationary copper ground plane, bonded to the floor or to the wall of the shielded room via copper bonding straps.

The standard specifies that the shielded enclosure must be sufficiently large to handle the EUT and the necessary test antennas. Since shielded enclosures introduce errors into the measurements (resulting from multipath reflections, enclosure

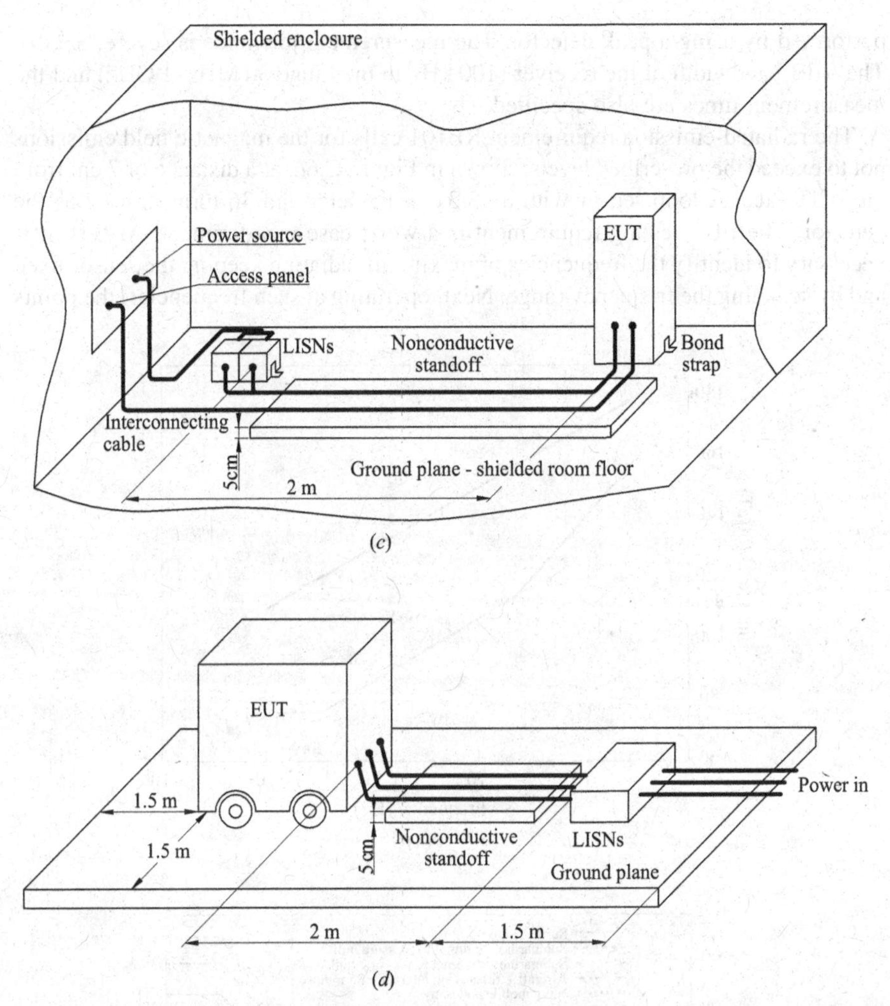

(*c*)

(*d*)

FIGURE C.5 (*Continued*)

resonances, and antenna loads), the standard suggests that RF absorber materials be used to reduce the reflections of EM energy and to improve accuracy and repeatability in electric-field radiated emissions or radiated-susceptibility testing.

To standardize the radiated (and conducted) emission behavior of the line providing input power to the EUT (so that measurements from different labs can be compared), the line impedance is controlled by a line impedance stabilization network (LISN). This is a combination of a voltage probe and a 50 Ω filter that isolates the power source from the test sample, standardizes the source impedance at 50 Ω, and allows the measurement receiver to be capacitively coupled to the test-sample power line.

In all the test procedures the standard requires that the entire measurement system be initially calibrated from the sensor to the display using standardized reference signals for each procedure, in order to reduce systematic errors. Measurements are performed by using a peak detector. The measurement tolerance is severe: ±2 dB. The 6 dB bandwidth of the receiver (100 kHz in the range 30 MHz–1 GHz) and the measurement times are also specified.

The radiated-emission requirement RE101 calls for the magnetic field emissions not to exceed the prescribed levels, shown in Figure C.6*a*, at a distance of 7 cm from the EUT face. A loop sensor with a 13.3 cm diameter and 36 turns is used as the detector. The EUT testing requirement is a worst-case measurement. At first, it is necessary to identify the frequencies of maximum radiation keeping the sensor fixed and by scanning the frequency range. Next, operating at such frequencies, the points

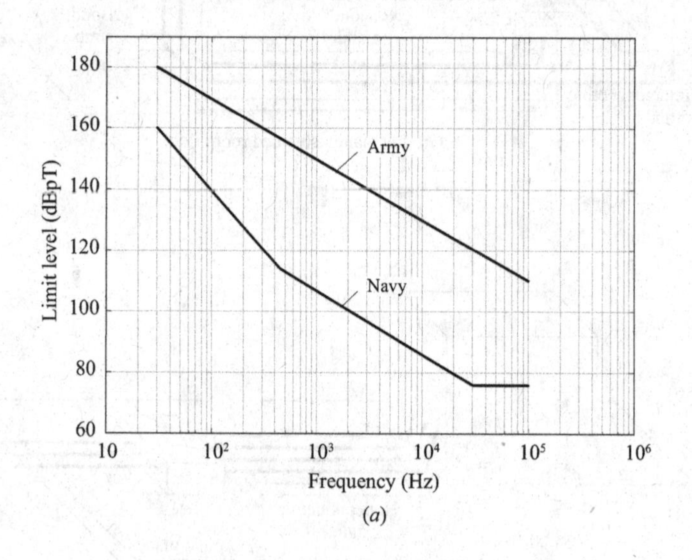

(*a*)

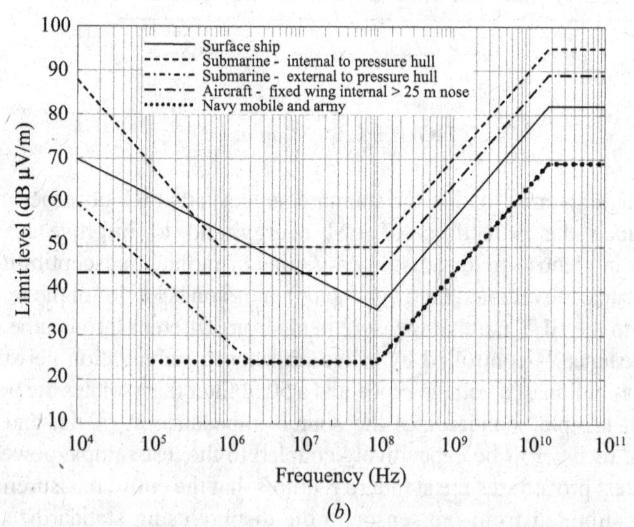

(*b*)

FIGURE C.6 MIL-STD 461E emission limits: (*a*) magnetic field; (*b*) electric field.

of maximum radiation have to be identified; this is accomplished by moving the sensor around the EUT and orienting the plane of the loop.

The RE102 requirement calls for the electric-field emissions not to exceed the prescribed levels shown in Figure C.6*b*, at a distance of 1 m. Above 30 MHz, the limits must be met for both horizontally and vertically polarized fields. Several receiving electric-field antennas are used: a 104 cm rod antenna (10 kHz–30 MHz), a biconical antenna (30 MHz–200 MHz), and a double-ridge antenna (200 MHz–18 GHz). The test procedure calls for the maximum emissions to be located at prescribed positions of the receiving antenna (which has to be placed at a distance of 1 m from the front edge of the EUT setup boundary) by scanning the measurement receiver for each frequency range and by orienting the antenna for both horizontally and vertically polarized fields.

The RS101, RS103, and RS105 requirements call for the EUT not to exhibit any malfunction, degradation of performance, or deviation from specified indications (beyond the tolerances indicated in the individual equipment or subsystem specification) when subjected to prescribed low-frequency magnetic fields, high-frequency electric fields, and transient EM fields, respectively. In the case of RS101, the radiating antenna is a loop with 12 cm diameter and 20 turns, placed at 5 cm from the EUT face. The test procedure involves scanning the considered frequency range to identify those frequencies, if any, where significant susceptibility effects take place. Then the transmitting loop is moved around the EUT to determine the locations of susceptibility. In the case of RS103, the transmitting antenna is placed 1 m from the test setup boundary at prescribed positions, depending on the boundary dimensions. The testing involves scanning the required frequency range to monitor the EUT performance for susceptibility effects, with the transmitting antenna both vertically and horizontally polarized. In the high-frequency range (200 MHz–40 GHz) the standard suggests an optional test in a mode-stirred reverberation chamber. The document calls for a number of prescribed tuner positions, minimizing the need for rotating the test sample and moving the transmitting antenna for monitoring susceptibility effects. The RS105 testing involves placing the EUT in a transverse electromagnetic cell (TEM-cell; see Figure C.7*a*), where it is exposed to a pulsed field with a prescribed waveform (shown in Figure C.7*b*).

MIL-STD 461E is a widely referenced standard. Through the years it has become the *de facto* EMC standard used by NATO and by several defense departments in countries around the globe. In any case some remarks about this standard are in order. When calling for the use of a shielded enclosure, the standard does not adequately address the problem of enclosure resonances: it only requires enough spacing to contain the setup. Although the addition of absorber materials is recommended, the semi-anechoic and anechoic chambers lead to some of the problems of metal-box enclosures such that the same product when tested in different-sized enclosures may yield different results. The LISN impedance is not stable across the entire measurement frequency range and must be calibrated at the test sample's equivalent load current. In addition significant errors can be encountered because of the impedance mismatch between the LISN and the measurement receiver. The prescribed calibration procedures do an excellent job in

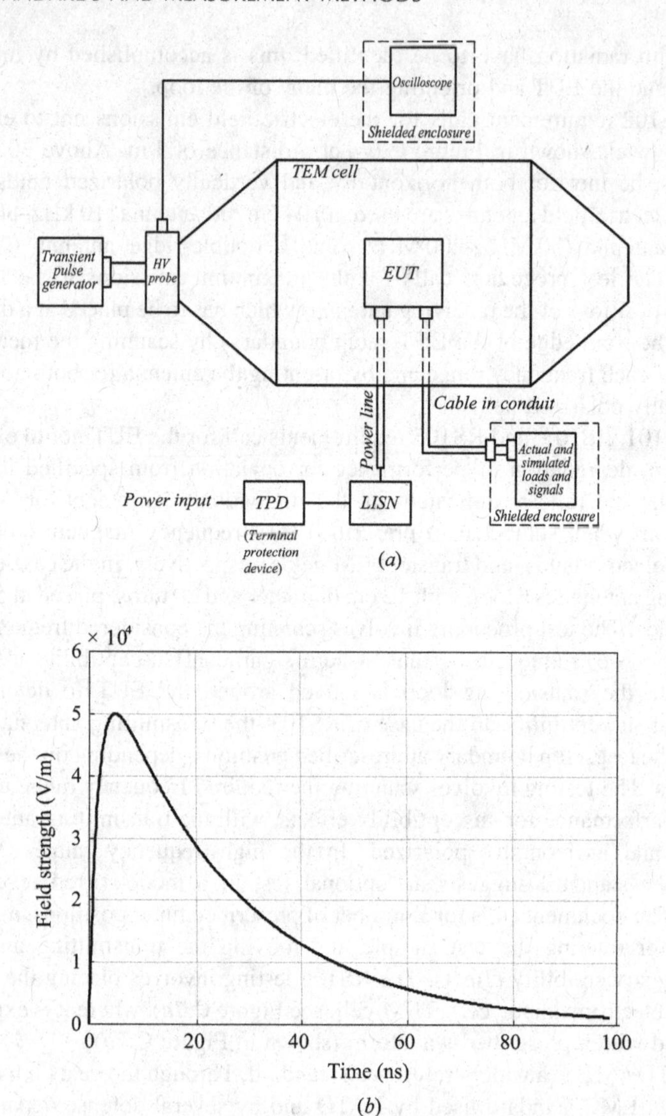

FIGURE C.7 RS105 setup (*a*) and pulsed field waveform (*b*).

minimizing the errors for conducted measurements, but the radiated measurements are nevertheless prone to errors resulting from the interaction of the radiated field with the surroundings. The standard calls for testing both a vertically and a horizontally polarized field. However, the measured signals are not necessarily polarized the same way as the antenna, so the arriving signal at the required measurement distance may be cross-polarized with respect to the antenna. The largest angular difference is 45°, which is equivalent to a −3 dB systematic error.

C.6 CODE OF FEDERAL REGULATIONS, TITLE 47, PART 15

Ever since the first poorly shielded home computers caused annoying herringbone patterns on neighbors' television screens, the Federal Communications Commission (FCC) has issued strict limits to the levels of RF emissions that a digital-logic product can spew into the environment. The requirements of the FCC are presently contained in the Code of Federal Regulations, Title 47, Part 15, "Radio Frequency Devices." FCC Part 15 [14] rules cover both unintentional and intentional radiators: the first are devices such as computers and television receivers, which may generate radio signals as part of their operation, though they are not designed to transmit such signals; the second are devices, such as garage-door openers, cordless telephones, and wireless microphones that depend on deliberate radio signals to function. Digital devices are categorized as class B devices (which are marketed for use in a residential environment) and class A devices (which are marketed for use in a commercial, industrial, or business environment). Examples of class B devices include (but are not limited to) personal computers, calculators, and similar electronic devices that are marketed for use by the general public.

For unintentional radiators, including digital devices, the FCC calls for measurements over the frequency range from the lowest RF signal generated or used in the device (without going below the lowest frequency for which a radiated-emission limit is specified) to an upper frequency depending on the highest frequency f_u used in the device: in particular, such an upper frequency is 30 MHz if $f_u < 1.705$ MHz, 1 GHz if 1.705 MHz $< f_u <$ 108 MHz, 2 GHz if 108 MHz $< f_u <$ 500 MHz, 5 GHz if 500 MHz $< f_u <$ 1 GHz, and 5th harmonic of the highest frequency or 40 GHz, whichever is lower, if f_u is above 1 GHz. Additional requirements are provided for inadvertent radiators, with the exception of digital devices with f_u less than 30 MHz and for receivers employing superheterodyne techniques. In the case of intentional radiators, the FCC calls for measurements over a frequency range from that of the lowest radio frequency (RF) used in the equipment cabinet (without going below 9 kHz) to an upper frequency that depends on the highest operating frequency of the used equipment. For equipments operating below 10 GHz, such an upper frequency is the 5th harmonic of the highest fundamental frequency or 10 GHz, whichever is lower; if the equipment operates between 10 and 30 GHz, it is the 10th harmonic of the highest fundamental frequency or 40 GHz, whichever is lower. If the equipment operates above 30 GHz, the upper frequency limit is the 5th harmonic of the highest fundamental frequency or 200 GHz, whichever is lower.

For unintentional radiators, except for class A digital devices, the field strength of the radiated emissions at a distance of 3 m must not exceed the values in Table C.2.

Also for intentional radiators, the field strength of the radiated emissions at a distance of 3 m must not exceed the values in Table C.2. For class A digital devices, the field strength of the radiated emissions at a distance of 10 m must not exceed the values in Table C.3.

As noted in Part 15, the testing can be according to the limits given either in the text of the regulations or in CISPR 22 [29], with the following points applying: the

TABLE C.2 FCC-Part 15 Limits for Intentional and Class B Unintentional Radiators

Frequency (MHz)	Distance (m)	Radiated (dBμV/m)	Radiated (μV/m)
30–88	3	40	100
88–216	3	43.5	150
216–960	3	46	200
Above 960	3	54	500

limits CISPR 22 must be used in their entirety, and the test procedures must be those specified in Part 15 and ANSI C63.4, not those in CISPR 22.

The measurements must be performed according to the measuring procedures FCC/OET MP-4 [15] and ANSI C63.4 [16]. Measurements are suggested to be made in an open area test site (OATS). However alternative test sites are possible, such as RF absorber-lined, metal test chambers, office or factory buildings, provided that they comply with the prescribed volumetric normalized site attenuation (NSA) requirements. The test environment must have both conducted and radiated RF ambient levels at least 6 dB below the specification. Radiation measurements made in a shielded enclosure are adequate only for determining emission frequencies of the EUT caused by multiple reflections, and such measurements can be used below the resonant frequencies of the enclosure (usually under 30 MHz). For frequencies below 1 GHz, the measurements are carried out using a CISPR quasi–peak detector function, whereas for frequencies above 1 GHz an average detector function is used. At radio frequencies, the prescribed 6 dB bandwidth of the measuring instrument is equal to that of the MIL-STD-461E: not less than 100 kHz in the range 30 MHz to 1 GHz and not less than 1 MHz for frequencies above 1 GHz. In the frequency range 30 to 1000 MHz, the receiving antenna must be a tuned half-wave dipole; other linearly polarized antennas can be used, provided that the measurement can be correlated with that made by means of a tuned dipole with an acceptable degree of accuracy. In the range of 1 to 40 GHz, calibrated, linearly polarized antennas must be used (double-ridged guide horns, rectangular waveguide horns, pyramidal horns, optimum gain horns, etc). Tests must be made in both horizontal and vertical planes of polarization over the frequency range. A reflecting ground plane must be installed on the floor of the radiated electric-field emission test site to provide a uniform,

TABLE C.3 FCC-Part 15 Limits for Class A Unintentional Digital Radiators

Frequency (MHz)	Distance (m)	Radiated (dBμV/m)	Radiated (μV/m)
30–88	10	39.08	90
88–216	10	43.5	150
216–960	10	46.44	210
Above 960	10	49.54	300

predictable reflection of radiated emissions measured at the site. Unlike MIL-STD-461E, the FCC tests are performed with the EUT placed on a turntable and rotated through 360°, while simultaneously raising and lowering the receiving antenna (from 1 to 4 m). For floor-standing EUTs, the turntable must be metal-covered and flush with the ground plane; for tabletop EUTs, the turntable can be nonmetallic and located on top of the reference ground plane.

It should be noted that acquiring a complete and exhaustive data set according to FCC (and MIL-STD-461E) requirements can take a long time. In the RF range, for each position of the turntable and each elevation of the receiving antenna, it is necessary to make for both horizontal and vertical polarizations a frequency span in the range 30 MHz to 1 GHz (970 MHz range) using a 100 kHz bandwidth.

C.7 ANSI\SCTE 48-3

The Society of Cable Telecommunications Engineers (SCTE), a nonprofit professional association, is the cable telecommunications organization accredited by the American National Standards Institute (ANSI). The International Telecommunication Union (ITU) also recognizes SCTE, allowing SCTE standards to be referenced by the ITU. SCTE submits standards to the ITU through the US Department of State and works in cooperation with the European Telecommunications Standards Institute (ETSI).

The standard ANSI/SCTE 48-3 [17] details the procedure for measuring the SE of a coaxial cable using a gigahertz transverse electro-magnetic (GTEM) cell (see Figure C.8 for the measurement setup). No performance requirement is indicated. In particular, this procedure applies to measuring the SE of the 75 Ω braided coaxial drop cables presently used within the broadband communications industry. The SE is calculated as the level difference (in dB) between the coupling-loss measurement of the unshielded and shielded device under test (DUT); the coupling-loss measurement is a measure of the voltage loss in dB between the RF voltage input to the GTEM cell and the RF voltage coupled into the DUT and received at the point where the cable exits from the GTEM cell. The measurements are conducted in the frequency range of 5 to 1000 MHz, and the average value across the entire range is considered as the SE of the DUT.

The unshielded calibration sample consists of a 1 m length of core (inner conductor and dielectric insulation only), with the appropriate coaxial cable connectors fitted to each end. The cable is installed within the GTEM cell, with the cable connectors attached to the adapter interfaces on the feed-through panel located in the center of the GTEM floor. One adapter is terminated with a 75 Ω load while the other is connected to the receiving port of the spectrum analyzer. The shielded sample consists of a 1 m length of cable; the coaxial connectors can be included or removed from the measurement. When the connectors have to be removed, two inches of jacket are removed from the cable in order to accommodate the feed-through adapters in the GTEM floor. The measurements are typically performed with the use of an amplifier.

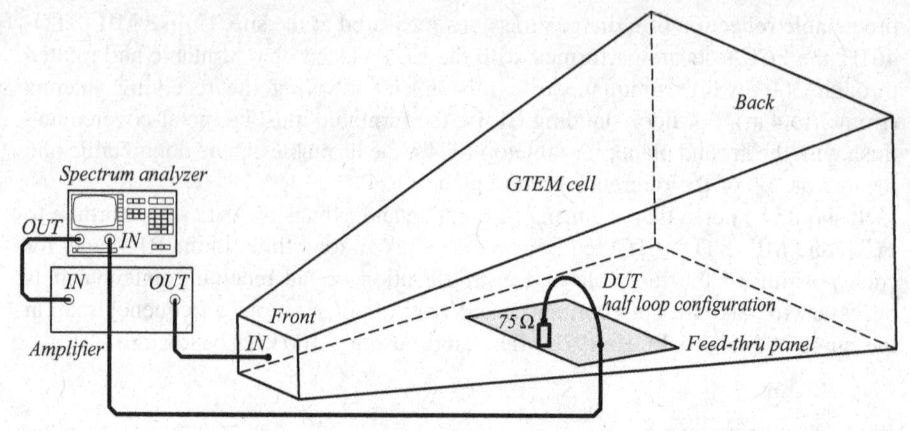

FIGURE C.8 ANSI/SCTE 48-3 measurement setup.

C.8 MIL-STD 1377

The MIL-STD 1377 [18], mandatory for use by the Department of the Navy, covers the methods of measuring the SE of weapon enclosures, cables, and cable connectors over the frequency ranges 100 kHz to 30 MHz and 1 GHz to 10 GHz. The document clarifies that it is unnecessary to measure the SE between 30 MHz and 1 GHz, due to the limited use that Navy makes of these frequencies. The standard provides the test setup and the necessary apparatus to characterize the shielding performances only of cables since it neglects weapon and connectors, which are very specific topics.

At frequencies below 30 MHz, the surface transfer impedance (STI) of the cable is measured according with the setup shown in Figure C.9*a*. With the output of a signal source (any RF signal generator) applied to the test cable (through a cable adapter), the center-conductor current and the voltage on the outer surface of the shield between the ends of the test cable are measured with an ammeter and an RF voltmeter, respectively. The STI is thus calculated by dividing the voltage measurement by the current measurement.

Above 1 GHz, the setup for measuring the SE of the cable shield is more complex, as shown in Figure C.9*b*. The main apparatus is a cabinet enclosure, with sufficient space to accommodate the cable and with a SE greater than 60 dB. Inside the enclosure, there are an input antenna and an output antenna, one opposite to the other, connected to bulkhead connectors mounted on the walls of the cabinet: they consist of a wire running from the connector to one corner of the cabinet, then diagonally across the end of the cabinet, and then along the length of the cabinet parallel to the edge. Inside the cabinet, a paddle wheel turner is installed too: it consists of three dipoles 8, 6.5, and 5.25 inches long and each 1 inch wide. In addition the setup requires a signal generator, impedance matching devices capable of matching the input and output impedance of the test enclosure to 50 Ω, a directional coupler capable of providing a signal proportional to the forward power, and power meters. The measurement procedure involves two steps: first a calibration

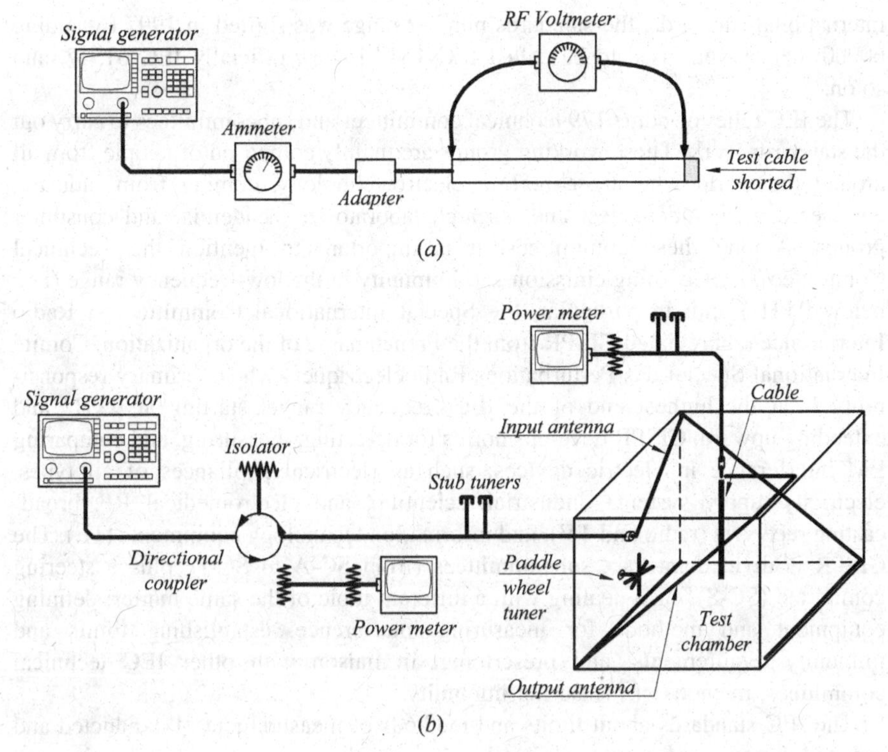

FIGURE C.9 MIL-STD 1377 measurement setup: STI low-frequency setup (*a*) and SE high-frequency setup (*b*).

is made with the output antenna connected to the receiving power meter, and then a measurement is made with the cable connected to the output bulkhead connector in place of the output antenna. The technique to be applied is the same: with the signal generator applied to the input of the cabinet, the matching devices are adjusted for maximum power as measured by the output power meter; then the paddle wheel tuner is rotated until the maximum output power is found; with the tuner fixed in this position, the final matching is performed and the maximum output and input powers are measured. The SE can be computed from the values of the forward input P_{IS} and output P_{OS} power for the shield-loss measurement and forward input P_{IC} and output P_{OC} powers for the calibration measurement, as $\mathrm{SE} = 10\log(P_{\mathrm{IS}}P_{\mathrm{OC}}/P_{\mathrm{OS}}P_{\mathrm{IC}})$.

C.9 IEC STANDARDS

The International Electrotechnical Commission (IEC) is an international organization that develops standards on electrical, electronic, and related technologies, sometimes jointly with the International Organization for Standardization (ISO). In order to distinguish standards published by the IEC numerically from other international standards, the standards number range was shifted in 1997 by adding 60000: hence what used to be called IEC 1547 is now officially IEC 61547, and so on.

The IEC relies on some 179 technical committees and subcommittees to carry out the standards work. These working groups are mainly composed of people from all around the world who are expert in electrotechnology, coming from industry, commerce, government, test and research laboratories, academia, and consumer groups. Among these committees, it is important to mention the Technical Committee 77 concerning emission and immunity in the low-frequency range (i.e., below 9 kHz) and, in particular, the Special International Committee on Radio Interference (abbreviated CISPR from the French name of the organization, Comité International Spécial des Perturbations Radioélectriques) whose primary responsibility is at the highest end of the IEC frequency range, starting at 9 kHz and extending upward. CISPR develops norms for detecting, measuring, and comparing EM interference in electric devices, such as electrical appliances of all types, electricity-supply systems, industrial, scientific, and electromedical RF, broadcasting receivers (radio and TV), and information technology equipment (ITE). The CISPR is divided into six subcommittees (from SC-A to SC-I), plus a steering committee (SC-S), each dealing with a different topic of the same matter: defining equipment and methods for measuring interference, establishing limits and immunity requirements, and prescribing, in liaison with other IEC technical committees, methods of measuring immunity.

The IEC standards about limits and methods of measurement of conducted and radiated disturbances are a huge number. Among them an immunity collection and an emission collection can be identified. The better known standards worthy to be mentioned are as follows:

Immunity Standards

- CISPR 14-2 [19], a product family standard about immunity requirements for household appliances, electric tools, and similar apparatus.
- CISPR 20 [20], a product family standard about limits and methods of measurement of immunity characteristics of sound and television broadcast receivers and associated equipment.
- IEC 61547 [21], a product family standard about immunity requirements for general-purpose lighting equipment.
- IEC 61000-6-1 [22], a generic standard about immunity requirements for equipment to be installed in residential, commercial, and light-industry environments.
- IEC 61000-6-2 [23], a generic standard about immunity requirements for equipment to be installed in industrial environments.
- IEC 61000-4-3 [24], dealing with testing and measurement techniques to be used in radiated RF immunity tests.

Emission Standards

- CISPR 11 [25], a product family standard about emission requirements for industrial, scientific, and medical (ISM) radio-frequency equipment.
- CISPR 13 [26], a product family standard about emission requirements for sound and television broadcast receivers and associated equipment.
- CISPR 14-1 [27], a product family standard about emission requirements for household appliances, electric tools, and similar apparatus.
- CISPR 15 [28], a product family standard about emission requirements for electrical lighting and similar equipment.
- CISPR 22 [29], a product family standard about emission requirements for information technology equipment.
- IEC 61000-6-3 [30], a generic standard about emission requirements for equipment to be installed in residential, commercial, and light-industry environments.
- IEC 61000-6-4 [31], a generic standard about emission requirements for equipment to be installed in industrial environments.

It is no small task to relate all the referred IEC standards. Indeed the task is so daunting that only the major characteristics will be described on the following pages.

As regards immunity, the specification IEC 61000-4-3 calls for monitoring a DUT for continuous proper operation as it is subjected to a RF radiation level. It identifies four test levels related to general-purpose devices: level 1 with a field strength of 1 V/m, level 2 with 3 V/m, level 3 with 10 V/m, and level 4 with 30 V/m. The standard does not provide any specification on which level has to be used because it is up to the product committees to select the appropriate level; it only instructs the technician not to use a single level over the entire frequency range. The test frequency range is from 80 MHz to 6 GHz: the tests are usually performed without gaps in the frequency range 80 MHz to 1 GHz for general-purpose devices, while in the ranges 80 MHz to 960 MHz and 1.4 to 6 GHz for RF emitting devices, the tests are limited to those frequencies where the device actually operates. The signal must have amplitude modulated by a 1 kHz sine wave with a modulation depth of 80%. The specification IEC 61000-6-1 fixes the test levels of the RF EM field (80 MHz–1 GHz) to 3 V/m for equipments to be used in residential and commercial environments, while the specification IEC 61000-6-2 fixes it to 10 V/m for industrial equipments (10 V/m 80–1000 MHz, 3 V/m 1.4–2 GHz, 1 V/m 2–2.7 GHz). Three performance criteria are specified: criterion A, when the apparatus continues to operate as intended during and after the test; criterion B, when the apparatus continues to operate as intended after the test while during the test it shows degradation of performance; criterion C, when the apparatus shows a temporary loss of function, which is in fact self-recoverable.

RF signal generators, power amplifiers, and antennas are needed to generate the required field strength. Any linearly polarized antenna (e.g., biconical, log-periodic, and horn antennas) can be used as the transmitting antenna if it satisfies the frequency range and power requirements (circularly polarized antennas are not allowable). The standard requires that the amplifier must be able to handle the required modulation without saturating during the test. Field strength can be monitored through field probes and field monitors. Forward power can be monitored through dual-directional couplers and power meters with sensors. The test facility is an anechoic or semi-anechoic chamber of a size adequate to EUT and test equipments (partially lined screened rooms and open area test sites are not alternative test facilities): the minimum distance between the antenna and the EUT is 3 m. A metallic ground plane is not required. The EUT can be floor-standing or table-top-standing: in the latter, it must be placed on a nonconductive 0.8 m high table. Wooden tables and supports that have been widely used in the test setups for years (because they are affordable and easy to make) cannot be used anymore above 1 GHz where they can be reflective. A low-permittivity material, such as rigid polystyrene, is necessary to satisfy the field-uniformity requirements for frequencies greater than 1 GHz.

For the immunity test the standard depends on the concept of a uniform-field area (UFA). The UFA is a hypothetical vertical plane whose field variations are acceptably small. The calibration procedure is aimed at showing the capability of the test facility and equipment to generate such a field, building at the same time a database of field strengths. The size of the UFA is at least 1.5×1.5 m, with the lowest edge at a height of 0.8 m above the floor. The calibration procedure is a tedious process, and it is impractical to do it without a PC and calibration software. The UFA is divided into 16 points on a 0.5 m grid. By choosing a point near the center as the reference point, the RF source is set to the low end of the test range and the power into the antenna is adjusted to produce the test level as monitored by a field-strength meter. The frequency is stepped from the low end of the desired spectrum to the high end, and the amplifier output is adjusted at each step to get the same field strength as that at the reference frequency. The operator then proceeds through the other 15 points, driving the antenna at the same level as before for each frequency step and recording the field strength. From the results the field is considered uniform if its magnitude at not less than 75% of all grid points (i.e., 12 points) is within the interval [0, +6] dB of the nominal value. The standard states that the level of any harmonic frequency generated by the power amplifier and measured in the UFA must be at least 6 dB below that of the fundamental frequency. The final setup is shown in Figure C.10.

As regards the emission, the measurement procedure is the same as that described in several CISPR standards. The requirements are also close to those of the ANSI C63.4 procedure. The measurements must be conducted with a quasi-peak measuring receiver in the frequency range 30 MHz to 1 GHz. The receiving antenna must be a balanced dipole resonant in length and must be adjusted between 1 and 4 m in height above the ground plane for a maximum meter reading at each frequency. Both the horizontal and vertical polarizations must be considered. The test site must be an open-area or an alternative site validated by an appropriate measurement of the provided attenuation. A conductive ground plane is required. The antenna-to-EUT azimuth must be varied during the measurement to find the

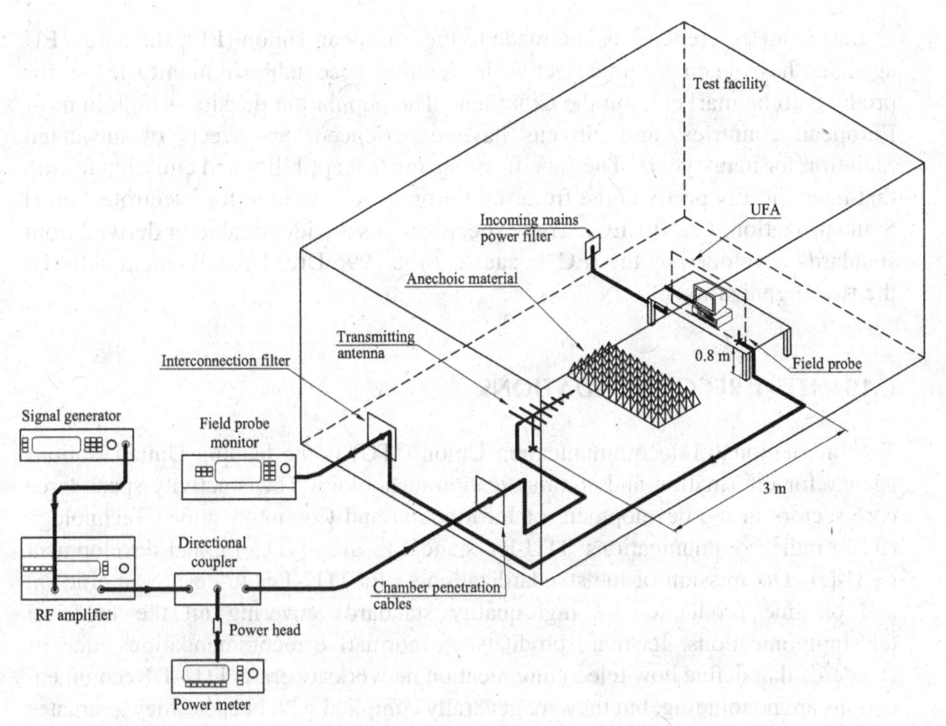

FIGURE C.10 IEC 61000-4-3 measurement setup for immunity tests.

maximum field-strength reading, either by rotating the EUT or by making the measurements around the fixed EUT (alternative choice).

As in the FCC Part 15, the devices are divided in two categories: class B equipment (for use in a domestic environment) and class A equipment (for use in an industrial environment). The emission limits are given in standards IEC 61000-6-3 and IEC 61000-6-4, respectively, and are summarized in Table C.4. For class A devices, the measurement can be conducted at a 10 m distance using the limits increased by 10 dB (thus meeting the previsions of CISPR 11). This way it is possible to see that the limits are very similar to those imposed by FCC Part 15.

TABLE C.4 IEC Emission Limits

Frequency (MHz)	Distance (m)	Radiated (dBμV/m)	Radiated (μV/m)
Class B (IEC 61000-6-3)			
30–230	10	30	31.6
230–1000	10	37	70.8
Class A (IEC 61000-6-4)			
30–230	30	30	31.6
230–1000	30	37	70.8

Last some reference must be made to the European Union (EU) standards. EU agencies have been the most active in defining acceptable immunity levels for products to be marketed on the Continent. The population density is high in most European countries, and citizens have experienced the effects of unwanted radiation for many years. The specifications for susceptibility and emission testing that have the EN prefix come from the European Committee for Electrotechnical Standardization (CENELEC). These specifications are identical to or derived from standards developed by the IEC, because of the 1996 Dresden agreement between the two organizations.

C.10 ITU-T RECOMMENDATIONS

The International Telecommunication Union (ITU) is the leading United Nations agency for information and communication technology. ITU's activity spans three core sectors in the development of Information and Communications Technology (ICT): radio-communications (ITU-R), standardization (ITU-T), and development (ITU-D). The mission of the standardization sector ITU-T is to ensure an efficient and on-time production of high-quality standards covering all the fields of telecommunications. Its main products are normative recommendations, that is, standards that define how telecommunication networks operate. ITU-T Recommendations are nonbinding, but they are generally complied with because they guarantee the interconnectivity of networks and enable telecommunication services to be provided on a worldwide scale. Currently there are about 3100 Recommendations in force on several topics, ranging from service definition to network architecture and security, from dial-up modems to Gbit/s optical transmission systems. Here it is worth reminding the K-series of the ITU-T Recommendations dealing with "protection against interference." The K-series Recommendations are a huge number; in the following the main recommendations dealing with topics related to the content of the book will be reported.

Recommendation K.42 [32] explains the basic principles on which EMC standardization is based in the ITU-T sector. It clearly states the need for collaboration with other international organizations (ISO and IEC) whose EMC standards must be considered as well. The above-mentioned IEC, ISO, and CISPR standards and publications are cited and the document describes procedures that are followed in the preparation of ITU-T Recommendations on EMC requirements about telecommunications equipments: environmental classification, emission, and immunity.

Recommendation K.34 [33] gives the details on the classification of EM environmental classes for telecommunication equipments covering all the relevant EM environmental parameters. It refers to the IEC 61000-2-5 [34]. The document defines four classes of environments for telecommunication equipments: major telecommunication centers, minor telecommunication centers, outdoor locations, and customer premises. The characteristic severities and some characteristics of the relevant parameters are stated for each environmental class. The parameters are

TABLE C.5 ITU-T K.43 Immunity Requirements

Frequency (MHz)	Test Level (V/m)	Performance Criterion
Equipment for telecom center		
80–800	1	A
800–1000	10	A
1400–2000	10	A
Equipment for customer premises		
80–800	3	A
800–1000	10	A
1400–2000	10	A

given according to the coupling path: signal lines entering the building, signal lines remaining within the building, ac power mains, dc power distribution and enclosure, such as in the coupling of EM fields to the internal equipments. As concerns the last parameter, the amplitude of the modulated RF EM field that can be expected in each environmental class is as follows: 1 V/m for major telecommunication centers, 3 V/m for minor telecommunication centers, 10 V/m for outdoor locations, and 3 V/m for customer premises in the frequency range 9 kHz to 2 GHz.

Recommendation K.43 [35] specifies the essential immunity requirements for equipments used within the public telecommunication networks and for terminal equipments connected to such networks. The document refers to the IEC 61000-4-3 standard: the performance criteria (A, B, C) are the same. Table C.5 summarizes the immunity requirements.

Emission from telecommunication networks are dealt with in Recommendation K.60 [36]. The document applies to wire-line telecommunication networks (e.g., all the telecommunication networks using telecommunication cables, their in-house cabling extensions, and connected telecommunications terminal equipments), all the telecommunication networks using the low-voltage ac mains network, and community antenna TV distribution networks. The document provides the methodology to measure in situ the disturbance emissions in the frequency ranges 9 kHz to 30 MHz and 30 MHz to 3 GHz, providing also the prescribed limits. As for other standards, in the low-frequency range a loop antenna must be used, whereas in the high-frequency range broadband dipole, biconical, log-periodic, or horn antennas (linearly polarized) are suggested. Emission from large systems can be accomplished according to Recommendation K.38 [37].

The last Recommendation worthy to be mentioned is Recommendation K.48 [38], which is a family Recommendation about telecommunication equipments such as switching, transmission, power, digital mobile base station, wireless LAN, and digital radio relay system, digital subscriber line, and supervisory equipments. The document specifies the emission and immunity requirements, mainly referring to CISPR 22 and ITU-T Rec. K.43 for methodologies and limits. The emission limits are the same as those of CISPR 22, while the immunity requirements are reported in Table C.6.

TABLE C.6 ITU-T K.48 Immunity Requirements

Frequency (MHz)	Test Level (V/m)	Performance Criterion
Equipment for telecom center and equipment for customer		
80–800	3	A
800–960	10	
960–1000	3	A
1400–2000	10	A

C.11 AUTOMOTIVE STANDARDS

Today automotive EMC standards are very dynamic. This is mainly due to the fact that the major companies and national standards bodies around the world are working to harmonize their standards with the international ones. International standards for automotive applications fall under two standards organizations: ISO (Technical Committee 22, Subcommittee 3, Working Group 3 for immunity concerns) and IEC (CISPR Subcommittee D for automotive and related products). In the United States, the American National Standards Institute (ANSI, which is the coordinating organization of US standards) has delegated the standards-writing activity to the Society of Automotive Engineers (SAE), which is dedicated to advancing mobility engineering worldwide, helping its members share information and exchange ideas.

The SAE EMC standards deal with both ground and aerospace vehicles. As concerns automotive regulations, the SAE carries out its work of standards-writing through two primary EMC committees, the Electromagnetic Immunity (EMI) and the Electromagnetic Radiation (EMR) Standards Committees, which have developed an extensive collection of EMC standards. The SAE EMI Standards Committee addresses immunity of automotive electrical and electronic systems in the vehicle and the modules or components of the vehicle, whereas the SAE EMR Standards Committee is primarily concerned with emissions from a vehicle and its modules or components that may cause radio-reception interference. As concerns emission and immunity of aerospace equipments, the EMC standards are written by the AE-4 Electromagnetic Environmental Effects Committee.

Now the main problem in the harmonization process is that SAE did not accept some test methods proposed by ISO (e.g., the stripline test method), while ISO did not adopt several of the long-accepted SAE test methods, because of much European pressure to limit the number of test methods in international standards.

The standards are huge in number. The main standards are listed in Table C.7. As regards ground automotive, the SAE standards (together with their corresponding ISO/IEC standards) can be broadly divided into two classes: standards applicable to vehicles or devices powered by an internal combustion engine or electric motor, and standards applicable to components. The first class is mainly composed of the standard family SAE J551 (with the counterpart family ISO 11451), and the second

TABLE C.7 SAE/ISO EMC Standards

SAE Number	Title	ISO Number	Title
J551-1	Performance Levels and Methods of Measurement of Electromagnetic Compatibility of Vehicles, Boats (up to 15 m), and Machines (16.6 Hz to 18 GHz)	11451-1	Road Vehicles–Vehicle Test Methods for Electrical Disturbances from Narrowband Radiated Electromagnetic Energy—Part 1: General Principles and Terminology
J551-5	Performance Levels and Methods of Measurement of Magnetic and Electric Field Strength from Electric Vehicles, Broadband (9 kHz to 30 MHz)		
J551-11	Vehicle Electromagnetic Immunity—Off-Vehicle Source	11451-2	Road Vehicles–Electrical Disturbances by Narrow-Band Radiated Electromagnetic Energy–Vehicle Test Methods—Part 2: Off-Vehicle Radiation Source
J551-12	Vehicle Electromagnetic Immunity–On-Board Transmitter Simulation	11451-3	Road Vehicles–Electrical Disturbances by Narrow-Band Radiated Electromagnetic Energy–Vehicle Test Methods—Part 3: On-Board Transmitter Simulation
J551-13	Vehicle Electromagnetic Immunity–Bulk Current Injection	11451-4	Road Vehicles–Electrical Disturbances by Narrow-Band Radiated Electromagnetic Energy–Vehicle Test Methods—Part 4: Bulk Current Injector (BCI)
J551-16	Electromagnetic Immunity–Off-Vehicle Source (Reverberation Chamber Method)—Part 16: Immunity to Radiated Electromagnetic Fields		
J551-17	Vehicle Electromagnetic Immunity–Power Line Magnetic Fields		
J1113-1	Electromagnetic Compatibility Measurement Procedures and Limits for Components of Vehicles, Boats (up to 15 m), and Machines (Except Aircraft) (16.6 Hz to 18 GHz)	11452-2	Road Vehicles–Electrical Disturbances by Narrow-Band Radiated Electromagnetic Energy–Component Test Methods—Part 2: Absorber-Lined Chamber

(continued)

TABLE C.7 (*Continued*)

SAE Number	Title	ISO Number	Title
J1113-4	Immunity to Radiated Electromagnetic Fields–Bulk Current Injection (BCI) Method	11452-4	Road Vehicles–Electrical Disturbances by Narrow-Band Radiated Electromagnetic Energy–Component Test Methods—Part 4: Bulk Current Injection (BCI)
J1113-21	Electromagnetic Compatibility Measurement Procedure for Vehicle Components—Part 21: Immunity to Electromagnetic Fields (30 MHz to 18 GHz), Absorber-Lined Chamber	11452-2	Road Vehicles–Electrical Disturbances by Narrow-Band Radiated Electromagnetic Energy–Component Test Methods—Part 2: Absorber-Lined Chamber
J1113-22	Electromagnetic Compatibility Measurement Procedure for Vehicle Components—Part 22: Immunity to Radiated Magnetic Fields	11452-8	Road Vehicles–Component Test Methods for Electrical Disturbances from Narrowband Radiated Electromagnetic Energy—Part 8: Immunity to Magnetic Fields
J1113-24	Immunity to Radiated Electromagnetic Fields; 10 kHz to 200 MHz–Crawford TEM Cell and 10 kHz to 5 GHz–Wideband TEM Cell	11452-3	Road Vehicles–Electrical Disturbances by Narrow-Band Radiated Electromagnetic Energy–Component Test Methods—Part 3: Transverse Electromagnetic Mode (TEM) Cell
J1113-27	Electromagnetic Compatibility Measurements Procedure for Vehicle Components—Part 27: Immunity to Radiated Electromagnetic Fields–Mode Stir Reverberation Method		
J1113-28	Electromagnetic Compatibility Measurements Procedure for Vehicle Components—Part 28: Immunity to Radiated Electromagnetic Fields–Reverberation Method (Mode Tuning)		
ARP 1173	Test Procedure to Measure the RF, Shielding Characteristics of EMI Gaskets		
ARP 1705A	Coaxial Test Procedure to Measure the RF Shielding Characteristics of EMI Gasket Materials		
ARP 5583	Guide to Certification of Aircraft in a High Intensity Radiated Field (HIRF) Environment		

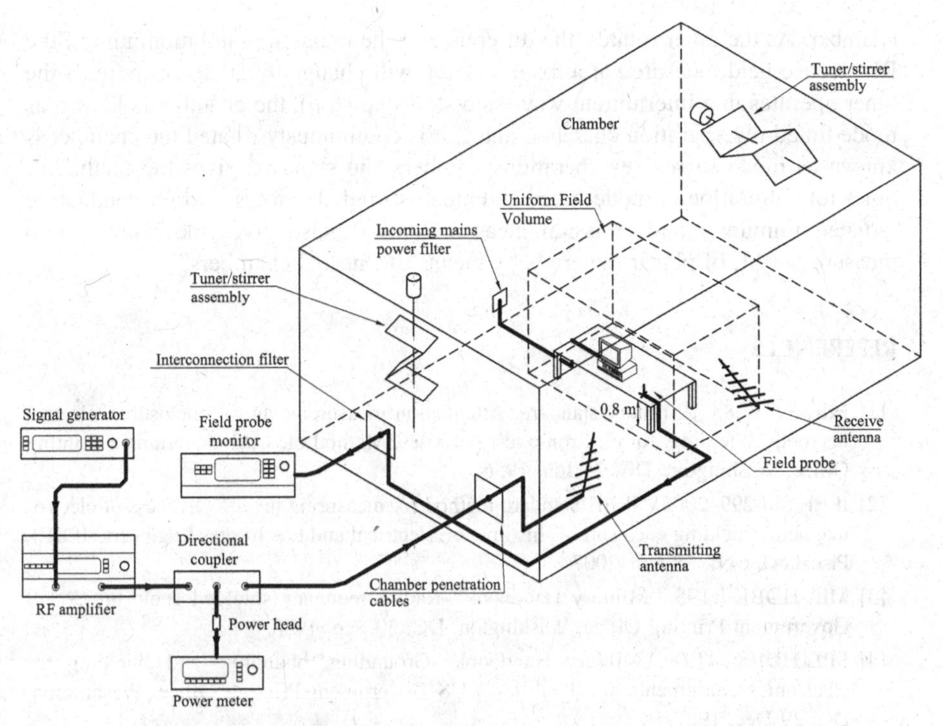

FIGURE C.11 IEC 61000-4-21 reverberation chamber setup.

class is composed of the standard family SAE J1113 (with the counterpart family ISO 11452).

The details of the standards in Table C.7 are not given. For more specific topics, the interested reader is referred to the mentioned standards. A point of interest to note here is that the SAE standards cover reverberation-chamber radiated immunity testing on vehicles and components, which is currently not covered by any ISO standard. However, the reverberation-chamber test methods are covered by the IEC 61000-4-21 [39], whose general setup is shown in Figure C.11.

The reverberation chamber is a shielded room, with a high-quality factor Q whose boundary conditions are changed via a rotating tuner. The tuner is a rotating metallic reflector that, in changing the boundary conditions as it rotates, moves the location of nulls and maximums of the field inside the reverberation chamber. The overall effect is a time-averaged uniform electric field inside the chamber, or better, in a portion of it (uniform-field volume). The IEC standard makes a detailed overview of the reverberation-chamber theory, adding several details on the chamber selection. For optimum chamber performance (especially at low frequencies), the volume of the chamber should be as large as possible, the room dimensions should not be integer multiples of one another (rooms with integer multiple dimensions will have degenerate modes and may not reverberate at all frequencies), and the tuner should be electrically large (greater than or equal to the wavelength at the lowest frequency of operation) to ensure its effectiveness in redistributing the energy inside the chamber. As the tuner rotates, the difference of the maximum and minimum of the RF electric field measured at a fixed location will change by 20 dB or more. If the tuner operates in a intermittent way (step-stop step-stop), the chamber is known as mode-tuned reverberation chamber, and if it is continuously rotated the chamber is known as mode-stirred reverberation chamber. The standard gives the methodologies for calibration of mode-tuned and mode-stirred chambers, and for conducting radiated immunity and emission measurements. It also gives the procedure to measure the SE of planar materials by means of "nested chambers."

REFERENCES

[1] MIL-STD 285. "Military standard. Attenuation measurements for enclosures, electromagnetic shielding, for electronic test purposes, Method of." US Government Printing Office, Washington, DC, 25 Jun. 1956.

[2] IEEE Std-299–2006. "IEEE Standard method for measuring the effectiveness of electromagnetic shielding enclosures." Institute of Electrical and Electronics Engineers (IEEE), Piscataway, NJ, 28 Feb. 2007.

[3] MIL-HDBK 1195. "Military Handbook—Radio frequency shielded enclosures." US Government Printing Office, Washington, DC, 30 Sep. 1988.

[4] MIL-HDBK 419A. "Military Handbook—Grounding, bounding, and shielding for electronics equipments and facilities." US Government Printing Office, Washington, DC, 29 Dec. 1987.

[5] MIL-HDBK 1857. "Grounding, bonding, and shielding design practices." US Government Printing Office, Washington, DC, 27 Mar. 1998.

[6] NSA 73-2A. "National Security Agency. Specification for foil RF shielded enclosure." National Security Agency, Fort George G. Meade, MD,15 Nov. 1972.

[7] NSA 65-5. "National Security Agency. RF shielded acoustical enclosures for communications equipment: General specification." National Security Agency, Fort George G. Meade, MD, 30 Oct. 1964.

[8] NSA 65-6. "National Security Agency. Specification for shielded enclosures for communications equipment: General specifications." National Security Agency, Fort George G. Meade, MD, 30 Oct. 1964.

[9] NSA 94-106. "National Security Agency. Specification for shielded enclosures." National Security Agency, Fort George G. Meade, MD, 24 Oct. 1994.

[10] ASTM E1851. "Standard test method for electromagnetic shielding effectiveness of durable rigid wall relocatable structures." American Society for Testing and Materials (ASTM), West Conshohocken, PA, 2004.

[11] ASTM D4935 (withdrawn). "Standard test method for measuring the electromagnetic shielding effectiveness of planar materials." American Society for Testing and Materials (ASTM), West Conshohocken, PA, Jun. 1999.

[12] MIL-STD 462D. "Military standard. Measurement of electromagnetic interface characteristics." US Government Printing Office, Washington, DC, 11 Jan. 1993.

[13] MIL-STD 461E. "Department of Defense Interface Standard. Requirements for the control of electromagnetic interface characteristics of subsystems and equipments." US Government Printing Office, Washington, DC, 20 Aug. 1999.

[14] Code of Federal Regulations, Title 47, Part 15. "Radio frequency devices" (Subpart A—General; Subpart B—Unintentional Radiators; Subpart C—Intentional Radiators). Federal Communications Commission (FCC), 10 Jan. 1998.

[15] FCC/OET MP-4. "FCC procedure for measuring RF emissions from computing devices." Federal Communications Commission (FCC)—Office of Engineering and Technology (OET), Jul. 1987.

[16] ANSI C63.4. "Methods of measurement of radio-noise emissions from low-voltage electrical and electronic equipment in the range of 9 kHz to 40 GHz." American National Standards Institute (ANSI), 30 Jan. 2004.

[17] ANSI-SCTE 48-3. "Test procedure for measuring shielding effectiveness of braided coaxial drop cable using the GTEM cell." American National Standard (ANSI) and Society of Cable Telecommunications Engineers (SCTE), Exton, PA, 2004.

[18] MIL-STD 1377. "Effectiveness of cable, connector, and weapon enclosure shielding and filters in precluding hazards of electromagnetic radiation to ordnance, measurement of." US Government Printing Office, Washington, DC, 1971.

[19] CISPR 14-2. "Electromagnetic compatibility–Requirements for household appliances, electric tools and similar apparatus—Part 2: Immunity–product family standard." International Special Committee on Radio Interference, 15 Nov. 2001.

[20] CISPR 20. "Sound and television broadcast receivers and associated equipment. Immunity characteristics—Limits and methods of measurement." International Special Committee on Radio Interference, 27 Nov. 2006.

[21] IEC 61547. "Equipment for general lighting purposes—EMC immunity requirements." International Electrotechnical Commission, 20 Sept. 1995.

[22] IEC 61000-6-1. "Electromagnetic compatibility (EMC)—Part 6-1: Generic standards–Immunity for residential, commercial, and light-industrial environments." International Electrotechnical Commission, 9 Mar. 2005.

[23] IEC 61000-6-2. "Electromagnetic compatibility (EMC)—Part 6-2: Generic standards–Immunity for industrial environments." International Electrotechnical Commission, 27 Jan. 2005.

[24] IEC 61000-4-3. "Electromagnetic compatibility (EMC)—Part 4-3: Testing and measurement techniques—Radiated, radio-frequency, electromagnetic field immunity test." International Electrotechnical Commission, 7 Feb. 2006.

[25] CISPR 11. "Industrial, scientific and medical (ISM) radio-frequency equipment—Electromagnetic disturbance characteristics–Limits and methods of measurement." International Special Committee on Radio Interference, 25 Jun. 2004.

[26] CISPR 13. "Sound and television broadcast receivers and associated equipment—Radio disturbance characteristics–Limits and methods of measurement." International Special Committee on Radio Interference, 13 Mar. 2006.

[27] CISPR 14-1. "Electromagnetic compatibility–Requirements for household appliances, electric tools and similar apparatus—Part 1: Emission." International Special Committee on Radio Interference, 13 Nov. 2005.

[28] CISPR 15. "Limits and methods of measurement of radio disturbance characteristics of electrical lighting and similar equipment." International Special Committee on Radio Interference, 17 Jan. 2007.

[29] CISPR 22. "Information technology equipment—Radio disturbance characteristics–Limits and methods of measurement." International Special Committee on Radio Interference, 20 Mar. 2006.

[30] IEC 61000-6-3. "Electromagnetic compatibility (EMC)—Part 6-3: Generic standards–Emission standard for residential, commercial and light-industrial environments." International Electrotechnical Commission, 17 Jul. 2006.

[31] IEC 61000-6-4. "Electromagnetic compatibility (EMC)—Part 6-4: Generic standards–Emission standard for industrial environments." International Electrotechnical Commission, 10 Jul. 2006.

[32] ITU-T K.42 Recommendation. "Preparation of emission and immunity requirements for telecommunication equipment—General principles." Telecommunication Standardization Sector of International Telecommunication Union (ITU), May 1998.

[33] ITU-T K.34 Recommendation. "Classification of electromagnetic environmental conditions for telecommunication equipment—Basic EMC recommendation." Telecommunication Standardization Sector of International Telecommunication Union (ITU), Jul. 2003.

[34] IEC 61000-2-5. "Electromagnetic compatibility (EMC)—Part 2: Environment—Section 5: Classification of electromagnetic environments. Basic EMC publication." International Electrotechnical Commission, 22 Sep. 1995.

[35] ITU-T K.43 Recommendation. "Immunity requirements for telecommunication equipment." Telecommunication Standardization Sector of International Telecommunication Union (ITU), Jul. 2003.

[36] ITU-T K.60 Recommendation. "Emission limits and test methods for telecommunication networks." Telecommunication Standardization Sector of International Telecommunication Union (ITU), Jul. 2003.

[37] ITU-T K.38 Recommendation. "Radiated emission test procedure for physically large systems." Telecommunication Standardization Sector of International Telecommunication Union (ITU), Oct. 1996.

[38] ITU-T K.48 Recommendation. "EMC requirements for telecommunication equipment—Product family recommendation." Telecommunication Standardization Sector of International Telecommunication Union (ITU), Sep. 2006.

[39] IEC 61000-4-21. "Electromagnetic compatibility (EMC)—Part 4-21: Testing and measurement techniques–Reverberation chamber test methods." International Electrotechnical Commission, 26 Aug. 2003.

Index

WILEY SERIES IN MICROWAVE AND OPTICAL ENGINEERING

KAI CHANG, Editor
Texas A&M University

FIBER-OPTIC COMMUNICATION SYSTEMS, Third Edition • *Govind P. Agrawal*

ASYMMETRIC PASSIVE COMPONENTS IN MICROWAVE INTEGRATED CIRCUITS • *Hee-Ran Ahn*

COHERENT OPTICAL COMMUNICATIONS SYSTEMS • *Silvello Betti, Giancarlo De Marchis, and Eugenio Iannone*

PHASED ARRAY ANTENNAS: FLOQUET ANALYSIS, SYNTHESIS, BFNs, AND ACTIVE ARRAY SYSTEMS • *Arun K. Bhattacharyya*

HIGH-FREQUENCY ELECTROMAGNETIC TECHNIQUES: RECENT ADVANCES AND APPLICATIONS • *Asoke K. Bhattacharyya*

RADIO PROPAGATION AND ADAPTIVE ANTENNAS FOR WIRELESS COMMUNICATION LINKS: TERRESTRIAL, ATMOSPHERIC, AND IONOSPHERIC • *Nathan Blaunstein and Christos G. Christodoulou*

COMPUTATIONAL METHODS FOR ELECTROMAGNETICS AND MICROWAVES • *Richard C. Booton, Jr.*

ELECTROMAGNETIC SHIELDING • *Salvatore Celozzi, Rodolfo Araneo, and Giampiero Lovat*

MICROWAVE RING CIRCUITS AND ANTENNAS • *Kai Chang*

MICROWAVE SOLID-STATE CIRCUITS AND APPLICATIONS • *Kai Chang*

RF AND MICROWAVE WIRELESS SYSTEMS • *Kai Chang*

RF AND MICROWAVE CIRCUIT AND COMPONENT DESIGN FOR WIRELESS SYSTEMS • *Kai Chang, Inder Bahl, and Vijay Nair*

MICROWAVE RING CIRCUITS AND RELATED STRUCTURES, Second Edition • *Kai Chang and Lung-Hwa Hsieh*

MULTIRESOLUTION TIME DOMAIN SCHEME FOR ELECTROMAGNETIC ENGINEERING • *Yinchao Chen, Qunsheng Cao, and Raj Mittra*

DIODE LASERS AND PHOTONIC INTEGRATED CIRCUITS • *Larry Coldren and Scott Corzine*

RADIO FREQUENCY CIRCUIT DESIGN • *W. Alan Davis and Krishna Agarwal*

MULTICONDUCTOR TRANSMISSION-LINE STRUCTURES: MODAL ANALYSIS TECHNIQUES • *J. A. Brandão Faria*

PHASED ARRAY-BASED SYSTEMS AND APPLICATIONS • *Nick Fourikis*

FUNDAMENTALS OF MICROWAVE TRANSMISSION LINES • *Jon C. Freeman*

OPTICAL SEMICONDUCTOR DEVICES • *Mitsuo Fukuda*

MICROSTRIP CIRCUITS • *Fred Gardiol*

HIGH-SPEED VLSI INTERCONNECTIONS, Second Edition • *Ashok K. Goel*

FUNDAMENTALS OF WAVELETS: THEORY, ALGORITHMS, AND APPLICATIONS • *Jaideva C. Goswami and Andrew K. Chan*

HIGH-FREQUENCY ANALOG INTEGRATED CIRCUIT DESIGN • *Ravender Goyal (ed.)*

ANALYSIS AND DESIGN OF INTEGRATED CIRCUIT ANTENNA MODULES • *K. C. Gupta and Peter S. Hall*

PHASED ARRAY ANTENNAS • *R. C. Hansen*

STRIPLINE CIRCULATORS • *Joseph Helszajn*

LOCALIZED WAVES • *Hugo E. Hernández-Figueroa, Michel Zamboni-Rached, and Erasmo Recami (eds.)*

MICROSTRIP FILTERS FOR RF/MICROWAVE APPLICATIONS • *Jia-Sheng Hong and M. J. Lancaster*

MICROWAVE APPROACH TO HIGHLY IRREGULAR FIBER OPTICS • *Huang Hung-Chia*

NONLINEAR OPTICAL COMMUNICATION NETWORKS • *Eugenio Iannone, Francesco Matera, Antonio Mecozzi, and Marina Settembre*

FINITE ELEMENT SOFTWARE FOR MICROWAVE ENGINEERING • *Tatsuo Itoh, Giuseppe Pelosi, and Peter P. Silvester (eds.)*

INFRARED TECHNOLOGY: APPLICATIONS TO ELECTROOPTICS, PHOTONIC DEVICES, AND SENSORS • *A. R. Jha*

SUPERCONDUCTOR TECHNOLOGY: APPLICATIONS TO MICROWAVE, ELECTRO-OPTICS, ELECTRICAL MACHINES, AND PROPULSION SYSTEMS • *A. R. Jha*

OPTICAL COMPUTING: AN INTRODUCTION • *M. A. Karim and A. S. S. Awwal*

INTRODUCTION TO ELECTROMAGNETIC AND MICROWAVE ENGINEERING • *Paul R. Karmel, Gabriel D. Colef, and Raymond L. Camisa*

MILLIMETER WAVE OPTICAL DIELECTRIC INTEGRATED GUIDES AND CIRCUITS • *Shiban K. Koul*

MICROWAVE DEVICES, CIRCUITS AND THEIR INTERACTION • *Charles A. Lee and G. Conrad Dalman*

ADVANCES IN MICROSTRIP AND PRINTED ANTENNAS • *Kai-Fong Lee and Wei Chen (eds.)*

SPHEROIDAL WAVE FUNCTIONS IN ELECTROMAGNETIC THEORY • *Le-Wei Li, Xiao-Kang Kang, and Mook-Seng Leong*

ARITHMETIC AND LOGIC IN COMPUTER SYSTEMS • *Mi Lu*

OPTICAL FILTER DESIGN AND ANALYSIS: A SIGNAL PROCESSING APPROACH • *Christi K. Madsen and Jian H. Zhao*

THEORY AND PRACTICE OF INFRARED TECHNOLOGY FOR NONDESTRUCTIVE TESTING • *Xavier P. V. Maldague*

METAMATERIALS WITH NEGATIVE PARAMETERS: THEORY, DESIGN, AND MICROWAVE APPLICATIONS • Ricardo Marqués, Ferran Martín, and Mario Sorolla

OPTOELECTRONIC PACKAGING • *A. R. Mickelson, N. R. Basavanhally, and Y. C. Lee (eds.)*

OPTICAL CHARACTER RECOGNITION • *Shunji Mori, Hirobumi Nishida, and Hiromitsu Yamada*

ANTENNAS FOR RADAR AND COMMUNICATIONS: A POLARIMETRIC APPROACH • *Harold Mott*

INTEGRATED ACTIVE ANTENNAS AND SPATIAL POWER COMBINING • *Julio A. Navarro and Kai Chang*

ANALYSIS METHODS FOR RF, MICROWAVE, AND MILLIMETER-WAVE PLANAR TRANSMISSION LINE STRUCTURES • *Cam Nguyen*

FREQUENCY CONTROL OF SEMICONDUCTOR LASERS • *Motoichi Ohtsu (ed.)*

WAVELETS IN ELECTROMAGNETICS AND DEVICE MODELING • *George W. Pan*

OPTICAL SWITCHING • Georgios Papadimitriou, Chrisoula Papazoglou, and Andreas S. Pomportsis

SOLAR CELLS AND THEIR APPLICATIONS • *Larry D. Partain (ed.)*

ANALYSIS OF MULTICONDUCTOR TRANSMISSION LINES • *Clayton R. Paul*

INTRODUCTION TO ELECTROMAGNETIC COMPATIBILITY, Second Edition • *Clayton R. Paul*

ADAPTIVE OPTICS FOR VISION SCIENCE: PRINCIPLES, PRACTICES, DESIGN AND APPLICATIONS • *Jason Porter, Hope Queener, Julianna Lin, Karen Thorn, and Abdul Awwal (eds.)*

ELECTROMAGNETIC OPTIMIZATION BY GENETIC ALGORITHMS • *Yahya Rahmat-Samii and Eric Michielssen (eds.)*

INTRODUCTION TO HIGH-SPEED ELECTRONICS AND OPTOELECTRONICS • *Leonard M. Riaziat*

NEW FRONTIERS IN MEDICAL DEVICE TECHNOLOGY • *Arye Rosen and Harel Rosen (eds.)*

ELECTROMAGNETIC PROPAGATION IN MULTI-MODE RANDOM MEDIA • *Harrison E. Rowe*

ELECTROMAGNETIC PROPAGATION IN ONE-DIMENSIONAL RANDOM MEDIA • *Harrison E. Rowe*

HISTORY OF WIRELESS • *Tapan K. Sarkar, Robert J. Mailloux, Arthur A. Oliner, Magdalena Salazar-Palma, and Dipak L. Sengupta*

SMART ANTENNAS • *Tapan K. Sarkar, Michael C. Wicks, Magdalena Salazar-Palma, and Robert J. Bonneau*

NONLINEAR OPTICS • *E. G. Sauter*

APPLIED ELECTROMAGNETICS AND ELECTROMAGNETIC COMPATIBILITY • *Dipak L. Sengupta and Valdis V. Liepa*

COPLANAR WAVEGUIDE CIRCUITS, COMPONENTS, AND SYSTEMS • *Rainee N. Simons*

ELECTROMAGNETIC FIELDS IN UNCONVENTIONAL MATERIALS AND STRUCTURES • *Onkar N. Singh and Akhlesh Lakhtakia (eds.)*

ELECTRON BEAMS AND MICROWAVE VACUUM ELECTRONICS • *Shulim E. Tsimring*

FUNDAMENTALS OF GLOBAL POSITIONING SYSTEM RECEIVERS: A SOFTWARE APPROACH, Second Edition • *James Bao-yen Tsui*

RF/MICROWAVE INTERACTION WITH BIOLOGICAL TISSUES • *André Vander Vorst, Arye Rosen, and Youji Kotsuka*

InP-BASED MATERIALS AND DEVICES: PHYSICS AND TECHNOLOGY • *Osamu Wada and Hideki Hasegawa (eds.)*

COMPACT AND BROADBAND MICROSTRIP ANTENNAS • *Kin-Lu Wong*

DESIGN OF NONPLANAR MICROSTRIP ANTENNAS AND TRANSMISSION LINES • *Kin-Lu Wong*

PLANAR ANTENNAS FOR WIRELESS COMMUNICATIONS • *Kin-Lu Wong*

FREQUENCY SELECTIVE SURFACE AND GRID ARRAY • *T. K. Wu (ed.)*

ACTIVE AND QUASI-OPTICAL ARRAYS FOR SOLID-STATE POWER COMBINING • *Robert A. York and Zoya B. Popović (eds.)*

OPTICAL SIGNAL PROCESSING, COMPUTING AND NEURAL NETWORKS • *Francis T. S. Yu and Suganda Jutamulia*

SiGe, GaAs, AND InP HETEROJUNCTION BIPOLAR TRANSISTORS • *Jiann Yuan*

ELECTRODYNAMICS OF SOLIDS AND MICROWAVE SUPERCONDUCTIVITY • *Shu-Ang Zhou*

Printed in the United States
By Bookmasters